Principles of Neural Design

Principles

Compute with chemistry

Compute directly with analog primitives

Combine analog and pulsatile processing

Sparsify

Send only what is needed

Send at the lowest acceptable rate

Minimize wire

Make neural components irreducibly small

Complicate

Adapt, match, learn, and forget

Principles of Neural Design

Peter Sterling and Simon Laughlin

The MIT Press
Cambridge, Massachusetts
London, England

First MIT Press paperback edition, 2017

MIT Press books may be purchased at special quantity discounts for business or sales promotional use. For information, please email special_sales@mitpress.mit.edu.

This book was set in Stone Sans and Stone Serif by Toppan Best-set Premedia Limited. Printed and bound in the United States of America.

Library of Congress Cataloging-in-Publication Data
Sterling, Peter (Professor of neuroscience), author.
Principles of neural design / Peter Sterling and Simon Laughlin.
    p. ; cm.
Includes bibliographical references and index.
ISBN 978-0-262-02870-7 (hardcover : alk. paper), 978-0-262-53468-0 (pb.)
I. Laughlin, Simon, author. II. Title.
[DNLM: 1. Brain—physiology. 2. Learning. 3. Neural Pathways. WL 300]
QP376
612.8'2—dc23

2014031498

10 9

For Sally Zigmond and Barbara Laughlin

Contents

Preface

Neuroscience abounds with stories of intellectual and technical daring. Every peak has its Norgay and Hillary, and we had imagined telling some favorite stories of heroic feats, possibly set off in little boxes. Yet, this has been well done by others in various compendia and reminiscences (Strausfield, 2012; Glickstein, 2014; Kandel, 2006; Koch, 2012). Our main goal is to evince some principles of design and note some insights that follow. Stories deviating from this intention would have lengthened the book and distracted from our message, so we have resisted the natural temptation to memoir-ize.

Existing compendia tend to credit various discoveries to particular individuals. This belongs to the storytelling. What interest would there be to the Trojan Wars without Odysseus and Agamemnon? On the other hand, dropping a name here and there distorts the history of the discovery process—where one name may stand for a generation of thoughtful and imaginative investigators. Consequently, in addition to forgoing stories, we forgo dropping names—except for a very few who early enunciated the core principles. Nor do the citations document who did what first; rather they indicate where supporting evidence will be found—often a review.

Existing compendia often pause to explain the ancient origins of various terms, such as cerebellum or hippocampus. This might have been useful when most neuroscientists spoke a language based in Latin and Greek, but now with so many native speakers of Mandarin or Hindi the practice seems anachronistic, and we have dropped it. Certain terms may be unfamiliar to readers outside neuroscience, such as physicists and engineers. These are italicized at their first appearance to indicate that they are technical (*cation channel*). A reader unfamiliar with this term can learn by Googling in 210 ms that "cation channels *are pore-forming proteins that help establish and control the small voltage gradient across the plasma membrane of all living cells . . .*"

(Wikipedia). So rather than impede the story, we sometimes rely on you to Google.

Many friends and colleagues long aware of this project have wondered why it has taken so long to complete. Some have tried to encourage us to let it go, saying, "After all, it needn't be perfect . . ." To which we reply, "Don't worry, it isn't!" It's just that more time is needed to write a short book than a long one.

Acknowledgments

For reading and commenting on various chapters we are extremely grateful to: Larry Palmer, Philip Nelson, Dmitry Chklovskii, Ron Dror, Paul Glimcher, Jay Schulkin, Glenn Vinnicombe, Neil Krieger, Francisco Hernández-Heras, Gordon Fain, Sally Zigmond, Alan Pearlman, Brian Wandell, and Dale Purves.

For many years of fruitful exchange PS thanks colleagues at the University of Pennsylvania: Vijay Balasubramanian, Kwabena Boahen, Robert Smith, Michael Freed, Noga Vardi, Jonathan Demb, Bart Borghuis, Janos Perge, Diego Contreras, Joshua Gold, David Brainard, Yoshihiko Tsukamoto, Minghong Ma, Amita Sehgal, Jonathan Raper, and Irwin Levitan. SL thanks colleagues at the Australian National University and the University of Cambridge. Adrian Horridge, Ben Walcott, Ian Meinertzhagen, Allan Snyder, Doekele Stavenga, Martin Wilson, Srini (MV) Srinivasan, David Blest, Peter Lillywhite, Roger Hardie, Joe Howard, Barbara Blakeslee, Daniel Osorio, Rob de Ruyter van Steveninck, Matti Weckström, John Anderson, Brian Burton, David O'Carroll, Gonzalo Garcia de Polavieja, Peter Neri, David Attwell, Bob Levin, Aldo Faisal, John White, Holger Krapp, Jeremy Niven, Gordon Fain, and Biswa Sengupta.

For kindly answering various queries on specific topics and/or providing figures, we thank: Bertil Hille, Nigel Unwin (ion channels), Stuart Firestein, Minghong Ma, Minmin Luo (Olfaction); Wallace Thoreson, Richard Kramer, Steven DeVries, Peter Lukasiewicz, Gary Matthews, Henrique von Gersdorff, Charles Ratliff, Stan Schein, Jeffrey Diamond, Richard Masland, Heinz Wässle, Steve Massey, Dennis Dacey, Beth Peterson, Helga Kolb, David Williams, David Calkins, Rowland Taylor, David Vaney, (retina); Roger Hardie, Ian Meinertzhagen (insect visual system); Nick Strausfeld, Berthold Hedwig, Randolf Menzel, Jürgen Rybak (insect brain); Larry Swanson, Eric Bittman, Kelly Lambert (hypothalamus); Michael Farries, Ed Yeterian, Robert Wurtz, Marc Sommer, Rebecca Berman (striatum); Murray Sherman,

Al Humphreys, Alan Saul, Jose-Manuel Alonso,Ted Weyand, Larry Palmer, Dawei Dong (lateral geniculate nucleus); Indira Raman, Sacha du Lac, David Linden, David Attwell, John Simpson, Mitchell Glickstein, Angus Silver, Chris De Zeeuw (cerebellum); Tobias Moser, Paul Fuchs, James Saunders, Elizabeth Glowatzki, Ruth Anne Eatock (auditory and vestibular hair cells); Jonathan Horton, Kevan Martin, Deepak Pandya, Ed Callaway, Jon Kaas, Corrie Camalier, Roger Lemon, Margaret Wong-Riley (cerebral cortex).

For long encouragement and for skill and care with the manuscript and illustrations, we thank our editors at MIT: Robert Prior, Christopher Eyer, Katherine Almeida, and Mary Reilly, and copy editor Regina Gregory.

Introduction

A laptop computer resembles the human brain in volume and power use—but it is stupid. Deep Blue, the IBM supercomputer that crushed Grandmaster Garry Kasparov at chess, is 100,000 times larger and draws 100,000 times more power (figure I.1). Yet, despite Deep Blue's excellence at chess, it too is stupid, the electronic equivalent of an idiot savant. The computer operates at the speed of light whereas the brain is slow. So, wherein lies the brain's advantage? A short answer is that the brain employs a hybrid architecture of superior design. A longer answer is this book—whose purpose is to identify the sources of such computational efficiency.

The brain's inner workings have been studied scientifically for more than a century—initially by a few investigators with simple methods. In the last 20 years the field has exploded, with roughly 50,000 neuroscientists applying increasingly advanced methods. This outburst amounts to 1 million person-years of research—and facts have accumulated like a mountain. At the base are detailed descriptions: of neural connections and electrical responses, of functional images that correlate with mental states, and of molecules such as ion channels, receptors, G proteins, and so on. Higher up are key discoveries about mechanism: the action potential, transmitter release, synaptic excitation and inhibition. Summarizing this Everest of facts and mechanisms, there exist superb compendia (Kandel et al., 2012; Purves et al., 2012; Squire et al., 2008).

But what if one seeks a book to set out principles that explain how our brain, while being far smarter than a supercomputer, can also be far smaller and cheaper? Then the shelf is bare. One reason is that modern neuroscience has been "technique driven." Whereas in the 1960s most experiments that one might conceive were technically impossible, now with methods such as patch clamping, two-photon microscopy, and functional magnetic resonance imaging (fMRI), aided by molecular biology, the situation has reversed, and it is harder to conceive of an experiment that can*not* be done.

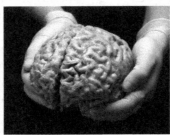

Figure I.1
How do neural circuits use space and power so efficiently? Computer: Image http://
upload.wikimedia.org/wikipedia/commons/d/d3/IBM_Blue_Gene_P_supercomputer
.jpg. Brain: Photo by UW-Madison, University Communications © Board of Regents
of the University of Wisconsin System.

Consequently, the idea of pausing to distill principles from facts has lacked
appeal. Moreover, to many who ferret out great new facts for a living, it has
seemed like a waste of time.

Yet, we draw inspiration from Charles Darwin, who remarked, "My mind
seems to have become a kind of machine for grinding general laws out of
large collections of facts" (Darwin, 1881). Darwin, of course, is incompara-
ble, but this is sort of how our minds work too. So we have written a small
book—relative to the great compendia—intending to beat a rough path up
"Data Mountain" in search of organizing principles.

Principles of engineering

The brain is a physical device that performs specific functions; therefore, its design must obey general principles of engineering. Chapter 1 identifies several that we have gleaned (not being engineers) from essays and books on mechanical and electrical design. These principles do not address specific questions about the brain, but they do set a context for ordering one's thoughts—especially helpful for a topic so potentially intimidating. For example, it helps to realize that neuroscience is really an exercise in "reverse engineering"—disassembling a device in order to understand it.

This insight points immediately to a standard set of questions that we suppose are a mantra for all "reverse engineers": *What does it do? What are its specifications? What is the environmental context?* Then there are commandments, such as *Study the interfaces* and *Complicate the design*. The latter may puzzle scientists who, in explaining phenomena, customarily strive for simplicity. But engineers focus on designing effective devices, so they have good reasons to complicate.[1] This commandment, we shall see, certainly applies to the brain.

Why a brain?

To address the engineer's first question, we consider why an animal should need a brain—what fundamental purpose does it serve and at what cost to the organism? Chapter 2 begins with a tiny bacterium, *Escherichia coli* which succeeds *without* a brain, in order to evaluate what the bacterium can do and what it cannot. Then on to a protozoan, *Paramecium caudatum*, still a single cell and brainless, but so vastly larger than *E. coli* (300,000-fold) that it requires a faster type of signaling. This prefigures long-distance signaling by neurons in multicellular organisms.

The chapter closes with the tiny nematode worm, *Caenorhabditis elegans*, which does have a brain—with exactly 302 neurons. This number is small in absolute terms, but it represents nearly one third of the creature's total cells, so it is a major investment that better turn a profit, and it does. For example, it controls a multicellular system that finds, ingests, and digests bacteria and that allows the worm to recall for several hours the locations of favorable temperatures and bacterial concentrations.

Humans naturally tend to discount the computational abilities of small organisms—which seem, well . . ., mentally deficient—nearly devoid of learning or memory. But small organisms *do* learn and remember. It's just that their memories match their life contexts: they remember only what they need to and for just long enough. Furthermore, the mechanisms that they evolved for these computations are retained in our own neurons—so we shall see them again.

The progression bacterium → protozoan → worm is accompanied by increasing computational complexity. It is rewarded by increasing capacity to inhabit richer environments and thus to move up the food chain: protozoa eat bacteria, and worms eat protozoa. As engineering, this makes perfect sense: little beasts compute only what they must; thus they pay only for what they use. This is equally true for beasts with much larger brains discussed in chapter 3.

Why a bigger brain?

The brain of a fruit fly (*Drosophila melanogaster*) is 350-fold larger than *C. elegans'*, and the brain of a human (*Homo sapiens*) is a million-fold larger than the fly's. These larger brains emerge from the same process of natural selection as the smaller ones, so we should continue to expect from them nothing superfluous—only mechanisms that are essential and pay for themselves. We should also expect that when a feature works really well, it will be retained—like the wheel, the paper clip, the aluminum beer can, and the transistor (Petroski, 1996; Arthur, 2009). We note design features that brains have conserved (with suitable elaborations) across at least 400 million years of natural selection. These features in the human brain are often described as "primitive"—reptilian—reflecting what are considered negative aspects of our nature. But, of course, any feature that has been retained for so long must be pretty effective.

This chapter identifies the core task of all brains: it is to regulate the organism's internal milieu—by responding to needs and, better still, by anticipating needs and preparing to satisfy them before they arise. The advantages of omniscience encourage omnipresence. Brains tend to become universal devices that tune all internal parameters to improve overall stability and economy. "Anticipatory regulation" replaces the more familiar "homeostatic regulation"—which is supposed to operate by waiting for each parameter to deviate from a "set point," then detecting the error and correcting it by feedback. Most physiological investigations during the 20th century were based on the homeostatic model—how kidney, gut, liver, pancreas, and so on work independently, despite Pavlov's early demonstration of the brain's role in anticipatory regulation (Pavlov, 1904). But gradually anticipatory control has been recognized.

Anticipatory regulation offers huge advantages.[2] First, it matches overall response capacity to fluctuations in demand—there should always be enough but not too much. Second, it matches capacity at each stage in the system to anticipated needs downstream, thus threading an efficient path between excess capacity (costly storage) and failure from lack of supplies. Third, it resolves potential conflict between organs by setting and shifting priorities. For example, during digestion it can route more blood to the gut and less to muscle and skin, and during exercise it can reverse this priority. This allows the organism to operate with a smaller blood volume than would otherwise be needed. Finally, it minimizes errors—which are potentially lethal and also cause cumulative damage.

Anticipatory regulation includes behavior

An organ that anticipates need and regulates the internal milieu by overarching control of physiology would be especially effective if it also regulated behavior. For example, it could reduce a body's need for physiological cooling (e.g., sweating—which costs energy and resources—sodium and water) by directing an animal to find shade. Moreover, it could evoke the memory of an unpleasant heatstroke to remind the animal to take anticipatory measures (travel at night, carry water). Such anticipatory mechanisms are driven ceaselessly by *memories* of hunger, cold, drought, or predation: *Pick the beans! Chop wood! Build a reservoir! Lock the door!*

The memories of danger and bad times that shape our behavior can be our own, but often they are stored in the brains of our parents and grandparents. We are reared with *their* nightmares—the flood, the drought, the famine, the pogrom. Before written history, which spans only 6,000 years, all lessons that would help one anticipate and thus avoid a lethal situation could be transmitted only by oral tradition—the memory of a human life span. Given that the retention of memories in small brains corresponds to their useful span, and that retention has a cost, human memory for great events should remain vivid with age whereas memories of lesser events should fade (chapter 14).

The most persistent dangers and opportunities, those extending far beyond a few generations, eventually become part of the neural wiring. Monkeys universally fear snakes, and so do most humans—suggesting that the response was encoded into brain structure before the lines split—on the order of 35 million years. But beyond alertness for predators, primate societies reserve their most acute observations and recall for relationships within the family and the troop. The benefit is that an individual's chances

for survival and reproduction are enhanced by the *group's* ability to anticipate and regulate. The cost is that the individual must continuously sense
the social structure—in its historical context—to receive aid when needed
and to avoid being killed or cast out (Cheney & Seyfarth, 2007).

Consequently, primate brains have expanded significantly in parts concerned with social recognition and planning—such as prefrontal cortex and
amygdala. Humans greatly expand these areas and also those for social
communication, such as for language, facial expression, and music. These
regions serve both the cooperative and the competitive aspects of anticipatory regulation to an awesome degree. They account for much of our brain
structure and many of our difficulties.

Flies too show anticipatory behavior—to a level consonant with their
life span and environmental reach. A fly need not wait for its blood sugar
to fall dangerously low, nor for its temperature to soar dangerously high,
before taking action. Instead its brain expresses prewired commands: *Find
fruit! In a cool spot!* Anticipatory commands are often tuned to environmental regularities that predict when and where a resource is most likely
to appear—or disappear. Thus, circadian rhythms govern foraging and
sleep. Seasonal rhythms, which broadly affect resource availability, govern
mating and reproduction. Consequently, specific brain hormones tuned
to season send orders to prewired circuits: *Court a mate! Intimidate a
competitor!*

What drives behavior?

To ensure that an organism will execute these orders, there are neural
mechanisms to make it "feel bad" when a job is undone and "feel good"
when it has succeeded. These are circuits whose activity humans experience, respectively, as "anxiety" and "pleasure." Of course, we cannot know
what worms or flies experience—but the same neurochemicals drive similar
behaviors. This is one wheel that has certainly been decorated over hundreds of millions of years, but not reinvented.

To actually accomplish a task is vastly complicated. Reconsider Deep
Blue's task. Each side in chess has 16 pieces—that move one at a time,
slowly (minutes), and only in two dimensions. Each piece is constrained to
move only in certain ways, and some pieces repeat so that each side has
only six different types of motion. This relatively simple setup generates so
many possible moves that to evaluate them requires a Deep Blue.

But the organ responsible for anticipatory regulation takes continuous
data from every sensory neuron in the organism—both internal and

external—plus myriad hormones and other chemicals. While doing so, it is calculating in real time—milliseconds—how to adjust every body component inside and out. It is flying the fly, finding its food, shade, and mate; it is avoiding predators and intimidating competitors—all the while tweaking every *internal* parameter to match what is about to be needed. Thus, it seems fair to say that Deep Blue is stupid even compared to a fruit fly. This defines sharply the next engineering question: what constrains the design of an effective and efficient brain?

What constrains neural design?

When Hillel was asked in the first century B.C.E. to explain the whole Torah while standing on one leg, he was ready: "That which is hateful to you, do not unto another. The rest is commentary—and now go study."

There is a one-leg answer for neural design: "As information rate rises, costs rise disproportionately." For example, to transmit more information by spikes requires a higher spike rate. Axon diameter rises linearly with spike rate, but axon volume and energy consumption rise as the diameter squared. Thus, the essence of neural design: "Send only information that is needed, and send it as slowly as possible" (chapter 3). This key injunction profoundly shapes the brain's macroscopic layout, as explained in chapter 4. We hope that readers will . . . go study.

If spikes were energetically cheap, their rates would matter less. However, a 100-mV spike requires far more current than a 1-mV response evoked by one packet of chemical transmitter. Obviously then, it is cheaper to compute with the smaller currents. This exemplifies another design principle: minimize energy per *bit* of information by computing at the finest possible level. Chapter 5 identifies this level as a change in protein folding on the scale of nanometers. Such a change can capture, store, and transmit one bit at an energetic cost that approaches the thermodynamic limit. Chapter 6 explains how proteins couple to form intracellular circuits on the scale of micrometers, and chapter 7 explains how a neuron assembles such circuits into devices on a scale of micrometers to millimeters.

It emerges that to compute most efficiently in space and energy, neural circuits should *nanofy*:

1. Make each component irreducibly small: a functional unit should be a single protein molecule (a channel), or a linear polymer of protein subunits (a microtubule), or a sandwich of monomolecular layers (a membrane).

2. Combine irreducible components: a membrane to separate charge and thus permit a voltage, a protein transporter to pump ions selectively across the membrane and actually separate the charges (charge the battery), a pore for ions to flow singly across the membrane and thus create a current, a "gate" to start and stop a current, an amplifier to enlarge the current, and an adaptive mechanism to match a current to circumstance.
3. Compute with *chemistry* wherever possible: regulate gates, amplifiers, and adaptive mechanisms by binding/unbinding small molecules that are present in sufficient numbers to obey the laws of mass action. Achieve speed with chemistry by keeping the volumes small.
4. For speed over distance compute *electrically*: convert a signal computed by chemistry to a current that charges membrane capacitance to spread passively up to a millimeter. For longer distance, regenerate the current by appropriately clustered voltage-gated channels.

Design in the visual system

Having discussed protein computing and miniaturization as general routes to efficiency, we exemplify these points in an integrated system—phototransduction (chapter 8). The engineering challenge is to capture light reflected from objects in the environment in order to extract informative patterns to guide behavior. Transduction employs a biochemical cascade with about half a dozen stages to amplify the energy of individual photons by up to a million-fold while preserving the information embodied as signal-to-noise ratio (S/N) and bandwidth. We explain why so many stages are required.

The photoreceptor signal, once encoded as a graded membrane voltage, spreads passively down the axon to the synaptic terminal. There the analogue signal is quantized as a stream of synaptic vesicles. The insect brain can directly read out this message with very high efficiency because the distance is short enough for passive signaling (chapter 9). The mammal brain can*not* directly read out this vesicle stream because the distance is too great for passive signaling. The mammal eye must transmit by action potentials, but the photoreceptor's analogue signal contains more information than action potentials can encode. Therefore, on-site retinal processing is required (chapters 10, 11).

Principles at higher levels

The principles of neural design at finer scales and lower levels also apply at larger scales and higher levels. For example, they can explain why the first visual area (V1) in cerebral cortex enormously expands the number and diversity of neurons. And why diverse types project in parallel from V1 to other cortical areas. And why cortex uses many specific areas and arranges them in a particular way. The answers, as explained in chapter 12, are always the same: diverse circuits allow the brain to send only information that is needed and to send it at lower information rates. This holds computation to the steep part of the benefit/cost curve.

Wiring efficiency

Silicon circuits with very large-scale integration strive for optimal layout—to achieve best performance given cost of space, time, and energy. Neural circuits do the same and thereby produce tremendous diversity of neuronal structure at all spatial scales. For example, cerebellar output neurons (*Purkinje cells*) use a two-dimensional dendritic arbor whereas cerebral output neurons (*pyramidal cells*) use a three-dimensional arbor. Both circuits employ a layered architecture, but the large Purkinje neurons lie *above* a layer of tiny neurons whereas the large pyramidal neurons lie *below* the smaller neurons. Cerebellar cortex folds intensely on a millimeter scale whereas cerebral cortex on this scale is smooth.

Such differences originate from a ubiquitous biophysical constraint: the irreducible electrical resistance of neuronal cytoplasm. Passive signals spread spatially and temporally only as the square root of dendritic diameter ($\sqrt{d}$). This causes a second law of diminishing returns: a dendrite, to double its conduction distance or halve its conduction delay, must quadruple its volume. This prevents neural wires from being any finer and prevents local circuits from being any more voluminous. In both cases conduction delays would grow too large. The constraint on volume drives efficient layout: equal lengths of dendrite and axon and an optimum proportion of wire and synapses. Chapter 13 will explain.

Designs for learning

All organisms use new information to better anticipate the future. Thus, learning is a deep principle of biological design, and therefore of neural design. Accordingly, the brain continually updates its knowledge of every internal and external parameter—which means that learning is also a brain function. As such, neural learning is subject to the same constraints as all other neural functions. It is a design principle that must obey all the others.

To conserve space, time, and energy, new information should be stored at the site where it is processed and from whence it can be recalled without further expense. This is the synapse. Low-level synapses relay short-term changes in input, so their memories should be short, like that of a bacterium or worm. These synapses should encode at the cheapest levels, by modifying the structure and distribution of proteins. High-level synapses encode conclusions after many stages of processing, so their memories deserve to be longer and encoded more stably, by enlarging the synapse and adding new ones.

A new synapse of diameter (d) occupies area on the postsynaptic membrane as d^2 and volume as d^3. Because adding synapses increases costs disproportionately, learning in an adult brain of fixed volume is subject to powerful space constraints. For every synapse enlarged or added, another must be shrunk or removed. Design of learning must include the principle "save only what is needed." Chapter 14 explains how this plays out in the overall design.

Design and designer

This book proposes that many aspects of the brain's design can be understood as adaptations to improve efficiency under resource constraints. Improvements to brain efficiency must certainly improve fitness. Darwin himself noted that "natural selection is continually trying to economize every part of the organization" and proposed that instincts, equivalent in modern terms to "genetically programmed neural circuits," arise by natural selection (Darwin, 1859). So our hypothesis breaks no new conceptual ground.

A famous critique of this hypothesis argues that useless features might survive pruning if they were simply unavoidable accompaniments to important features (Gould & Lewontin, 1979). This possibility is undeniable, but if examples are found for neural designs, we expect them to be rare because each failure to prune what is useless would render the brain less efficient—more like Deep Blue—whereas the brain's efficiency exceeds Deep Blue's by at least 10^5.

So what do we claim *is* new? The energy and space constraints have been known for a while, as have various principles, such as "minimize wire." The present contribution seems to lie in our gathering various rules as a concise list and in systematically exemplifying them across spatial and functional scales. When a given rule was found to apply broadly with constant explanatory power, we called it a "principle." Ten are listed as a round number. As with the Biblical Commandments and the U.S. Bill of Rights, some readers will find too many (redundancy) and others too few. We are satisfied to simply set them out for consideration.

Some readers may object to the expression "design" because it might imply a designer, which might suggest creationism. But "design" can mean *"the arrangement of elements or details,"* also *"a scheme that governs functioning."* These are the meanings we intend. And, of course, there *is* a designer— as noted, it is the process that biologists understand as natural selection.[3]

Limits to this effort

Our account rests on facts that are presently agreed upon. Where some point is controversial, we will so note, but we will not resort to imagined mechanisms. Our goal is not to explain how the brain *might* work, but rather to make sense of what is already known. Naturally what is "agreed upon" will shift with new data, so the story will evolve. We gladly acknowledge that this account is neither complete nor timeless.

We omit *so* much—many senses, many brain regions, many processes— and this will disappoint readers who study them. We concentrate on vision partly because it has dominated neuroscience during its log growth phase, so that is where knowledge goes deepest at all scales. Also we have personally concentrated on vision, so that is where our own knowledge is deepest. Finally, to apply principles across the full range of scales, but keep the book small, has required rigorous selection. We certainly hope that workers in other fields will find the principles useful. If some prove less than universal and need revision, well, that's science. The best we can do with Data Mountain really is just to set a few pitons up the south face.

1 What Engineers Know about Design

During the Cold War, the Soviets would occasionally capture a U.S. military aircraft invading their airspace, and with comparable frequency a defecting Soviet pilot would set down a MiG aircraft in Japan or Western Europe. These planes would be instantly swarmed by engineers—like ants to a drop of honey—with one clear goal: to "reverse engineer" the craft. This is the process of discovering how a device works by disassembling and analyzing in detail its structure and function. Reverse engineering allowed Soviet engineers to rather quickly reproduce a near perfect copy of the U.S. B-29 bomber, which they renamed the Tu-4. Reverse engineering still flourishes in military settings and increasingly in civilian industries—for example, in chip and software development where rival companies compete on the basis of innovation and design.

The task in reverse engineering is accelerated immensely by prior knowledge. Soviet engineers knew the B-29's purpose—to fly. Moreover, they knew its performance specifications: carry 10 tons of explosive at 357 mph at an altitude of 36,000 feet with a range at half-load of 3,250 miles. They also knew how various parts function: wings, rudder, engines, control devices, and so forth. So to grasp how the bomber must work was straightforward. Once the "how" of a design is captured, a deeper goal can be approached: what a reverse engineer really seeks is to understand the *why* of a design—*why* has each feature been given its particular form? And *why* are their relationships just so? This is the step that reveals principles; it is the moment of "aha!"—the thrilling reward for the long, dull period of gathering facts.

Neuroscience has fundamentally the same goal: to reverse engineer the brain (O'Connor, Huber, & Svoboda, 2009). What other reason could there be to invest 1 million person-years (so far) in describing so finely the brain's structure, chemistry, and function? But neuroscience has been somewhat handicapped by the lack of a framework for all this data. To some degree we resemble the isolated tribe in New Guinea that in the 1940s encountered a crashed airplane and studied it without comprehending its primary function. Nevertheless, we can learn from the engineers: we should try to state the brain's primary goal and basic performance specifications. We should try to intuit a role for each part. By placing the data in some framework, we can begin to evaluate how well our device works and begin to consider the why of its design. We will make this attempt, even though it will be incomplete, and sometimes wrong.

Designing de novo

Engineers know that they cannot create a general design for a general device—because there is no general material to embody it.[1,2] Engineers must proceed *from* the particular *to* the particular. So they start with a list of questions: Precisely what is this machine supposed to accomplish? How fast must it operate and over what dynamic range? How large can it be and how heavy? How much power can it use? What error rates can be tolerated, and which type of error is most worrisome—a false alarm or a failure to respond? The answers to these questions are design specifications.

Danger lurks in every vague expression: "very fast," "pretty small," "power-efficient," "error free." Generalities raise instant concern because one person's "very" is another's "barely." To a biologist, "brief" is a millisecond (10^{-3} s), but to an electronic engineer, "brief" is a nanosecond (10^{-9} s), and the difference is a millionfold. Engineers know that no device can be truly instantaneous or error free—so they know to ask how high should we set the clock rate, how low should we hold the error rate, and at what costs?

The engineer realizes that every device operates in an environment and that this profoundly affects the design. A car for urban roads can be low slung with slender springs, two-wheel drive, and a transmission geared for highway speeds. But a pickup for rough rural roads needs a higher undercarriage, stouter springs, four-wheel drive, and a transmission geared for power at low speeds. The decision regarding which use is more likely (urban or rural) suffuses the whole design. Moreover an engineer always wants to quantify the particular environment to estimate the frequencies of key features and hazards.

One assumes, for example, that before building a million pickups, someone at Nissan bothered to measure the size distribution of rocks and potholes on rural roads. Then they could calculate what height of undercarriage would clear 99.99% of these obstructions and build to that standard.

Knowing the frequencies of various parameters allows rational consideration of safety factor and robustness: how much extra clearance should be allowed for the rare giant boulder; how much thicker should the springs be for the rare overload? Such considerations immediately raise the issue of expense—for a sturdier machine can always be built, but it will cost more and could be less competitive. So design and cost are inseparable.

Of course, environments change. Roads improve—and then deteriorate—so vehicle designs must take this into account. One strategy is to design a vehicle that is cheap and disposable and then bring out new models frequently. This allows adaptations to environmental changes to appear in the next model. Another strategy is to design a more expensive vehicle and invest it with intrinsically greater adaptive capacity—for example, adjustable suspension. Both designs would operate under the same basic principles; the main difference would lie in their strategies for adaptation to changes in demand. In biology the first strategy favors small animals with short lives; the second strategy, by conserving time and effort already invested, favors larger animals with longer lives. As we will see, these complementary strategies account for many differences between the brains of tiny worms, flies, and humans.

Design evolves in the context of *competition*. Most designs are not de novo but rather are based upon an already existing device. The new version tries to surpass the competition: lighter, faster, cheaper, more reliable—but each advance is generally modest. To totally scrap an older model and start fresh would cost too much, take too long, and so on. However, suppose a small part could be modified slightly to improve one factor—or simply make the model prettier? The advance might pay for itself because the device would compete better with others of the same class. A backpacker need not outrun the bear—just a companion—and the same is true for improvements in design. The revolutionary Model T Ford was not the best car ever built, but it was terrific for its time: cheaper and more reliable than its competitors.

How engineers design

An engineer takes account of the laws of physics, such as mechanics and thermodynamics. For example, a turbine works most efficiently when the pressure drop is greatest, so this is where to place the dam or hydro-tunnel. Similarly, power generation from steam is most efficient at high temperatures, which requires high pressures. But using pressure to do work is most efficient when the pressure change is infinitesimally small—which takes infinitely long. There is no "right" answer here, but the laws of physics govern the practicality of power generation and power consumption—and thus affect many industrial designs.

Similarly, a designer is aware of unalterable physical properties. Certain particles move rapidly: photons in a vacuum (3×10^8 m in a second). In contrast, other particles move slowly: an amino acid diffusing in water (~1 μm in a millisecond)—a difference of 10^{14}. So for a communications engineer to choose photons to send a message would seem like a "no brainer"— except that actual brains rely extensively on diffusion! This point will be developed in chapters 5 and 6.

Designers pay particular attention to the interfaces where energy is transferred from one medium to another. For example, an automobile designed for a V-8 engine needs wide tires to powerfully grip the road. This is the final interface, tire-to-road, through which the engine's power is delivered; so to use narrow, lightly treaded tires would be worse than pointless—it would be lethal. More generally it is efficient to match components—for their operating capacities, robustness, reliability, and so on. Efficient designs will match the capacities of all parts so that none are too large or too small.

Matching may be achieved straightforwardly where the properties of the input are predictable, such as a power transformer driven by the line voltage, or a transistor switch in a digital circuit. But the engineer knows that the real world is more variable and allows for this in the design—by providing greater tolerances, or adjusting the matches with feedback. And to estimate what tolerances or what sorts of feedback are needed, the engineer—once again—must analyze the statistics of the environment. Chapters 8–12 will do this for vision.

What components?

Having identified a specific task, its context and constraints, a designer starts to sketch a device. The process draws on deep knowledge of the available components—their intrinsic properties (both advantageous and problematic), their functional relationships, robustness, modifiability, and cost. A mechanical engineer draws from a vast inventory of standard bolts, gears, and bearings and exploits the malleability and versatility of plastics and metal alloys to tailor new parts to particular functions. For example, Henry Ford, in designing his 1908 Model T, solved the mechanical problem of axles cracking on roads built for horses by choosing a tougher, lighter steel alloyed with vanadium.[3] An electrical engineer solves an electronic problem by drawing on a parts catalog or else drawing on known properties and costs to design a new chip. Consequently, as models advance, the number of parts grows explosively: the Boeing 747 comprises 6 million parts.

In these respects the genome is a parts catalog—a list of DNA sequences that can be transcribed into RNA sequences ("messengers") that in turn can be translated into amino acid sequences—to create proteins that serve signaling in innumerable ways. This extensive genetic parts list is not the end, but rather a start—for there are vast opportunities for further innovation and tailoring (see chapter 5). An existing gene can be duplicated and then modified slightly, just as an engineer would hope, to produce an alternative function. For example, the protein (*opsin*) that is tuned to capture light at middle wavelengths (550 nm) has been duplicated and retuned in evolution by changing only a few amino acids out of several hundred to capture longer wavelengths (570 nm). This seemingly minor difference supports our ability to distinguish red from green.

At the next level, a single DNA sequence can be transcribed to produce shorter sequences of messenger RNA that can be spliced in alternative patterns to produce subtle but critical variants. For example, alternative splicing produces large families of receptor proteins with subtly different binding affinities—which give different time constants. Other variants desensitize at different rates. How these variations are exploited in neural design will be discussed, as will the capacity to further innovate and tailor the actual proteins by binding small ions and covalently adding small chemical groups (posttranslational modification). In short, with 20% of our genome devoted to coding neural signaling molecules, plus the additional variation allowed by duplication, alternative splicing, and posttranslational modification, the brain draws from a large inventory of adaptable parts. The versatility of these components, as explained further in chapter 5, is a major key to the brain's success.

At a still higher level biological design builds on preexisting structures and processes. Where a need arises from an animal's opportunity to exploit an abundant resource, natural selection can fashion a new organ from an old one that served a different purpose. Famously, for example, the panda's "thumb" evolved not from the first digit that humans inherited from earlier primates but from a small bone in the hand of its ancestors that served a different purpose (Gould, 1992). Thus, efficient designs can be reached via natural selection from various directions and various developmental sequences. This was recognized a century ago by a key founder of neuroscience, Santiago Ramón y Cajal (1909):

We certainly acknowledge that developmental conditions contribute to morphological features. Nevertheless, although cellular development may reveal how a particular feature assumes its mature form, it cannot clarify the utilitarian or teleological forces that led developmental mechanisms to incorporate this new anatomical feature. (edited for brevity)

Neatening up

A designer's list of needed functions might also reveal several that could be accomplished with a single, well-placed component. This has been termed "neatening up," which Ford certainly did for the Model T. For example, rather than manufacture separate cylinders and bolt them together (the standard method), he cast the engine as a solid block with holes for the cylinders. Moreover, rather than use a separate belt to drive the magneto (which provides spark to initiate combustion), he built magnets into the engine's flywheel—thereby reducing parts and weight. The sum of his efforts to improve components (vanadium steel) and design (flexible suspension) and neaten up (engine block, magneto) produced a model that was 25% lighter and delivered 25% more horsepower per pound than the competition, such as the Buick Tourabout.

Brain design reflects this process of neatening up. For example, one synapse can simultaneously serve two different pathways: fast and slow; ON and OFF. One neuron can alternately serve two different circuits: one during daylight and another during starlight (chapter 11). But this strategy must not compromise functionality.

Complicate but do not duplicate

Scientists are constantly lashed with the strop from Occam's razor. That is, we are forcefully encouraged to keep our explanatory models and theories *simple*. So the following design principle seems, not merely surprising, but actually counterintuitive: *if one design is simple and another complicated, choose the complicated* (Glegg, 1969; Pahl et al., 2007). Here is the reasoning: when one part is forced to do two jobs, it can do neither well. An example is the two-stroke engine.

The operating cycle of a four-stroke automobile engine involves four sweeps of the piston through the cylinder. One draws in the fuel, the next compresses it, the third delivers power as combustion drives the piston outward, and the fourth sweeps out the exhaust. The two-stroke engine discharges the exhaust with the same stroke that draws in fuel at the bottom of the combustion stroke and the beginning of the compression stroke. This serves well for a lawn mower or a power saw because the simpler design avoids the need for a valve gear to separately port fuel and exhaust, giving a better ratio of power to weight. However, the four-stroke engine's more complicated design delivers much more power per liter of fuel and runs smoother and quieter. Moreover, its more efficient combustion discharges fewer pollutants (French, 1994).

There are general advantages to providing a separate part for each task. First, each part can be independently tuned for speed, sensitivity, and so forth—without compromise. Second, each part can be regulated independently. Third, more parts provide more opportunities for further refinement, innovation, and improvement—simply because there are more starting points.

Complicate! is such an important principle for neural design that it seems justified to give one example. The vertebrate retina might have used only one type of photoreceptor but instead it uses two: rod and cone. The rod photopigment is more stable but slower to regenerate, so it serves best in dim light. The cone photopigment is less stable but faster to regenerate, so it serves best in bright light. Having complicated the retina's cellular architecture with two cell types, each type has developed its own molecular refinements—specialized versions of the transduction molecules and of *their* regulatory molecules—all tuned for different light intensities. To take full advantage of these refinements, the two cell types have developed different circuits within the retina. However, just before the retinal output, there is a neatening up: rod and cone circuits merge to share a set of excitatory synapses onto a set of common output cells (ganglion cells). Further explanation is to be found in chapters 8 and 11.

There is another way to complicate a design: include several parts that appear to serve the same function. For example, a neuron may express several enzymes that produce the same product. And neighboring cells may express different versions of a similar protein; for example, axons and astrocytes (glial cells) in optic nerve both express a sodium/potassium pump but with subtly different properties. Also one region may connect to another via multiple pathways: dorsal spinocerebellar tract, ventral spinocerebellar tract, spino–reticulo–cerebellar tract, spino–olivo–cerebellar tract, and so on. These parallel features might once have been regarded as "redundant"— to increase reliability and protect against failure. But now most biologists appreciate that multiple pathways generally serve different roles and thus are not truly redundant.

Indeed, engineers try avoid redundancy, and for good reason. A part waiting for a function occupies space, adds weight, and costs extra. So, this consideration raises suspicion that the multiplicity of intracellular phosphatases, sodium/potassium pumps, spinocerebellar tracts, and so on represent complexity of the good kind.

Choosing materials

Engineers can choose from diverse materials. However, they must try to select the least costly material that is appropriate for the task. For a sailboat mast, wood was traditional, but it is heavy. Graphite can be equally stiff for less weight, but it is brittle; titanium gives the best physical performance, but it is costly. So the choice depends on whether the boat is a dinghy for weekend sailing or a 12 meter yacht for the America's Cup.

Brain design is forced to select from a far narrower set of materials. For example, biological membranes are composed of lipids and proteins. Although mechanisms for regulating the passage of substances and ions *across* the membrane in either direction are myriad (ion channels, pumps, cotransporters, antiporters, flippases, etc.), the intrinsic properties of the membrane itself are relatively constant. In particular, the membrane's specific capacitance is fixed at around 1 $\mu F\ cm^{-2}$. Neurons generate electrical signals by opening and closing channels in the membrane that allow ions to move down their electrochemical gradient and carry charge in and out of the cell.

The time constant of this electrical response is the product of the membrane resistance and capacitance, but capacitance is fixed. Therefore, to speed up an electrical process, a neuron of given surface area must reduce its membrane resistance by opening more channels, thus allowing more ions to cross the membrane. The cost of restoring these ions so as to maintain the electrochemical gradient is high—in fact, it is the human brain's major energy cost: more than 60% goes for pumping ions, making this a key constraint upon design. Thus, for the brain, as for 12 meter yachts and automobiles, speed comes at a premium—and the brain is forced to use it sparingly. This theme will recur.

Integrating across systems

Engineers look for trade-offs among individual components to improve overall performance. For example, because a truck's suspension reduces shock, investment in better springs and shock absorbers can be traded for weight and strength in the axles. So designers evaluate the whole system to discover where investment in one component is more than compensated by savings in others. This integrated approach to design extends the principle of matching components to include cost.

An example relevant to neural design is the mobile telephone (Mackenzie, 2005). Like many animals, it is small and roams on limited power. Models compete fiercely, and success depends upon performance, beauty, and energy efficiency. One notable innovation provides the phone's "brain," its tiny internal computer, with a *turbo code* that extracts wireless signals from environmental noise. This code employs an algorithm for *belief propagation* that is computationally expensive. But the investment pays because the code eliminates noise so effectively that the efficiency of wireless communication approaches the theoretical limit defined by Shannon's equation (chapter 5).

Optimizing efficiency allows the phone to reduce the amplitude of its output signals. These consume the highest proportion of the phone's power because more energy is needed to transmit radio signals long distances in all directions than to send electrical pulses along short connections in a tiny computer. Consequently, the energy invested in the phone's brain for turbo coding produces much larger savings in the heavy work of signal transmission. By analogy, an animal's small brain saves energy by efficiently directing the activities of large, power-hungry muscles.

To understand the design of an integrated system requires teamwork. When no single person can grasp the details of every component and process, designers team up. Specialists integrate their detailed knowledge of each particular into an efficient whole. It was a team of specialists that reverse engineered the B-29. They needed to combine expertise in aerodynamics, structural engineering, materials science, fluid mechanics, control systems, and so on. Neuroscientists are reaching the same conclusion and forming teams that integrate specialized knowledge to reverse engineer their systems. Brains are integrated systems because they evolved to integrate, so how else can we understand them?

How to proceed and a caution

To consider brain design as a problem of reverse engineering, we must begin with an overview of its main tasks, establish some basic measures of performance, and then see how these relate to the investment of resources in particular mechanisms (chapters 2 and 3). Having established some basic principles, we select one important system—vision—and treat each stage of processing in the framework of design. We present the environmental context, then the circuit structure and some "hows" of its functioning. Then for each stage we will point out some "whys" of the design and note where other neural systems use similar principles.

The design principles evinced here do not explain everything. In fact, principles cannot explain how anything works—not the B-29, not the Model T, and certainly not the brain. What, then, is their use? Design principles deepen our understanding of why things work the way they do, and armed with this deeper understanding, we can reverse engineer more efficiently. Of course, applying inappropriate or misguided principles would slow us down. Thus, principles derived theoretically, without real objects and mechanisms to illustrate them, are not yet of much use. So we attempt to balance the insights that come from principled explanations against the doubts that come from overdoing them.

2   Why an Animal Needs a Brain

Essentially only one thing in life interests us: our psychical constitution. The considerations which I have placed before you employ a scientific method in the study of these highest manifestations in the dog, man's best friend.
—Ivan Pavlov, Nobel lecture, 1904 (edited for brevity)

Brain books generally begin at the lowest levels—neurons, axons, synapses, and ion channels. But that approach ill suits our goal of reverse engineering. One cannot explain a B-29 by starting with the nuts and bolts. So we postpone the parts lists and detailed schematics to consider first a larger question: why do we *need* a brain?

One's first thought, of course, is that we need it for the magical activities and feelings it confers: art, music, love . . . consciousness. But although these features arouse intense curiosity—as Pavlov emphasized—we shall see that they are merely baroque decorations on the brain's fundamental purpose and should not be mistaken for the purpose itself. What we identify here as the brain's purpose, especially because we are seeking principles, should apply not only to humans but as well to the nematode worm, *C. elegans*, and to flies. The deep purpose of the nematode's brain of 302 neurons, the fruit fly's brain of 10^5 neurons, and our own brain of 10^{11} neurons (Azevedo et al., 2009) must be the same. By identifying the basic purpose, we set a context for later considering the "decorations." We expect that research on the mammalian cerebral cortex will not reveal many new principles—rather it will elaborate the core ones. In general, it should be easier to discover them in simpler brains.

The brain's purposes reduce to regulating the internal milieu and helping the organism to survive and reproduce. All complex behavior and mental experience—work and play, music and art, politics and prayer—are but strategies to accomplish these functions. Sharing these fundamental tasks, the brains of worms, flies, and vertebrates show significant similarities—which will be discussed. But first, consider that a tiny bacterium, *E. coli*, and a much larger single-celled protozoan, *Paramecium,* manage these two tasks quite well without a brain. How?

Lives of the Brainless

A bacterium foraging

E. coli is miniscule (1×3 µm) and thrives in a nutritive soup—adrift in the intestinal digests of a large animal (figure 2.1; Alberts et al., 2008). The microbe is equipped with "taste" receptors, a battery of proteins each of which specifically binds an attractant (such as an amino acid or sugar) or a repellant. These receptor proteins cluster on the surface membrane and form signaling complexes within which they cooperate to increase sensitivity and response speed. The largest cluster is at the forward end ready to taste what comes as the bacterium ploughs through the soup. Although each cluster comprises thousands of molecules—to increase the chance of catching a taste—there are only five types of receptor molecule, each responding to a range of related compounds.

The first function of these receptors is to evaluate the *soup du jour*. Each potential nutrient (amino acid, sugar, etc.) requires its own specific transporter (*permease*) for uptake into the bacterium, plus a particular enzyme or even a whole set of enzymes to process it for energy and materials for growth. It would be uneconomical to maintain high levels of all possible transporters and processing enzymes when only a subset is needed at a given moment. Therefore, a cell refrains from synthesizing proteins for uptake and digestion until a taste receptor binds the target molecule. A receptor's binding affinity determines the concentration at which protein synthesis becomes economical.

For its default fuel *E. coli* uses glucose. But when glucose is off the menu, it can use lactose. This requires lactose detectors to call for two proteins: a permease to admit lactose and an enzyme, galactosidase, to split it. The genes coding these proteins are adjacent in *E. coli*'s DNA, comprising an *operon* (genes that work together). Their expression is blocked by a repressor protein that binds to this stretch of DNA and blocks the entry of RNA polymerase, the molecular machine that transcribes DNA to RNA to initiate protein synthesis (figure 2.2). The repressor *is* the lactose detector which, upon binding allolactose (an isomer that always accompanies lactose) changes shape and releases from the DNA. This allows RNA polymerase to move off and transcribe the operon (figure 2.2; Phillips et al., 2009).

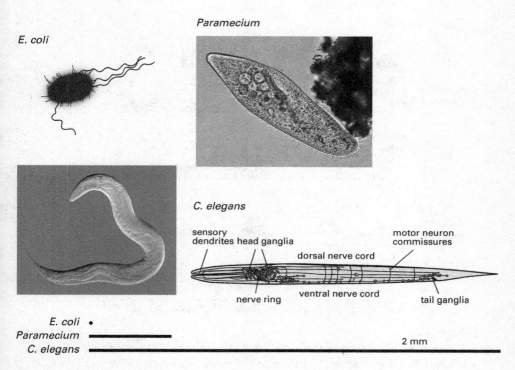

Figure 2.1
Three organisms of increasing size: bacterium, protozoan, and a nematode worm.
Note the different scales: micrometers to millimeters. Body lengths are drawn to the same scale at the bottom of the diagram. *Paramecium caudatum* and *C. elegans* photos are light micrographs of live specimens. Diagram of worm indicates the positions of neurons that form the brain. Light micrographs from Wiki commons. *C. elegans* from Wikimedia Commons, CC BY-SA 3.0 / Bob Goldstein, UNC Chapel Hill, http://bio.unc.edu/people/faculty/goldstein/. *Paramecium* by Alfred Kahl, public domain, from Wikimedia Commons.

In effect, the lactose receptor *predicts* for the organism what it will need to exploit this new resource. By encoding the permease and the digestive enzyme together, one sensory signal can evoke all necessary components in the correct ratios. Thus, a given level of lactose in the soup calls for the proper amount of permease which is matched by the proper amount of galactosidase. This design principle—matching capacities within a coupled system—is a key to the organization of multicellular animals where it is called "symmorphosis" (Weibel, 2000). We see here that symmorphosis begins in the single cell.

The lac operon

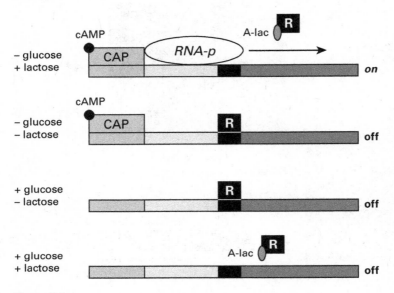

Figure 2.2
The lac operon: a molecular mechanism that discriminates between patterns of input and determines action. To transcribe the lac operon's genes, RNA polymerase (*RNA-P*) must bind to its site and move into the operon's DNA. Its movement is blocked by the repressor R, but R cannot bind and block when holding a molecule of allolactose (A-lac). To start moving, *RNA-p* must be activated by the protein CAP. This activator protein only binds to its site on the DNA when it is binding cAMP, and cAMP is eliminated in the presence of glucose. Thus, *RNA-p* only transcribes the lac operon when glucose is absent and lactose is present.

On occasions, such as when its host has eaten an ice cream, *E. coli* is presented with both lactose *and* glucose. Now the bacterium need not metabolize lactose and so need not build machinery to process it. To block this futile activity, there is a second molecular switch. RNA polymerase, to step along the DNA transcribing the lac operon, must be activated by the protein CAP, and CAP must be binding a small signaling molecule, cAMP. Biochemical pathways couple the production of cAMP to the concentration of glucose. As glucose rises, cAMP falls; this turns off the RNA polymerase (figure 2.2), and *E. coli* stops producing unneeded machinery.

Thus, a molecular control system combines information from two inputs to compute the correct conditions for processing lactose: IF lactose AND NO glucose, then GO; IF lactose AND glucose, then NO GO. The chemical network controlling the lac operon enables a single cell to detect specific patterns of events and to mount concerted patterns of response that promote survival and reproduction. Of course, this is what a brain does on a larger scale, and in doing so it builds upon the capacities for executing logic that reside in the molecular control systems of single cells (Bray, 2009).

E. coli does more than just taste the soup and reprogram its digestive enzymes. The taste receptors also direct the cell to forage, that is, to discover and migrate to regions of higher nutrient concentration. To execute this process, *chemotaxis*, the bacterium propels itself with flagella, which are helical screws that rotate at 6,000 rpm. Their beating sends it tumbling off in random directions for brief periods, each followed by a short, straightish run. A surface receptor, sensing the instantaneous concentration of a nutrient, compares it to the past concentration—"past" lasting 1 s. If the new concentration is higher, the motor apparatus holds the forward course for a bit longer.

This search strategy (*biased random walk*, figure 2.3) resembles the party game where an object is hidden and a searcher is simply told "warmer . . . cooler . . . warmer, warmer. . ." The mechanism can sum signals from several attractants—maintaining the direction of motion for a longer time. Or, it can sum antagonistic signals (attractant + repellent) and change direction sooner. Thus, with a sensor, plus a "working memory" that controls a propeller, a microbe's wandering eventually delivers it to a greener pasture (Berg, 1993).

A microbe's memory

E. coli's working memory is simple: it is imprinted on the receptor protein by means of a negative feedback loop. The activated taste receptor causes an enzyme to attach methyl groups to the receptor complex, decreasing its sensitivity. The number of methyl groups on a receptor indicates how strongly it has been activated, and because the feedback loop is sluggish, the record stretches back into the bacterium's frantic past—1 s. The mechanism, by using the past to set receptor sensitivity, determines the bacterium's response in the present—a reasonable definition of memory. Thus, a single cell can store information cheaply through chemistry—by covalently modifying a signaling molecule.

In accomplishing the basics (preserve internal milieu and reproduce), this single cell uses mechanisms that are either optimal or highly economical: just the right number and distribution of taste receptors, just the right ratios of transporters and digestive enzymes, just the right levels of protein expression to match costs versus resources, plus the smallest signaling

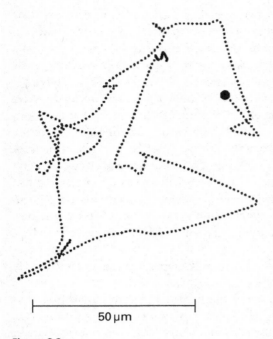

50 μm

Figure 2.3
E. coli's **biased random walk.** By moving forward more and turning less, as the concentration of attractant increases, *E. coli* approaches the attractant's source. Tracing shows 26 runs over about 30 s with a mean speed of 21.2 μm/s. Reprinted with permission from Berg and Brown (1972). For videos of *E. coli* swimming see http://www.rowland.harvard.edu/labs/bacteria/index_movies.html/.

network for chemotaxis that could provide sufficiently robust performance. Moreover, its working memory suffices to steer the motor toward food and mates. Although a memory lasting only 1 s may not seem impressive, realize that to store a long history of lactose concentrations would be pointless— because they are themselves evanescent. Given its lifestyle, the bacterium's memory is just about as long as it *should* be.

This microbe easily lives like a Zen master—in the moment. Feed the cell, and in an hour it is gone, divided among its progeny. But once an organism becomes large enough for a brain, the Zen injunction—"Live in the moment"—itself becomes a Zen koan. A brain provides the organism with a more significant individual past and a more extended future with which to exploit it. But so equipped, staying in the moment becomes as unimaginable as the sound of one hand clapping.

Limitations to life as a microbe

Given that bacteria accomplish the basics so well, one must consider the limitations. First, their ability to respond to environmental challenge resides largely in genetic memory. A *population* thrives by reproducing rapidly and exchanging genetic material—so that when the environment changes, at least one individual in the population will contain a gene to deal with it. Thus, a population can "learn" to exploit new resources—such as potentially delicious industrial waste. However, an individual microbe, suddenly losing glucose in a lactose-rich medium, can respond only if its genome already contains the lac operon.

Second, an individual microbe cannot actively move very far. It can neither return to the site of its last meal nor deliberately transfer to a new host. This confines each species of microbe to the restricted environment for which it has specialized: a termite's gut or the skin of a human inner elbow (Grice et al., 2009)—where the bacterial genome is prepared for what it will likely encounter, and where surprises are relatively few. But this leaves a wider world unexplored and thus unexploited.

To explore would certainly increase the chances of encountering a more favorable medium—but there is a limiting challenge: size. For such a miniscule object, water is tremendously viscous. Top speed for *E. coli* is 30 μm per second, and when its effort ceases, there is insufficient inertia to carry it forward, so it abruptly stops within 0.01 nm (chapter 5; Purcell, 1977; Nelson, 2008). For a human it would be like swimming in thick molasses—agonizingly slow and energetically expensive. Consequently, to move over long distances, bacteria have evolved other methods, for example, by being sticky and hitching rides on animals.

In short, a bacterium inhabits a tiny universe—barely a few centimeters—where the critical factors are beyond its control. When transportation relies on random, energetically expensive self-propulsion or the kindness of strangers, life is precarious. A cell that could propel itself more rapidly and cheaply could forage more widely, but to overcome the effects of Brownian buffeting and high viscosity it must enlarge. And it need not get very large before motor coordination becomes an issue—as we now explain.

Protozoa: bigger and faster but still brainless

Paramecium, the familiar single-celled protozoan, measures up to 350 μm × 50 μm. Being 300,000-fold larger than *E. coli*, it is less subject to viscous forces. *Paramecium* propels itself with cilia that cover its surface and coordinate their beating to send synchronous waves from head to tail. Cruising speed can reach roughly 1,400 μm per second, 50-fold faster than *E. coli* and with lower relative energy cost. In human terms this is the difference between exploring on foot at 4 mph and racing a car at 200 mph. Consequently, *Paramecium* can explore relatively enormous volumes of pond water and harvest bacteria by sweeping them into its "mouth." This microshark is guided by a variety of taste receptors to approach sites where bacteria proliferate, for example, clumps of rotting vegetation. It also has nociceptors to detect toxic sites, such as overripe sludge contaminated with hydrogen sulfide.

In its cluttered environment *Paramecium* inevitably encounters immovable obstacles, and to avoid the futility of continual ramming, *Paramecium* has evolved a useful response (figure 2.4; Jennings, 1904; Eckert, 1972). At the first bump it throws its cilia into reverse and backs off by a few millimeters. Then it does a quick twiddle, switches to forward, and sets off in a new direction. This avoidance response is fast—completed within a fraction of a second—and it has to be. Futile activity wastes time and energy; moreover, the immovable object might be a predator!

E. coli's chemical signaling systems could not trigger and coordinate this rapid response. Diffusion suffices for *E. coli* because the distance is short—a small intracellular messenger molecule diffuses throughout the bacterium in about 4 ms. But diffusion time increases as the distance squared (Nelson, 2008), so for a *Paramecium* that is 100-fold longer than *E. coli*, diffusion from "head" to "tail" would be 10,000-fold slower, about 40 s. Obviously, this is far too slow for receptors at the head to call "*Reverse!*" to the tail cilia. Electrical signals spread much faster: a change in membrane voltage initiated at the head reaches the tail in milliseconds.

Electrical signaling for this avoidance response requires several new components. First, a mechanoreceptor is needed to detect the bump. This involves a specialized cation channel inserted into the cell membrane. Stretch on the membrane deforms the channel, opening it to sodium ions that rapidly depolarize the membrane (<100 μs). Depolarization opens voltage-sensitive calcium channels that admit a rush of calcium ions— further depolarizing the membrane, opening still more calcium channels, and so on. This positive feedback produces a robust response that recruits calcium channels across the entire membrane (figure 2.4). They open briefly, then close and inactivate. Thus, the two components—stretch-gated sodium channel plus voltage-gated calcium channel—cooperate to deliver a synchronous pulse of calcium over the cell's entire surface.

The reason to spread the electrical signal via a calcium channel, rather than a voltage-gated sodium channel (such as used by nerve and muscle), is

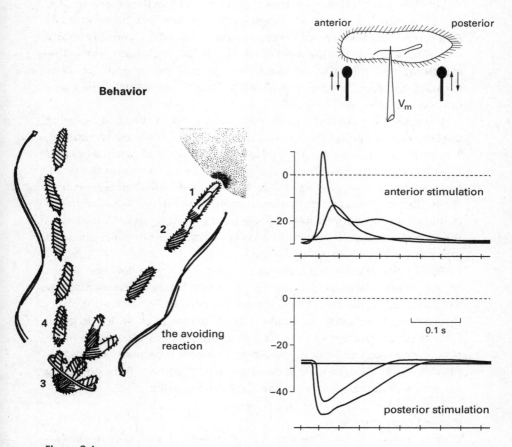

Figure 2.4
Paramecium's **avoidance response: behavior and electrical mechanism. Left**: The four
stages of behavior. (1) Bumps up against immovable object, (2) backs off by reversing
cilia, (3) gyrates while cilia switch from reverse to forward, and (4) sets off in a new
direction. **Upper right**: Measuring electrical response to mechanical stimuli. Intracellular microelectrode records membrane potential and probes prod the membrane.
Middle right: Membrane potential recorded following stimulation with anterior
probe. A weak prod depolarizes membrane for 300 ms (lower trace). A strong prod
generates a short calcium action potential followed by longer depolarization (upper
trace). **Lower right**: Posterior prod hyperpolarizes. The response to the weaker prod
is smaller and has a longer latency. Adapted from Eckert (1972), with permission.

that a calcium ion can also serve intracellularly as a chemical messenger. In this case the chemical message arrives synchronously at the base of all cilia, saying "*Reverse beat*," and their simultaneity adds power to the reversal. As *Paramecium* backs up, calcium pumps in the membrane vigorously reduce the calcium level, allowing patches of cilia to slip back into "forward"— explaining the indecisive twiddle. Once most of the calcium has been extruded and all cilia again beat forward, *Paramecium* heads off in a new direction (figure 2.4).

The system is polarized. The stretch channels are at the head, ensuring that the calcium pulse that reverses the cilia will also reverse the animal. The decision to reverse is structured as a simple threshold: when a bump is sharp enough, stretch channels open sufficiently to depolarize the membrane smartly enough to kick the calcium channels into their regenerative cycle. The numbers and sensitivities of stretch channels are adjusted to discriminate a truly immoveable obstacle from a yielding one. Conceivably, they are even tuned by experience via the attachment of some chemical group as with *E. coli*'s working memory.

Finally, the twiddle that sets *Paramecium* off in a new direction occurs because some patches of cilia enter forward gear before others, perhaps by the molecular noise in calcium pumps (chapter 6). Whatever the exact mechanism, the twiddle generates a random direction—which is good. Lacking distance receptors, *Paramecium* cannot predict which search direction is most likely to be best, so random behavior is optimal (Reynolds & Rhodes, 2009). Also, random motion prevents a predator from predicting *Paramecium*'s next move, thus making it harder to catch.

Where brains emerge

Despite the advantage of its fast control system for locomotion, *Paramecium*'s behavioral repertoire is limited. One impediment to richer behavior is that there is only one cell membrane and thus only one line for fast (electrical) communication. But more deeply, the cell is still so small that locomotion must be slow, and the environment remains so evanescent that richer behavior and longer memory offer no advantage. *Paramecium*'s exploitable world remains sufficiently restricted that one communication channel is plenty. Multicellularity can pay—but only when an animal becomes slightly larger and lives slightly longer in an environment where clues to food and danger persist.

The crossover—where multicellular animals arise and dominate (eat the unicellular)—occurs at a size of around 1 mm and a lifetime of days.[1] Then cells specialize and associate to form tissues, tissues form systems, and systems cooperate to form a more versatile organism. Thus, multicellularity follows the engineering principle *complicate* (Glegg, 1969/2009a). The many tasks performed by a single cell are now divided among many specialized components. Naturally, coordination is required at each level (cell, tissue, organ, system, and organism) and across levels.

Coordination demands some mechanism with an overview that enables it to weigh alternatives, set priorities, and then exert ultimate authority to execute. Fortunately, the multicellular design that demands such integration also provides a special class of cells to accomplish it. These cells—neurons—now do what *Paramecium* could not: provide multiple fast lines for communication. In short, for a multicellular organism a brain becomes necessary, possible, and profitable.

Worm with tiny brain

The nematode worm, *C. elegans*, measures about 1×0.1 mm (figure 2.1) and in its predominant hermaphroditic form comprises exactly 959 somatic cells (Herman, 2006). It lives close to the soil surface and feeds on bacteria in rotting vegetable matter. Unlike *Paramecium*'s pond water, chemicals in soil and humus are not swept away by convective currents—they move by diffusion and capillarity through a matrix, so traces persist (Félix & Braendle, 2010). The matrix and surface film provide firmer substrates for locomotion, and these allow the worm's sinuous crawl to open up whole new continents for exploitation.

The worm's enlarged territory and its locomotion through a labyrinthine matrix with persistent chemical traces warrant an upgrade. The worm improves the chemotaxis system and adds diverse sensors (of current state, opportunity, and danger), plus a larger repertoire of behavioral responses and a longer memory (de Bono & Maricq, 2005). Because bacteria-rich patches are oases where many species compete, the worm's success requires that it move smartly across a patch to efficiently find and exploit the productive regions, meet, mate, and lay eggs.

Improved foraging must be matched by more efficient systems for digestion, absorption, metabolic storage, and elimination. And as the behavioral repertoire expands, there is more need to evaluate and prioritize. For example, upon encountering a good hunting ground, how much heat or acidity should it tolerate? Upon encountering two chemical traces, which should it follow? When to search and when to graze? When to mate and when to be stilled by "satisfaction"? In short many of the choices posed for humans by Ecclesiastes arise even for this apparently simple worm—which decides with its tiny brain.

The worm's brain may be small, but its 302 neurons plus 56 glial and support cells comprise nearly 40% of its body's entire complement. The figure in humans is close to 1%. So we first consider some behavioral advantages that justify its immense investment. Then we consider the brain's design, noting the features shared with larger brains that suggest they are governed by principles of neural design.

Locomotion

Grazers must keep on the move. The worm moves forward by bending just behind the head and then propagating the bend toward the tail. Driven by this sinusoidal wave, it threads its way through soil and rotting vegetable matter, swims through pools of fluid, and crawls across moist surfaces (e.g., decaying fruits, agar plates in laboratories). A worm travels fastest when rigid objects are regularly spaced at 0.5 mm (figure 2.5), and if this spacing is changed by just 10%, their forward speed halves. A worm seems designed to cope best with the average particle size in its preferred habitat, like a pickup truck designed for rough roads (Park et al., 2008).

But *C. elegans* is both truck and driver, continually adapting its propulsion to cope with changing conditions. When the worm goes from swimming in a pool to crawling across a wet surface, the surface tension increases viscous forces 10,000-fold, and the worm adjusts its undulations accordingly (figure 2.5). Frequency falls tenfold, wavelength shortens threefold, and more muscular power is transferred to the viscous medium. The worm continuously adjusts its drive train over a wide range of conditions, maintaining the wave's angle of attack at an efficient value, close to 45° (figure 2.5). To understand how, we must examine the integrated locomotor system: brain, muscles, body, and substrate.

A sequence of muscular contractions produces the moving wave (Sengupta & Samuel, 2009). Muscle cells on the upper side of the body contract to bow out the lower side, and when the upper cells relax, the body springs back, driven by an internal hydrostatic pressure of 0.5 atmospheres. The wave is propagated by sending two opposite bends along the body, one after the other (figure 2.6), and this sequence repeats at the frequency of undulation. When the head leads the tail, the wave moves down the worm, pushing it forward, and when the tail leads, the worm moves backward. The head also wags from side to side, and when the worm decides to

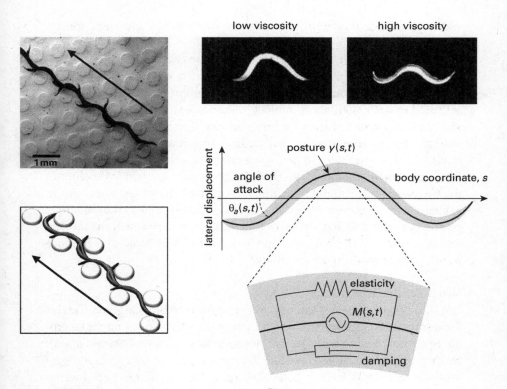

Figure 2.5
C. elegans locomotion matches the terrain and adapts to viscosity. Spacing of soil
particles affects forward speed, as shown when worm crawls through a regular array
of agar posts of given spacing. **Upper left:** Superposition of 10 photos taken at 200-
ms intervals as a worm traversed the array in which it moved forwards at maximum
speed. **Lower left:** Tracings of five of the above photos, taken at 400-ms intervals,
show why speed is maximum: body wavelength matches post spacing to distribute
thrust efficiently. **Upper right:** The wavelength of undulation is longer in a low-
viscosity medium and shorter in high viscosity. **Middle right:** Body posture is described by $y(s,t)$, the lateral displacement, y, changing with position along body, s,
and time, t. The angle of attack at a given position and time, $\theta_a(s,t)$, is critical for
determining thrust against the substrate. **Lower right:** The factors determining body
posture and its dependence on viscosity. These vary with position along the body,
s, and change with time t. In a simple biomechanical model the muscle force $M(s, t)$
interacts with body elasticity and viscous damping by the medium, to determine lateral displacement $y(s, t)$ and the angle of attack $\theta_a(s, t)$. Left reprinted with permission
from Park et al. (2008). Right reprinted with permission from Fang-Yen et al. (2010).

suddenly change direction, it bends the whole body and then springs back—a good tactic for evasion and escape.

These four distinct patterns (forward, reverse, wag, and turn) are produced by 75 motoneurons that control 95 muscle cells. Each muscle cell receives input from one excitatory and one inhibitory neuron which are activated in strict alternation (Bullock, Orkand, & Grinnell, 1977). To bend the head, an excitatory motor neuron on one side of the body activates a muscle, and an inhibitory motor neuron suppresses the corresponding muscle on the other side. To propagate the bend as a wave, motor neurons activate sequentially along the body. Their output frequency determines the frequency of the undulation, and their phase determines its waveform. Excitatory motoneurons on one side activate with inhibitory motoneurons on the opposite side and alternate with excitatory motor neurons on that side (figure 2.6). Where should one look for the oscillators that produce these cycles of motor neuron activity?

Search for the oscillators

Early studies of animal locomotion were fraught with bitter argument about the origins of cyclical activity—such as stepping. Oscillations might be produced within the nervous system by local circuits (*central pattern generators*). Or they might be produced outside the nervous system by cycling sensory feedback (Marder & Bucher, 2001; Goulding, 2009). The feedback mechanism was proposed early for vertebrate stepping. One set of motor neurons excites muscles that extend the limb. This activates sensors that inhibit the extensor motor neurons and excite the flexor motor neurons, thus retracting the limb. Flexion activates sensors that inhibit the flexor motor neurons and excite the extensor neurons, and so on.

Many animals combine the two mechanisms. A central pattern generator sends cyclical commands to the motor neurons, and sensory feedback adjusts their phase, frequency, and amplitude to match changes in external load (Burrows, 1996). But the worm's circuitry seems not to use a central pattern generator. No intrinsically oscillating neurons have been found, nor does the brain's wiring diagram (see below) show the typical oscillatory circuit—a small group of neurons that send signals around a closed loop. Worms are capable of making central pattern generators—some of their cells use internal biochemical oscillators to control the rhythmical movements of ingestion, defecation, and copulation. That the worm can make central pattern generators but does not do so for locomotion suggests that it might have found a better way. Rather than relying on a pattern generator in its brain, the worm exploits its body.

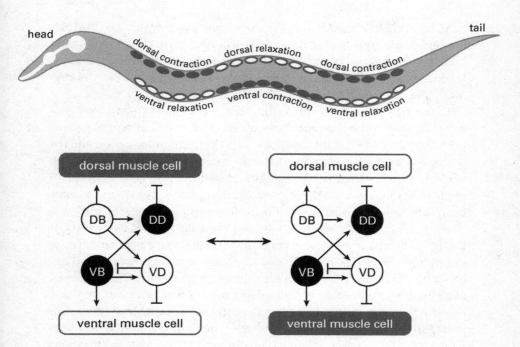

Figure 2.6
Neural circuit that bends the worm. Excitatory motor neurons (DB, VB) alternately cause dorsal and ventral muscles to contract, whereas inhibitory motor neurons (DD, VD) alternately cause them to relax. The excitatory motor neuron on one side drives the inhibitory neuron on the other side so that the body bows downward (DB and VD active), or upward (VB and DD active). This cross-inhibitory circuit repeats along the worm to promote a traveling wave. Modified from Sengupta & Samuel (2009), with permission.

Cycling with the body

The worm builds its oscillator by combining feedback with body mechanics. A burst of activity in motor neurons drives the muscles on one side. Their contraction bends the body and tensions the body's intrinsic spring—internal hydrostatic pressure. Sensors excited by these forces feed back to inhibit motor neurons, whereupon the muscles relax and the body springs back. This terminates the negative feedback, allowing the motor neurons to reactivate and start a new cycle (figure 2.6). Because the spring is damped by viscous forces (figure 2.5), the oscillation is well behaved. Also, it automatically adjusts to changes in viscous load, smoothly shifting the worm's gait to match operating conditions.

So by using its biomechanics the worm can dispense with a central pattern generator, thus freeing up brain space. Here, then, is a useful design principle for motor systems: lighten the brain's load by using the body. Engineers call this *embodied computation* (also embodied intelligence or cognition; Pfeifer & Bongard, 2006).

In the early days of robots, crawling and stepping movements were generated by an all-powerful central computer—an omniscient central pattern generator. This artificial intelligence collected sensory information and fed it into a complicated program that, by modeling the robot's mechanics, worked out the necessary commands and sent them to slavish limbs. To implement this top-down design required the robot to drag around a heavy computer, which, in turn, meant thicker limbs and stronger actuators—the result, a power-hungry behemoth. It was eventually realized that the robot and its limbs *are* a computer, an analogue computer that runs its mechanics in real time (Brooks, 1990). This analogue computer comes for free and can be set up to process information for control by, for example, being part of an oscillator. This insight inspired a new generation of small, efficient, and adroit stepping machines that blew away the behemoths. Thus, the worm exemplifies embodied computation with a neuromechanical system that matches and integrates a few basic components to meet specifications efficiently.

Neural circuits coordinate patterns of movement

Despite the contribution of body mechanics to the oscillator, neural circuits are still essential—they close the loop inside the worm. The neural circuits must be correctly configured and tuned to work with the biomechanics. Sensors must give the right feedback to motor neurons, and motor neurons must send the right signals to the right muscles with the right timing. Circuits are constructed to make this happen by ensuring that as muscles on one side of the body contract, the antagonistic muscles on the other side relax: motor neurons on one side inhibit the excitatory motor neurons for the antagonists and also excite their inhibitory motor neurons (figure 2.6). Here, then, is a circuit motif, *reciprocal inhibition* (Sherrington, 1906), that is widely employed in brains because it simply and effectively solves a common problem.

Changing direction

The brain produces motor rhythms for "forward" and "backward" using two separate sets of motor neurons. Each set has its own circuit: one works with the biomechanics to send the undulatory wave head-to-tail and the other works to send the wave tail-to-head. This is not a popular design. Most animals use a single set of motor neurons as the final common pathway for all commands to muscle. Using two independent sets, each with a full complement of connections and synapses to muscles, seems wasteful, so why does the worm do this? We speculate that for a small brain with neuromechanical oscillation, two sets of motor neurons are cheaper than a complicated central pattern generator.

Directing action

Like *E. coli* and *Paramecium*, the worm acts to improve its chances of completing its production on the ecological stage. Equipped to move further and faster, its costs are higher and the risks greater, but so are the opportunities and rewards. So the acts must be directed appropriately (de Bono & Maricq, 2005; Lockery, 2011).

The simplest acts are aversive responses, similar in purpose and effect to *Paramecium*'s avoidance response. Tap the worm's head, and it immediately wriggles backward; tap its tail, and it wriggles forward. Two simple circuits generate this behavior (figure 2.7). Mechanosensory neurons at the front drive interneurons that activate the "backward" set of motor neurons, and mechanosensory neurons at the rear drive interneurons that activate the "forward" set. The two sets have cross connections to prevent their working in opposition.

Just as the purpose of *E. coli*'s actions is laid out in chemical circuits in a single cell, so the purpose of the worm's behavior is laid out in the connections between neurons. Naturally a brain with many neurons can generate richer behavior because, by forming connections between cells, it makes more circuits. How has the worm's brain harnessed this potential and moved its behavior beyond the simple reactions of *E. coli* and *Paramecium*?

Brain and behavior

Like the single-celled organisms the worm retreats from noxious chemicals, but its decision is more finely judged. A single sensor, the neuron labeled ASH in the brain's wiring diagram, controls this behavior by driving a "retreat" command interneuron, AVA, which shuts down the "forward" motor neurons and activates the "backward" motor neurons (figure 2.8). The sensor ASH expresses molecular receptors and detectors for a variety of potential threats, such as heavy metals, detergents, acids, or high temperature. Each input contributes to ASH activity, and when their sum suffices to trigger the command neuron, the worm backs off. Thus, a single neuron ASH serves as lawyer, jury, judge, and enforcer. It defines what constitutes

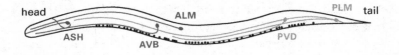

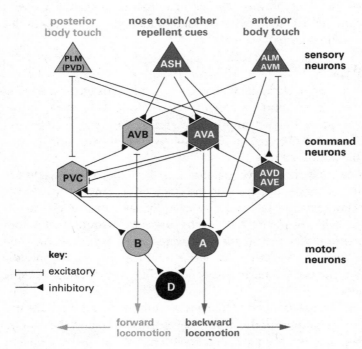

Figure 2.7
The circuit for aversive behavior. Mechanosensory neurons in the nose and in other anterior parts of the body drive command neurons for "backward" motor neurons. Mechanosensory neurons at the posterior end drive command neurons for "forward" motor neurons. These two pathways cross inhibit at the levels of command neurons and motor neurons. Adapted from de Bono & Maricq (2005), with permission.

evidence by selecting which receptors to express on its surface, collects the evidence, weighs it, judges if it warrants escape, and mandates the decision. The worm has several such sensory neurons, collecting other lines evidence for other actions.

Finding warmth, food, and mates
The worm seeks congenial places to feed, grow, and mate. *C. elegans* thrives and reproduces in a fairly narrow range of conditions: dim light,

temperature 13°–25° C, oxygen concentration 7%–14%, moderate pH, ample bacteria, and so on. To find these conditions, the worm needs a signal to warn it of imminent departure from the range—"bacteria depleted," "temperature dropping," and so forth. This search signal activates forward crawl. Foraging now for bacteria by taste and smell, the first whiff activates gradient ascent. Upon reaching favorable conditions, the worm needs a stop signal to announce "satisfaction"—what was sought is found. This signal activates a sequence of turns that places the worm in graze mode. But the worm remains vigilant. If at any moment sensors for noxious conditions are activated, they suppress the forward movement and turning, and they activate reverse.

The worm retains *E. coli*'s basic strategy for moving up or down a gradient, the biased random walk. As conditions improve, the worm turns less and runs ahead more; as conditions worsen, it turns more and runs ahead less. The mechanism is also similar: molecular receptors that drive the forward run adapt, and the decay of their output signals allows a turn. Stronger signals decay more slowly, prolonging the run.

However, with multicellularity comes an advance: ascending the gradient with paired sensors. For salt, a sensor on the right side of the head is excited by *increasing* salt, and a sensor on the left side is excited by *decreasing* salt. The right sensor excites the "forward" circuit and inhibits turning. Once the worm finds the peak concentration, this cell falls silent. If the worm moves off the peak, the left cell, excited by decreasing salt, reduces forward motion and excites turning. This search pattern, brief forward motion followed by turning, continues until the concentration starts to rise again.

The worm uses head wagging to expose both sensors to new territory and combines this action with forward thrust. This exemplifies a motor output modulated by sensing. This system also provides a case where two communication channels collect *identical* information, by sensing the same gradient, but extract different patterns and use them to drive opposite motor responses. Here is something else that a brain offers—new forms of pattern recognition that improve foraging.

Improved sensing and control are needed because *C. elegans* is to *E. coli* as a supertanker is to a rowboat. To steer a whole organism in random directions with gradual correction works on a small scale, but on a larger scale it becomes wasteful. Better for the worm to be more discriminating, to search with its *head* and inform the body once a course can be plotted. In still larger animals the sensors themselves are motorized—an insect antenna, a cat external ear, a human eye (chapter 4).

Because most worms use the same foraging circuits, they accumulate at the same sites—like undergraduates at a good café. And the subtext is similar: a place to feed is also a place to find mates. Moreover, the worms, unlike most undergraduates, are commonly hermaphroditic, so doubling their chances of a satisfying encounter. Even so, many worms enhance their attractiveness by releasing a pheromone to which intrinsically social worms are attracted. Movement toward the pheromone is controlled by a single neuron, RMG, a network hub that collects and integrates inputs from a suite of sensors and pheromones and drives the appropriate command interneurons (figure 2.8). A worm's degree of sociality is adjusted by a particular peptide released within the brain in response to changing conditions. The peptide, one member of a class of *neuromodulators*, binds to receptor proteins on specific neurons to change their activity—and hence behavior (Bargmann, 2012).

Stick and carrot
When local conditions begin to deteriorate, some definite signal is needed for the worm to move on. One such signal is the neuromodulator, octopamine. When food reserves fall, certain neurons release octopamine, which binds to receptors on particular target neurons, modifying their excitability and changing their synapses. This inhibits turning and activates the forward motor pattern. Thus, a single agent, released in response to a change in conditions, acts on specific neurons to alter circuits and switch the worm's program from "graze" to "roam."

When food is found, roaming stops and grazing resumes. This involves a second neuromodulator, dopamine. In mammalian brain, dopamine signifies (among other things) that a reward has exceeded its expected value. In worm, dopamine is released by the presence of food when, for example, mechanosensors touch particles the size of bacteria. Dopamine binds to receptors on target neurons, turning off the octopamine receptors and restoring the circuit to its previous configuration. This switches the worm from roaming to collecting its food reward. Thus, two neuromodulators, octopamine and dopamine, provide this tiny brain with a primordial stick and a primordial carrot to mediate, as they do in larger brains, "anxious" searching and "pleasurable" repetition (de Bono & Maricq, 2005).

Imminent starvation is not the only stress. Others include low oxygen, high CO_2, acidity and overcrowding. All suggest an exhausted patch—time to move on. As with humans, stress increases urgency. A comfortable worm

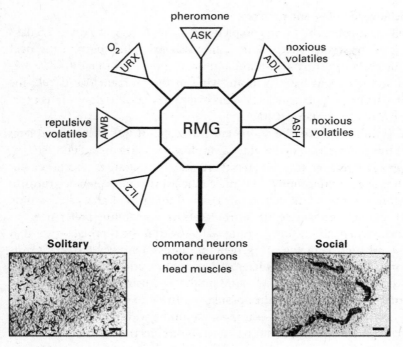

Figure 2.8
C. elegans. **Spoke and hub circuit controls solitary versus social behavior (dispersed vs. huddled). Upper:** The neuron RMG integrates social cues sensed by particular sensory neurons, ASK etc., and drives neurons that implement behavior. **Lower:** Social behavior. Solitary worms disperse and keep apart. Social worms huddle in groups. Each worm appears as a dark speck. Diagram adapted from Sokolowski (2010). Solitary and social worms from de Bono & Bargmann (1998), with permission; scale bars 1 mm.

moves leisurely up a promising chemical gradient, but a worm subjected to low oxygen for several hours ascends quickly. To change from stroll to rush, neuromodulators reconfigure the circuit for gradient ascent (Bargmann, 2012). For example, the sensors ADF and ASG respond to low oxygen by releasing another neuromodulator, serotonin.

Just as "carrot and stick" oversimplify human motivation, so it is for the worm. Competing for limited resources requires many factors to be weighed in deciding whether to roam or graze. A rich suite of neuromodulators allows the worm's brain of 302 neurons to evaluate contextual factors, such as nutritional status, food availability, crowding, and social signals, and then reconfigure accordingly.

Associative learning and memory

When life is good, the worm completes its life cycle (egg to egg) in 3.5 days and lives for several weeks. With a life span extending beyond the next mitotic cycle, allowing a past and a future, it now pays to recall what was good and what was bad. Far from living in the moment like *E. coli*, the worm uses its brain to associate events over time and thus draw on its experience (Ardiel & Rankin, 2010).

A worm remembers the temperature at which it was well fed and later seeks this temperature by moving up or down a thermal gradient. Finding the preferred temperature, it hangs there, searching along the isotherm. But dopamine decays promptly, so if the cupboard is bare, preference turns to aversion and the worm crawls off. Upon finding food and thus earning another shot of dopamine, the worm resets its temperature preference.

The mechanism for this learning resides within the thermal sensor that drives oriented crawling. This neuron senses changes of $0.003°$ C. Its response is minimal at the preferred temperature and rises on either side. The temperature for this minimum is reset by adjustments to the neuron's internal signaling; this requires protein synthesis and takes several hours. This learning process—chemical reprogramming within a single neuron—changes protein molecules but not synaptic connections.

Chemical preferences can also become associated with particular signals. For example, NaCl (salt) normally attracts worms, but when a worm has been starved in the presence of salt for only 10 minutes, it later avoids salt. A particular neuron downstream from the salt sensor releases another neuromodulator (insulin) that feeds back to an insulin receptor on the salt sensor to activate an internal signaling pathway (involving PIP3-kinase) to suppress attraction. Again, reprogramming a signaling pathway *within* a neuron allows experience to change the balance between attraction and repulsion. This mechanism also serves odorants. *C. elegans* even learns to avoid odorants from a particular pathogenic strain of bacteria that has made it sick.

These memory traces promote survival by extending the time over which an animal can identify and use patterns. The number of trials needed to establish an association is modest, five to ten repeats over 20 minutes. This makes sense in an environment where conditions are sufficiently shifty that to be useful, an association must establish rapidly and decay rapidly. In short, the worm's behavior demonstrates its reliance on information from three distinct sources: outside, inside, and the past. Its brain integrates these streams to select behaviors that, reflecting a wider context, improve the worm's vitality and reproductive success.

Some design aspects of this tiny brain

C. elegans' brain may be small, but it is not simple. To achieve its panoply of behaviors, the worm draws on a large catalog of molecular parts. This includes diverse proteins for intracellular chemical and electrical signaling, plus numerous parts for processing information at synapses. For example, signaling proteins occupy 20% of the worm's genome, and its 300+ synaptic parts amount to one third the number for mammals (Emes et al., 2008). In fact the worm brain uses many of the same components present in larger brains. Since parts are shared, one might expect some design rules to be shared as well. If some rules were not shared, that would also be instructive, for it might suggest costs and benefits of scaling up.

Here then are some design features gleaned from considering the worm's brain and what they might imply for bigger brains.

Computes as much as possible within a single cell

This feature is exemplified by the worm's *receptors* and their *sensors*. We distinguish these terms: "receptor" refers to an individual *protein molecule* that responds to a specific event—like stretch, temperature, protons, or chemical binding; "sensor" refers to an individual *neuron* that expresses one or more types of receptor. Although neuroscientists understand this difference perfectly well, for historical reasons they often use "receptor" for both the molecule and the neuron. We use different terms to reduce confusion for readers unfamiliar with the jargon, and also because they raise two design problems.

First, a single receptor molecule is subject to stochastic fluctuations, such as thermal noise. Therefore a neuron might need to improve the signal-to-noise ratio of signals conveyed by one receptor by averaging over a population of the same type. This raises the following design question: How many receptors of the same type should be expressed by each sensor? The answer will be given in chapter 6.

Second, receptors are more diverse than the sensor neurons that express them. Therefore, how should diverse receptor types be apportioned among sensors? For this problem *C. elegans* has a rule. If a set of receptors all lead to the same final action, they share a common sensor. For example, the sensor ASH collects signals from various types of receptor for noxious stimuli that require an aversive response; ASH couples its output to a single neuron that executes a command: *Scram!*

This rule explains receptor grouping generally. The worm uses more than 1,700 different types of receptor molecule for chemoreception (taste and olfaction). This considerably exceeds the 800 or so used in mammals, but unlike mammals where each receptor type is typically assigned its own sensor, the worm provides only about 30 separate sensor neurons. Like sensors of noxious stimuli, each chemosensor sends its signal to a specific command neuron. So the signals from 1,700 different input channels (receptors for taste and olfaction) are assembled for action, not by circuits higher in the brain, but by a few dozen sensory neurons.

Computing *within* a cell economizes on neuron numbers. The worm meets all basic requirements for behavior (sensory pattern recognition, sensorimotor integration, and motor control) with small numbers of neurons. Thirty-eight sensors connect to 82 interneurons (whose processes are confined within the brain) that contact 119 motor neurons (cells whose processes leave the brain to contact the worm's 100 muscle cells). This reserves about 70 neurons for internal regulation and mating.

Yet there is a downside to performing several operations in a single cell. A cell's capacity to handle information is limited by factors such as internal noise, dynamic range, and energy supply. So a sensor that processes inputs from several types of receptor compromises its ability to handle the information from any one receptor type. A dedicated sensor can devote more receptors to its particular modality and thus improve sensitivity and signal-to-noise. This is the engineer's principle from chapter 1: to prevent one component from doing two tasks suboptimally, complicate.

Complication goes up the line. Better sensors warrant better sense organs: eyes for vision, ears for hearing, and so on. To benefit from these more accurate and discriminating sense organs, specialized sensory systems evolve in larger brains, each devoted to processing a single modality. The conclusion is obvious: as brains scale up to improve behavior, neurons specialize. Chapter 3 will suggest how and why, but now we consider a related question, how does a worm's tiny brain manage to compute efficiently?

Uses chemistry wherever possible

Many worm neurons use internal molecular circuits to perform functions that in larger brains use a circuit of several neurons. For example, a single sensory neuron, AFD, determines the worm's temperature preference by adding new proteins to its intracellular signaling network. Another neuron, AWC^{ON}, changes a behavioral response to suit the situation. When an odorant is present *without* food, AWC^{ON}'s molecular receptors adapt and chemotaxis declines. However, when the same odorant is present *with* food, its receptors are sensitized, and chemotaxis increases (de Bono & Maricq, 2005). These competing responses are controlled by an intracellular mechanism that switches the connection between sensor and behavioral output to reverse the control of chemotactic turning behavior (Pereira & van der Kooy, 2012).

These examples show that chemical computing by circuits *within* a neuron can manage behavior. Moreover, this can be very efficient because chemical signals are orders of magnitude cheaper than electrical signals (chapters 5 and 6). Chemical diffusion is slow for long distances, but the worm *is* small and slow. Thus, the worm's size and speed well suit its reliance on cheap chemical signaling. In addition, chemical signals can be broadcast to specific targets, which brings us to another design feature.

Uses neuromodulators to switch behaviors

Three neuromodulators were mentioned (octopamine, serotonin, and dopamine) that switch the worm's behavior in response to stress or the prospect of reward. But this is just page one from the parts catalog since the worm expresses 250 small peptides with known neuromodulatory functions. Their diversity and ubiquity is understandable because neuromodulation is so ingenious (Harris-Warrick & Marder, 1991). A neuromodulator can be broadcast widely yet still act locally and specifically, affecting only neurons that express an appropriate receptor. The receptors often couple to a protein that modulates intracellular signaling, so in effect a neuromodulator uses *trans*cellular chemistry to modulate *intra*cellular chemistry.

A neuromodulator's reach is further enhanced because its receptor diversifies into multiple subtypes that couple to different intracellular signaling networks. Consequently, one small molecule can retune and reconfigure a whole neural circuit without altering the anatomical connections. This allows every circuit to always be doing something and then to be recruited for something else as required. Thus, neuromodulators allow the brain to use components to their fullest.

Conserves synapses

The worm brain makes only about 6,400 chemical synapses. This is roughly the number that in a mammal contact a single retinal ganglion cell or a single cortical pyramidal cell. How can a worm operate with so few synapses? The neurons are far smaller and therefore can be driven by fewer synapses. But since a single synapse is unreliable, how can so few synapses signal reliably?

One answer is: *slowly*—a neuron can improve reliability by averaging over time. This can be tolerated because, compared to many animals, the worm lives in the slow lane. For example, its olfactory sensor uses a chemical amplifier, a G protein signaling cascade that integrates for more than 20 s (chapter 5, figure 5.6). This sensor drives a synapse that integrates over several minutes. By comparison, a fly's olfactory system acts in less than 1 s. Locomotor waves descend the worm's body at 1 Hz, but an insect moves its legs faster than 10 Hz. So the worm can prosper with few synapses because it is slow. This suggests another feature: *send information as slowly as possible* because this uses fewer synapses, smaller cells, and less energy. Later chapters explain more.

Uses stereotyped components

Efficient design gives every component a definite task. Once all components are optimized for their tasks and optimally fitted together, it is efficient to repeat them across individuals. Similarly, every neuron in *C. elegans* has a definite role optimized by natural selection to meet a specified level of performance. Correspondingly every neuron is "identified," meaning that it exhibits a stereotyped morphology, chemistry, and location in every animal (White et al., 1986). The circuits are also identified, meaning that the synaptic connections are essentially identical across animals. This was established by reconstructing the entire nervous system from thousands of electron micrographs of serial sections—to produce the worm's *connectome* (figure 2.9). Identified neurons and circuits are consistently found in small brains: worm and water flea, leech and lobster, and so on.

Minimizes wiring costs

The layout of *C. elegans'* neural wiring suggests that all 302 neurons are located as near as possible to the sites where they are needed (Varshney et al., 2011). Chemical and thermal sensors concentrate at the head; tactile sensors that guide locomotion distribute along the body axis; motor neurons that propel the worm forward distribute along the rear half of the body, and motor neurons for reverse locomotion distribute along the front half (figure 2.6). But does the layout approach the optimum sought by chip designers—the unique set of placements that minimizes the total length of connections in the brain?

Designers of silicon chips have developed algorithms to optimize component placement. Their rule: place the most densely interconnected components close together and the more sparsely connected components further apart (figure 2.9). This algorithm applied to the worm's brain shows that 90% of neurons are optimally positioned (Cherniak, 1995; Chen et al., 2006; Pérez-Escudero & de Polavieja, 2007). The 10% of neurons not in their optimal position suggests competing needs. For example, neurons

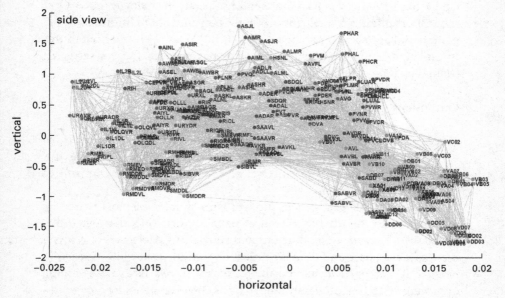

Figure 2.9
C. elegans connectome reconstructed from serial sections photographed in the electron microscope. Each neuron is identified, and its synaptic connections are shown in gray. At the time of writing this is one of the most complete wiring diagrams established for any part of any brain (the other is the fly lamina cartridge, figures 9.2 and 9.3). Careful estimates suggest that this worm connectome is 93% accurate. Such are the technical difficulties of tracing neurons' thin connections that, after two decades of work on 302 neurons, 7% of connections are "missing." Reprinted with permission from Varshney et al. (2011).

that communicate most frequently with each other may be placed closer together to save energy and reduce conduction delays between them. Although layouts in larger brains certainly reflect this, conduction delay may be less relevant for *C. elegans* because the distances are so short, and the worm is so slow. "Short and slow" suggests another design feature.

Favors analogue over pulsatile
Because electrical signals in the worm travel less than a millimeter, neurons can conduct passively, as graded (analogue) changes in electrical potential. The brief, sharp, energy-intensive action potentials that dominate long-distance signaling in larger brains are unneeded, so the worm can rely solely on analogue computations, which are direct and energy efficient

(Sarpeshkar, 1998). Even its motor neurons operate in analogue mode. Over these short distances, analogue signaling transmits more information per neuron and at lower cost (chapter 7). So firmly does *C. elegans* hold to this feature that it has abandoned the gene that encodes the voltage-gated sodium channel used by larger, faster species to produce spikes.

Conclusions

Three organisms of ascending size, *E. coli*, *Paramecium*, and *C. elegans*, show why an animal needs a brain to process information on a larger scale. It is to increase opportunities for survival and reproduction in a competitive and variable environment.

The small single cell, *E. coli*, survives with surface receptors that relay information to the internal chemical signaling networks that determine metabolism, growth, reproduction, and movement. However, *E. coli* is a mere speck in space and time with most opportunities beyond its reach. A larger cell, *Paramecium*, moving more briskly travels farther, expanding opportunities, but is ultimately limited by its chemical signaling networks—diffusion and internal communication by intracellular motors are both too slow. Voltage-gated ion channels added to the cell membrane allow fast electrical signaling, but trapped in a viscous world, a single cell can only do so much.

The multicellular worm, *C. elegans*, overcomes viscosity by enlarging, and it moves faster and farther by specializing cells. This leads it to more opportunities and dangers—richer sources of information to be gathered and processed that finally need a brain. The key innovation is the neuron, a cell type specialized to collect, process, and communicate. Each neuron links its rich web of internal chemical communication to the electrical network at the surface membrane and thence to other neurons via synapses. Neuromodulators retune selected neurons to reconfigure whole circuits. Thus, a brain of only 302 neurons extends the worm's horizon by providing a behavioral repertoire that adapts to changing contexts.

The worm accomplishes the same tasks as a bacterium or protozoan—finds growth conditions and mates while avoiding unproductive or toxic sites. And it does so with similar behaviors, such as gradient ascent by biased random walk and avoidance. But with its brain *C. elegans* can cover more territory, and with its longer lifespan (weeks instead of minutes), it can adapt to nasty surprises as an *individual* rather than as a miniscule part of an adapting *population*.

Here emerges another design principle. Life span and lifestyle are related to the appearance of particular types of memory and particular decay times. Nothing should be remembered that is unlikely to enhance survival and reproduction. Nor should memories exceed the typical time constants of useful correlations—because when correlations decay, memory ceases to predict anything useful. But it *is* useful to establish the memory trace rapidly before it is outdated—and that seems to occur—few trials, closely spaced. This suggests that the longest and deepest human memories are not mere decoration but serve to shape character over a lifetime, promoting survival in our complex social fabric (chapter 14).

Finally, given that *C. elegans* does so well with only 302 neurons, one might look critically at an assumed truth—that it is better to have a bigger brain. So why *have* animals evolved still bigger brains?

3   Why a Bigger Brain?

This chapter will explain why, despite the worm's success with 302 neurons, brains expand. The mouse cerebral cortex contains about 10^7 neurons. This seems like a lot until you consider that the cortex of the macaque monkey, a key experimental model, is larger by 100-fold, and that human cortex is 10-fold larger still (Herculano-Houzel, 2011). Despite this huge range of scales, one feels comfortable generalizing about the "mammalian brain"—because every part identified in mouse can also be identified in macaque and human (figure 3.1; Kaas, 2005).

Consider also the fly brain. It has 500-fold fewer neurons than the mouse brain, but 500-fold more neurons than the worm brain, plus a rich structure—so warranting a slot in the "large brain" category. Insect and mammal brains share many similarities. For example, both gather their neurons into clusters and their axons into cables (*tracts*). Both employ special structures to accomplish the same broad tasks: store high-level input patterns, generate low-level output patterns, and retrieve patterns using reduced instructions. Of course, there are differences, given the differences in body design and behavior. Yet, despite half a billion years of evolutionary opportunity to diverge, brain designs in insect and mammal seem to have followed the same rules.

For designs to have persisted across this immensity of time and spatial scale implies that they are neither arbitrary nor accidental. Rather, they must have emerged as responses to some broad constraint. That is what elevates the shared responses to the status of *principles*. This chapter will identify the key constraint and indicate how it leads to three principles that govern the organization of larger brains.

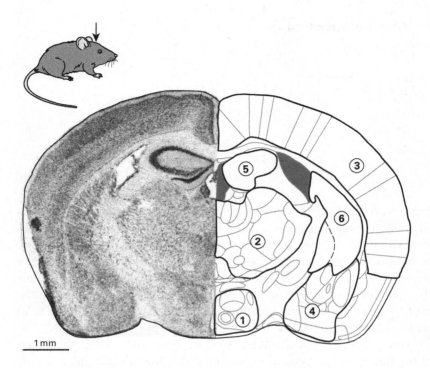

1 mm

① Generate patterns for wireless signaling and appetitive behaviors.
② "Preprocessing" to shape signals for higher processing.
③ High-level processing: assemble larger patterns, choose behaviors.
④ "Tag" high-level patterns for emotional significance.
⑤ Store and recall.
⑥ Evaluate reward predictions.

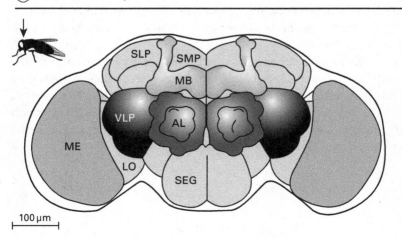

100 µm

Figure 3.1
Mammalian and insect brains share many broad aspects of design. Upper: Cross section through mouse brain; inset indicates plane of section. **Left:** Fine dots are neurons; dark regions are neuron clusters; bright regions are myelinated tracts (chapter 4). **Right:** Numbered regions dedicated to core tasks: (1) *hypothalamus*; (2) *thalamus*; (3) *cerebral cortex*; (4) *amygdaloid complex*; (5) *hippocampus*; (6) *striatum*. Reprinted with modifications and permission from Franklin and Paxinos (1996). **Lower:** Cross section through fly brain; inset indicates plane of section. Brain is built of more than fifty clusters, each specialized for particular tasks. Depicted here are ME, medulla—detect and map local visual patterns; LO, lobula—assemble local visual patterns into larger patterns; AL, antennal lobe—preprocess olfactory signals for pattern recognition; VLP, ventrolateral protocerebrum; SLP, superior lateral protocerebrum; SMP, superior medial protocerebrum—all involved in high-level integration; MB, mushroom body—store and recall; SEG, subesophageal ganglion—integrate information for wired and wireless output to body.

A brain's core tasks

As animals emerge from the soil to a wider, less viscous world, the possibilities for foraging expand immensely. A worm explores mainly in two dimensions over an area of 0.01 m^2 whereas a honeybee typically covers an area of nearly 10^7 m^2, and a fly somewhat less. So foraging area expands by 10^9 (1 billionfold). Add the third dimension, and the volume to be explored becomes astronomical. Larger animals, such as fish, birds, and mammals, may migrate and thus forage over thousands of kilometers—thus millions of square kilometers.

Such gigantic territories contain immense resources and, of course, harbor innumerable dangers. For an animal to find the one and avoid the other requires it to rapidly gather vast amounts of information from the environment. To calibrate "vast" with one example, the eye sends the brain about 10 megabits per second, roughly the rate of an Ethernet connection (Koch et al., 2006). All sense data reach the brain in the form of tiny patterns—evanescent pieces of a dynamic jigsaw puzzle—and to be of any use, they require assembly to reveal a larger pattern. So if gathering information is to be at all rewarding, the brain must commit resources to assembling larger patterns on spatial and temporal scales that are relevant to behavior.

Yet, even a larger pattern might be useless until it is compared to a library of stored patterns where it can be identified: *edible/toxic, friend/foe,* or *search item not found.* Either outcome provides a basis for behavioral choice. A match allows confident choice: eat or decline, approach or flee. A non-match suggests caution and need to gather more data. Thus, the brain requires "pattern comparators," and these must couple to mechanisms that select behaviors: *feed, fight, copulate, investigate*. These, in turn, couple to mechanisms for detailed motor patterns to drive muscles for moving limbs or wings.

Any given motor behavior *might* match exactly the action that was ordered: the arrow might strike the exact point at which it was aimed. But often there are errors due to environmental or neural perturbations, and these need to be identified, so that performance can progressively improve. Thus, a brain needs mechanisms to evaluate the mismatch between the orders it gave and the actual motor performance. So, in addition to sensing and processing patterns to discover "what's important out there," the brain also devotes considerable resources to sensing and processing its own motor errors, and other errors of internal "intentional" signaling in order to improve the accuracy and efficiency of the next round. This is "motor learning."

Behaviors are subject to another important class of errors. Every action has both costs and consequences. The costs are partly energetic: how much energy was spent? But also there are "opportunity costs": could the return have been greater and the risk less for some different action? Every behavior, even when perfectly executed, needs to be evaluated from this perspective: wise or foolish? repeat or not? These evaluations of *reward prediction*, like those for motor errors, are used to update stored knowledge in order to improve the next round of predictions. The nematode worm already shows this type of evaluation to some degree, but animals in the wider world allot it major neural resources.

In sum, to succeed in the wider world, an animal must exchange larger amounts of information with its external environment and also evaluate the costs and consequences of its actions. The seven core tasks that every brain must accomplish are summarized in figure 3.2. What the brain does for the external environment it also does for the internal environment which has also expanded and complexified. Moreover, the mechanisms for managing the internal and external environments need to couple closely in order to serve each other (figure 3.2).

Why the internal milieu needs a brain

To support richer external behaviors, an animal requires specialized internal tissues and organs. Some digest the bounty foraged from the outer

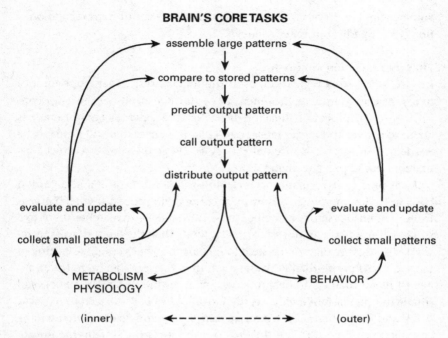

BRAIN'S CORE TASKS

Figure 3.2
Large brains accomplish the same broad tasks. Note that inner and outer tasks couple to serve *each other* (↔).

world; others store metabolites and energy-rich compounds for release upon demand. Still others regulate ionic balance and cleanse the internal milieu, or distribute oxygen and metabolites to hungry tissues. Specialized organs of immunity protect against infectious agents and parasites. Organs couple to form systems, and systems cross-couple to optimize overall function.

The standard idea is that the internal systems more or less take care of themselves. Each parameter is supposed to have a set point, like a thermostat, from which deviations trigger feedback to correct the mismatch (*homeostasis*). Internal regulation also employs *autonomic nerves*—so termed because they are in some sense independent of voluntary control—thus, autonomous. We cannot "will" our heart to beat faster or our blood pressure to decrease. However, we can accomplish these shifts by recalling or imagining the appropriate scene. This implies the existence of neural pathways from pattern stores to pattern generators for autonomic circuits. Thus, although the autonomic nerves are generally supposed to serve emergencies ("fight or flight"), they actually serve continuous regulation—not just for panic, but for efficiency.

Efficient regulation anticipates

In fact, all internal regulation, even the mildest sort, is far from autonomous. As the external environment presents opportunity or cause for concern, internal processes must predict what the external environment is about to deliver and must prepare particular responses that will probably be needed in support. For internal processes the goal is not to correct mismatches but to prevent them.

Such predictive regulation was demonstrated for feeding and digestion by Ivan Pavlov more than a century ago: the brain processes small patterns from the outside (sight or smell of some substance) and matches them to a stored pattern that identifies a particular food. Then the brain triggers secretions all along the digestive system to prepare for what's coming, starting in the mouth (if bread, then amylase; if fat, then lipase), then on to the stomach (if meat, then acid plus protease), the intestine (if fat, then bile), and finally the circulation (if glucose, then insulin). All of these secretions occur *before* and *during* the meal, triggered *predictively*—anticipating what will be coming down the gastrointestinal tract—thus preparing systems for absorption and uptake in order to prevent deviations that would need correction by negative feedback (Fu et al., 2011).

Modern work extends this point: as the stomach releases its contents to the next stage, it also signals the brain to prepare for the next bout of foraging. The brain responds by tuning up sensitivity of the olfactory receptors and by increasing the rate of sniffing (Julliard et al., 2007; Tong et al., 2011). Thus, the stomach warns the brain "Prepare to forage again"—well before the body has begun to deplete its reserves. Moreover, as fat reaches the small intestine, the gut can predict confidently the approach of satiety. Therefore, the gut warns the brain "cease feeding and proceed to the next activity"[1] (Fry et al., 2007).

Each "next activity" requires the brain to predict continuously, and in timely fashion, the need for a particular blood pressure. Consider the record of mean arterial pressure over 24 hours (figure 3.3). In early afternoon, as the subject attends a lecture, his brain anticipates reduced demand and allows him to doze: pressure falls. Startled awake by the jab of a pin, the brain predicts danger: pressure spikes; then, identifying a prank, the brain directs the nap to resume: pressure falls. At midnight the subject has sexual intercourse: pressure spikes, but then falls profoundly and stays low during sleep. Come morning, the brain predicting a busy day, restores the pressure.

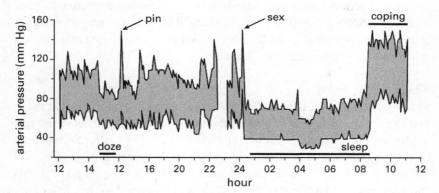

Figure 3.3
Internal systems match behavior. Arterial pressure fluctuates with demand. Each shift in pressure is accompanied by parallel shifts in hormonal and neural signaling that follow the broad catabolic and anabolic patterns. Redrawn from Bevan et al. (1969) and reprinted from Sterling (2004b).

Such anticipatory tuning requires coordinated action of multiple organs and organ systems. To raise pressure, the heart accelerates and vessels constrict. Also the kidney expands blood volume by pumping more salt water into the circulation. The kidney also signals the brain that the body will soon need more supplies of salt and water. Thus, like the gastrointestinal tract, the kidney alerts the brain well in advance of an upcoming need to resupply. Each contribution operates on a different timescale: faster for heart and vessels, slower for kidney's pumping, and still slower for the brain's rise of salt appetite and thirst. These contributions to internal regulation are all initiated simultaneously—and largely by the same signals.

In short, every move we make is matched by a corresponding cardiovascular and renal pattern. Of this we are generally unaware. Yet if the motor command ("Arise!") slightly precedes the internal command ("Tighten vessels!"), blood flow to the head drops, and we faint. That this experience, *postural hypotension*, occurs rarely attests to the rigorous coupling between the cardiovascular pattern and muscular patterns on a 100-ms timescale. On a slower timescale "Arise!" increases by eightfold a signal to the kidney to save water.[2]

Note that matching blood pressure to environmental context requires all of the brain's broad tasks as diagramed in figure 3.2—the collecting and assembling of patterns, the comparison to stores, and so forth. How else to decide if the jab is from a friend or enemy? Moreover, every high-level call to external action is delivered simultaneously to multiple internal organs. Thus, collecting patterns and distributing patterns are both thoroughly coupled between inner and outer worlds. Where and how the brain effects this coupling will be treated in chapter 4.

Adapt, match, trade

Although this book concerns efficient neural design, we must keep in mind that the brain comprises only 2% of the body's mass and 20% of its energy. So the body also needs to operate efficiently. Each organ should match its capacity to the anticipated need of the organ downstream. Too little and the system will fail; too much and capacity is wasted. So each organ needs constant tuning to anticipate the next demand (figure 3.4). But what happens when a need exceeds the capacity to supply? This problem is solved by arranging various short-term "trade-offs." Such cooperation enhances the range of performance while greatly reducing average excess capacity (figure 3.4).

For example, the "resting" heart pumps 6 L of blood per minute through the respiratory system and then out to the general circulation. Resting skeletal muscle uses about 20% of the oxygenated blood—matched to its modest need for maintaining posture. During peak exercise, muscle must increase its supply by nearly 20-fold, but the pulmonary and systemic circulation can increase their outputs only fourfold. Therefore, the body must either reduce its peak capacity for exercise or increase its peak pulmonary and cardiovascular capacity by fivefold—imagine the chest! Or it can borrow.

Indeed, during peak exercise the splanchnic circulation (gut and liver) and the renal circulation (kidney) both reduce their shares by four- to fivefold, enough to pay part of muscle's bill for exercise. During digestion, when the splanchnic circulation needs more blood, it borrows from muscle and skin—unless skin needs blood for cooling. The brain neither makes loans nor allows overdrafts that might cause it to overheat. Anyone who has eaten and then exercised in the sun will recall how these conflicting demands from muscle, gut, and skin are resolved: by corrective motor commands to internal systems ("Vomit!") and to external systems ("Lie down!"). Moreover, the experience receives a strongly negative evaluation that updates the knowledge store ("Do not repeat!").

This example illustrates three key rules for efficient regulation: (1) adapt response capacity to changes in input level, (2) match response capacities across the system, and (3) trade between systems. Regulatory responses begin promptly—as soon as there is sufficient statistical evidence to predict

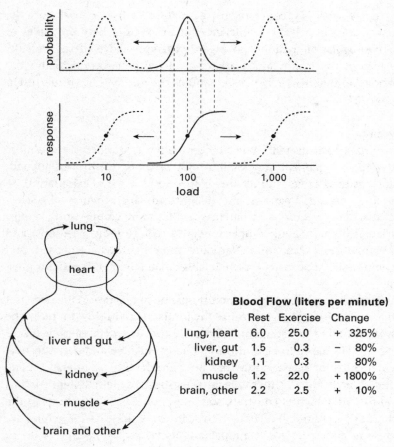

Blood Flow (liters per minute)

	Rest	Exercise	Change
lung, heart	6.0	25.0	+ 325%
liver, gut	1.5	0.3	− 80%
kidney	1.1	0.3	− 80%
muscle	1.2	22.0	+ 1800%
brain, other	2.2	2.5	+ 10%

Figure 3.4
Adapt, match, trade. Upper: Adapt response capacity to load. Every system confronts some distribution of probable loads (bold). As conditions shift, so does the distribution (dashed). The response curve (bold) is typically sigmoid with its most sensitive region (steep part) matched to the most probable loads. As a sensor detects a statistically reliable change in the distribution, it prepares the effectors by shifting their response curves to match the new distribution (dashed). Each sensor also adapts its own sensitivity. Reprinted from Sterling (2004b). **Lower:** Organs and organ systems couple efficiently by matching loads to capacities. Trade-offs allow better performance while reducing unused capacity and enhancing "portability." Blood flow pattern changes with exercise: total flow quadruples, but that is insufficient for muscle. To meet the full need, blood is routed from liver, gut, and kidney, temporarily reducing their performance but eventually benefiting from what the muscular effort has accomplished. Data from Weibel (2000).

a new target level. By comparison, self-regulation by feedback to a set point would be hopelessly inefficient. But to execute these principles of predictive regulation requires an organ with knowledge of the outside, knowledge of the inside, and knowledge of the past to anticipate what the whole animal will need over various timescales—the whole brain (Sterling, 2012).

Bigger brains

We seem to have answered "Why a bigger brain?" In a wider world, a more effective brain expands the possibilities for behavior. Control of append-ages such as fins, wings, and legs lends speed and scope to exploration, so that vastly more small patterns are encountered which then require selec-tion and assembly. More large patterns require more comparisons, requir-ing a larger library; more comparisons also require more decisions, and these require more evaluation. Naturally, more neurons are needed, and since neuronal components are irreducibly small (chapter 7), a brain must enlarge.[3]

The larger brain, to be effective, must operate in real time. One need not watch a sloth for very long to realize the limits to life in slow motion. The larger, faster brain must still remain portable and also metabolically afford-able. So a brain needs to be both functionally effective and cost-effective. These demands for speed, portability, and affordability all interact; there-fore, individually and together they raise questions of brain design. We turn now to the fundamental constraint on any brain design that leads to the first three design principles. Then, in the context of these few principles, we discuss some actual designs (mammal and insect).

Design constraints

The fundamental constraint on brain design emerges from a law of physics. This law governs the costs of capturing, sending, and storing *information*. This law, embodied in a family of equations developed by Claude Shannon, applies equally to a telephone line and a neural cable, equally to a silicon circuit and a neural circuit. This law constrains neural design at all scales and cannot be avoided any more than a B-29 bomber can avoid the law of gravity. But, though the brain is fundamentally an organ that manipulates information, few neuroscientists are familiar with this law or aware of its value for understanding brain organization. We explain it briefly here and give more detail in chapters 5 and 6.

What *is* "information"?

Information is *the reduction of uncertainty about some situation X associated with observing any variable Y that is causally correlated with X*. Uncertainty defines the standard measure: one *bit* is the information needed to decide between two equally likely alternatives. Information depends on causality because, to reduce uncertainty, a message must be reliably relatable to its source, the event that caused it. Any factor that reduces the reliability of this connection, such as noise, increases uncertainty and destroys information.

Reduction of uncertainty succinctly describes the brain's purpose. A spike in an ON ganglion cell reduces the brain's uncertainty that a brighter than average object is located in a particular region of the visual field (chapter 11). And when the brain matches the sensory pattern coded by a patch of ganglion cells to a stored pattern, it reduces a key uncertainty: "Friend or foe?" The answer helps to select the next behavior and implement it. To this end, a motor neuron spike decreases the uncertainty that its target muscle fibers will contract and help the animal move in the appropriate direction. In short, to achieve its core purpose, the brain uses physical devices (neurons and circuits) that represent and manipulate information. So now we must ask: how much information can a neuron represent, and what constrains its capacity?

A neuron's information capacity

To convey information, a neuron must represent the state of its input as a distinct output (input and output must be causally related). It follows that a neuron's capacity to convey information is limited by the number of distinctly different outputs that it can generate. The number of different outputs a spiking neuron can generate in a given time is the number of distinctly different spike trains that it can produce in that time. This depends on two factors, mean firing rate (R spikes per second) and the precision of spike timing (Δt seconds). The upper bound on firing rate is set by spike duration plus the period following a spike when a neuron is refractory (cannot spike). Certain neurons reach this limit during brief bursts, but most neurons operate far below this limit. Precision is limited by channel noise and membrane time constant. Here biophysics limits information capacity.

What is the relation between spike rate, timing precision, and the number of different spike trains a neuron can produce? When a neuron transmits for 1 s, it produces R spikes with a timing precision of Δt (Rieke et al.,

1997). The number of different spike trains, M, is the number of ways the neuron can place its R spikes in $T = 1/\Delta t$ intervals (figure 3.5). Deriving M is a standard exercise in calculating combinations that is often set to students in quaint terms, such as placing peas in pots. The solution is

$$M = T!/(R!(T - R)!), \tag{3.1}$$

where ! denotes factorial and $(T - R)$ is the number of empty (spikeless) intervals.

The number of different messages, M, that a neuron can generate in 1 s converts to information rate. According to Shannon, the information, H, is given by

$$H = \log_2(M). \tag{3.2}$$

Substituting for M using (3.1) gives

$$H = \log_2(T!/(R!(T - R)!)) = \log_2(T!)—\log_2(R!)—\log_2((T - R)!). \tag{3.3}$$

Because Shannon used a logarithmic scale, a message lasting twice as long conveys twice as much information. And, because he used log base 2, information is in bits. Thus, H, the information that a neuron can transmit with messages 1 s long, is its information capacity in bits per second (figure 3.5).

With this expression we can "follow the money." That is, using a standard currency (bits) we can ask like good engineers: how fast does a neuron send information (bits per second) and how efficiently (bits per spike)? And at what cost in space (bits per cubic millimeter) and energy (bits per molecule of adenosine tri-phosphate)? This molecule, abbreviated *ATP*, is the standard intracellular molecule for transferring energy.

Information costs energy and space
Information rate increases with spike rate and with spike timing precision, that is, reduction in Δt. However, for any given precision, information rate increases sublinearly with spike rate (figure 3.5). Consequently, as spike rate rises, bits per spike should fall, and this theoretical decline in bits per spike is observed experimentally (figure 3.5).

There is another way to explain why more frequent spikes carry less information. A symbol that occurs less frequently is more surprising and so more informative (chapter 4, equation 4.2). This effect, which Shannon called *surprisal*, makes a code with fewer spikes more efficient. For example, a code that distributes spikes sparsely among a population of neurons conveys more bits per spike (chapter 12; Levy & Baxter, 1996).

This simple law—infrequent spikes carry more bits—profoundly influences neural design because, following the money, one finds that spikes are

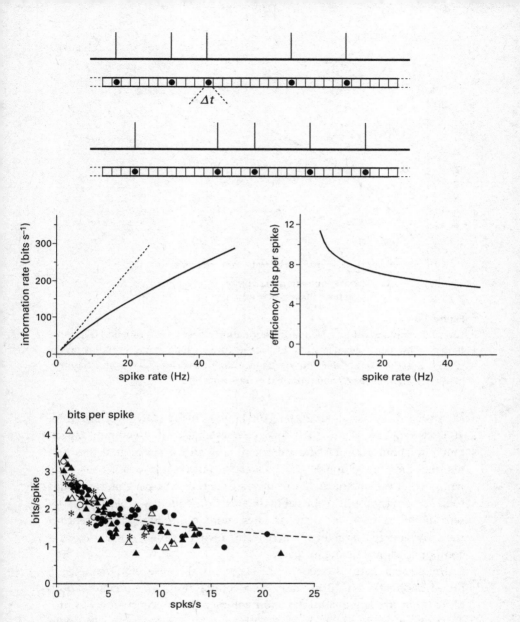

Figure 3.5
Mathematics and biophysics govern the representational capacity of signal trains.
Upper: Distinct sequences of spikes in time intervals Δt represent different inputs.
Middle left: Theory predicts information rate to increase sublinearly with spike rate,
with the consequence shown at **middle right:** Increasing spike rate reduces the information transmitted per spike. These theoretical curves were calculated using the
standard approximation for signal entropy at low spike rates (Rieke et al., 1997, equation 3.22). In general neurons do not achieve their theoretical capacity because of
noise and redundancy; consequently, measured values of bits/spike are lower (figure
11.25). **Lower:** Measured bits per spike falls as mean spike rate increases. Data pooled
from several classes of guinea pig retinal ganglion cell. Reprinted with permission
from Balasubramanian & Sterling (2009).

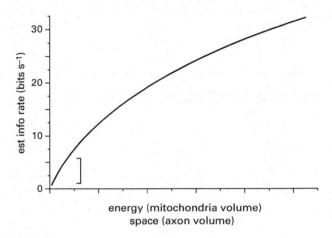

Figure 3.6
Law of diminishing returns. Doubling information rate of retinal ganglion cells more than doubles space and energy costs. Consequently, neural designs try to stay on the steep region (bracket) of this empirically measured curve. Modified from Balasubramanian & Sterling (2009) and reprinted with permission.

expensive. They use about 20% of the brain's energy (Attwell & Laughlin, 2001; Sengupta et al., 2010). A spike charges a neuron's membrane capacitance by about 100 mV, and the membrane area is substantial due to a neuron's local branching. Higher mean spike rates require a larger cell body with greater membrane area; this increases energy cost per spike and adds to the cost of transmitting bits at high rates. Consequently, where spikes are sent sporadically and at low mean rates, more information can be sent for the same energy—more bits per ATP. This saving in energy by low rates is compounded by a saving in space.

Higher spike rates also require thicker axons.[4] Because axon diameter, d, increases directly with firing rate, axon volume rises as d^2; therefore, doubling the firing rate quadruples axon volume. The concentration of mitochondria, an indicator of energy cost, tends to be constant with axon diameter; therefore, as volume quadruples, so does the energy supply (Perge et al., 2009, 2012). In summary, there is a *law of diminishing returns*: cost per bit, both in energy and space, rises steeply with bit rate (figure 3.6).

Three principles of neural design
The inescapable cost of sending any information and the disproportionate cost of sending at higher rates lead to three design principles: *send only what is needed; send at the lowest acceptable rate; minimize wire, that is, length and*

diameter of all neural processes. This last principle seems obvious, but it actually reflects a subtle point that arises from the constraint on rate.

Designs should reduce wire, of course, because wire uses space and energy. But wires also use *time* for transmission, and that is time lost to processing and action (Howarth et al., 2012). The constraint is particularly onerous for neural wires because they transmit more slowly than copper wire. Neural conduction velocity is 100 millionfold lower and, for biophysical reasons, faster conduction requires thicker wires (chapter 7). Thus saving time by sending at higher information rates (bits per second) and higher conduction velocities (meters per second) requires thicker axons, which, as noted, involves disproportionate costs in energy and space (Wen & Chklovskii, 2005). Thus, the only economical way to save time is to rigorously shorten wires. This principle shapes brain design across all scales, from an axon's branching and the microscopic design of local circuits, to the overall layout (chapter 13).

With these few principles we can now consider how the mammalian and fly brains are organized on a scale of about 1 mm and why. This macro-organization cannot explain the actual computations because those occur mostly on a finer scale. Nor do we claim that every feature represents the best of all possible designs. Others might work just as well—but they have not been tested. All we can say is that these three principles illuminate the layout of real brains—across a millionfold range of scale and half a billion years of evolution.

4   How Bigger Brains Are Organized

I sensed the earth's slow turning into the dark. The shadow of night is drawn like a black veil across the earth, and since almost all creatures, from one meridian to the next, lie down after the sun has set, one might in following the setting sun, see on our globe nothing but prone bodies, row upon row, as if leveled by the scythe of Saturn.

—W. G. Sebald, paraphrasing Sir Thomas Browne (edited for brevity)

The preceding chapter established that for the brain to send information requires energy and space. Moreover, higher rates (more bits per second) require disproportionately more energy and space because they need thicker axons—for which both space and energy rise as the diameter *squared*. Consequently, the most efficient designs will send only information that is essential and will send it at the lowest rate allowable to serve a given purpose. If information can be sent without any wire at all, that is best. If wires are absolutely needed, they should be as short and as thin as possible. These principles allow substantial insight into how bigger brains are organized.

One design decision is so ubiquitous as to require immediate mention. Brains segregate the wires that interconnect local circuits with each other and with distant circuits. The reason is simple and fundamental: to mingle the wires with the circuits increases total wire length and thickness— violating the principle minimize wire (chapter 13). In mammals axons segregate if they travel beyond a few millimeters. The reason is that increasing distance requires increasing conduction speed to avoid computing delays, and this requires thicker axons. When axon diameter exceeds about 0.5 μm, the axon becomes wrapped in *myelin*, which increases conduction speed by about 6 mm ms^{-1} for every 1-μm increase in diameter (chapter 7). Because myelin in the living brain glistens white, extended sheets of myelinated axons are termed *white matter*.

Saturn's scythe sets brain design

The most profound condition for all life on Earth, the one that uniquely shapes every cell in every organism, is the daily rotation of our planet about its axis. This motion shifts the intensity of arriving solar radiation over the course of 24 hours by a factor of 10^{10}. The impact of this motion is so profound that for many cultures it opens the story of Creation. One familiar example waits only until line 4: "... *and God divided the light from the darkness ... and there was evening and there was morning, one day"* (Genesis 1:4–5).

Animals can certainly survive without light (e.g., in caves), but those with access to light generally choose a particular time of day to forage and thus a particular range of light intensities. The basic choices are diurnal, nocturnal, and crepuscular (dawn and dusk).[1] This decides their investment in sensors: fine spatial vision with color versus acute hearing, possibly with echolocation, versus olfaction plus whiskers. Foraging period also decides their strategies to deal with predators occupying the same slot: camouflage, evasive flight, or skulking behavior.

During its active period the body expends chemical energy to support external behaviors, such as foraging, and internal activities, such as digestion and absorption. Some needs rely on both internal and external actions, for example, thermoregulation. Thus, the active phase involves a broad metabolic pattern, *catabolism*: (1) disassemble large polymeric molecules (proteins, fats, carbohydrates, nucleic acids) into their monomeric building blocks (amino acids, fatty acids, sugars, nucleotides); (2) distribute monomers to metabolically active tissues; (3) convert monomers into energy-bearing molecules, such as ATP, that drive cellular processes; and (4) use an aerobic (oxygen requiring) pathway to produce ATP because it is sixteenfold more efficient (ATP per glucose monomer) than the anaerobic pathway.

During its *in*active period, the body shifts to a broad pattern of renewal, *anabolism*: (1) assemble new polymers for growth, repair, remodeling, and immunity, and (2) replenish reserves by storing residual monomers as resynthesized polymers. Thus, liver converts spare glucose to the storage polymer glycogen; fat cells convert excess glucose to monomeric fatty acids which are then used to build the storage polymer, fat. Because catabolism and anabolism involve opposing sets of biochemical reactions, it would be inefficient to run them simultaneously. Thus, natural selection has separated internal processes into complementary patterns for different segments of the daily cycle.

The brain itself participates in the catabolic/anabolic cycle. During wakefulness it collects, processes, and distributes immense amounts of information. During sleep, the brain switches over to anabolism via a specific regulatory enzyme and uses this phase to store recently acquired information (Dworak et al., 2010). This involves remodeling local circuits by retracting certain synapses and adding new ones and, in some cases, generating new neurons (chapter 14).

The obligatory alternation between catabolism and anabolism involves throttling down one set of biochemical pathways and revving up another—both of which take time. Consequently, each pattern needs to anticipate the environmental shift—in order to optimally match the key time windows for sleep and foraging. Thus, the pattern seen in figure 3.3, where blood pressure falls with sleep and rises with waking, is completely general: all processes in body and brain move through this cycle. So it is efficient for them to share the same broad signals, and although some processes cease during darkness and others during light, all must follow Saturn's scythe.

Brain clock

Many somatic cells contain an intrinsic clock, established by oscillations of interacting proteins, with a period of approximately 24 hours (*circadian*). But without a mechanism to trim them up, these clocks would soon drift out of phase. So a master clock is needed to track the day, including its continual shift, due to Earth's axial tilt, during its annual revolution about the sun. The master clock comprises a discrete cluster of neurons (about 8,600 in human), the suprachiasmatic nucleus (*SCN*).[2] One subgroup of SCN neurons contains a circadian clock that resets daily based on signals from the retina that track the slow shifts of light intensity across the day and season (figure 4.1).

The master clock requires neither color, nor spatial, nor fine temporal information—only slow intensity changes. Therefore, following two design principles, the retina sends as little as needed and sends as slowly as possible. It uses just a small fraction of retinal output neurons (0.2%), types that cover the retina sparsely and fire at very low rates, a few Hertz averaged over the day (Crook et al., 2013; Wong, 2012). SCN neurons themselves fire between about 8 Hz (day) and about 1 Hz (night; Häusser et al., 2004). To follow another principle, minimize wire, the SCN locates exactly where the optic tracts join the brain (see figure 4.1). But how does the master clock govern patterns across the entire body and the brain as well?

The SCN's relatively few neurons, about 10^4 in rat, could not conceivably contact all other cells directly (Güldner, 1983). Nor should they because their job is not to micromanage every cell but mainly to keep the time. Except for time, the SCN is fairly ignorant—largely unaware of internal

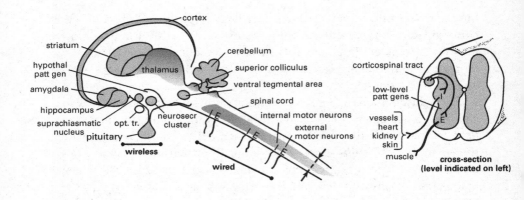

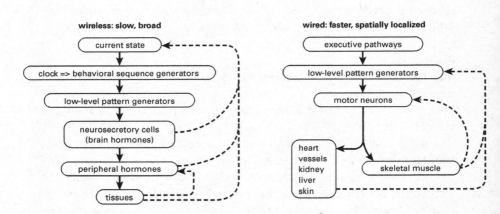

Figure 4.1

Brain's master clock (suprachiasmatic nucleus) informs a network of high-level pattern generators in hypothalamus. These coordinate internal physiology and external behavior. Their network selects a behavior, plus the endocrine and autonomic patterns needed to support it, and communicates the orders to low-level pattern generators, both wireless and wired. **Upper left**: Longitudinal section (diagrammatic) shows the spatial layout of this hierarchy. Opt. tr., optic tract; hypothal patt gen, hypothalamic pattern generator. **Upper right**: Cross section through spinal cord indicates that corticospinal tract tells local pattern generators to match internal physiology to external behavior. **Lower**: Schemes for wireless and wired control.

physiology and external behavior. Therefore, it could not responsibly tell either the body or the brain when to shift the broad pattern. For example, a rat normally forages at night, but what if food becomes sparse at night and plentiful at noon? Were the SCN to directly instruct a command center for foraging, it might send the rat to sleep without its supper.

Coupling clock to behavior: A hypothalamic network

Instead, the SCN couples to an adjacent region, the hypothalamus, that for its comparatively small extent is extremely well informed (figure 4.1; Saper et al., 2005; Thompson and Swanson, 2003). This region monitors myriad internal parameters, including temperature, blood levels of salt, and metabolites, hormonal signals for satiety, hunger, thirst, pain, fear, and sexual state. Some of its neuron clusters express their own endogenous oscillators, and at least one of these responds to changes in food availability (Guilding et al., 2009). This territory also monitors stored patterns—such as best places and times to forage and past dangers. And it monitors the external environment using every sense. Integrating all these data, plus SCN clock time, this region calculates which needs are urgent. Then, balancing urgency against opportunity and danger, it tells the rat whether to forage, mate, fight, or sleep. To execute, it does not micromanage but instead calls the appropriate pattern of behavior (Saper et al., 2005; Thompson & Swanson, 2003).

Hypothalamic circuits, designed to anticipate impending needs, generate signals that elicit various "motivated behaviors," that is, foraging for food, or drink, or sex in response to these integrated signals. As these motivating signals are broadcast to other brain regions, there arises a subjective component that we (among other animals) experience as desire. If one area can be considered as the wellspring of unconscious desires, this is it. It seems amazing that such a small region could access and integrate so much information and evoke such a variety of core behaviors. How could there be sufficient space for hypothalamic neurons to do so much?

Part of the answer is that this well-informed region dictates *sequences* of low-level patterns. For example, feeding behavior requires the sequence: sniffing → biting → chewing → swallowing. These components are programmed in detail by dedicated pattern generators located down in the brain stem near their effector muscles. The local pattern generators manage the exact timings of muscle contraction required for coordinated behavior. The broad sequence that smoothly calls each component into play can be dictated to local pattern generators with a reduced instruction set—something like a music conductor following a score to call forth a Beethoven symphony from 80 low-level players with nothing but a slender baton. The analogy does not explain the magic in either case, but it does emphasize the design principle: send simple instructions and compute the complex details locally (Büschges et al., 2011).

This economical design allows the hypothalamic region to accommodate a dedicated circuit for each behavioral pattern. These are sufficiently compact that a fine electrode can stimulate them separately, revealing that each circuit evokes a full behavioral pattern, plus the appropriately matched visceral pattern (Hess, 1949; Bard & Mountcastle, 1947). For example, a cat with an electrode placed to evoke "angry attack" arches its back, hisses, and strikes with bared claws and teeth (somatic pattern). Simultaneously it dilates pupils, raises hackles, and increases cardiovascular activity (visceral pattern; Büschges et al., 2011; Hess, 1949).[3] Moving the electrode by a few millimeters can activate circuits for other behaviors: feeding or drinking or copulating or curling up to sleep. In short, many circuits fit in a small space because their output messages are simple.

Each behavior circuit is demonstrably guided by a rich set of input signals. For example, a cat electrically stimulated to feed will attack a ball of cotton that mimics a mouse, but only briefly, whereas it persistently attacks a real mouse until the current stops. If the mouse is replaced by a substantial rat, the cat retreats to its home corner. Evidently the feeding circuit is modulated by inputs that identify prey, distinguish true prey from false, and recognize dangerous prey—all based on comparison to stored patterns. Moreover, each behavior is imbued with a motivational component— apparent when an animal stimulated to feed will seek hidden food and work to obtain it (press a lever).

How does this small region, the hypothalamus, access the brain's core systems for perception, spatial memory, danger, economic value, and urgency? Again, it relies on details computed elsewhere and delivered only as conclusions: time from the SCN; integrated physiological data from myriad sources that define internal state; selected memories of location and danger from hippocampus and amygdala; recent history of reward value from the striatal system; high-level analysis of choices from prefrontal cortex. Because these inputs to the hypothalamic region all send summaries, they can use low information rates and thus fine fibers, thereby greatly conserving space (figure 4.1). Energy is also conserved, allowing this crucial region to have among the lowest metabolic rates (Sokoloff, 1977).

This strategy allows a major organ for memory, the hippocampus, to access key aspects of an animal's life history but send only modest clips to guide a particular behavior. This might explain why its output tract (*fornix*)

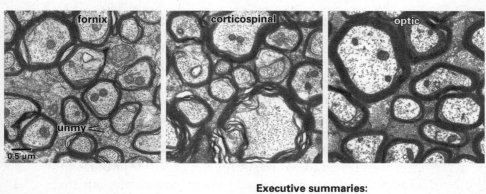

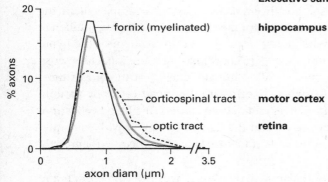

Executive summaries:

hippocampus

motor cortex

retina

Figure 4.2
Fiber tracts that transmit summaries share an economical design. Their axon diameters distribute log-normally, with many thin axons and fewer thick ones. Halving the diameter reduces space and energy costs by fourfold. These "summary tracts" use low mean firing rates (see figure 4.6). Reprinted with permission from Perge et al., 2012).

can manage with mostly fine fibers, resembling the optic nerve, which itself sends strongly edited summaries from the retina (chapter 11). An apparently similar strategy allows sensorimotor areas of the cerebral cortex to lend speed and agility to motor behaviors via an output tract (*corticospinal tract*) of similar fine structure (figure 4.2; Quallo et al., 2012). In short, the hypothalamic network is designed to receive executive summaries as input and deliver broad memoranda as output (Perge et al., 2012).

Resurrection
To be awakened from a deep sleep feels horrible. And no wonder: every cell in the body and brain struggles to function according to its catabolic phase—against all central instructions to remain in anabolic phase. But

when anabolism has gone to completion—when the body has replenished stores, healed wounds, rebuilt muscles and immune systems, and when the brain's sorting mechanisms have punched "delete" or "save"—then all the cells and tissues finally wake up more or less simultaneously.

The SCN signals "dawn" to the hypothalamic network—which then decides, based on many factors, whether it is auspicious to awaken.[4] If so, the network signals a nearby cluster of neurons (comparable in size to SCN) to secrete the peptide transmitter *orexin*. The orexin neurons project widely over the brain to activate a cascade of systems that regulate arousal (Sakurai, 2007). Because orexin neurons couple the clock to the brain's arousal system, an animal lacking orexin tends to collapse unpredictably into sleep.

The orexin cluster specifically awakens olfactory sensors, enhancing their sensitivity, and it awakens motor mechanisms for foraging (Julliard et al., 2007). Informed by the master clock, the orexin cluster uses the hypothalamic pattern generator network to coordinate alertness, olfactory sensitivity, and the sense of hunger—all to initiate foraging at the proper time. Now it is time for brain signals to reinstate the broad catabolic pattern: mobilize energy stores from liver and oxygen carriers (red blood cells) from spleen and bone marrow; re-expand the vascular reservoir with salt water from the kidney. And it is time to *de*mobilize anabolic processes for growth, repair, and immunity.

In summary, the hypothalamic network manages the whole brain and all of its functions—without micromanaging. But now, what about micromanaging? A conductor is all well and good, but someone must play the bassoon. So how are the processes that do involve micromanaging governed by the design principles considered here?

Distributing output patterns

Wireless signaling

Design principles dictate that the slowest processes should be governed by the slowest effectors and the least wire. Where signals can be sent with zero wire, that is best. Consequently, the effectors for micromanaging the broad catabolic and anabolic patterns are endocrine glands. For example, the adrenal gland secretes a steroid hormone that enhances the kidney's uptake of sodium and a different one that enhances catabolism, mobilizing energy and suppressing growth and repair. Testis secretes anabolic steroids that enhance muscle, and liver secretes a hormone that stimulates red blood cell production. What coordinates these low-level effectors? Higher-level endocrine signals from the pituitary gland, which is in turn governed by hormones from the brain. Wireless regulation of two particular functions, blood pressure and muscle contraction, is summarized in figure 4.3.

Brain hormones are secreted directly into the circulation by neurosecretory neurons whose clusters lie adjacent to the hypothalamic network of pattern generators. The pattern generators deliver their well-informed but simple orders via very fine, very short wires (figure 4.1). Each node in the hypothalamic network can call a particular pattern of brain hormones for release into the blood just upstream of the pituitary, thus stimulating it to release its own hormones into the general circulation. The whole endocrine network reaches every cell in the body within seconds. Not blazingly fast, but on the other hand, the messages are broadcast without any wire at all and with zero energy cost above what the heart is already doing.

The genius of this wireless system lies partly with the receivers. Although all somatic cells are exposed to all hormones, only certain cell types download a given message. To do so, they produce a specific molecular receptor that binds a particular hormone and triggers a particular intracellular response. Thus, information broadcast diffusely to the whole body can be read out by a restricted number of cell types—whose responses to the signal are thereby coordinated. The molecular mechanism and reasons why it is so economical are described in chapter 5.

Another clever feature is that receiver cells can express different subtypes of the molecular receptor. Each subtype can couple within the cell to a particular *second messenger* with its own stereotyped action. For example, one messenger can greatly amplify the hormonal signal and use it to either activate or suppress some intracellular process. Thus, a single message broadcast wirelessly can evoke complex response patterns among different tissues that include negative as well as positive correlations.

For example, skeletal muscle acts rapidly on the outer world via fast signals over thick wires. Yet, it is also a tissue within the body and is thus regulated wirelessly by various hormones, including anabolic steroids, insulin, growth hormone, and thyroxin (figure 4.3, lower panel). Thus, wireless signaling helps the brain to efficiently couple inner and outer worlds.

Wireless collecting

The brain also uses wireless receivers, a small set of *circumventricular organs* that locate at specialized interfaces between brain and blood vessels. There the normal barrier between blood and brain parts, thus exposing neurons to circulating chemicals. These neurons select just what they need by expressing the appropriate molecular receptors. For example, the *subfornical organ* locates near the hypothalamic pattern generators that regulate

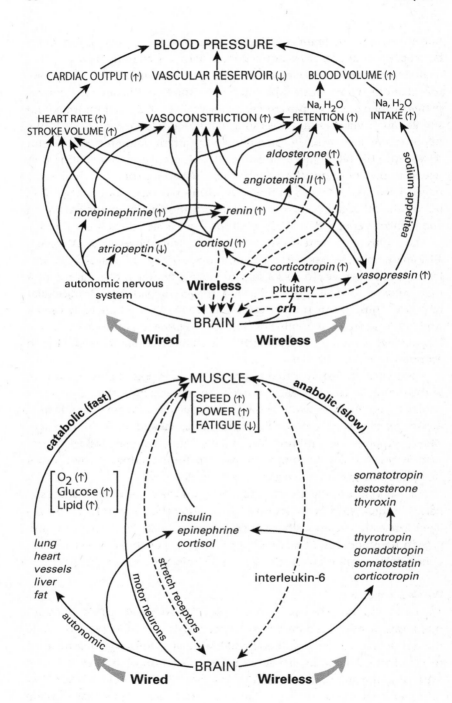

Figure 4.3
Wireless regulation broadcasts slow signals to efficiently couple inner and outer worlds. Upper: To adjust blood pressure rapidly and locally, the brain uses wires (autonomic nerves). But to shift pressure slowly and broadly, it uses wireless signals (hormones) (*italicized*). Dashed lines indicate wireless feedbacks to brain. Feedbacks by wire are used by certain sensors, such as for oxygen and pressure, but are not shown. *CRH, corticotropin releasing hormone.* **Lower:** Catabolism in muscle activates rapidly to support contraction; so to rapidly activate catabolism, the brain uses wires. But anabolism in muscle is slower, so the brain activates those processes with wireless signals (*italicized*).

appetite for salt and water (figure 4.4). The neurons sense the blood's sodium level, plus levels of hormones (*angiotensin II* and *aldosterone*) that tell the kidney to conserve sodium (figure 4.3, upper panel). Thus, this wireless receiver closes the loop for anticipatory regulation: the brain sends instructions to kidney regarding salt and water, and the brain's subfornical organ wirelessly receives information about the current state[5] of sodium balance.

Need for wires: Faster, spatially directed signaling

Neurosecretions spread slowly (over seconds) and modulate target cells slowly because the packets of hormone molecules released into the voluminous vascular system become greatly diluted (to concentrations $\sim 10^{-9}$ M). Therefore, molecular receptors need high affinity to capture the hormone, and thus their *un*binding is slow (chapter 6). However, this delay is inconsequential because the intracellular processes that they are regulating typically span minutes or hours. Thus, the slow rhythms of wireless signaling match their targets, physiological processes that rise and fall slowly.

Where faster responses are needed, the hormone is released into a *portal vessel* leading directly to a target downstream. Because the hormone is less diluted, it can be captured by lower affinity receptors, which unbind faster, and operate on the steep limb of the binding/response curve. For example, the brain hormone corticotropin-releasing hormone is secreted into portal vessels leading to the pituitary; the adrenal cortex secretes steroid hormones into portal vessels leading to the adrenal medulla. Yet certain internal process must proceed still more smartly, and that needs wire.

For example, for the brain to initiate a change in body posture, it must alter the pattern of muscle contraction. This will require a change in the distribution of oxygen and thus an altered vasomotor pattern to redistribute blood. Furthermore, active muscle will need to take up glucose, and that will require triggering insulin secretion from pancreatic cells. These vascular and endocrine adjustments need to be initiated along with the muscle activity, and these faster, spatially localized signals demand wires.

This need is served by autonomic neurons whose axons contact every internal organ and blood vessel. Their mean firing rates are less than 1 Hz, and thus in Shannon's sense they transmit at low information rates. This seems intuitive, since a message—"Secrete some insulin" or "Constrict this vessel"—goes somewhat beyond "yes" or "no" (one bit), but not by much, and thus it can be accomplished with few spikes. Signals that transfer at rates below 1 Hz use the finest, cheapest axons.

What manages these autonomic effectors? *Answer*: low-level pattern generators located in the brain stem and spinal cord near the output clusters (figure 4.1, right). The latter form two subsystems (*sympathetic* and *parasympathetic*), which employ different transmitters. Each transmitter couples to several receptor types, which in turn couple to different second messengers. Consequently, the autonomic effectors can generate rich internal patterns. They are the orchestral players—ready and waiting for the conductor to select the next pattern and tempo.

What manages the muscles that change the body's posture? Again the answer is low-level pattern generators located near the motor neuron clusters. These pattern generators must increase force from certain muscles and decrease it from others—in just the right amounts and at just the right instants. Sharp timing requires large currents, rapid integration (short time constants), and high mean firing rates (chapter 7). Therefore, these pattern generators need large neurons with thick dendrites and thick axons.[6] To reduce costs, they locate near their effectors. This lengthens the descending pathways that supervise them, but as noted, those are cheaper (figure 4.2).[7]

Motor control requires rapid feedback. The fastest signals from skin and joint receptors travel at about 50 m s^{-1}, and those from muscle receptors travel at about 100 m s^{-1}. These velocities require very thick, myelinated axons, 8–17 μm in diameter.[8] These fibers are 10-fold thicker than for the descending tracts and thus 100-fold greater in volume. Were pattern generators located higher in the brain, for example, nearer to the hypothalamic pattern generators, feedback would be delayed, even though these axons are huge. Thus, the combined needs for fast output and fast feedback constrain the low-level generators of motor patterns to locate near their effectors, the motor neurons (figure 4.5).

Arrangement of effector clusters
The neurosecretory clusters locate adjacent to the hypothalamic network, which can thus modulate them with very little wire (figure 4.1). But the

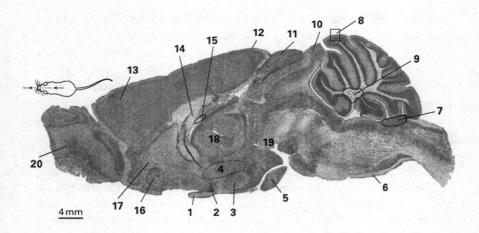

1. optic nerve
2. suprachiasmatic nucleus (clock)
3. hypothalamic neuroscretory cluster (brain hormones)
4. hypothalamic pattern generators (high-level)
5. pituitary gland (wireless signals → periphery)
6. corticospinal tract (summaries from motor cortex to low-level pattern generators)
7. area postrema (monitor blood chemistry)
8. cerebellar cortex (correct errors of intention)
9. cerebellar output clusters (integrates cerebellar output)
10. inferior colliculus (early auditory processing)
11. superior colliculus (orient head and eyes toward key information sources)
12. primary visual cortex (far from long-term storage sites)
13. frontal cortex (near long-term storage sites)
14. fornix (summaries from hippocampus to hypothalamic pattern generators)
15. subfornical organ (monitor blood sodium and related hormones)
16. amygdala (tag high-level patterns for storage)
17. striatum (evaluate predictions of reward)
18. thalamus (process signals for economical transfer to cerebral cortex)
19. ventral tegmental area (dopamine neurons → frontal cortex + striatum)
20. olfactory bulb

Figure 4.4
Longitudinal section through rat brain. This section shows relative size and location of various structures discussed in this chapter. From http://brainmaps.org/ajax-
viewer.php?datid=62&sname=086&vX=-47.5&vY=-22.0545&vT=1 © The Regents of
the University of California, Davis campus, 2014.

autonomic and somatic motor neuron clusters lie far from the hypothalamic network, distributing from the midbrain down through the spinal cord. This extended distribution allows space for their low-level pattern generators. The total volume of the autonomic effectors and their pattern generators, summed over the length of the spinal cord, is about 100-fold greater than that of the hypothalamic network.[9] This need for space easily justifies extending the brain tailward and helps explain why this design has been conserved. Moreover, the extension allows additional efficiencies.

Neurons that share input from the local-pattern-generator should cluster close together. Thus, the autonomic effector neurons that regulate internal organs and endocrine cells align in a column, allowing them to share input from the columnar low-level generator of autonomic patterns. Somatic motor neurons also align in columns—parallel to the autonomic column and near it; therefore, circuits for internal physiology and external behavior can be coordinated locally via short wires (figure 4.1).

Because low-level pattern generators for internal physiology and behavior locate together, descending tracts can regulate them together with no extra wire. For example, the corticospinal tract sends a reduced instruction set from motor cortex to low-level pattern generators for muscle (Yakovenko et al., 2011) and also to adjacent autonomic pattern generators for kidney (see figure 4.1). Thus, the descending message, "Arise!" can be sent efficiently to both effectors (Levinthal & Strick, 2012).

Somatic motor neurons extend this design for efficient component placement to a still finer level (figure 4.5). Motor neurons for a given muscle often fire together, implying shared inputs, so they cluster. Motor neurons for muscles that act synergistically across a joint also often fire together, also implying shared inputs, so their clusters stay close. Motor neurons for muscles that cooperate across multiple joints also fire together, but less often, so their clusters are further apart, distributing longitudinally with separations roughly corresponding to their frequencies of coactivation. Finally, motor neurons for antagonistic muscles tend to fire reciprocally, flexors excited/extensors inhibited. This reciprocity depends on a shared circuit (cross-inhibition, like the worm), so the clusters of antagonistic motor neurons also stay close—in parallel columns that run down the spinal cord (Sterling & Kuypers, 1967; figure 4.5).

In short, somatic motor neurons distribute according to a broad design rule: *neurons that fire together should locate together.*[10] This rule also governs sensory maps and all the brain's orderly topographic connections (chapters 12 and 13).

Design for an integrated movement

The placement of motor neurons in longitudinal columns allows a pattern generator to economically evoke an integrated limb movement (Bizzi & Cheung, 2013). The task is to excite contractile units in dozens of muscles across several joints and suppress their antagonists (Sherrington, 1910; Creed & Sherrington, 1926). The key is for motor neurons to send their dendrites longitudinally within a column for long distances (about 1 mm) so that dendrites of synergists overlap. Then, an input axon can coactivate synergists simply by branching as a T within the column and distributing synapses at regular intervals. Strong synergists will greatly overlap their dendrites and thus share more input than weaker synergists that overlap less (figure 4.5, lower). All inputs to the motor neuron columns follow this rule, including axons from sensory receptors, axons from local pattern generators, and axons from cortex (figure 4.5, lower). This design uses less wire than any other conceivable geometry, and thus it is optimal (chapter 13).

Pattern-generator neurons use thick, myelinated axons to synchronously activate motor neurons at different levels of the motor neuron column. To do this while least disturbing the synaptic circuitry, the axons are routed into the white matter where upon reaching the appropriate levels, they reenter the motor column and connect (figure 4.5).

One benefit of this architecture is that different sensory receptors from the same location can efficiently evoke opposite responses. Here, pressure receptors from the foot connect to the extensor pattern generator, so as weight shifts to that foot, all the extensors are excited to support the limb. Pain receptors connect to the flexor pattern generator, so as weight shifts to that foot, all the flexors are excited (and extensors inhibited) to withdraw the limb. These alternative decisions are accomplished at the lowest level, thereby avoiding the costs in time, space, and energy of consulting higher levels. The corticospinal tract delivers "executive summaries" from motor cortex to the pattern generators. So a corticospinal axon can simply say "Flex!" and local circuits do the rest (Bizzi & Cheung, 2013).

Collecting input patterns

Different senses, different costs

The wider world that makes a larger brain such a good investment contains a seeming infinity of patterns carried by diverse forms of energy: electromagnetic (light), heat, mechanical vibration of air (sound), direct

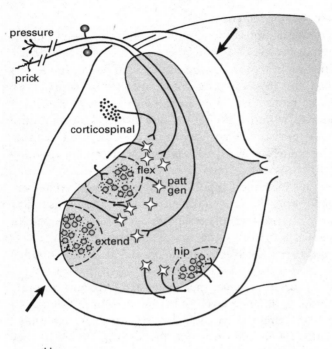

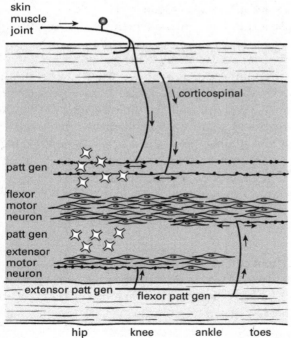

Figure 4.5
Efficient wiring for integrated movement. Upper: Cross section through the spinal cord. Flexor and extensor motor neurons for the leg form separate clusters, which locate near each other and also near to the pattern-generator neurons that reciprocally excite and inhibit them. The flexor and extensor clusters form parallel columns extending over several segments of spinal cord. Each column is structured as a motor map: motor neurons for thigh muscles locate at higher spinal levels, then in descending order: knee, ankle, and toes. Within a column, the motor neuron dendrites extend longitudinally for about 1 mm in both directions; consequently motor neuron dendrites for synergistic muscles overlap. Their overlap allows a pattern-generator axon to excite motor neurons for synergistic muscles simply by spreading its axon arbor longitudinally within the dendritic plexus. This uses the least possible wire to excite motor neurons for several muscles. The longitudinal dendrites appear in this plane as dots scattered within the motor neuron clusters. Motorneuron clusters for hip muscles locate separately, near the midline. Patt gen, pattern generator. **Lower:** Longitudinal section through spinal cord in the plane indicated by arrows in upper diagram. This plane reveals the motor neurons' longitudinal dendritic plexus that spans the motor map from hip to toe. This plane shows the pattern generator axons leaving the white matter to enter a flexor or extensor dendritic plexus where they encounter overlapping dendrites of synergistic motor neurons. The pattern generator neurons do not orient longitudinally and thus do not overlap. Consequently, a sensory axon or a corticospinal axon, coursing longitudinally within the pattern-generator columns, can efficiently access a discrete subset of pattern-generator neurons and thus a subset of motor neurons for a particular integrated limb movement.

mechanical contact, volatile molecules (odorants), molecules in solution, electrical patterns, magnetic fields, and gravity. Animals evolve mechanisms to collect information carried by all these forms—and use them to find food and mates, to avoid predators, and to orient in space and time. The challenge is to decide which forms to invest in and how much. Some are intrinsically cheap whereas others are intrinsically costly. Yet for certain lifestyles, cheap won't work and expensive is well rewarded. So an animal selects from the universe of patterns according to how it makes a living and during what phase of the planet's daily rotation.

Animals that forage by day invest heavily in photoreceptors sensitive to wavelengths between 300–700 nm. Animals that forage by night invest heavily in other receptors. Snakes that hunt mice use temperature receptors to extend their range to the infrared[11] (about 800 nm). Moths and frugivorous bats invest heavily in olfactory receptors, but certain bats prefer moths over fruit and so invest heavily in sonar systems that produce, detect, and process ultrasound (frequencies up to 180 kHz).

Fish that inhabit clear water invest in photoreceptors and, because the spectral content shifts with depth toward blue, those that inhabit deeper waters shift their peak photosensitivity correspondingly. Fish residing in caves *dis*invest in photoreceptors and are essentially blind. Certain fish inhabiting rich, but turbid tropical rivers invest in electrosensory systems that interrogate their surroundings by emitting brief electrical pulses or sinusoidal waves up to 2 kHz, and measuring the electrical field with electroreceptors.

Sensors differ greatly in cost. Olfactory sensors are slow and relay information at low mean rates, so their axons are extremely fine, approaching the limit set by channel noise (chapter 7). Vision is faster, so retinal ganglion cell axons (optic nerve) fire at higher mean rates and are somewhat thicker; and hearing is still faster, so auditory axons are far thicker (figure 4.6). This progression of axon calibers corresponds to a linear progression of firing rates (figure 4.6). However, since space and energy costs rise steeply with diameter and firing rate, the thickest auditory axon costs 100-fold more than an olfactory axon (Perge et al., 2012).

Systems for sensing at the skin follow similar design rules. Mechanosensors employ various mechanisms to transduce and filter pressure and touch. Some sense high frequencies (vibration) and transmit via thick axons (figure 10.3); other mechanosensors sense lower frequencies and transmit via finer axons. Sensors for pain and temperature send at the lowest spike rates and use the finest axons. Centrally, the fast and slow systems are processed in parallel and to a large degree arrive at their thalamic relay over separate tracts (Willis & Coggeshall, 1991; Maksimovic et al., 2013; Boyd & Davey, 1968).

Of course, these costs of collecting primary patterns are merely down payments. Auditory patterns arriving at high rates must be *processed* at high rates—so their initial central circuits use thick wires and fast (expensive) synapses (Carr & Soares, 2002). The most expensive parts in a mammalian brain are those devoted to early auditory processing, for example, the *superior olivary nucleus and inferior colliculus* (see figure 4.4; Mogensen et al., 1983; Borowsky & Collins, 1989). Thus, the ultrasonic imaging system of an insectivorous bat is intrinsically more expensive than the olfactory system of a frugivorous bat.

For fish that use electrical signaling, the cost is tremendous. One set of neurons needs to produce high-frequency pulses; another needs to detect them and signal the brain. Then, as for the insectivorous bat, processing is expensive. The computations required by this system are executed by cerebellar circuits, so the cerebellum greatly expands (figure 4.7). Consequently,

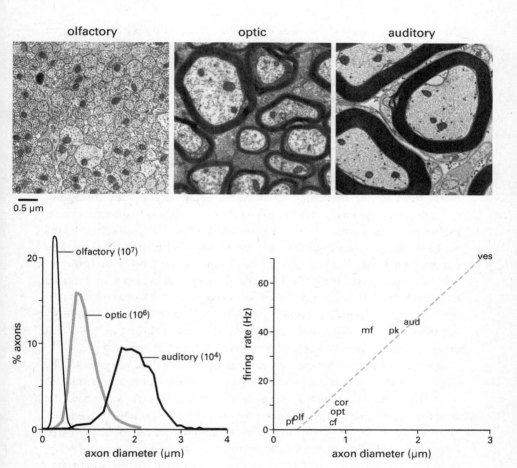

Figure 4.6
Unit cost of sending information differs greatly across senses. Upper row: Electron micrographs of cross sections through the olfactory, optic, and cochlear nerves shown at the same magnification. **Lower left:** Distributions of axon diameters. The auditory axons are nearly sevenfold thicker than the olfactory axons, so their unit volume and energy cost are nearly 50-fold greater. In parentheses are the number of axons serving that sense. The relation is reciprocal: low unit cost allows a many-unit design (olfactory) whereas high unit cost restricts the design to fewer units (auditory). **Lower right:** Higher mean firing rates require thicker axons. Vestibular axon unit cost is 100-fold greater than that unit cost of an olfactory axon. Reprinted with modifications and permission from Perge et al., 2012. ves, vestibular; aud, auditory; pk, Purkinje; mf, mossy; cor, corticospinal; opt, optic; olf, olfactory; pf, parallel fiber.

the brain of a mormyrid fish that uses electrical signaling is huge compared to a trout of comparable body size (figure 4.7) and requires 60% of the resting animal's energy budget! This emphasizes that the purpose of brain design is not necessarily to operate on the cheap—for that would limit functionality. Rather, it is to ensure that the brain's investment pays off.

Design and usage of sensor arrays

In the mammalian ear each auditory hair cell is tuned to a particular range of frequencies—with the cells mapped along the cochlea's basilar membrane from lowest frequency (20 Hz in human) at the apex to highest (20,000 Hz) at the base.[12] The axons serving the highest frequencies fire at higher mean rates and are roughly threefold thicker than those for the lowest frequencies. Consequently, they use nearly 10-fold more volume and energy (figure 4.8). For humans the most critical frequencies are those for speech—which peak below 500 Hz and decline gradually out to 3500 Hz (figure 4.8); From the perspective of brain economy it is fortunate that natural selection has placed human speech at the lower end of the auditory nerve's frequency range, which is the most economical (figure 4.8). This design decision also saves costs downstream for central processing.

It turns out that music uses the same frequencies as human speech. The most frequent intervals in music correspond to the greatest concentrations of power in the normalized spectrum of human speech. Moreover, the structure of musical scales, the preferred subsets of chromatic scale intervals, and the ordering of consonance versus dissonance can all be predicted from the distribution of amplitude–frequency pairings in speech (Schwartz et al., 2003). Thus, music's tonal characteristics match those of human vocalization, which are the predominant natural source of tonal stimuli. This match seems understandable given that music serves to express and communicate emotions. It seems that the blues evoke sadness because those are the sounds that ancient humans uttered in communicating *their* sadness (Bowling et al., 2012; Han et al., 2010).

Music is processed by auditory areas in the right hemisphere, the side specialized for perceiving and expressing emotion; language is processed by corresponding areas on the left. It might seem redundant to analyze sounds with the same frequencies and structure in both hemispheres, but the computations are quite different, so it is economical to separate the circuits. What is the payoff for investing such substantial neural resources? Human survival and reproduction requires social cooperation—which depends upon communicating emotionally as well as cognitively. In short, music

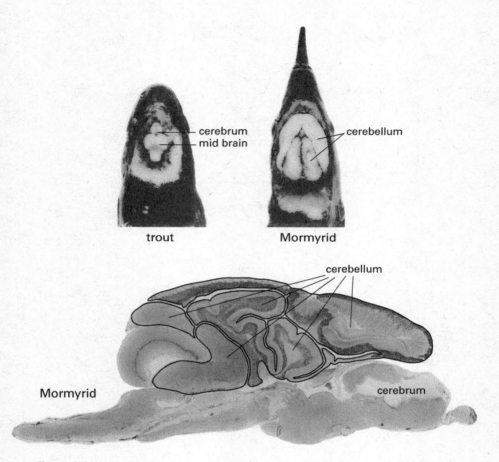

Figure 4.7
Mormyrid brain greatly expands cerebellar structures. Upper: Electrosignaling Mormyrid from turbid waters resembles trout in body size but requires a far larger brain, most of which is a highly elaborated cerebellum. **Lower:** Longitudinal section shows that the cerebellum (outlined) occupies most of the brain, completely obscuring the cerebrum. Central processors of high temporal frequencies often use a cerebellar-like design, including, in mammals, the dorsal cochlear nucleus (Oertel & Young, 2004; Bell et al., 2008). Reprinted with permission from Nieuwenhuys & Nicholson (1969).

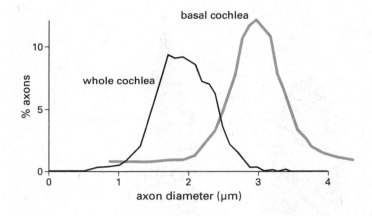

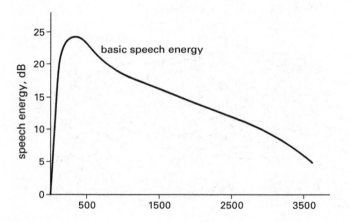

Figure 4.8
Speech uses lower frequencies and thus finer axons. Upper: Axons from the high-frequency end of cochlea (basal) are thicker and cost more space and energy than axons from the low-frequency end. **Lower:** Human speech occupies mostly frequencies below 500 Hz—the cheaper end. Upper, reprinted with permission from Perge et al., 2012); lower, after Freeman (1999).

helps communal life, made difficult by large brains, to be at least tolerable and occasionally joyous (Chanda & Levitin, 2013).

A sensor array must be fine enough to resolve the details that are critical to its task. For example, human vision resolves a spatial pattern of 60 cycles per degree, and this requires 120 cones per degree (Nyquist's rule). In two dimensions this amounts to 200,000 cones mm^{-2} (Packer et al., 1989). Again, this is just the down payment—for to *preserve* this spatial resolution, the communication line from each cone must remain separate all the way up to the visual cortex. All design must foresee the subsequent costs.

The general solution is to sample densely with a small part of the array and more sparsely with the rest. Therefore, our retina packs half of all its cones densely in a tiny patch (*fovea*), which occupies only 1% of the retinal surface. In this design the visual cortex devotes half of its volume to processing what the fovea delivers—thus allowing a fine analysis without unacceptably expanding the cortex.

For this strategy to work, it is often necessary to make the sampling array mobile—so that it can be trained on any feature of potential importance. Therefore, a fovea requires a system of muscles to move the eye, plus a control system to direct its constant exploration, and a higher-level system to select an object to be tracked. The effect is to stabilize the object on the fovea, allowing it to be sampled at high spatial resolution.[13] Stabilization confers an additional economy: it reduces the range of temporal frequencies on the fovea, allowing foveal neurons (and their subsequent processors) to operate at lower information rates, that is, on the steep segment of the rate-versus-cost curve for space and energy (figure 3.6).

This strategy also works for the tactile sense—dense distributions of sensors to fingertips, lips, and tongue—and explains the distorted *homunculus* in maps of human cortex, also the *barrel fields* in mouse cortex that represent the whiskers (Pammer et al., 2013) and the bizarre countenance of the star-nosed mole (figure 4.9).

Motorizing the sensors

The strategic choice of a fine, mobile sampler raises two other design issues: first, how to point the sensor where it is needed and, second, how to tell the brain that the sensor is *being* pointed. Both design issues require a dedicated part, the *superior colliculus* (figure 4.4).

The mechanism that chooses where to point the sensor needs visual input. When a retinal region outside the fovea senses a moving object, retinal signals drive a motor mechanism to smartly move the fovea onto that object and track it. The superior colliculus does this efficiently by placing a

Sensory periphery   **Cortex**

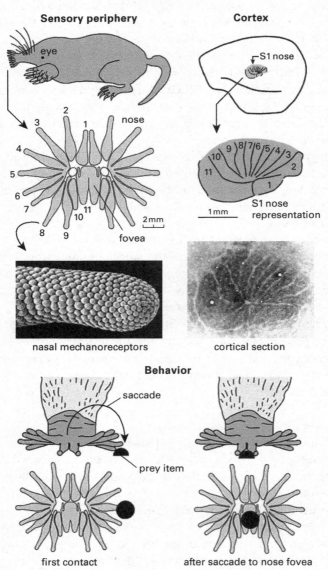

nasal mechanoreceptors   cortical section

Behavior

saccade

prey item

first contact   after saccade to nose fovea

Figure 4.9
Design of sampling arrays. Fine sampling required for spatial acuity requires large areas of cortex. Shown here is the mechanosensory system of the star-nosed mole. **Upper:** Frontal view shows tip of nose surrounded by 22 fleshy appendages. **Middle left:** Each nostril surrounded by 11 appendages, all covered by mechanoreceptors. No. 11 bears the densest distribution of receptors and thus serves as a mechanosensory fovea. **Middle right:** Each appendage is represented separately in somatosensory cortex (S1), with no. 11 occupying the greatest area. **Lower:** When a lateral appendage contacts an object of interest, the nose shifts to touch it with no. 11, the foveal appendage. Reprinted with permission from Sachdev and Catania (2002).

retinal map in register with a motor map so that each retinal point, using extremely short axons (~0.1 mm) can excite the corresponding point in the motor map and drive the eyes toward that location. Other senses also couple to the same motor map so that any of them—a flash, a bang, a slap—can announce which region of space needs the brain's immediate attention.[14]

Of course, we also attend to milder stimuli that match some stored pattern, especially when aroused by an internal signal of desire (food, sex). So the collicular mechanism for orienting the sensors needs to be informed by many issues. Decisions regarding where to look are made at the cortical level, which requires the cerebral cortex to communicate with the superior colliculus.

The upper collicular layers receive visual patterns and relay upward for further processing by cortical areas concerned with motion, and they receive signals from the same areas (Berman & Wurtz, 2010). The deeper collicular layers collect signals from the highest executive levels—frontal and parietal cortex—which convey a highly informed decision regarding where to look. The computations needed to reach that decision are extensive, involving much of the brain. But the decision can be relayed to the superior colliculus via a rather modest tract that requires only 6% of the corticocollicular pathway (Collins et al., 2005).

In short, the deeper layers of the colliculus know *where* to direct the eyes—that circuit is hardwired between the motor map and the low-level pattern generators that coordinate eye muscles. The deeper layers learn *whether* to move the eyes and *when*, by integrating raw-ish sensory inputs[15] with processed signals descending from cortex. The integrated output delivers instructions regarding vector and timing to pattern generators in brainstem that micromanage eye movements, and to those in the upper spinal cord that micromanage head movements. Thus, the descending collicular tract resembles various other tracts, such as fornix, hypothalamic, optic, and corticospinal, in being organized to send minimal instructions.

The motor stratum of the superior colliculus represents an intermediate-level pattern generator. It is tweaked by succinct executive decisions from above and delivers succinct instructions to low-level pattern generators. But it must also fulfill one more responsibility—to inform higher levels that its order: "Look!" has been sent.[16] This signal, termed *corollary discharge*, informs frontal cortex that the sensor is being repositioned. Why is this signal needed?

Corollary Discharge

When the retina is swept passively across a scene, the scene appears to move. The reader can confirm this by closing one eye and jiggling the other with a forefinger (gently!). However, when the superior colliculus *orders* the eye to sweep actively, the scene appears stable. The trick to stabilizing the scene when the brain moves the eye is to relay the order: "move!" to brain regions where the smaller patterns have finally been assembled into large, coherent patterns—corresponding to integrated perceptions. These areas, lying anteriorly in the parietal and prefrontal cortex (*frontal eye field*), know where the eye is looking—but they also need to know where the eye is *about to look*, so that they can compensate in advance before the motion occurs. This prediction, by allowing compensation, stablilizes perception—when we move our eyes, the world appears to remain stationary, as it should (Sommer & Wurtz, 2008; Wurtz et al., 2011).

The anterior frontal cortex is as far away from the superior colliculus as it is possible to be, so one might wonder why spend so much wire? One reason is that large patterns are assembled step-wise by cortical areas that press ever forward (chapter 12). By the stage where behaviorally relevant patterns have been assembled, compared to stores, and readied for use in selecting an action, the anterior frontal lobe is pretty much the last bit of available real estate. Moreover, because this cortical region decides where to look, it is precisely the site that needs corollary discharge to compensate for self-motion.

Another reason to control eye movements from the anterior frontal lobe is that, beyond their aid to sensing, eye movements also serve social communication. When someone looks us in the eye (or fails to), we notice. Even a dog notices and becomes aggressive when stared down by an unfamiliar human. Thus, as the cortical areas for social communication expand in the frontal and temporal lobes (chapter 12), they require a mechanism for sending executive summaries down to the superior colliculus. So design again economizes by using long pathways to send modest messages: "Look here!" or *"Don't look here!"*—skipping long, expensive explanations.

In summary, to build efficient sensors, the brain makes them mobile. It also compensates for self-induced motion, targeting the highest levels where choices and actions are being selected. These high-level mechanisms can then efficiently direct the low-level circuits that generate stereotyped patterns of movement. This *motif* drives orienting movements: the primate's eyes, the cat's external ear, and the rodent's whiskers and sniffing. These circuits use modest tracts to govern low-level pattern generators located near the relevant motor neuron clusters. This is the same motif that regulates internal systems and behavior.

Processing and storage of input patterns

Patterned inputs encounter the same constraints as patterned outputs, and to economize, they follow the same principles. First, the inputs deliver what can be computed locally; second, they relay upward only what is needed to assemble larger patterns. Each successive stage of processing sheds unneeded information. These principles also apply to storage: save only what is needed, for as long as it is needed, and in the most compact form.

Compute locally

Economy begins with sensory transduction. Because sending information at high rates costs more (figure 3.6), sensors use separate lines for different rates. For example, certain mechanosensors in the skin are wrapped in an onion-like capsule that filters out slow changes and delivers the fast ones to a mechanosensitive cation channel in the nerve terminal at the onion's core (figure 10.3). Other types with different capsules locate at different depths within the skin to help filter out the fast changes and capture slower ones. Skin sensors of temperature, noxious pressure, and noxious chemicals operate still more slowly—which allows still lower spike rates and finer axons. Consequently, the distribution of fiber diameters from sensory nerves resembles that of central tracts: many fine fibers and fewer thick ones.

Exemplifying the rule, *compute locally*, are two types of pressure receptor located on the foot. Each demands a prompt behavioral response without waiting 200 ms and expending more wire to consult higher processors. The responses are opposite: one extends the limb to support the body; the other flexes the limb to remove it from contact with the ground.

For example, pressing your bare foot on a smooth surface activates an array of low-frequency pressure receptors that excites the pattern generator for limb extension to support your weight. But pressing your foot on a sharp point activates higher frequency pressure receptors that excite the pattern generator for limb flexion to withdraw your weight and for limb extension on the opposite side to support your weight. This occurs faster than you can *feel* "Ouch!" because the higher frequency pressure responses travel over thick, fast-conducting wires that couple directly to the local pattern generator (figure 4.1).

Such direct functional connections between specific sensory inputs and specific motor outputs were historically termed *reflexes* (Sherrington, 1906). By now the design is seen as coupling each receptor type to the appropriate pattern generator. This design saves time, wire . . . and grief.

Relay to cortex
The small pattern carried by a single sensory axon resembles a piece of jigsaw puzzle to be assembled with other pieces into a larger pattern of sufficient quality for comparison to stored patterns. Assembly is a task for the cerebral cortex, but to reach that level, input arrays require serial "preprocessing" to reduce firing rates by stripping away redundancy and unneeded information. This requires that slow and fast signal components that were transduced separately maintain their separation via *parallel pathways* all the way to cortex. Thus, skin sensors signaling pain and temperature with low mean rates are processed by one set of circuits near their entry points (spinal cord and lower brainstem) whereas sensors signaling joint angle, muscle length, and whisker deflection with high mean rates are processed by different circuits[17] in lower brainstem.

For most sensors the spike rates are still too high for direct relay to cortex, so a central integrator (*thalamus*) is interposed to concentrate the message, that is, more bits per spike (figure 3.5C). This allows a two- to fourfold reduction in mean spike rate on the path to cortex. The thalamus is also used by other brain regions, such as cerebellum, *striatum*, and superior colliculus, for the same function (Bartlett & Wang, 2011; Sommer & Wurtz, 2004).[18] The computational strategy and synaptic mechanisms to achieve this function are described in chapter 12. The exceptions to this design are the olfactory sensors which signal at such low rates that, following a single stage of preprocessing in the *olfactory bulb*, they are allowed to skip the thalamic relay and ascend directly to cortex (Friedrich & Laurent, 2001).

Cortex finds larger patterns
The task of sensory cortex is to rapidly capture correlations of higher order from the array of local correlations relayed from thalamus. This proceeds by

stages, first across layers of each primary area (*V1, S1, A1*) and then across successive areas, until single neurons eventually report patterns of clear behavioral relevance that identify an object by sight, touch, or sound (figure 12.11). Such patterns emerge in specialized patches where most neurons respond only to that pattern and not to the fragments that comprise it, thus an area for faces, objects, scenes, and so on (chapter 12).

A reader might worry that the world's infinity of categories would require a corresponding infinity of cortical areas, but actually, the number only needs to match categories that matter most deeply to the animal. Smaller brains operate with fewer categories, so the whole mouse cortex divides into about 20 areas, whereas human cortex has about 200 (Kaas, 2008). As areas attain higher levels of abstraction, each contains less information and thus requires less space. So the early cortical areas, which first process thalamic input, are large, whereas later areas for high-order patterns are small (figure 12.11)

This design—many small areas operating in parallel—continues the principles of economy. Resources can be assigned according to what matters most to the animal. Processing can proceed at lowest acceptable rates and at lowest acceptable spatial resolution. For example, an *object-grasp area* that needs only coarse patterns can download them at an earlier stage than an *object identification area* that needs more detail (Srivastava et al., 2009; Fattori et al., 2012). Wire is saved by locating areas that assemble the patterns near to the areas that use them (chapters 12 and 13). For example, face areas locate anteriorly in the temporal lobe on the path toward areas that evaluate facial expression. An object-grasp area locates posteriorly in the parietal lobe—on the path toward motor cortex that guides grasping. Thus, the overall processing scheme for cortex reflects the three design principles seen at lower levels: send only what's needed; send slowly as possible; minimize wire.

Storing Patterns

To store small, evanescent patterns encoded by an array of thalamic neurons, would be costly. If patterns were all stored at this level, high-level images could in principle be reconstructed. However, with the optic nerve delivering 10 Mbit s^{-1} to the thalamus, storage needs would soon exceed any conceivable capacity. Moreover, if data were stored raw, it could only be filed by order of arrival—so to retrieve images from stored fragments would be a computational nightmare and impractically slow. So an animal should store high-level patterns and only *particular* ones that can improve future behaviors.

Each species stores patterns critical for its economic strategies. For example, a nutcracker jay living at high altitude caches nuts at numerous sites in autumn and descends to a valley for winter. Returning in spring, it recalls myriad cache locations to sustain itself until the summer brings fresh groceries. For humans, what matters most is our ability to rapidly recall a face, along with any historical significance that we can attach to the face we are facing. This allows the best chances for selecting an appropriate behavior.

Yet we must not store every face encountered on a stroll through the park—only ones likely to prove significant. So a potentially important face needs to be tagged—cognitively and affectively—and then filed. Upon reencounter, the original image is retrieved and held in "working memory" for comparison to the current image. These various processes require cooperation between several neural structures. The main cortical face area connects with the amygdala, which "stamps" the image from its catalog of innate emotional expressions. To further annotate the image, the striatal system for reward prediction connects to the face area via a long loop and to the amygdala (Middleton & Strick, 1996). Then, they all connect to sites for working memory and behavioral choice in prefrontal cortex.

These organs for pattern recognition, storage, evaluation, and behavioral choice interconnect strongly; therefore, by locating near each other, wire is reduced. Their location anteriorly in temporal and frontal lobes is no mystery: the posterior regions are already occupied by areas concerned with pattern assembly. Thus, in mammals where higher degrees of sociality require the brain to enlarge, the expansion occurs disproportionately in anterior regions for cognition and emotional expression (Dunbar & Shultz, 2007). Thus, although human and macaque collect similar amounts of sensory information (e.g., their retinas are nearly identical),[19] humans greatly expand the number and size of cortical areas for assembling the higher order patterns. This occurs especially in forward regions that include amygdala, prefrontal cortex, and hippocampus.

Correcting errors

Evaluating behavior: Two kinds of prediction error

The parts of the motor system that directly generate and distribute final output patterns (behavior) require only a small fraction of total brain volume. However, the adjective "final" is slightly misleading. Each motor act is also a beginning: it is a provisional answer to some predicted need. Since needs recur, output patterns might be improved if their effectiveness could be evaluated. Therefore, the brain invests heavily in several systems for evaluation and error correction.

One system asks, "How precisely did the actual output pattern match the intended pattern?" This system computes the difference between the intended pattern and the actual pattern; then it feeds the error back to command structures that gradually improve performance. This serves *motor learning*—what is gained from practicing the piano or the golf swing. Mindful repetition improves speed and accuracy—and also efficiency—since a motion that begins awkwardly eventually gains grace and saves energy (Huang et al., 2012). This system also serves cognitive and affective processes: it compares intended cognitive and emotional patterns to what actually occur and then feeds back to improve subsequent performance. Thus, motor learning is subset of *intention learning*.

Another system asks, "Was the act, however well performed, worth the energy and the risk?" This system compares the expected payoff from a particular act to what was actually gained. The neural mechanism rewards a better outcome by releasing a pulse of dopamine at key brain sites and punishes a poorer outcome by reducing dopamine and enhancing other chemical signals. This is *reward-prediction learning*, and one can easily imagine its myriad ramifications. Reward-prediction learning evaluates every choice and thus charts the course of our lives: cereal or toast; law or medicine; choice of mate, friends, and retirement fund (chapter 14).

Intention learning and reward-prediction learning employ different brain structures, and both are large (Doya, 2000). The organ for intention learning is the cerebellum, and the organ for reward-prediction learning is the striatum (figure 4.10). Neither structure directly modulates the final output: they do not send wires to the low-level pattern generators. Rather, they return error signals to particular high-level organizers of behavior. For example, the cerebellar region that serves motor learning (*anterior lobe*) returns its updating signal to motor cortex. Cerebellar regions that serve perceptual, cognitive, and affective learning return their updates to cortical areas for pattern recognition in temporal and parietal cortex and to areas for behavioral choice, such as prefrontal cortex (Strick et al., 2009; Schmahmann & Pandya, 2008).

Cerebellar and striatal output tracts both use high spike rates that require thick axons. In fact, the striatum is named for striations due to bundles of thick, myelinated axons (figure 4.10). High spike rates should be reduced before the messages are broadcast. Both circuits do this, as noted, via a thalamic relay. Cerebellar and striatal design will be considered further in chapter 13.

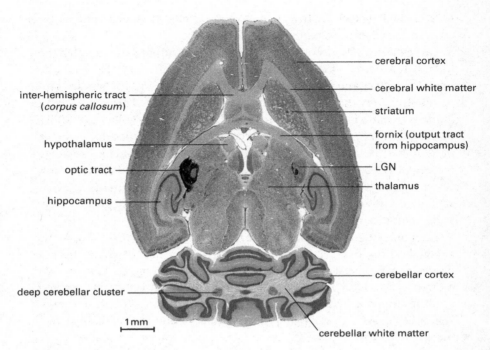

Figure 4.10
Rat brain in horizontal section. Note that striatum lies nearest to the anterior cerebral
cortex. Striatum contains dense bundles of myelinated axons (pale) whose large caliber reflects their high spike rates. Note also the deep cerebellar clusters which reduce
the number of high-rate axons before projecting to thalamus where rates are reduced
before relay to cerebral cortex. Left optic tract is dark because a protein tracer injected
into the eye was taken up by ganglion cells and transported inside their axons to the
brain. Tracer is visualized here by a specific chemical reaction. Image courtesy of H. J.
Karten and reprinted with permission; © The Regents of the University of California,
Davis campus, 2014.

Conclusions regarding organization of mammal brain

This chapter has sketched how three principles (*send only what is needed*;
send at the lowest acceptable rate; *minimize wire*) shape brain design to
accomplish its seven broad tasks (see figure 3.2). The layout explained here
extends upward from a scale of millimeters. It does not explain design of
local circuits that analyze and integrate input patterns or generate output
patterns. Those compute on a scale of nanometers to micrometers and
are topics for chapters 7–11. This chapter also does not explain the brain's

striking structural diversity on the scale of micrometers to millimeters, such as the different structures for cerebellar versus cerebral cortex and the specialized substructures of cerebral cortex. These are treated in chapter 13.

Insect Brain

We consider now the insect brain, emphasizing *Drosophila*, because of its importance for genetic analysis—like mouse. But we also include other insects, such as locust, wasp, cricket, and bee that share various broad features of somatic and neural design and are profoundly specialized for particular lifestyles and habitats (Burrows, 1996; Strausfeld, 2012). Just as we referred to "mammalian" brain in preceding sections, we will refer to "insect" brain in this section.[20]

The first point is that the insect brain needs to accomplish the same basic tasks as the mammalian brain (figure 3.2). Second, it is governed by the identical constraints: the law of diminishing returns (figure 3.6), plus the need to minimize wire (figure 4.1). Third, the insect brain also predictively regulates the internal environment and efficiently couples the internal organs (see figure 3.4). Finally, the insect brain couples the inner and outer worlds (figures 3.2 and 4.3) and, following the scythe of Saturn, encounters the same types of information, which it must analyze and integrate to satisfy similar behavioral demands. So we should expect similarities of macro-organization. Indeed, they are numerous and striking (figure 4.11).

The insect brain, like the mammalian, is organized into defined neural clusters with locally dense connections plus distinct tracts for more distant connections (Chiang et al., 2011). Brain outputs include a rich system for wireless signaling, starting with two neurosecretory bodies at the back of the brain (*corpora cardiaca* and *corpora allata*) whose neurons secrete neuromodulators and hormones into the circulation (analogous to the hypothalamic neurosecretory clusters). These neuromodulators and hormones, which include over 50 neuropeptides, govern the insect's internal milieu by acting on energy metabolism, salt and water balance, growth/molting, and reproduction. Autonomic neurons cooperate with these hormones to coordinate visceral function with behavior (Cognigni et al., 2011). For example, gut neurons interact with hormones to increase intestinal throughput to fuel egg production and also to control appetite. These concerted actions of wireless and slow wire processes in insects resemble those of the vertebrate hypothalamo-pituitary and autonomic systems, and there appears to be a common evolutionary origin (Arendt, 2008).

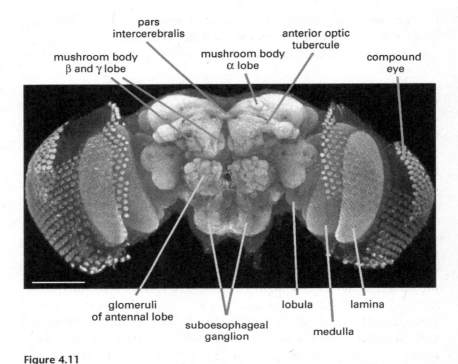

Figure 4.11
Frontal view of fly brain shows prominent areas devoted to specific functions. *Vision:*
Compound eye, hexagonal array of optical sampling units passes information sequentially to lamina—collect and sort inputs, medulla—detect local patterns, lobula
(and lobula plate not seen in this view)—assemble small patterns into larger patterns,
anterior optic tubercule—associate larger patterns. *Olfaction:* Glomeruli in antennal
lobe—collect and sort inputs and project to mushroom bodies, which identify patterns. *Learning:* mushroom bodies—integrate diverse information, learn patterns
and associate with punishment and reward. *Integration:* Pars intercerebralis connects
two sides of brain. *Distribution:* Suboesophageal ganglion—integrate information for
wired and wireless output to body. View of a three-dimensional reconstruction of a
Drosophila brain stained with antibody for synapses to show areas where processing
takes place. Image courtesy of Ian Meinertzhagen. Reconstruction can be rotated and
viewed from different angles at http://flybrain.neurobio.arizona.edu/Flybrain/html/
contrib/1997/sun97a/.

Insect brain uses these systems to coordinate visceral, behavioral, and immune responses to stress, but instead of the vertebrate's adrenalin (*epinephrine*), insects use octopamine (Verlinden et al., 2010). Thus, during emergencies, "fight or flight," octopaminergic neurons raise octopamine concentration in hemolymph (like adrenalin in vertebrate blood), which acts broadly on endocrine cells and fat body (similarities with vertebrate liver) to mobilize energy reserves, on muscle to increase power, and on sensory receptors and circuits to increase sensitivity and response speed. Octopamine neurons also directly contact endocrine glands, heart, muscle, and certain brain regions for specific purposes. For example, in locust 40 identified neurons (*DUM*) innervate flight muscles to regulate fuel supply (Burrows, 1996). At rest the neuron fires steadily, maintaining the supply of "fast burning" sugars needed for takeoff. During steady flight the DUM is silenced, and energy supply switches to the larger reserves of slower burning fats. The DUM's low mean firing rate, 0.5–1 Hz, resembles mammalian autonomic nerves.

Insect brains also have clocks set by light—indeed the molecular mechanism of animal clocks was first determined in *Drosophila* (Weiner, 1999). *Drosophila*'s roughly 150 clock neurons form a distributed system that governs catabolic/anabolic phases, including a sleep phase for the consolidation of neural processing (Allada & Chung, 2010; Crocker & Sehgal, 2010). Some clock neurons form small clusters, mini-SCNs, that collect specific entraining inputs from the compound eye, the simple eyes (*ocelli*), and a pair of photoreceptor cells within the brain. Other clock neurons express their own photopigment and so can collect photons through translucent cuticle. Thus, the fly's clocks locate anywhere they are needed. We speculate that this distributed design saves wire in a small brain.

Collecting patterns

Investment in sensors to collect patterns is strongly tuned to social and economic strategies. *Drosophila*'s compound eye is relatively small, and the photoreceptors gather information at a low rate—good enough for hovering over decaying fruit. However, *Coenosia*, a close relative with similar body size, is an aerial predator and, to resolve and track its prey, requires a threefold larger eye and photoreceptors with fourfold higher bit rates (Gonzalez-Bellido et al., 2011). In accordance with the law of diminishing returns for photoreceptors, *Coenosia*'s high-rate eye costs more space and energy per bit (chapter 8).

To identify rotting fruit and detect *pheromones* (secreted chemical factors that trigger social responses), *Drosophila* invests in about 50 types of olfactory receptor. These are more than are used by the louse that parasitizes humans (10) but fewer than are used by the honeybee (160) and fire ant (400) for their extensive foraging and chemical communication. Certain insects also invest in mechanical apparatus to improve their efficiency at pheromone detection. For example, male moths commonly use broad antennae as molecular sieves, which they push through the air to trap molecules of female attractant.

Meanwhile, *Drosophila*'s antennae specialize to register not the aroma of courtship but its music. Both sexes sing to each other. The vibrations, reaching 500 Hz, are received via the antenna and transmitted to its base to activate about 500 mechanosensors (*Johnston's organ*). These are equipped with mechanical feedback to boost the gain, like hair cells in the mammalian cochlea, to operate near the sensitivity limit set by Brownian noise (Immonen & Ritchie, 2011).

Moths are hunted by bats using echolocation. So the moth invests in a pair of simple ears, each with only one or two sensors, and couples their outputs to a simple pattern generator for evasive flight. When the sensors detect a bat's ultrasonic chirp, evasive flight is engaged, and the moth dives to the ground (Roeder, 1967). This system provides a cheap answer to the bat's high-tech, super-expensive sonar.

Insect sensor arrays, like mammalian sensor arrays, are subject to the sampling theorem (Nyquist's rule). To achieve high resolution at acceptable cost, they too combine broad, coarse sampling with local, fine sampling— both in space and time. For example, a male housefly pursuing an evasive female at high angular velocities is aided by his visual *lovespot*. The forward-facing photoreceptors pack especially densely to improve spatial resolution, and they produce especially fast electrical responses to improve temporal resolution—both needed to track the speedy female (Burton & Laughlin, 2003). But the lovespot, like a mammalian fovea, must not be too broad, because it is expensive, so the fly uses the same solution: motorize the sensor. During pursuit, a dedicated tracking system controls head and body movements to keep the lovespot centered on the target.

In short, insects invest in sensors according to need and locate the sensors where they will be most useful: olfactory and auditory sensors on the antennae that project into the air stream; auditory sensors on crickets' forelegs to space them widely (thereby improving sound localization), taste sensors on the landing gear (feet), mechanosensors on the wing. Each sensory system is used to inform the others; for example, an odor that attracts *Drosophila* increases the accuracy with which its visual system guides its flight (Chow et al., 2011)—cross-modal interactions that are also used by mammals (Burge et al., 2010).

Processing and storage

Insect sensory processing resembles mammalian processing in that small patterns collected by sensors are filtered and then assembled into larger patterns. To assemble visual patterns, the fly identifies spatial and temporal correlations via successive neural layers (figure 4.12). First, the lamina sums correlated inputs and removes redundancy associated with the level of illumination. Then the medulla identifies local features which the next layers (*lobula* and *lobula plate*) use to detect larger and more complicated patterns. Then their outputs distribute to various smaller regions (*optic glomeruli*) where they are processed before projecting forward to integrative centers in the *protocerebrum*. Each optic glomerulus collects inputs from a particular ensemble of neurons in the lobula, suggesting that higher order patterns are being segregated.

The architecture of the fly visual system resembles in several respects that of mammal. The fly preserves spatial continuity of the retinal image by mapping the output from one layer, point by point, onto the next layer—across the many stages of processing. However, at the final stage, the optic glomeruli abandon retinotopic organization, thus shedding "where" information while sorting out "what," reminiscent of the *ventral stream* of the mammalian cortical pathway (chapter 12).

The layers and maps of vision's earlier stages are computationally efficient because all parts of an object represented in the retinal image are spatially and temporally continuous. These properties of the input allow local features (local motion, local edges) to be extracted and mapped at the lower levels and then assembled at higher levels to define objects and scenes. Extracting all local features first, as with insect medulla and mammalian visual cortex, provides a communal data set to be shared by various higher order mechanisms, and this conserves space and energy. Local processing, mapping, and the orderly projections from each layer to the next also save wire, as do the orderly maps of different modalities within a tract (Niu et al., 2013; chapter 13).

Despite an efficient architecture, visual processing for form, motion, and color is computationally demanding. The visual system uses 70% of the fly's neurons, of which most are in the medulla, which extracts local features using about 150 different types of identified neuron. Thirty-five types, replicated in each of the medulla's 800 retinotopic columns, interrogate the image for local features. In this respect the fly's medulla is analogous to the mammal's primary visual cortex, also the largest visual area (chapter 12).

The olfactory system is structured differently (figure 4.12). Whereas vision assembles patterns stepwise across four layers, olfaction uses just two (Masse et al., 2009). The first layer (*antennal lobe*) collects input from 45

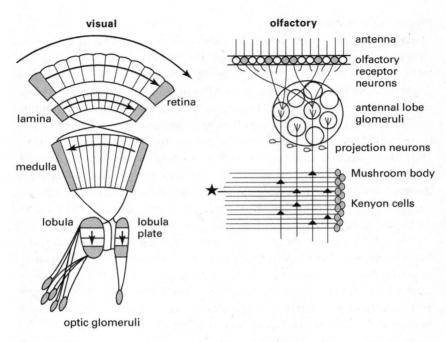

optic glomeruli

Figure 4.12
The visual system is deep and maps spatial position. The olfactory system is shallow and processes globally, without reference to spatial position. Left: Fly visual system processes retinal image in four successive layers. Lamina assembles and sums correlated inputs and reduces redundancy (chapter 9); medulla extracts local features; lobula and lobula plate assemble larger patterns (lobula—color, form and motion; lobula plate—motion). The first three layers map retinal image (arrow) across columns of neurons. The last layer, optic glomeruli, does not map, it generalizes. Each glomerulus collects from all neurons coding the same pattern, irrespective of spatial position. **Right:** Fly olfactory system processes information in just two layers. First the antennal lobe assembles and sums correlated inputs from receptor neurons. A glomerulus collects from neurons with same olfactory receptor and filters to reduce redundancy. Then 2500 Kenyon cells in mushroom body extract from all 45 glomeruli the patterns that define odors. Each Kenyon cell associates synaptic input (triangles) from small subset of 10 glomeruli to form an efficient sparse code (2 associated synapses shown on starred Kenyon cell). Diagrams simplified and not to scale. Visual based on Strausfeld (2012); olfactory based on Masse et al. (2009).

types of olfactory receptor on the antenna, collects each type in a separate synaptic glomerulus, sums these correlated inputs to reduce noise and filters to reduce redundancy. The results are relayed to the second processing layer, residing in the mushroom body, which is the insect's seat of learning (see below). The second-stage neurons compare all 45 olfactory inputs and learn by association the unique patterns of glomerular input that define particular odors. The mammalian olfactory system employs a remarkably similar structure (Wilson & Mainen, 2006). It uses an olfactory bulb with glomeruli, one for each receptor type, and after filtering, it projects straight to cortex for association and learning.

Two-stage processing works for olfaction because, unlike vision, there are no local features. The molecule or mixture that characterizes an odor arrives in a volume of air for a certain time, but there are no higher order spatial correlations to help identify it. The correlations that identify an odor are distributed across receptors: each type binds a spectrum of molecular species, each with different affinity. Thus, an odorant, whether from a single molecular species or a mixture, activates several receptor types to different degrees, to produce a correlated pattern of receptor activations—and that defines an odor.

The pattern from the array of glomeruli transfers to the mushroom body (Laurent, 2002; figure 4.12). Because each odorant stimulates several receptors, and each receptor contributes to the coding of many odorants, the mushroom body's task is to find correlations across receptor inputs—the pattern that defines a particular odor. When a new and significant odor is encountered, the new pattern is learned. To optimize the number of different patterns that can be represented by the mushroom body's 2500 Kenyon cells, the information is coded sparsely with few spikes (Jortner et al., 2007).

In short, there are profound differences across sensing systems within an animal, and profound similarities for a given sensing system across animals (insect vs. mammal). Olfactory and visual designs differ because the small patterns that they collect present different statistics and thus require different processing. Olfactory designs are similar because the input statistics for insect and mammal are the same and thus require similar processing. The same goes for visual designs.

Nonetheless, insect and mammalian designs are not identical, probably because they are differently constrained. For example, the fly visual system lacks a thalamus, which the mammal needs to reduce spike rates. Many fly visual neurons connect centrally over distances less than 0.5 mm, which means that signals can travel passively in graded (analogue) form. This saves space and energy in two ways: analogue can transmit high

information rates cheaply (chapter 5), and can avoid costly analogue →
pulsatile and pulsatile → analogue conversions. Thus the insect brain uses
a more efficient design that cannot be implemented in a larger brain.

Assembling patterns and choosing an action

A fly assesses its current state from sensory patterns, compares this state to
stored patterns to learn how its state is changing, and adjusts behavior
accordingly. For example, it may steer flight to maintain a constant bearing
with respect to the sun or change course to approach a rewarding object or
avoid an aversive one. The *central complex*, a compact modular structure
strategically placed deep in the brain, plays a pivotal role in these processes
of assessment, decision, and direction (Strausfeld, 2012; Strauss et al., 2011).

The central complex links sensory patterns to motor commands within
a framework of body orientation (figure 4.13). Its three largest structures,
the *protocerebral bridge*, the *fan-shaped body*, and the *ellipsoid body* are linear
arrays of neural modules that map the angle of azimuth (compass bearings
on a horizontal plane) around the fly. The 16 modules of the protocerebral
bridge map 16 sectors, eight on the fly's left and eight on its right (figure
4.13), and project to eight modules in the fan-shaped body. Each fan-shaped
body module accepts input from a protocerebral bridge module on the left
side, and from its opposite number on the right side. This convergence
establishes eight horizontal axes that pass through the center of the fly.

The eight fan-shaped body modules then connect straight to the eight
modules of the ellipsoid body which, in turn, connect to the lateral accessory lobes. Here the outputs from the central complex contact the descending neurons that drive motor pattern generators in the segmental ganglia.
In short, by explicitly linking signals to azimuthal bearings (horizontal
lines of sight from the fly's cockpit), the central body relates the position of
a sensory pattern to the body's orientation and direction of movement.

Information on sensory patterns and stored patterns project across the
directional modules via *horizontal neurons*. Some horizontal neurons establish memory traces, and this allows generalization. Information gathered
from a pattern observed in one direction is distributed so that an object
learned in one location can be recalled in another. The horizontal projections stratify the fan-shaped body, and two of its layers have been linked to
specific components of visual patterns: one layer to the orientation of visual
contours and the other to the elevation of an object above the horizon.

Some patterns processed by the central complex are used for navigation.
The central complex serves as a sky compass that enables locusts and monarch butterflies to fly on a constant bearing by maintaining the body at a

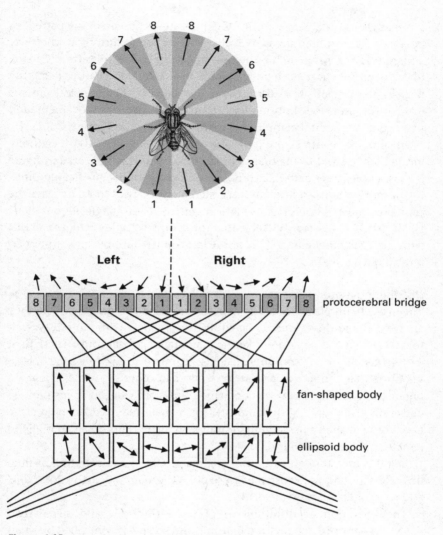

Figure 4.13
Central complex maps horizontal lines of sight. Protocerebral bridge's 16 modules map 16 sectors viewed from head, 8 on insect's left and 8 on its right. Projection to fan-shaped body's 8 modules connects opposite sectors (e.g. Left 1 and Right 8) to establish and map axes that pass through centre of head. This map is projected to ellipsoid body's 8 modules, for output to neurons that select and control motor patterns. The central complex then sends information about position of stimuli with respect to the head to neurons that control body orientation and the direction of locomotion. Figure based on Strausfeld (2012) and Strauss et al. (2011). Fly image from http://openclipart.org/image/800px/svg_to_png/120457/HouseFly2_.png.

given angle to the solar azimuth. When the sun is obscured, the pattern of polarized light in blue sky is used instead. For bees returning to the hive, and especially for monarch butterflies migrating 3,000 miles from Canada to Mexico, it is important to maintain the same true bearing (e.g., 185 degrees South Southwest) throughout the day. For this, the sky compass mechanism uses clock information to correct for the sun's movement, and neurons in the central complex are involved (Heinze & Reppert, 2011).

In short, the central complex is aptly named because it is both centrally located and central to the brain's broad tasks that were indicated in figure 3.2 (assemble larger patterns, compare to stored patterns, predict a promising output pattern, and call an integrated output). Thus in many ways the central complex is homologous to the mammal's basal ganglia (Strausfeld & Hirth, 2013). It seems remarkable that the central complex achieves all this with less than 600 neurons (592 at the last count). But how are output patterns implemented?

Distributing motor patterns

The insect brain places its motor neurons in the body segments where they are needed and drives their detailed firing sequences with pattern generators located at the same site (like the mammalian spinal cord). These final pattern generators are coordinated across segments (e.g., three pairs of legs) via fibers that connect to pattern generators in other segments. These are organized into complex behaviors which the brain can call or restrain via descending neurons. Most famously, for a male mantis to copulate, he needs only to shed a descending restraint—which occurs when an obliging female . . . bites off his head.

Significantly, though perhaps anticlimactically, the distribution of fiber diameters in the connecting tracts resembles the mammal: many fine axons and fewer thick ones (figure 4.14).

The activation or disinhibition of some rhythmic and stereotyped behaviors—for singing, mating, fighting, and so on—is controlled by small numbers of command neurons that activate dedicated networks (Hedwig, 2000). To an observer, these behaviors appear quite complex and plastic— for example, Google "drosophila aggression" and watch a YouTube film that resembles a professional boxing match. Complex behaviors can be evoked from larger insect brains by electrical stimulation of single command neurons—recalling the complex behaviors evoked with fine electrodes from the mammalian hypothalamus.

The insect brain, like the mammal, needs to distinguish activity created by its own motor commands from activity originating in the environment,

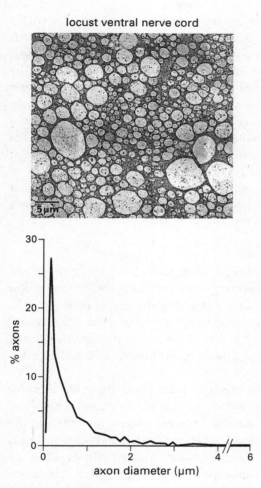

Figure 4.14
Distribution of fiber diameters in insect nerve cord. This distribution resembles many long pathways in mammalian brain. Reprinted with permission from Perge et al. (2012).

that is, it needs mechanisms for corollary discharge. For example, a cricket producing loud chirps risks desensitizing its own auditory system, which would prevent it from detecting softer external sounds (Poulet & Hedwig, 2007). To avoid desensitization, the small motor circuit that generates the chirp drives a single neuron that directly blocks inputs from the two ears (figure 4.15). This simple circuit shuts down auditory inputs for precisely the duration of a chirp, leaving the cricket free to listen for responses between chirps. This precise blanking-out of disruptive input resembles the suppression of visual inputs during a saccadic eye movement. The point here is that for most tasks that a mammalian brain needs to accomplish, so too must an insect brain. Moreover, the insect brain often uses similar strategies—but benefits from the smaller scale: fewer neurons and shorter distances (Chittka & Niven, 2009).

Correcting errors: Motor learning

The prominence of the cerebellum in mammalian brain might predict an obvious insect analogue, but there is no structure totally dedicated to motor learning. The suggestions are that motor learning is one of many tasks assigned to the mushroom bodies and to the central complex (Farris, 2011; Strauss et al., 2011). Indeed, with fewer body segments to coordinate, stiffer mechanics, and a body that is not continually growing, an insect arguably has less need for motor learning.

Nonetheless, some motor learning is essential. For example, flies improve their motor performance with practice (Wolf et al., 1992). Normally when a fly (or any animal) turns in one direction, the visual scene moves in the opposite direction. If this relationship between action and consequence is reversed by placing the fly in a flight simulator, the fly adjusts within 24 hours. Now when it wants to approach a promising target, it turns *away* from the target, and *voilá*, the target enters its field of view. This resembles Kohler's famous experiment: after students wore inverting spectacles for a day or two, the world appeared to be right-way up, but when they removed the spectacles, it appeared upside down. Why do flies need this motor learning? Motor learning is built into their flight control system to cope with changes of body mass (feeding, defecating, growth, and laying eggs) and damage to the wings.

Reward-prediction error

Insect brains are wired for associative learning and employ a system for computing reward-prediction error that follows the same basic learning rules as in mammals. The internal reward system uses dopamine and

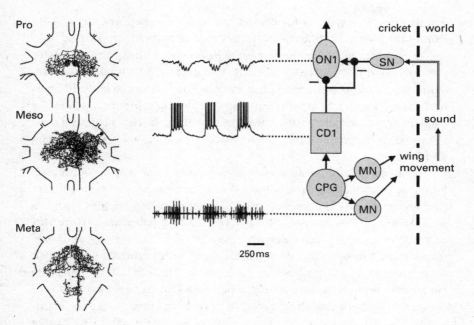

Figure 4.15
CD1, the neuron that prevents a cricket being deafened by its own chirp. Right:
CD1's circuit. Central pattern generator (CPG) drives motononeurons (MN) rhythmically to produce wing movements that generate chirps. Each chirp excites sensory
neurons (SN) in cricket's ear. SN output synapses excite omega neuron (ON), which
conveys auditory information to brain. CPG also drives CD1, which inhibits ON1
and output synapses of SN, thereby blocking signal to brain while chirp is being
produced. **Middle:** Recordings of signals within circuit. Bottom trace: Extracellular
recording of spikes in MN, driven rhythmically by CPG. Middle trace: Intracellular recording from CD1. Excitatory synapses from CPG depolarize CD1 to produce
bursts of spikes that follow CPG rhythm. Top trace: Intracellular recording from ON.
Inhibitory synapses made by CD1 produce rhythmical bursts of IPSPs that block ON1
output during chirps. **Left:** Morphology of CD1 revealed by intracellular dye injection. Axon connects dendritic arbors in the three thoracic ganglia, meta-, meso-, and
pro-. Mesothoracic dendrites receive excitatory synapses from CPG. Prothoracic dendrites make inhibitory synapses onto omega neuron, ON1, and all sensory neurons,
SN. Vertical scale bar: 20 mV for CD1; 5 mV for ON1. Reproduced from Poulet and
Hedwig (2006) with permission.

octopamine. The systems for computing reward-prediction error and for storing the lessons both reside in the mushroom body (figure 14.11).

The mushroom body, like mammalian cerebral cortex, participates in olfactory learning, associative learning, spatial learning, visual pattern recognition, attention, and sensory integration. The mushroom body, like cortex, shapes its circuit architecture to view multiple inputs, looking for coincidences to associate with reward or punishment. This suggests a multipurpose cross-correlator that can be wired to evaluate a variety of associations and store the lessons.

As with other computing devices, new models allow new opportunities. Primitive parasitic wasps (early model) use elaborate mushroom bodies to find and store the locations of grubs hidden at particular sites within a plant (Farris & Schulmeister, 2011). Social wasps (later model) use this capacity to recognize each colony member by its distinctive face and body markings and to store this information along with knowledge of its position in the dominance hierarchy (Sheehan & Tibbetts, 2011). Thus, the later model supports a complex social behavior that confers the benefits of communal foraging and the division of labor. Social insects, like social primates, build upon the low-level sensors, adding brain parts that enable social behavior. The parts that expand are those that recognize patterns, store them, and evaluate them via the system of reward prediction.

What a honeybee can do with a brain of 10^6 neurons seems prodigious. A bee learns to break camouflage, to navigate a maze via symbolic cues (blue, turn left; yellow, turn right), and to associate a flower with the time of day during which that particular species produces nectar. Bees can also perform delayed match-to-sample and symbolic match-to-sample tasks[21] that were thought, until recently, to be confined to monkeys, human, dolphin, and pigeon (Srinivasan, 2010; Menzel, 2012). In short, absolute numbers of neurons seem not to be everything. What seems most important is that design takes full advantage of small size.

Efficiencies of small size

Because an insect is small, it can use an external skeleton. Small body and exoskeleton both allow a smaller brain, which is intrinsically more efficient. A small brain uses disproportionately less wire than a larger brain, so it can locate cell bodies at the brain's margins, out of the way of wires and tracts (chapter 13). As well as saving space, this also saves energy because a distant cell body reduces load on a neuron's electrical circuit (chapter 7). Shorter wires allow more analogue signaling (e.g., worm; chapter 2), and what spikes are needed can travel at lower velocities on thinner axons.

Furthermore, a small brain allows a compact neuron to coordinate the activities of an entire system (figure 4.15) or to spread its dendrites broadly enough to extract a pattern from an entire sensory field.

An insect brain economizes by relaxing the specifications for workaday behavior. A low-mass insect, clad in tough exoskeleton, sustains less damage in a collision or a stumble, so it can tolerate accident rates that would for humans be criminally negligent. The exoskeleton also lessens the burden of motor control. Shorter limbs with stiffer joints and viscous damping are easier to manage, and the ability to place sensors in the exoskeleton to measure the most informative forces reduces the need to compute at higher levels. Mammals use the same strategy (see above), but an exoskeleton provides insects with more opportunities for sensor construction and placement. Insects do require some high-performance control systems; it would be impossible for a fly to fly without one, but in many respects the insect body is less demanding and more adaptable.

The exoskeleton provides opportunities to reduce demands on the brain through embodied computation. For example, to beat its wings at 200 Hz, *Drosophila* builds an oscillator from its flexible exoskeleton and muscles that, when excited, contract in response to stretch (Dickinson & Tu, 1997). To kick start, a dedicated neural circuit excites an auxiliary muscle to contract sharply and stretch the muscles that elevate the wings. As the elevators contract, they stretch the muscles that lower the wings. Coupled by the resonant exoskeleton, the antagonists pull back and forth, beating the wings. To keep the muscles excited, the brain need only deliver spikes at less than 10 Hz. Thus, an intermittent, low-rate input from the brain produces a high-rate, patterned output from the body, significantly reducing computational load. The kick-start muscle also yanks the legs straight, thrusting the fly upward as the wings start to beat, a case of "neatening up" (chapter 1).

The brain can further reduce its computational load by taking shortcuts. Challenging problems are solved with simpler solutions that, while inexact, work well enough, and many animals, including humans, use these efficient *heuristics* (Gigerenzer, 2008). Insects often use them to judge the sizes of much larger objects (Wehner, 1987)—an egg to be parasitized by a tiny wasp, a chamber in which to build a whole ants' nest, a target of given angular diameter—is it small and close by or big and far off?

A big problem for any animal is how to find its way in the world and return. The honeybee uses the sun as a compass to set its bearings from the hive to a productive clump of flowers. When the sun is obscured, the bee infers the sun's position from the pattern of polarized light in patches of blue sky. To relate a fragment of this polarization pattern to its stored map seems difficult, but the bee employs a shortcut: it reduces the two-dimensional sky map to a one-dimensional map of *e-vector* versus bearing to the sun, ignoring the sun's arc as it travels across the sky (Rossel & Wehner, 1982). As expected, this extreme simplification produces serious errors (up to 30 degrees depending on the time of day), but these are of little consequence because the bees all use the same map. Thus, when a scout returns to the hive on a bearing that, according to her faulty map is 50 degrees from the sun, she communicates this bearing to food gatherers. When these foraging bees set off, they head in the right direction because they use the same faulty map to set a bearing of 50 degrees.

In short, what insect designs demonstrate to an astonishing degree is the advantage of specialization. If a task is specified for a modest range of conditions, then it can be done with a highly specialized design. This is the significance of J.B.S. Haldane's famous remark that God seems to have had an "inordinate fondness for beetles." Their primordial design apparently allowed them to specialize enormously—so each could do with great efficiency what its niche required. A brain comprising small, specialized areas will, like an ecosystem of interacting specialists, be complicated.

Conclusions

Mammalian and insect brains accomplish the same core tasks and are subject to the same physical constraints, so both are designed to *send at the lowest acceptable rate* and *minimize wire*. Both brains regulate the body's internal milieu via slow, wireless (endocrine) signals, plus thin wires with extremely low firing rates (autonomic). Both send long-distance signals via tracts with mostly thin axons. Both arrange their sensors and brain regions in similar positions and use similar structures to perform similar computations. These designs operate at or above the level of the single neuron. But lower levels—molecules and intracellular networks—are subject to similar constraints and therefore follow similar principles, as described next in chapter 5.

5 Information Processing: From Molecules to Molecular Circuits

Chapter 3 explained that information is transmitted when a signal reduces uncertainty about the state of a source. It further explained that in transmitting information by pulses, the information rate (bits/s) depends on the pulse rate and timing precision. That chapter noted a law of diminishing returns: as pulse rate rises, there is less information per pulse (figure 3.6). Moreover, higher information rates (i.e., higher pulse rates and greater timing precision) use disproportionately more space and energy, both of which are limiting resources. These resource constraints directly suggested three principles for efficiency in transmitting information: *send only what is needed*; *send at the lowest acceptable rate*; *minimize wire*. Chapter 4 showed that these principles shape many aspects of brain design on a spatial scale of centimeters down to micrometers.

Yet, as pulses transfer information over distance, they are mainly reporting results. The actual processing of information occurs mostly on a 1,000-fold finer spatial scale, the scale of molecules. There information is processed by chemical reactions: molecules diffuse, bind, exchange energy, change conformation, and so on. The key actors at this level are single protein molecules (~6 nm). They are targets for diverse inputs, such as small "messenger" molecules that, upon binding to a receiver protein, reduce its uncertainty about a source. Protein molecules also provide diverse outputs that, for example, alter the energy or concentration of other molecules, thereby reducing their uncertainty.

These processes not only operate at different scale, they often use a different format. Rather than being pulsatile, molecular signals are often graded, that is, analogue. Despite the change in format, the task remains the same: to reduce uncertainty. Therefore, the same principles for communicating information still apply. Chapter 5 explains how information is processed by single molecules. It identifies constraints on the information capacity of a single protein molecule, and the irreducible cost of registering one bit. A logical place to begin is where information from an electrical pulse is forced to change format to a chemical concentration.

When one neuron sends a pulse to another neuron, there is a problem. The source wire that delivers it is separated physically from the receiver neuron by a gap of 20 nm. When a signal manages to cross that gap, there is another formidable barrier, a double layer of hydrophobic membrane about 5 nm thick. How to cross both barriers and finally deliver information to the receiver? The membrane is equally a problem for wireless signals (chapter 4): how can a hormone outside the cell deliver its information to the inside? The solution in both cases is for the message to change format. This presents boundless opportunities to process information and also opportunities to lose it.

Information from a pulse crosses the gap as a puff of small molecules—appropriately termed *transmitter*. Information finally enters a receiver neuron when one or more transmitter molecules bind to a protein molecule that spans the cell membrane. Binding triggers the protein to change conformation, and that carries information into the cell. A wireless messenger (hormone) works the same way—binds to a transmembrane protein to change its conformation.[1] *Thus, most transfer of information from a source neuron to a receiver neuron occurs via chemistry (concentrations, binding reactions) and physics (changes in molecular structure).*

Information can enter a cell in myriad ways. The change in protein conformation may open a channel through the membrane to admit ions that carry electrical current. Or it may cause a protein's cytoplasmic tail to release a small molecule that binds and alters other proteins. An altered protein may search out targets by random walk (diffusion). To save time its search may be reduced from three dimensions to two by allowing the altered protein to skate with little feet along the membrane's inner surface.

Such mechanisms accomplish much of the brain's information processing. They amplify, perform logical operations, store and recall, and so on. Although these mechanisms may be triggered by an all-or-none pulse, they themselves are generally graded: small molecules vary in concentration, activated proteins vary in number, ionic currents vary in amplitude, and so on. The information content of these analogue signals, as for the pulse code, can be usefully analyzed by Shannon's formulas. A very few equations, all intuitive, can explain fundamentally: (1) what constrains information processing by signals; (2) what reduces their information; and (3) why higher information rates are more expensive.

Shannon communication

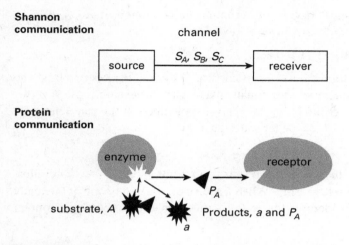

Protein communication

Figure 5.1
Shannon's general communication system maps onto communication between two protein molecules. When Shannon's source is in states *A*, *B*, or *C*, it transmits signals S_A, S_B, or S_C, so eliminating the receiver's uncertainty about the state of the source. The protein source is an enzyme that, upon encountering substrate *A*, produces two products P_A and *a*. The protein receiver is a receptor that specifically binds P_A. P_A's presence or absence at the receptor's binding site establishes the state at the source, namely, that *A* is present or absent.

The reward is the same as for pulses (equation 3.3): with these formulas one can "follow the money" and thereby discover how constrained neural resources are spent. Moreover, following the money at this nanoscale leads to all the remaining principles of neural design. So now we explain how Shannon calculated the amount of information needed to specify a source and how much information a signal can carry (figure 5.1; Shannon & Weaver, 1949).

How much information is needed to specify a source?

The information needed to specify a source increases with the number of states that the source might occupy. Where there is only one state, there is no uncertainty, so no information is required and signals indicating this known state are *redundant*. Efficient designs will reduce redundancy to satisfy the principle *send only what is needed*.

If there are two equally likely states, *A* and *B*, then by definition 1 bit of information eliminates uncertainty by identifying *A* or *B* (e.g., *A* = 0; *B* = 1).

Increase the source's degrees of freedom from two to four states, A, B, C, D, and the probabilities are lower, for example,

$p(A) = p(B) = p(C) = p(D) = 0.25.$

Now the situation is more uncertain, and to decide requires two bits. The first bit decides between two equally likely pairs, for example, (A, B) versus (C, D), and the second bit decides between members of the pair. These two bits constitute a 2-bit code for states, such as

$A = 00; B = 01; C = 10; D = 11.$

The fact that 1 bit specifies two states and 2 bits specifies four states illustrates a general relationship. When a source can be in any one of U equally likely states, to identify the state of the source a receiver must obtain at least

$I = \log_2(U)$ bits.   (5.1)

Note that, as expected, the quantity of information needed to define the state of a source increases with the complexity of the situation—here the number of possibilities, U.

Most sources in nature have states whose likelihoods differ, and this affects the quantity of information needed to specify a state. For example, when we change the probability distribution of the four states, A, B, C, D to

$p(A) = 0.125; p(B) = 0.5; p(C) = 0.25; p(D) = 0.125,$

all four states can be identified by a 2-bit code: ($A = 00; B = 01; C = 10; D = 11$), but a 3-bit code is more efficient (figure 5.2). The first bit decides if the state is B, the second if it is C, and the third if it is D or A. Note that each choice is binary and equiprobable—1 bit. When used repeatedly, this 3-bit code is, on average, more efficient than the 2-bit code. On 50% of occasions the 3-bit code needs just 1 bit to identify the correct state, $p(B) = 0.5$. On 25% of the occasions, it needs 2 bits to identify the correct state, $p(C) = 0.25$), and on 25% it needs three bits to identify the correct state, $p(A) + p(D) = 0.25$. With usage so distributed, the average number of bits per determination of state is

0.5×1 bit $+ 0.25 \times 2$ bits $+ 0.25 \times 3$ bits $= 1.75$ bits.

Thus, the 3-bit code is 12.5% more efficient than a 2-bit code. This illustrates one of Shannon's discoveries: it is efficient to match a coding scheme to the statistical distribution of the states being coded. The brain got there first (chapter 9).

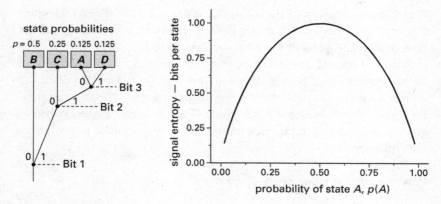

Figure 5.2
Two ways to improve efficiency with which signal states represent information. Left:
This decision tree implements a 3-bit code to represent four states that have different probabilities. An alternative would be to assign 2 bits to every state, but 3-bit code is more efficient because half the signals transmitted (those for state *C*) use only 1 bit, and this more than compensates for giving the least frequent states (*A*, *D*) 3 bits. **Right:** A limited number of signal states is used most efficiently when all states are used equally often. In this two-state system the condition $p(A) = p(B) = 0.5$ maximizes information capacity at 1 bit per state. Left reprinted from Laughlin (2011).

The 3-bit code also reminds us that the number of bits needed to specify a state increases with the state's uncertainty. One bit specifies the most likely state, $p(B) = 0.5$; two bits the next most likely, $p(C) = 0.25$; and three bits the least likely, $p(A) = p(D) = 0.125$. In general, when the probability of encountering state x is $p(x)$, the information required to specify x is

$$I_x = \log_2(1/p(x)) = -\log_2(p(x)) \text{ bits.} \tag{5.2}$$

This relationship is consistent with equation 5.1: when there are U equally likely states, $p(x) = 1/U$.

The four-state source explains the basics, but how does information theory apply to the riotous possibilities of the real world? For practical applications, such as the design of his employer's telephone network, Shannon derived a general equation. The number of bits needed to specify the state of *any* source is

$$H(x) = -\sum_{1}^{x} p(x)\log_2(p(x)). \tag{5.3}$$

This quantity, $H(x)$, takes the information per state, as defined by its probability $p(x)$ in equation 5.2, multiplies it by the proportion of time the state is used, $p(x)$, and sums this quantity across all states.

Shannon named this quantity, $H(x)$, *entropy* because its equation (5.3) has the same form as Boltzmann's equation for the entropy of a thermodynamic system. Indeed, the two entropies derive similar quantities. Boltzmann's entropy quantifies a system's total disorder. Shannon's entropy quantifies a system's total uncertainty, and it enabled him to answer our next question.

How much information can a signal carry?

The number of bits carried by a signal is given by the entropy equation, here the entropy of signal states. We start with a signal's ability to specify a source. When every source state is allotted its own signal state (a 1:1 mapping of source onto signal), the signal can carry all of the information needed to specify the source because it can always represent each and every state of the source, and from equation 5.3 this information is the *source entropy*. This equality suggests a general method to calculate the information carried by a signal. Identify the signal's states and use them to calculate the signal's entropy in bits. The calculation obviously holds when source states map 1:1 onto signal states, but is it valid when the source and signal states greatly differ? For example, is it valid when analogue signals from a microphone are transferred to the digital format of a CD or when analogue synaptic potentials trigger trains of action potentials?

Shannon proved mathematically that entropies equate across formats. Thus, it is always possible to devise a mapping whereby a signal with entropy H bits specifies the states of a source with an entropy H bits. Thus the information from a meandering source with many rare states, such as sounds in a telephone conversation, can be compressed into snappier codes that use fewer states more often, such as high frequency radio signals or bits in a digital network.[2] In short, to quantify how much information a signal can carry, just calculate its Shannon entropy using equation 5.3. Having done so, one can consider design issues for the signals that couple a neural source to a neural receiver.

Entropy sets the upper bound to a system's information capacity, but communication systems generally and neural systems in particular are unable to fill that capacity. The first constraint is noise because, when noise enters a system, information is lost. Thus, we must consider how noise affects the design of neural circuits. The second constraint is redundancy because, repeating a signal reduces a system's capacity to send *new* information. However, when noise is present, repetition can enhance the system's ability to specify the source. Consequently, noise and redundancy in every real communication system are complementary.

How noise destroys information

Noise (random fluctuation that does not correlate with changes in signal state) destroys information by introducing uncertainty. In a noise-free system, the receiver can associate a given signal state with a source state with total confidence; however, when noise is present, is a change sensed by the receiver signal, or is it noise? The quantity of information destroyed by noise depends on the uncertainty introduced by noise and, because bits resolve uncertainty, this is also the number of bits required to describe the noise—its Shannon entropy (equation 5.3). It follows that the information carried by a signal in the presence of noise is the signal entropy minus the noise entropy. Because entropy tends to increase logarithmically with the number of states (equation 5.1), and subtracting logarithms is equivalent to division, information increases as the logarithm of the ratio between signal and noise; $\log_2(S/N)$.

Redundancy

Redundancy (signal state that represents something already known) carries no information. Redundancy comes in two forms. The first is a less extreme form of repetition—states are no longer completely correlated; they are partially correlated. When state A correlates with state B, receiving A increases the probability of receiving B, thus reducing the uncertainty associated with B, and hence B's information content. Circuits commonly use lateral and self-inhibition to remove this form of redundancy in order to *send only what is needed*, information (chapters 9 and 11).

In the second form of redundancy, the signal states are carrying less information than they might because they are used too frequently or too rarely. Consider a binary signal with two states, A and B. The information carried by these two states depends upon the signal entropy,

$$H = -p(A)\log_2(p(A)) - p(B)\log_2(p(B)),\tag{5.4}$$

and H peaks at 1 bit per state when $p(A)$ is equal to $p(B)$ (figure 5.2). This optimum coding strategy, use states equally often, generalizes to systems with many states, and is widely employed in systems where the number of available signal states is severely limited by power restrictions and noise (e.g., satellites, mobile phones). Retinal neurons face similar limitations and use this same strategy as do many other cell signaling systems (chapters 9, 11; Bialek, 2012).

Now we know: (1) how much information is needed to describe a set of events (equation 5.3); (2) how much information a signal conveys about these events (also equation 5.3); and (3) how these quantities depend on redundancy and noise. This leads to another question: when events change rapidly, how can the brain keep up? To answer this, we derive an expression for information rates in bits per second.

Calculating the information rates of continuously changing signals

The rate of information transfer depends upon the amount of information conveyed by each signal state and the rate at which these states evolve over time. Chapter 3 gave the information rate for action potentials by calculating their entropy. This quantity depends on a physical property of the signal: discrete pulses timed with a given precision, a property that makes low rates cheaper. Other formats have different properties, and these impose different constraints on relationships between signal quality, bit rate, and efficiency.

Much of the brain's information is represented by analogue signals that, by definition, change continuously. These include changes in concentration of messenger molecules, changes in the number of receptor proteins activated by a ligand, and changes in the electrical potentials generated across neural membrane by ion channels. As an analogue signal varies, it runs through a series of signal states (figure 5.3). These states deliver information at a rate that is the number of bits conveyed per state multiplied by the rate at which states change. The number of discriminable states is the range of response covered by signal and noise, $(S + N)$, divided by the noise (figure 5.3). Thus, from equation 5.2, each state delivers $\log_2(1 + S/N)$ bits. The analogue signal can change level in time Δt (figure 5.3). Thus, states are delivered at a rate $R = 1/\Delta t$, and when successive signal states are uncorrelated (i.e., no redundancy in the input), the information rate is

$$I = R.\log_2(1 + S/N) \text{ bits s}^{-1}. \tag{5.5}$$

In many practical systems, calculating rate is more complicated. Equation 5.5 assumes that redundancy is zero, that is, there is no correlation between signal states. To achieve this, signal states must change randomly. To be truly random, the signal must be able to jump from any one state to

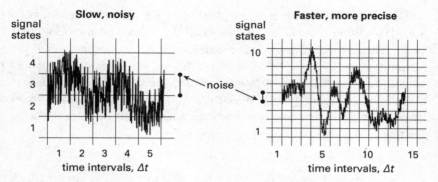

Figure 5.3
Signal range, noise, and response dynamics determine the information rates of analogue signals. Noise divides a waveform's signal range into discriminable states, and states can change at time intervals Δt. The faster, more reliable waveform obviously conveys more details of the signal. From equation 5.5, it also has a higher information rate because it has a higher S/N and, with shorter Δt, changes level at a higher rate.

any other, but this ability is constrained by the time needed to make the jump. For example, an enzyme generates a product at a finite rate, so it requires time to change the product's concentration in a compartment of given volume; similarly an electrical current supplied through a resistor requires time to charge a capacitor. Thus, the number of different states to which a signal can jump in one time interval, Δt, is limited by the rate at which the signal can change, but given sufficient time, it can move to any state. This time dependency complicates the calculation of information rates.

Shannon solved this problem by using the Fourier transform to convert the continuous analogue signal and noise into their frequency components. Each frequency component is independent, in the sense that changing the amplitude or phase of one frequency component has no effect on any other frequency; consequently, every frequency carries its own information. It follows that the total information carried by the signal is the sum of the information carried by each of its component frequencies.

$$I = \int_0^{co} \log_2[1 + S(f) / N(f)] \cdot df , \qquad (5.6)$$

where I is bits per second, $S(f)$ and $N(f)$ are the power spectra[3] of signal and noise, and co, the signal's cutoff frequency, defines its bandwidth.

There are two provisos to this derivation of information rate (equation 5.6). The system must be linear, and both the signal and the noise must vary randomly with Gaussian distributions so that the frequencies being transmitted are uncorrelated. These conditions are reasonably well met when systems are driven with low-amplitude Gaussian inputs (e.g., Rieke et al., 1997). Note that when the spectrum of $S(f)/N(f)$ is flat, the sum across frequencies reduces to equation 5.5, with a bandwidth of $1/2\Delta t$ replacing the rate R,

$$I = (bandwidth)\log_2(1 + S/N). \tag{5.7}$$

This relationship between *bandwidth*, S/N, and information rate, I, affects neural design because to transmit information at higher rates, a neuron needs a wider bandwidth (faster responses) plus higher S/N, and these require extra materials and energy. Thus, we have a trade-off between resources and performance that, as we will see, profoundly influences neural design.

Information in any real system must be embodied physically or chemically. The brain uses *signaling proteins* to process information, so we now examine their physics and chemistry.

How protein molecules transmit and process information

A protein acquires its specific function by folding to reduce its free energy

A protein molecule is formed from a linear chain of amino acids linked in a genetically specified sequence (Alberts et al., 2008). The linear sequence becomes a useful molecule as follows. The chain is flexible, so it bends and folds to reduce its free energy by minimizing potential energy and maximizing entropy (Williamson, 2011; Dror et al., 2012). The charged amino side groups attempt to form pairs of attractive opposites (+ with –) and to avoid repellant likes (+ with +) or (– with –). To increase entropy, the hydrophobic side chains avoid polar groups and coalesce into oily cliques. All of this jostling for position must be achieved within packing constraints.

Buffeted by thermal energy, yanked up and down potential gradients, exchanging order for disorder, the protein molecule constantly changes its three-dimensional structure (*conformation*) until it falls into a local minimum free energy that is deep enough to resist thermal motion. The protein molecule has reached a stable conformation (figure 5.4).

This stable conformation determines the protein molecule's physical and chemical properties (Williamson, 2011). A typical protein, with several hundred amino acids, folds into a 5- to 10-nm structure to adopt a form that supports its function: long fiber to make a hair, globular block with attachment knobs to build the cytoskeleton, part of a stepping leg to move materials, and so on. A protein may locate a subset of amino acids where they can bind and interact with a specific molecule. Such a binding site enables the protein to collect and send information.

Binding specificity allows information transfer

Recall that information transfers when a change at the receiver can be associated with the state of the source. Chemical binding satisfies this requirement. For example, when an enzyme molecule reacts with its substrate to produce its product, the enzyme only binds the substrate, and the receiver, a receptor protein, only binds the product (figure 5.1). Thus, the source tells the receiver "substrate present" by using a diffusible messenger, the product. If the enzyme and/or the receptor were to relax their binding specificities, other molecules in the cytoplasm would also bind. Such cross talk would reduce the probability that the receiver is responding to the presence of one particular substrate at the source. Thus, binding specificity enables information transfer.

Once a protein's binding site receives information, how can it be further processed? By *allostery*. This is a protein's ability to respond to a specific input, such as binding a messenger, by switching to a new stable conformation.[4]

How allostery works

Consider the protein molecule continuously changing conformation as it descends to its lowest available free energy level. This progression is, in effect, a voyage across an energy landscape (figure 5.4) in which the map coordinates represent the protein's conformation and the altitude represents its free energy.

The descent follows gradients in the energy landscape, and thermal jiggles push it over bumps. Thus, the protein explores a locale and finds a path to lower regions. When the protein enters a valley too deep for thermal forces to boost it out, it is trapped, and the conformation becomes confined to a small region (figure 5.4). Here the protein may shuttle between a small set of functionally distinct conformations, or it may remain centered on one stable conformation (figure 5.4). Thus confined, the molecule assumes a role dictated by its conformation.

Consider now what happens when an external factor alters the energy landscape. An external input could be a change in pH or electrical potential, it could be binding or releasing a specific molecule, or it could be an

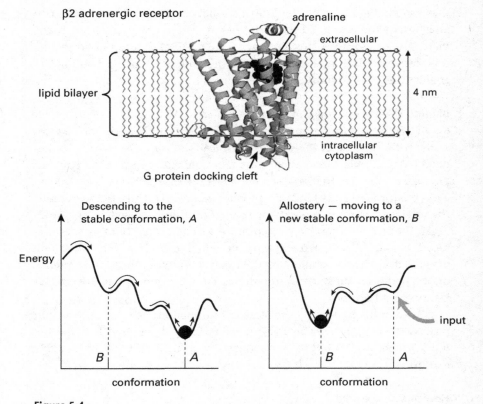

Figure 5.4
Protein structure, conformational state, energy landscape, and allostery. Upper:
The β2 adrenergic receptor protein spans the cell membrane's lipid bilayer. Here
it is shown in the conformation where binding an adrenaline molecule at a site
on the outside has opened a cleft for binding a G protein molecule on the inside.
Note the prominent helices crossing the membrane. **Lower left:** Section through a
protein molecule's energy landscape. During folding, the protein descends the energy landscape and adopts the stable conformation A. **Lower right:** An external input changes the energy landscape and the protein moves to conformation B. This
is allostery. Upper adapted from http://en.wikipedia.org/wiki/Beta-2_adrenergic_
receptor#mediaviewer/File:2RH1.png.

injection of energy via the attachment of a high-energy phosphate group to an amino side group. Such inputs alter the protein's energy landscape, depressing some regions and elevating others (figure 5.4). The protein responds by moving, within microseconds to milliseconds (Williamson, 2011; Dror et al., 2012), to a new stable conformation. The new conformation differs physically and chemically from the previous one, so the molecule reacts differently to chemical and physical inputs. This change enables it process information.

How a protein uses allostery to process information

A finite-state machine[5] processes information by running through a well-defined sequence of state changes (transitions), each triggered by a particular condition, such as the presence or absence of an input, or a conjunction of inputs. This is allostery. As a protein molecule runs through a sequence of state changes, each conditional upon a particular input, it produces an output conditional upon those inputs (Huber & Sakmar, 2011). Thus, allostery enables a single protein molecule to compute (Bray, 1995). For example, a single molecule is easily programmed to perform the Boolean operation, AND (figure 5.5).

The rest of this chapter treats one particular finite-state machine that comprises a pair of interacting proteins. The receptor protein accepts the wireless signal, adrenalin, a hormone that prepares an organism to fight or flee, then relays the information ("Adrenalin present!") across the cell membrane. There it transmits to receiver proteins on the membrane's inner face that amplify and broadcast the information within the cell. Both proteins then reset for the next signal. The receptor protein is the β2 adrenergic receptor, and the receiver protein is a G protein.

We choose this example for several reasons. First, the β2 adrenergic receptor and its G protein represent a broad, ubiquitous class of finite-state machines (chapters 2, 7, and 8). The human genome specifies more than 800 different receptor proteins that couple to a G protein and more than 100 different G proteins. Second, this example indicates the spatial scale used by most neural computations. Third, it exemplifies computation by amplifying, and in doing so illustrates molecular solutions to a broad design problem, overcoming noise. Fourth, it clarifies the reason to compute at this spatial scale: high efficiency in space and energy. The energy cost of 1 bit in this system, as will be explained, approaches the theoretical lower limit to within a factor of about 30.

The final reason to choose this example over other possibilities is that the sequence of conformational changes, triggered by adrenalin's

A single molecule AND gate: C if A and B

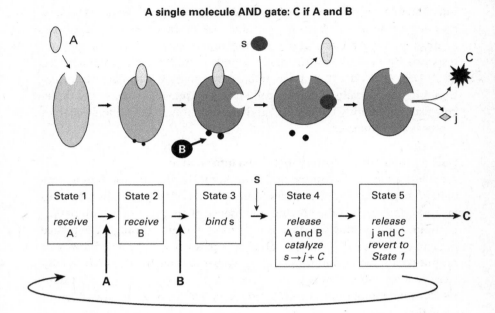

Figure 5.5
The allosteric protein as a finite-state machine. How a sequence of stimulus-evoked changes in allosteric state could enable a single protein molecule to perform a simple computation, here a logical AND on the two inputs A and B. Ligand A binds to the protein, exposing two sites to be phosphorylated by kinase B. The pair of attached phosphates alters the protein's conformation, exposing a catalytic site that digests the substrate s to produce products j and C. Bottom row gives the corresponding program of state transitions.

binding to the receptor and completed by the release of activated G proteins, has been documented at the atomic scale, by x-ray diffraction (Rasmussen et al., 2011; Chung et al., 2011; summarized in Schwartz & Sakmar, 2011).

Allostery in action

The system is ready to receive when the receptor's conformation exposes its adrenaline binding site on the cell membrane's outer face and masks the G protein's binding site on the inner face (figure 5.6). G proteins diffuse on the inner face, colliding with receptors, but encounter no signal. When adrenalin binds to the receptor, the protein changes conformational state

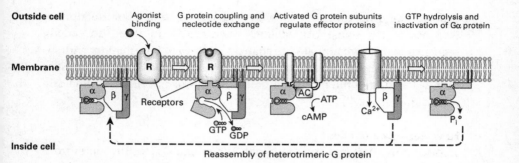

Figure 5.6

β2 adrenergic receptor and its G protein use allostery to operate as a finite-state machine. Receptor receives a wireless signal outside the cell and, by changing conformation, relays it across the membrane to G protein. G protein dissociates and α subunit broadcasts signal to effector proteins by diffusing on inner surface of the membrane. α subunit hydrolyses bound GTP and reverts to conformation that the binds the other subunits. G protein is reconstituted, ready to signal again. Further details in text. Figure adapted from summary diagram from the definitive study of structural changes that pass information through these two molecules (Rasmussen et al., 2011), with permission.

(figures 5.4 and 5.6). One of the seven helical coils that span the membrane (coil number 6) moves 1.4 nm and others move shorter distances. Together they open a cleft in the receptor molecule at the inner face to expose the G protein's binding site. At the next collision, a G protein engages this site with a special knob and docks securely (figure 5.6).

This coupling changes the energy landscape of both molecules. The G protein embarks on a sequence of conformational changes (figure 5.6). Two of its three subunits, β and γ, detach and diffuse into the cytoplasm. The α subunit responds to the loss of its partners by swinging apart two large sections at their hinge. This motion, spanning more than 110° and requiring several hundreds of microseconds, reveals, like an oyster showing its pearl, a small molecule, guanosine diphosphate (*GDP*), bound deep within the protein. The exposed GDP promptly exchanges with a molecule from the cytoplasm, guanosine triphosphate (*GTP*), whose additional phosphate gives it higher energy.

GTP's binding transfers energy to the α-subunit, again changing the landscape. The hinged gates swing closed, retaining the high-energy GTP that is fueling the sequence of state changes. The knob retracts, thereby uncoupling the α subunit from the receptor and freeing it to diffuse on the membrane's inner face. Now another binding site on the α subunit is exposed for other proteins to bind and change *their* conformation in response to the signal "Adrenaline!" (figure 5.6). In short, an orderly sequence of conformational state changes has carried information, "Adrenaline!," across the cell membrane, and by releasing an activated GTP-α subunit, it has started the process of broadcasting this information wirelessly within the cell.

How allostery amplifies

This form of allostery easily amplifies. When one GTP-α uncouples from the activated receptor protein, another docks in its place, is activated, then is released, and so on. The rates vary from 10–500 per second, depending mainly on the density of G proteins on the membrane—for this sets their frequency of encountering a receptor protein. The number of G proteins activated and released by a receptor increases with time as the cleft stays open. The amplification (*gain*) varies across systems, ranging from 4 in a system with short time constant, such as a fast fly photoreceptor (chapter 8) to 100 in systems with long time constant, such as a slow-acting hormone.

Amplification is a form of redundancy since each copy simply repeats a message without adding new information. Thus, multiple G proteins activated by the β2 receptor simply repeat, "Adrenalin!," "Adrenalin!" . . . Yet this redundancy is essential for two reasons. To produce a concerted response to adrenalin, the signal must reach many parts of the cell in good time, hence the activation of several G proteins. Second, the system must guard against noise. Because a thermal bump occasionally activates a single G protein molecule, the receptor must activate several molecules to generate a reliable message. Thus, when amplification protects information from noise, it also introduces inefficiency in the form of redundancy. An efficient design will strike an appropriate balance by matching the gain of amplification to the level of noise (chapter 6).

Although the β2 receptor and its G protein have worked together to amplify and broadcast the signal "Adrenaline!," the process is incomplete. This finite-state machine, which turned on in order to signal danger, must turn off when the warning has been sent. Then the machine must reset to be ready once again to deliver the message.

How allostery terminates the message and resets the system

Turnoff and reset are accomplished by continuing to move the receptor and the G proteins it activated through their sequences of conformational states. As for the all preceding steps of activation, each transition for deactivation serves a specific purpose. To inactivate the β2 receptor, an enzyme (*kinase*) accepts a high-energy phosphate group from an ATP molecule and attaches it covalently to a particular site on the β2 receptor. The phosphorylation of several such sites raises the receptor's energy level sufficiently to change its conformation, now exposing a binding site for a different protein molecule, *arrestin*. When arrestin binds, it blocks access to the G protein's docking cleft, thus preventing transmission.

Something is needed to protect unoccupied β2 receptors from being inactivated while they are in the receptive conformation, waiting for adrenalin. The receptor is engineered so that the receptive conformation hides the phosphorylation sites, and they become exposed only in the conformation triggered by binding adrenalin. Something is also needed to give time for an activated receptor to amplify, that is, to activate and release several G proteins. To achieve this, the kinases that attach high-energy phosphates are designed to work slowly. Moreover, by modulating this rate of phosphorylation, both the gain and time constant of amplification are adjusted for no extra space and little extra energy.

Once arrestin blocks transmission to the G protein, the β2 molecule resets—by continuing its journey through conformational states. The adrenalin molecule, whose initial binding to the receptor opened a cleft for docking the G protein, eventually *un*binds adrenalin, and this closes the docking cleft. This allows a *phosphatase* enzyme to remove the added phosphates, releasing arrestin, and restoring the receptor to its initial state. Its *signaling cycle* is complete: it has received, transmitted, and reset.

But what prevents the activated α subunit from continuing its diffusive search for partners? This subunit is also an enzyme that removes the high-energy phosphate from its own bound GTP (figure 5.6), and this provides an automatic cutout. Withdrawing the high-energy phosphate from the α subunit triggers its final sequence of conformational state changes. It rebinds the βγ units and once more protrudes its docking knob. Now the G protein has reset to the inactive αβγ-GDP form and is again ready to dock with an adrenalin-bound receptor.

In summary, this molecular finite-state machine uses two parts, receptor and G protein. It exploits three properties of a protein molecule—binding specificity, allostery, and freedom to diffuse—to execute a program of state changes. The program receives a signal at the cell surface and transmits it *mechanically* across the cell membrane. The program then amplifies the signal, broadcasts it within the cell, and resets. This computational device, the G-protein-coupled receptor (*GPCR*), being ubiquitous, will be discussed further (chapters 6–8). But here we explain another invaluable property of signaling proteins—how their energy efficiency approaches the thermodynamic limit.

Energy efficiency of protein devices

Why must molecular devices consume energy to process information?

A protein's signaling cycle starts and finishes at the same point in the energy landscape. If every conformational state within the cycle had the same free energy, the cycle could be completed without consuming energy. However, the protein would then depend on random thermal fluctuations to change states. Moreover, if free energy were constant, each transition would be reversible—with equal probabilities of moving forward or backward. To complete the cycle would be theoretically possible: a signal could be delivered without expending energy. However, such a lossless system would be impractical because, relying on a chain of improbable and reversible events, the receiver would wait for long and indeterminate times (Bennett, 1982, 2000).

Energy eliminates this intolerable wait by driving the protein through the conformational state transitions in the intended direction. Moreover, the effect is progressive: adding more energy speeds the cycle. But what about the lower bound: what is the least energy that can deliver information usefully?

Lower bound to energy cost in signaling

Thermodynamics suggests a minimum, the energy required to register one bit of information (Landauer, 1996; Schneider, 2010),

$$\Delta E = k_B T \ln(2) \approx 0.7 \; k_B T \text{ joules} \approx 3 \times 10^{-21} \text{ joules per bit,} \tag{5.8}$$

where k_B is Boltzmann's constant and T is temperature in degrees Kelvin. ΔE is tiny,[6] but single protein molecules are also tiny and so approach this thermodynamic limit to energy efficiency.

The signaling cycles of the β2 adrenergic receptor and its G protein can each register a bit by switching from OFF to ON and then resetting to OFF. Each protein draws energy from the cell's standard currency, the high-energy molecule, ATP. Hydrolysis of one ATP delivers 25 $k_B T$ joules, and the receptor uses at least three ATP molecules when it is phosphorylated (figure 5.5). This gives an efficiency of 75 $k_B T$ joules per bit, which is two orders of magnitude above the thermodynamic limit (equation 5.8). The G protein consumes the equivalent of 1 ATP when it hydrolyzes its GTP to GDP (figure 5.6), giving an efficiency of $25 \ k_BT$ joules per bit, between one and two orders of magnitude above the thermodynamic limit.

Thus, both proteins process a bit of information for less than the cost of a covalent bond (~100 k_BT). This seems plausible because a protein is a soft device, more like a machine made from jelly than a rigid clockwork (Williamson, 2011). Indeed, the free energy to stabilize a protein (folded vs. unfolded) is less than a quarter of the free energy to form a covalent bond and is about equal to the energy delivered by ATP.

What prevents these two protein molecules from operating closer to the thermodynamic limit? Realize that the 0.7 k_BT limit is the cost of simply registering a bit as a change of state. It does not include transmitting the bit. To send a bit across the membrane, the β2 receptor moves its helix number 6 by 1.4 nm, and to relay the bit into the cytoplasm, the G protein opens its large hinged section by 110°. Both movements require work (Howard, 2001), and work consumes energy. Energy is also used to drive the cycle at a rate appropriate for the function—recall that the β2 receptor signals "Emergency!" Considering that the energy cost of transmission by the GPCR includes these extra tasks, protein signaling appears astonishingly close to the thermodynamic limit. An order of magnitude is a reasonable guess.

Energy and the design of efficient signaling molecules

The receptor and G protein turn on and off abruptly and reliably—like a mechanical switch. The latter avoids accidental tripping by using an energy barrier. Some of the energy needed to trip it is recycled so that once triggered, the change goes quickly. Where safety is critical, the energy barrier is high, but where it is less critical, the barrier can be lowered to save energy. Likewise, a protein's energy landscape seems engineered to require just the right energy input for each state transition. The design also involves trade-offs between speed, reliability, and energy. For example, were viscous forces within a protein to increase with switching rate, the energy cost per transition would increase disproportionately, making lower rates more efficient. Thus, a design principle observed at the microscopic level for axons, *send at the lowest acceptable rate* (chapter 3), may also hold at the nanoscopic level for protein molecules, albeit for different reasons.

Summary

The signaling systems established by protein molecules receive and transmit information, as defined by Shannon, using different physical and chemical processes from the ones that Shannon originally treated. Three physical and chemical properties of proteins support the transmission and processing of information. Binding makes specific connections between molecules, enzymatic activity provides a potent means of generating and amplifying signals, and allostery enables information to pass through single molecules. Allostery also equips a single protein molecule to compute by operating as a finite-state machine. By running through a well-defined program of state changes, triggered by specific inputs, the molecule completes a program only when it encounters a specific combination of inputs. These properties equip proteins to form circuits of molecules that compute.

Circuits built from proteins satisfy two design principles. First, if we rule out quantum computation, these circuits are irreducibly small, and this saves space and materials. Protein circuits also save energy because protein molecules operate near the thermodynamic limit of energy efficiency. Moreover, protein chemistry allows energy to be delivered efficiently in just the amounts needed to meet the circuit's need for speed and accuracy. Thus, the performance of components in protein circuits can be matched to their tasks to gain economies that come with sending at the lowest rate.

These advantages—compactness, energy efficiency, and ability to adapt and match—all suggest the principle *compute with chemistry*. It is cheaper. But to realize the savings, protein circuits must support the brain's core tasks. Chapter 6 now explains how proteins equip molecular circuits to meet a brain's requirements for information processing.

6   Information Processing in Protein Circuits

Chapter 5 explained that information is encoded whenever a source's change in state registers as a change in state at a receiver. The primary mechanism at the nanometer scale is a protein's ability to connect specific inputs to specific outputs by, for example, binding molecules, catalyzing reactions, and changing conformation. These reactions are employed universally in biology and have two advantages for brains—energy efficiency and compactness. As noted in the previous chapter, the energy used by a protein molecule to register 1 bit approaches the thermodynamic minimum. Also, for changing conformation, its unique task, a protein is irreducibly small. Smaller would be better since a moderate-sized protein molecule (100 kDa) spans about 6 nm and occupies about 100 nm^3. But although a smaller peptide can serve as a ligand, it lacks a protein's rich possibilities for stable folds, pockets, and allostery that are essential to its receiving and processing information.

Chapter 5 noted that a protein molecule can compute. For example, it can amplify (one adrenalin bound to one β2 receptor protein activates several G proteins), and it can do logic (e.g., compute the Boolean AND; figure 5.5). However, one logical operation doesn't make a brain. A brain needs to do a lot more math than that. For starters, it needs mechanisms on the nanometer scale to calculate the four linear arithmetical operations (+, -, ×, ÷) and various nonlinear operations such as $\log(x)$ and x^n. It also needs switches (where an input causes a step change in output), filters (to remove certain frequencies and attend to particular timescales), correlators (to associate events), and so on.

For such nanometer-scale computations, the genome serves as a parts catalog—listing the codes for thousands of protein structures, each specified for some particular input/output (I/O) function. But executing an orderly sequence of operations that computes something requires something more: a specific subset of I/O components that link correctly. A cell's

Cascade amplifier

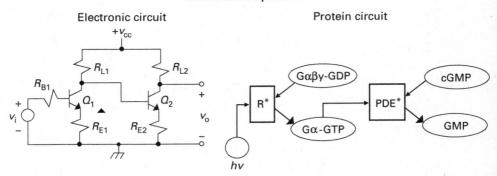

Figure 6.1
Circuit for cascade amplifier: silicon versus protein. In silicon, an input voltage, v_1, drives the first transistor Q_1, which amplifies the signal. Q_1's output drives transistor Q_2, which amplifies the signal again and generates the output v_o. In protein, a photon (hv) activates one molecule of a receptor protein (R), changing its conformation to (R*). Like the β-adrenergic receptor (figure 5.6), R* amplifies by catalyzing 20 G proteins to change from Gαβγ-GDP to Gα-GTP. Each Gα-GTP activates a molecule of the enzyme phosphodiesterase (PDE), which again amplifies by catalyzing the hydrolysis of 100s of messenger molecules of cGMP to GMP. Both silicon and protein amplifiers multiply the input by the product of the gains of the two amplification stages. Electronic circuit from http://en.wikipedia.org/wiki/Cascade_amplifier. Protein circuit for phototransduction in rods is described in chapter 8. GDP, guanosine diphosphate; GTP, guanosine triphosphate; R*, the photosensitive molecule rhodopsin, activated by a photon.

internal mechanism ensures that this occurs—that the right proteins are delivered to the right places at the right times (Alberts et al., 2008). In both respects—using components with specific I/O functions and linking them correctly—protein circuits resemble electronic circuits (figure 6.1).

To understand neural computing at the nanometer scale, one must consider what shapes a protein's I/O function. What determines, for example, whether it will take a sum or a logarithm, whether it will switch or filter? These functions emerge from a protein's three-dimensional structure, through its ability to react chemically, mechanically, and electrically, and to change state in response to these inputs—allosterically.

One must also consider how a sequence of I/O functions should couple to make a useful circuit. For example, should a protein couple directly to its target, should it diffuse, should it anchor and send a small messenger, or should it communicate electrically via the cell membrane? Here the broad answers are simple: diffusion slows as the square of molecular weight, and proteins are heavy, so the best choice for coupling depends on the required distance and allowable time. Diffusion time increases as distance squared and concentration decays exponentially. Thus, molecular size and concentration, plus the laws of diffusion, shape protein circuit design. Consequently, when distances are large and time is short, circuits use electrical signals. This chapter will explain further with some simple examples, starting with ligand binding. The concepts and principles introduced here will be exemplified more thoroughly in all subsequent chapters.

I/O functions emerge from the kinetics of chemical binding

I/O functions from a single binding site

A ligand diffuses under thermal bombardment to a specific site on a protein and binds. That is, it sticks for a time, and then comes off. While the ligand is bound, the protein adopts an active conformation in which it produces its *output*, for example, it is able to bind a downstream protein or catalyze a chemical reaction. Thus, the protein's *output* is proportional to the fraction of time it binds the ligand, and this is determined (Phillips et al. 2009, chapter 6, "Entropy Rules!"; Bialek, 2012) by the ligand concentration [ligand] and rate constants for unbinding (k_{OFF}) and binding (k_{ON}):

$$output/output_{max} = [\text{ligand}]/(k_{OFF}/k_{ON} + [\text{ligand}]). \tag{6.1}$$

This I/O function is *hyperbolic*; it rises steeply at first, and then tapers off as the binding site approaches saturation, $output_{max}$ (figure 6.2). The ratio k_{OFF}/k_{ON} is the dissociation constant k_D, and equals the ligand concentration required to produce a half maximal *output*. The same binding kinetics apply to protein–protein binding, so what is here explained for ligand–protein binding applies also to protein–protein binding.

The hyperbolic I/O function computes. It can perform, depending on input, three analogue operations:

1. At lower inputs levels (those causing < 0.25 maximum output), the function is linear (figure 6.2), so small inputs add.
2. At medium input levels (those causing 0.25 to 0.75 maximum output), the function is approximately logarithmic (figure 6.2). This reduces the sensitivity of the output to the absolute level of the input and scales the inputs proportionally such that a constant fractional change in input, $\Delta[\text{ligand}]/[\text{ligand},]$ causes a constant change in output, $\Delta output$. This type of scaling exists at the behavioral level for many categories of sensory

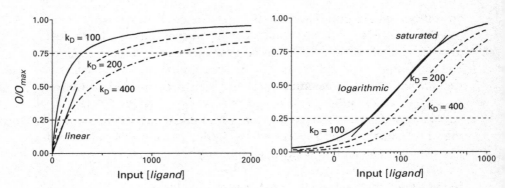

Figure 6.2
**Input/output (I/O) function generated by binding kinetics performs the same com-
putations across widely different input ranges by altering dissociation constant, and
hence binding affinity.** Left: Output (normalized to its maximum) is plotted against
input, [ligand], for three dissociation constants, k_D. When the output is small (<0.25
max), the I/O function is linear and adds. **Right:** Output plotted against log([ligand]).
When the output is medium (0.25–0.75 max) the function is logarithmic. In saturated regime (0.75 max – max) function's slope approaches zero.

discrimination (*Weber–Fechner law*). Thus, a computation that serves behavior starts with chemical kinetics at the nanometer scale.
3. When large, sudden increases in input drive the response from zero to
maximum, the function is a step and thus can serve as an ON/OFF switch
for Boolean operations.

Sensitivity depends on the protein's affinity for the ligand. Higher affinity
(tighter binding) decreases the OFF rate, thus reducing the k_D. The effect is
to reduce the concentration of ligand needed to cause a half-maximum
output. By adjusting k_D, a given I/O function can execute the same set of
computations across a wide range of mean ligand concentrations (figure
6.2). All that is needed is to tweak the protein's binding site to match its
affinity to the level of ligand by changing the protein's conformation
slightly. This can be executed stably in the genome, by changing the codons
that specify influential amino acids, to produce a different *isoform* of the
protein, or it can be done dynamically as the protein operates—for example, by using a kinase to add an energetic phosphate.

This capacity of a protein to implement its I/O function with altered
binding affinity serves in innumerable ways. For example, at low affinity
(high k_D) a protein can receive information from its ligand across a
short distance at high concentration, in a brief time, for example,

neurotransmitter diffusing across a 20-nm synaptic cleft. At high affinity (low k_D) the protein can receive information from the same ligand at 1,000-fold lower concentration over a much longer time, for example, a circulating hormone. These capacities are implemented for adrenalin by adrenergic receptors, probably by different isoforms. Dynamic adjustments to affinity can be used for physiological adaptation—to match the I/O function to changes in mean concentration of ligand (figure 3.4).

Protein molecules with different binding affinities transmit different temporal frequencies. High-affinity receptors cannot transmit high frequencies because they do not release their ligand quickly. Consequently, they maintain the same level of output for some time after the input ligand concentration falls. Thus, a high-affinity receptor acts as a low-pass filter—for example, at retinal synapses (chapter 11). By comparison, low-affinity receptors release their ligands promptly, so they transmit high frequencies as well as low, and this gives them a wider bandwidth.

Temporal filtering by a single protein molecule can be modified by *desensitization*. This property curtails the output even while the input ligand remains bound, so allowing a protein with sufficient binding affinity for a low mean concentration of ligand to cut off its response faster than the ligand can unbind. Now, the protein is a high-pass filter. For example, upon binding synaptic transmitter, a protein receptor changes conformation to open an ion channel, but conformational change continues and closes the channel long before the ligand comes off. Speed of desensitization is designed into a protein as part of its energy landscape (Sun et al., 2002), and its use in temporal filtering will be exemplified in chapter 11.

Steeper I/O functions from cooperative binding

A protein's hyperbolic I/O function is steepened by adding more binding sites for the ligand and requiring that several bind to generate the output (Koshland et al., 1982). When n sites have to cooperate, the I/O function follows the nth power of the ligand concentration:

$$output/output_{max} = [ligand]^n/(k_D + [ligand]^n). \tag{6.2}$$

Now the I/O function's lower region (figure 6.3) approximates a power function: $output/output_{max} = [ligand]^n$, and its logarithmic midregion (figure 6.3) is n times steeper: $output/output_{max} = \log([ligand]^n) = n \log([ligand])$. By adjusting both binding affinity and *cooperativity*, an I/O function's position and slope can be matched to the distribution of its input levels (figure 3.4)—which in the fly visual system optimizes coding efficiency (figure 9.10; Laughlin, 1981; Nemenman, 2012).

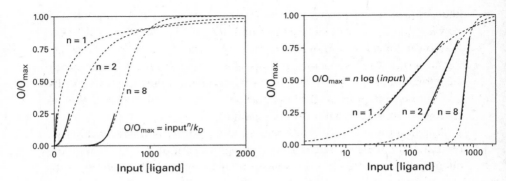

Figure 6.3
Cooperativity changes the input/output (I/O) function generated by binding kinetics to provide different computations. I/O functions are plotted with cooperativities $n = 2$ and $n = 8$ and, for comparison, without cooperativity ($n = 1$). k_D is constant. **Left:** Cooperativity implements the power function $output = input^n/k_d$ with small outputs (<0.25 max). It also shifts the I/O function to higher input values without losing sensitivity (the slope remains steep). In the extreme, for example, $n = 8$, cooperativity creates a switch. **Right:** Cooperativity implements the function $n\log(input)$ in the medium output range (0.25–0.75 max).

A high cooperativity provides a steeper I/O function for digital switching (figure 6.3, $n = 8$) which, by thresholding, can prevent input noise from passing further along a protein circuit. For example, in the protein circuit that releases a synaptic vesicle (chapter 7), a critical step is triggered by the protein synaptotagmin binding calcium ions at several sites. This cooperativity shifts the I/O function to higher concentrations (figure 6.3), so that noisy fluctuations in a cell's baseline calcium concentration rarely release a vesicle. Cooperativity also narrows the range of calcium concentrations that trigger release by increasing the I/O function's slope. Thus, when a voltage-gated calcium channel releases a puff of calcium, synaptotagmin responds promptly, and this increases the temporal precision of release.

Chemical circuitry supports analogue processing
In addition to the functions implemented by binding, proteins' chemical reactions support analogue processing with a rich repertoire of primitives. In brief, simple chemical circuits have equivalent electronic circuits (Sarpeshkar, 2010; figure 6.1) and are capable of implementing procedures used in analogue electronics, namely, amplify, oscillate (Tyson et al., 2003), differentiate, and integrate (Oishi & Klavins, 2011). As well as taking logs (figure 6.2) and raising to powers (figure 6.3), chemical circuits support the

arithmetic operations add, subtract, multiply, and divide (figure 6.4). Small chemical circuits also have the ability to perform more complicated functions—for example, take nth roots (Buisman et al., 2008), compute polynomials, and solve quadratic equations. Whether the brain explicitly implements this more advanced algebra in small chemical circuits[1] is an open question, but the point is made. Chemical circuits support Turing's Universal Computation (Hjelmfelt et al., 1991), which means that they can in principle be configured to compute any function.

Chemical circuits cover the time domain

Not only does chemistry compute, it equips the brain to compute over the range of timescales observed in animal behavior—from the microseconds of the electric sense and hearing to a century of memory. Binding and conformational change take microseconds to seconds. Sequences of reactions executed by protein circuits take from milliseconds (phototransduction, chapter 8) to days (the circadian clock, chapter 4). In chapter 14 we describe how memories that are first laid down by the modification of synaptic receptor proteins are then consolidated for years by the chemical synthesis of new proteins and the assembly of new structures.

What makes a protein circuit efficient?

Computation by circuits built from protein molecules is efficient for several reasons. It is efficient in energy because binding and conformational change approach the thermodynamic limit (chapter 5). It is efficient in space because a single molecule computes. Moreover, computation at this level proceeds directly—that is, by implementing "analogue primitives" (Sarpeshkar, 1998; 2014). Analogue computation typically needs fewer steps than digital to complete a basic operation. For example, analogue multiplies directly, but digital takes $PR^{1.585}$ steps, where PR is the numerical precision in bits (Moore & Mertens, 2011), so even with a low precision of 4 bits, eight steps are saved.

Transmission within a chemical circuit is wireless, so space for wires also reaches an absolute minimum and circuits share space seamlessly. Wireless transmission distributes signals with a minimum of equipment. Once a messenger molecule is broadcast, it can be received by any protein with the appropriate binding site. Thus, wireless transmission makes it easier to reconfigure circuits to change behavior—in the short term by sculpting circuits with neuromodulators (chapter 2) and in the long term by evolving new connections (Katz, 2011). Nor is additional energy needed for wireless

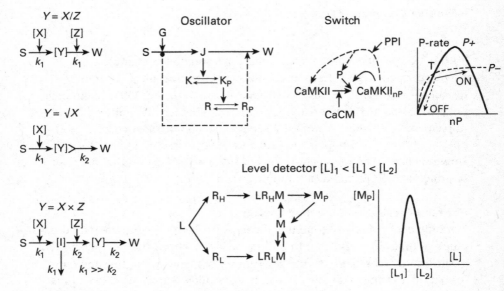

Figure 6.4
Computation by chemical circuits. Left: Circuits that divide, calculate square root, and multiply. The steady-state concentrations of enzymes [X] and [Z] determine the steady-state concentration [Y]. The substrate S is replenished to maintain its high concentration, and the waste product, W, is eliminated so that neither limit reaction rates. k_1 and k_2 are rate constants. In the square-root circuit, two molecules of Y react to form W. In the multiplication circuit, the enzyme X produces an intermediate I. Adapted from Buisman et al. (2008). **Upper middle:** Oscillates when enzyme G is activated. J builds up rapidly and also activates two delayed negative feedback loops (dashed line) by promoting the slower buildup of K_P and R_P. R_p depresses J by catalyzing its removal of J and blocking its production. As J falls, K_P and R_P convert back to K and R, negative feedback ceases, and the next cycle starts with the production of J. Adapted from Novák and Tyson (2008). **Upper right:** Autocatalytic switch implicated in synaptic memory storage (chapter 14). The switch protein, CAM Kinase II (CAM-KII) has 12 phosphorylation sites. If two sites are phosphorylated by the input, the calcium binding protein CaCM, then CAM Kinase II becomes autocatalytic and attaches more phosphates to itself. Rate of phosphate attachment, $P+$, increases steeply with nP, the number of attached phosphates, but then declines at high nP as more phosphorylation sites are occupied. The rate of phosphate removal, $P-$, by the phosphatase PPI increases with np and saturates at a medium nP. Consequently, when CaCM is strong enough to drive CAM Kinase II phosphorylation to the trip point, T, where $P+ > P-$, autocatalysis drives nP to the ON position. Here $P+ = P-$ and the switch can remain ON indefinitely. When CaCM fails to drive the system to T, PPI wins out and removes all phosphates—the switch remains OFF. Adapted from Miller et al. (2005). **Lower middle/right:** Level-detector circuit responds by generating M_p when concentration of [L] lies between $[L_1]$ and $[L_2]$. Two receptor types bind L, high-affinity R_H and low-affinity R_L. LR_H phosphorylates M to active M_p, but LR_L just binds M reversibly At low [L] only the high-affinity LR_H binds, and M_p production increases with L. At high [L] the low-affinity R_L also binds; it outcompetes LR_H for M, so M_p production falls. Adapted from Bray (1995).

transmission. Once the messenger is synthesized and concentrated, it diffuses down its gradient, agitated by thermal bombardment (Brownian motion).

The thermal bombardment that aids diffusion also randomizes movement, and this limits efficiency by introducing noise. Each messenger molecule that reaches a binding site has done so independently of all other messenger molecules; moreover, it has arrived *accidentally* by random walk (figure 2.3). It is the same for a protein designed to deliver information by skating on the membrane (chapters 5 and 8): it finds a receiver by random walk in two dimensions. Moreover, the processes that pass information *through* a protein—binding, allosteric state-transition, catalysis, and release—are also randomized by thermodynamic fluctuations. Therefore, chemical computation in molecular circuits has an associated degree of noise that, as noted in chapter 5, destroys information. Such thermodynamic noise cannot be eliminated, so it must be managed, as we now explain.

Managing noise in a protein circuit

Following the principle *send only what is needed*, a circuit should generally avoid sending noise.[2] Where noise is inevitable, it should be minimized before transmission, so most neural designs try to prevent noise or reduce it at early stages.

Where proteins remain tightly bound in small complexes, signals go directly, thereby avoiding Brownian noise. For more extensive circuits, molecules must move more freely. Now Brownian motion introduces uncertainty. This is reduced by placing proteins close to each other, on the membrane or attached to the cytoskeleton, and by confining diffusible messengers to small compartments. Small compartments also reduce costs—less messenger need be made to produce a signal of given concentration.

By reducing diffusion distances, complexes and compartments shorten delays and lower noise. This occurs where proteins are held together by a protein scaffold—for example, on both sides of a chemical synapse (chapter 7). A presynaptic complex of at least five different proteins (Eggermann et al., 2012) binds a synaptic vesicle and attaches it to the membrane, ready for release. When activated by a surge of calcium, the proteins run through their finite-state routines within 100 µs, to release the vesicle with a minimum of Brownian noise. Postsynaptically, a larger complex of protein species couple to each other and to the membrane. When the vesicle's transmitter molecules cross the 20-nm synaptic cleft and bind a receptor protein, the change in state triggers a host of postsynaptic protein pathways. This complex occupies a 25- to 50-nm layer beneath the postsynaptic membrane (figure 7.3). Compartments and complexes are used in all chemical synapses, in dendrites (chapter 7), in photoreceptors (chapter 8), and indeed in all cells, to promote economy, to speed responses, and to reduce noise.

Some of the noise associated with changes in a protein's conformational state can be prevented by elevating the barriers on the molecule's energy landscape (chapter 5). Although this reduces reaction rates and hence bandwidth, these can be restored by injecting more energy to drive the process. Thus, there are trade-offs between energy consumption, response speed (bandwidth), and reliability (S/N). This sort of intramolecular noise can also be removed by thresholding with a molecular switch (figure 6.3), but there are three penalties: (1) the high energy cost of having a system full ON when only partial ON would do; (2) the low information capacity of a binary system; and (3) the loss of analogue's ability to process directly. But despite complexes, small compartments, and binary switches, some noise remains. What then?

Noise reducer of last resort

There is another way to reduce noise, or more precisely, to improve S/N. The trick is to replicate a noisy signal, then send the replicates in parallel through multiple components, and sum their outputs. The amplitude of the transmitted signal increases linearly with the number of components, but because their noise is uncorrelated, noise increases as the square root. Thus, with an array of M identical components generating noise independently, the output S/N increases as $\sqrt{M}$. Such a parallel array can increase its S/N to arbitrarily high levels by adding more components. However, the solution must be used as a last resort, and then judiciously, because it is expensive.

The dependence of S/N on $\sqrt{M}$ imposes a law of diminishing returns. Cost rises in proportion to M, but benefit rises as $\sqrt{M}$, so efficiency falls as $1/\sqrt{M}$. Here then is the downside of molecular processing. A single molecule can process near the thermodynamic limit to energy efficiency, but that molecule suffers thermodynamic fluctuations. This noise can be countered with a parallel array of the self-same molecules, but the additional resources consume some of what was saved by operating near thermodynamic limit. Therefore, the best a circuit can do is maximize the efficiency of its parallel array, and this it does by matching the size of the array (M) to the costs associated with the array, and to the S/N of the input.

Maximizing efficiency in a parallel array

To evaluate costs and benefits in the design of a parallel array, we use a general measure of performance, information capacity (Schreiber et al., 2002). An array's information capacity depends on S/N (chapter 5) and increases as $\log_2 (1 + S/N) = \log_2 (1 + \sqrt{M})$. However, the energy cost of passing the signal through the array increases as M. Thus, as M increases, the array's efficiency falls—unavoidably—because the array is redundant: all components try to transmit the same signal. Therefore, efficiency is maximum when $M = 1$. Unfortunately the signal generated by one protein molecule is usually too weak and noisy to be useful.

A more practical optimum emerges upon including the *fixed cost* of building and maintaining the circuit that contains the array. Then, as M increases, information per unit cost of signaling falls through redundancy, but information per unit fixed cost rises. An optimum occurs where these two competing tendencies balance. Consequently, a higher ratio of fixed cost to signaling cost gives a larger optimum array (figure 6.5, inset). The optimum array size also depends on the costs in other parts of the circuit. Where expensive components generate a high S/N and then couple to cheaper components, the cheaper array should enlarge beyond its optimum to retain the hard-won benefit. In general, good design distributes investment among components to maximize performance across the entire system (Alexander, 1996; Weibel, 2000).

A good design does not necessarily optimize an array's efficiency. Initially information capacity and efficiency both rise steeply with M (figure 6.5). But then the capacity curve starts to flatten, and an optimum is reached for given fixed cost where efficiency peaks (figure 6.5). As M rises above the optimum, capacity continues to increase, but efficiency declines, albeit more gradually than it rose. Consequently, an array should set M somewhat above the optimum to reduce the possibility of losing both efficiency and information when unexpected perturbations force it to operate below the optimum. Thus operating at the exact optimum may not be best. Robustness is important, too (Schreiber et al., 2002; Sterling & Freed, 2007).

But what *is* a protein circuit's fixed cost? Given that a circuit's viability requires the whole animal, must one count all vital functions? Although the far end to fixed costs looks hazy, the beginning is certainly clear: it is the cost of making a circuit's protein molecules. The average cost to synthesize an amino acid and insert it into a protein is approximately 5.2 molecules of ATP (chapter 5; Phillips et al., 2009), so to build a typical protein of 300 amino acids costs about 1,700 ATP molecules. Protein delivery and installation are extra. By comparison, the cost per signaling cycle is one to

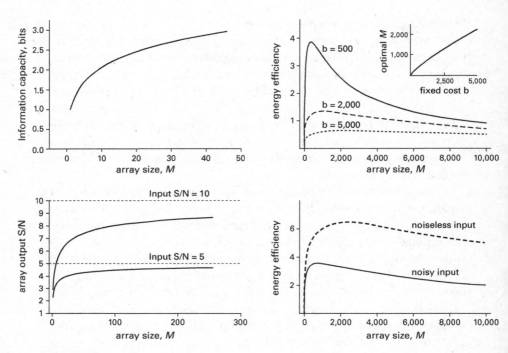

Figure 6.5
Optimizing the noise reducer of last resort—an array of *M* identical components.
Upper left: Increasing an array's size increases its information capacity with diminishing returns. **Upper right**: Energy efficiency (information capacity/energy cost) is
optimized at an array size, *M*, that depends on the fixed cost, *b*. Efficiency is in
arbitrary units, *b* is in units of signaling cost. Inset shows how optimal array size increases with fixed cost. **Lower left**: With a noisy input the output S/N cannot exceed
the input S/N (dashed lines). Lowering this ceiling reduces the advantage of larger
arrays. **Lower right**: Reducing input *S/N* reduces the size of the optimum array. Upper
and lower right redrawn from Schreiber et al. (2002). Upper and lower left calculated
using their formulae.

five ATP molecules (chapter 5), and this suggests a rule of thumb: the cost
of operating a protein molecule (signaling cost) will exceed its construction
cost when the molecule has completed 500–1,000 signaling cycles.

Returning to efficiency, the S/N of an input profoundly affects the array's
optimum size. The array cannot reduce input noise but can only let noise
cancel by averaging. Consequently, input noise imposes a ceiling to be
approached by the array's S/N. This reduces the efficacy of a large array at
low input S/N (figure 6.5) and the size of the most efficient array (figure

6.5). In other words, because an input with low S/N contains less information, and a smaller array has a lower information capacity, the optimum array matches its capacity to its input.

The matching of array size to input S/N follows the principle of symmorphosis (Weibel, 2000), whereby capacities match within a system to avoid waste. What was illustrated for flow of oxygen through lungs, heart, vessels, and muscle (figure 3.4) applies equally to the flow of information through an array of protein molecules. We will see that symmorphosis also holds for parallel arrays of ion channels in a membrane (below), for photoreceptors in a retina (chapter 8), for synapses in a neural circuit (chapter 9), and for neurons in a pathway (chapter 11).

Summary: Pros and cons of computing with chemical circuits

A chemical circuit processes information efficiently on several counts. Operating near the thermodynamic limit it is energy efficient, and its molecules makes efficient use of space and materials. Chemical computation is direct (analogue), which uses fewer steps than digital. Chemistry is wireless, which reduces space and energy for transmission and, by making it easier to form new connections, facilitates behavioral plasticity and evolutionary innovation. A downside is noise, which is handled in four ways. Some Brownian noise is avoided by coupling proteins in complexes and small compartments; some thermodynamic noise is avoided by raising intramolecular energy barriers; and some noise is removed by molecular switches. Unavoidable noise can be mitigated by signaling with parallel, redundant components that add n signals linearly and noise as the square root.

The cost of signaling increases with the concentration of the messenger. Therefore, efficiency might seem to favor high-affinity receptors that bind at low concentration. Yet, there is a penalty and, hence, a trade-off. High-affinity receptors decrease signal bandwidth by slowing the rate at which a signal decays. Low-affinity receptors need higher concentrations, which cost more but, releasing the ligand faster, provide higher bandwidths (Attwell & Gibb, 2005). Thus, speed and bandwidth consume materials and energy, making it advisable to send at the lowest rate.

Despite the advantages of chemical computing, there remains the important proviso *compute with chemistry wherever possible*. Chemistry is fast at the nanometer scale, but because diffusion slows and dilutes signals, chemistry beyond a few microns is too slow to coordinate immediate behavior. Thus, as for *Paramecium* (chapter 2), the need for speed over distance forces a more expensive option—protein circuits that process information electrically.

Information processing by electrical circuits

How electrical circuits meet the need for speed over distance

Electrical current in a silicon device is carried by electrons, but in a biological device it is carried by ions. The cell membrane, comprising a bilayer of nonpolar lipid, is impermeant to ions, so it separates charge, sustains a voltage difference across it, and has a capacitance of about $1 \ \mu F \ cm^{-2}$. Charging the membrane's capacitance constrains the speed of electrical signaling. The membrane's time constant, τ, is its resistance times its capacitance, RC, so τ can be shortened to speed up the signal by shrinking the membrane area and by reducing its resistance to the passage of ions.

An ion passes through the membrane via a channel (Hille, 2001); a large protein molecule assembled as a ring of subunits to form an aqueous pore in the membrane (figure 6.6). The pore is constructed to selectively pass particular ion species in single file by adjusting its width and strategically positioning charged amino acid side groups. A typical sodium ion channel is 10 times more permeable to sodium than to either calcium or potassium, and a potassium channel is more selective still—100-fold more permeable to potassium than to sodium, and almost totally impermeable to calcium.

The channel's energetically stable conformation sets it either closed or open. And thus it remains until a specific input, such as a ligand binding or a change in membrane potential, and/or thermal fluctuations cause the channel to open or close, allosterically. Any net transfer of charge through a channel changes the voltage across the membrane. This voltage signal transmits further and faster along the membrane than chemical diffusion allows, millimeters in milliseconds. But although allostery allows a cheap input, a channel's ionic current is an expensive output, as we now explain.

To charge the membrane quickly, ions must be driven through channels at high rates. The primary driving force is a concentration gradient maintained across the membrane by ion pumps (figure 6.7). Most important is the sodium–potassium pump, which maintains low sodium concentrations and high potassium concentrations inside the neuron. This pump is a molecular machine, a protein complex spanning the membrane which hydrolyzes one ATP molecule to export three sodium ions and import two potassium ions. This asymmetrical exchange generates an outward current of one positive charge per pump cycle and sets up the two concentration differences, $[K]_{in} > [K]_{out}$ and $[Na]_{in} < [Na]_{out}$. These two gradients power most of the brain's electrical circuits. Consequently, the sodium–potassium pump consumes 60% of the brain's energy (Attwell & Laughlin, 2001).

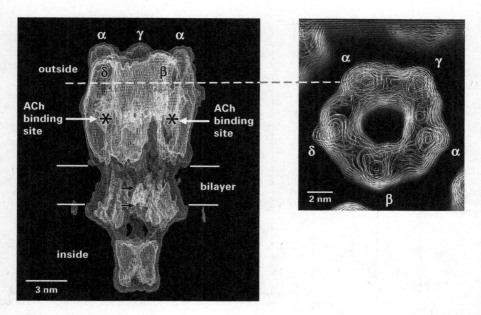

Figure 6.6

An ion channel is a large protein with a pore that conducts ions across the membrane. Ligand gated channel from the electric organ of a torpedo ray opens to admit sodium ions and potassium ions when it binds two molecules of the neurotransmitter acetylcholine, Ach. **Left:** Channel imaged side-on. The channel is formed by a ring of five protein subunits, two αs, β, γ, and δ. All contribute to the extracellular vestibule, the narrower pore that crosses the membrane's lipid bilayer, and the intracellular domain. Asterisks show binding sites for neurotransmitter acetylcholine on the two α subunits. When both bind the channel opens and passes sodium ions and potassium ions. Large intracellular domain has phosphorylation sites for modulating channel's sensitivity. **Right:** Cross section through channel at level indicated on left by dashed line. Three-dimensional structure of channel reconstructed from electron micrographs of crystalline channel arrays, with a resolution of 0.4 nm. Image courtesy of Nigel Unwin. Further details in Unwin (2013).

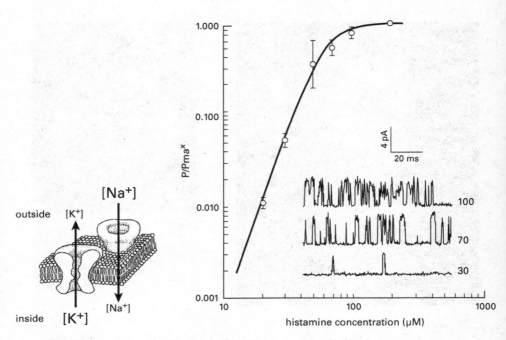

Figure 6.7
Concentration gradients drive ions through channels that open and close rapidly in response to a specific input. Left: Sodium and potassium ions cross the membrane through ions channels, driven by concentration gradients. **Right:** A chloride ion channel opens to pass ~4 pA of current when it binds the neurotransmitter histamine. Currents recorded from a single channel, by patch clamp, at three histamine concentrations: 30, 70, and 100 μM. The open probability increases with histamine concentration according to the binding equation, 6.2, with cooperativity $n = 3$. Channel recorded in membrane of a large monopolar cell from the fly lamina (chapter 9). Left, after Hille (2001). Right modified and reprinted with permission from Hardie (1989).

The concentration gradient is equivalent to a battery whose voltage drives ions through the channel at the same rate (figure 6.8). The battery's voltage is given by the Nernst equation, which converts the chemical potential of the concentration difference into an equivalent electrical potential. Thus, for ionic species, x, its battery's voltage is

$$E_x = RT/(zF) \ln([X]_o/[X]_i) = 2.303\ RT/(zF) \log([X]_o/[X]_i), \tag{6.3}$$

where $[X]_o$ and $[X]_i$ are the concentrations of ion x outside and inside the cell, z is its charge, R is the universal gas constant, T is the temperature in Kelvin, and F is Faraday's constant.

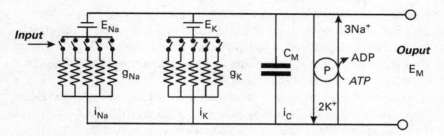

Figure 6.8
The simple resistor–capacitor (RC) circuit formed by ion channels in the neuronal membrane. The input opens sodium channels, and the output is the membrane potential, E_M. A bank of potassium channels, each with conductance g_K, passes outward current i_K, driven by the potassium ion battery E_K. Without input, the potassium channels maintain a *resting potential* of E_K. Input opens sodium channels, g_{Na}, which, driven by the sodium battery, E_{Na}, pass inward current, i_{Na}. To change the output, E_M, the membrane capacitance, C_M, is charged and discharged by the capacitative current, i_C. Sodium-potassium pump, P, keeps batteries charged using energy obtained from hydrolysis of one molecule of ATP to ADP to export 3 sodium ions and import 2 potassium ions, thereby generating an outward pump current.

The two ionic batteries that dominate electrical signaling, potassium with $E_K \sim -85$ mV and sodium with $E_{Na} \sim +50$ mV, provide a dynamic range of about135 mV. A neuron exploits this to the fullest when it generates its fastest signal, an action potential. Before the action potential the neuron is at rest. Mainly potassium channels are open, and the membrane potential sits close to E_K. Here a sodium ion experiences its maximum force, pulled inward by a membrane potential of −85 mV, and pushed inward by a concentration difference equivalent to +50 mV. So when a sodium channel opens to initiate an action potential, sodium ions surge in, driven by 135 mV, and their powerful current helps meet the need for speed.

Less than a millisecond later, when the action potential peaks close to E_{Na}, a potassium ion experiences its maximum force, so when a potassium channel opens to return the membrane to rest, potassium ions surge in, driven by 135 mV. Again, this helps meet the need for speed by increasing the power of the potassium current.

To improve power delivery, a channel's bore is designed to transmit rapidly: ions pass at rates up to 10^8 s^{-1} (Williamson, 2011). These are the highest output rates known for protein molecules (Hille, 2001). By comparison, the fastest chemical output by an enzyme (carbonic anhydrase) is 20-fold slower, and most enzymes are 100-fold slower (Williamson, 2011).

Chemical signaling by molecules, such as ligand-binding receptors and G proteins, operate slower than an ion channel by 4 to 7 orders of magnitude. With its exceptional output rate, a voltage-gated sodium channel opening for 1 ms admits 6,000 Na^+ ions. This 1 pA ionic current delivers $2.4 \times 10^4 k_B T$ joules, giving a power rating of 200 fW.

Fast processing also requires molecules that switch quickly. Channels are structured to open or close in tens of microseconds (figure 6.7)—near the limits of allosteric state change (Chakrapani & Auerbach, 2005). The energy used to open a channel, ~25 $k_B T$ joules (Chowdhury & Chanda, 2012), is 35 times the thermodynamic minimum for a bit (chapter 5), high enough above to be reliable, but low enough not to put too much of a brake on processing speed. With an input energy of 25 $k_B T$ joules and an output of $2.4 \times 10^4 k_B T$ joules, a sodium channel opening for 1 ms has a power gain ×1,000. Thus, a channel's combination of sensitivity, fast switching, and gain satisfies the need for speed. But as noted, it comes at a price.

The price is paid to keep ionic batteries fully charged. An ion passing through a channel drops its battery's voltage by reducing the concentration gradient (equation 6.3). The gradient is restored by pumping the ion back across the membrane, so when a sodium channel opens for 1 ms and admits 6,000 Na^+ ions, sodium-potassium pumps hydrolyze 2,000 ATP molecules to ADP to pump these ions back. The efficiency of the conversion of the chemical energy supplied by ATP to the electrical energy delivered by the channel is reasonably high, 50%.[3] Nevertheless, a channel's signaling cycle (open, admit ions for a millisecond, close, restore ions) uses 2,000 times more ATP than a G protein's cycle. This is the price paid for speed over distance.

In summary, an ion channel changes a neuron's membrane potential rapidly by operating as a power transistor that is irreducibly small and operates close to thermodynamic limits. Engineers seek similar efficiency savings by developing their version of a single molecule power transistor. Biology evolved this device over a billion years ago and solved the not inconsiderable problem of connecting its molecular "transistors" to form circuits.

How circuits built from ion channels operate electrically

Ion channels naturally form electrical circuits because they connect two lower resistances (extracellular space, cytoplasm) across an insulating membrane. Consider the simplest circuit, two types of ion channel working against each other to code an analogue input as an analogue output, namely, a change in membrane potential, E_M (figure 6.8).

The circuit's behavior is captured by an electrical model in which each channel is a switched resistor, connected to its battery (figure 6.8; Koch, 1999). The resistor represents the channel's conductance, g, (conductance = 1/resistance) and the switch opens the channel. For a channel that passes ions of species x, the current, i_x, is given by Ohm's law:

$$i_x = g_x (E_x - E_m), \tag{6.4}$$

where E_m is membrane potential, E_x is the electromotive force (EMF) of the ionic battery (equation 6.3), and g_x is the single-channel conductance for ion x. Note that when $E_M = E_x$, there is a tipping point where the direction of current reverses. This point is used to determine E_x experimentally, so it is often called the *reversal potential*.

For ion channels to change the membrane potential, they must charge and discharge the membrane's capacitance (~ 1 μF cm^{-2}), represented in the model by the capacitor, C_M. The fourth component, the sodium–potassium pump, P, hydrolyzes ATP to keeps the ionic batteries charged. Because the rate at which the pump exchanges three sodium ions for two potassium ions is effectively independent of membrane potential, it is treated as a constant current source.

This RC circuit model describes how the membrane potential changes when channels open and close. Applying Kirchoff's law,

$$i_{Na} + i_K + i_C + i_P = 0, \tag{6.5}$$

where i_C is the capacitative current and i_P is the pump current. Substituting for the currents flowing through the channels and the capacitor,

$$(E_{Na} - E_M)N_{Na}g_{Na} + (E_K - E_M)N_K g_K + C_M \, dE_M/dt + i_P = 0, \tag{6.6}$$

where N_{Na} and N_K are the numbers of open sodium channels and open potassium channels. Because the pump maintains the concentration gradients for sodium and potassium, $i_P = 0.5 \, i_K$, giving

$$(E_{Na} - E_M) \, N_{Na}g_{Na} + 3/2(E_K - E_M) \, N_K g_K + C_M \, dE_M/dt = 0. \tag{6.7}$$

This current-balance equation captures the biophysics of electrical signaling across a neural membrane and easily extends to include other channels (including ones that depend on time and voltage), other pump currents, and currents generated by ion exchangers. Consequently, an equation of this form is the core of the many more complicated models of electrical interactions in neurons (Hodgkin & Huxley, 1952; Koch, 1999). One insight is that this irreducibly simple circuit is inherently *self-shunting*. That is, current driven through a channel pushes the membrane voltage toward the channel's reversal potential, thereby progressively diminishing the current passed per channel as more channels of this type open. This nonlinear behavior shapes the circuit's I/O function and supports information processing.

I/O function of the basic circuit

To explain the circuit's I/O function we drive it with an input that opens sodium channels. Sodium ions enter, pushing E_M toward the positive potential of the sodium battery. This shift in voltage encodes the input intensity, I, as an output. To derive the relationship between input and output, assume that the input acts linearly, so the number of open sodium channels is

$$N_{Na} = aI, \tag{6.8}$$

where a is channel gain, in open channels per unit input. Thus, the sodium conductance is

$$G_{Na} = g_{Na}N_{Na} = g_{Na}aI. \tag{6.9}$$

The opposing potassium conductance is held constant, $G_K = g_K N_K$, where g_K is the conductance of a single potassium channel and N_K is the number of open potassium channels.

The circuit's I/O function now follows. Without input, $G_{Na} = 0$, and the circuit rests with $E_M = E_K$. A step rise in I opens aI sodium channels whose inward current charges the membrane capacitance to a new steady voltage with a time constant

$$\tau_M = C_M R_M, \tag{6.10}$$

where R_M, the membrane resistance, is $1/(G_{Na} + G_K)$. This steady state is reached long before pump currents change because they are slow (see below) whereas τ_M is typically milliseconds; consequently $i_C = i_P = 0$. Solving the circuit's current balance equation gives the new steady-state membrane potential

$$E_M = (G_{Na}E_{Na} + G_K E_K)/(G_{Na} + G_K). \tag{6.11}$$

Dividing through by G_K, we see that E_M depends on the conductance ratio, G_{Na}/G_K,

$$E_M = (E_{Na}G_{Na}/G_K + E_K)/(G_{Na}/G_K + 1). \tag{6.12}$$

This relationship is simplified by expressing the voltage output relative to a baseline of zero input so that $output = E_M - E_K$, then normalizing output to its maximum, $output_{max} = E_{Na} - E_K$. Note that the setting of E_K to zero simply

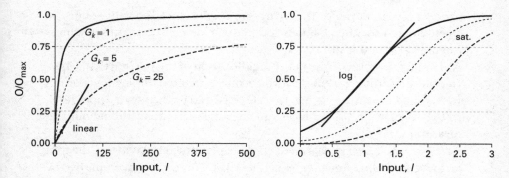

Figure 6.9
Input/output (I/O) function generated by the basic electrical circuit allows the same computations across different input ranges by changing the shunting conductance G_K. Normalized output, O/O_{max}, is plotted against input, I, for three different shunting conductances. **Left:** When output is small (<0.25 max), the I/O function adds. **Right:** When the output is medium (0.25–0.75 max), the function is logarithmic. In saturated regime (0.75 max – max) function's slope approaches zero. Note similarity with I/O function produced by chemical binding (figure 6.2).

shifts the voltage scale without altering the EMFs experienced by ions, so response amplitudes are unaffected. Now

$$output/output_{max} = (G_{Na}/G_K)/(G_{Na}/G_K + 1). \tag{6.13}$$

Substituting aIg_{Na} for G_{Na}, we obtain a simple form of the circuit's I/O function

$$output/output_{max} = kI/(kI + 1), \tag{6.14}$$

where the gain factor $k = ag_{Na}/G_K$. The electrical circuit's I/O function is hyperbolic (equation 6.14; figure 6.9), like the I/O function for chemical binding, because it too saturates. And like the chemical circuit, the electrical circuit's hyperbolic I/O provides operators for processing information (Koch, 1999; Silver, 2010).

An electrical circuit's hyperbolic I/O supports six operators
1. *Addition* (A + B) occurs when the circuit operates in the bottom quartile of the I/O function where it is approximately linear (figures 6.2 and 6.9), When inputs A and B open the same species of ion channel, they add.
2. *Subtraction* (A – B) also occurs in this linear region when input A opens an ion channel that carries current inward and B opens a channel that

carries a current outward. The changes in conductance and voltage must be small enough for the channels to approximate constant current sources driving a constant load.

3. The *log* transform occurs in the middle region of the I/O function, where *output ~ k*log *I* (figure 6.9B). As with chemical circuitry, this log transform is widely used in sensory circuits to scale responses to changes in input level, so that a constant $\Delta I/I$ produces equal changes in output throughout this logarithmic range.

4 & 5. *Multiplication* (×) *and division* (÷) are performed by changing the gain factor, *k*, in the I/O function (equation 6.14). This can be accomplished by altering the channel gain (*a*) and/or the potassium conductance (G_K). For example, increasing G_K shunts the input from G_{Na}. This mechanism is widely used for multiplicative gain control and divisive normalization (chapters 8 and 12), procedures that optimize coding and facilitate the extraction of patterns (Koch, 1999; Carandini & Heeger, 2012). Changing channel gain, *a*, does not, strictly speaking, multiply and divide within the circuit, but it has this effect on the I/O function. The important distinction for design is that increasing G_K increases both signal quality (S/N, band-width) and energy consumption by increasing the number of open chan-nels, whereas reducing *a* reduces signal quality and energy consumption by reducing the number of open channels.

6. *Exp* (inverse of log) is implemented by installing cooperativity in ion channels—for example, by requiring that *n* binding sites be occupied to open a ligand-gated channel. As in chemical circuits, cooperativity raises the output to the *n*'th power of the input, so steepening the I/O curve and shifting it to higher input levels. Cooperativity is used at blowfly photore-ceptor output synapses to match a neuron's coding function to the range of input levels (figure 3.4). The neurotransmitter, histamine, must occupy 3 binding sites to open a postsynaptic chloride channel. This steepens the I/O function (figure 6.7) to help achieve a match with the probability distribu-tion of input signals (figure 9.10).

How electrical circuits support analogue processing
Ion channels implement the four elements of analogue electrical circuits, resistance, R; capacitance, *C*; inductance, *L*; and memristance, *M* (Chua, 1971). Resistance and capacitance are obvious (figures 6.7 and 6.8), but the uses of inductance and memristance need explanation. With an induc-tance the voltage is proportional to the rate of change of current. Thus, when the current is increasing more rapidly, the voltage is larger, and this

advances the phase of the response to a sinusoidal input. Voltage-gated potassium channels advance phase by means of delayed negative feedback (Koch, 1999).

A memristor changes its resistance in proportion to the quantity of charge it has conveyed and then holds this resistance when charge stops flowing (Strukov et al., 2008). This resistance with memory is provided by a channel that couples electrical signaling to chemical signaling. For example, take an ion channel that passes mostly sodium with a little calcium. This calcium provides a measure of the total charge flowing through the channel. Arrange that calcium binds to the mechanism that opens the channel, and alters its open probability. Now one has a memristor in which charge entry couples to the channel's effective conductance. Photoreceptors use this mechanism to control their gain (chapter 8).

How voltage-gated channels meet a need for speed over distance

A voltage-gated channel opens or closes allosterically, in response to membrane potential. Thus, a voltage-gated channel can be activated within milliseconds by channels opening millimeters away. In addition, a voltage-gated channel amplifies an electrical input. By virtue of these properties, voltage-gated channels can produce a larger signal that transmits more quickly and reliably than the signals generated by ligand-gated channels—most notably an action potential (figure 6.10).

A typical action potential, an approximately 100-mV pulse lasting about 1 ms (figure 6.10), is produced by a large and sudden influx of sodium ions followed by a similar efflux of potassium ions. These currents are produced by sodium channels and potassium channels (figure 6.10) that, gated by depolarization, generate the action potential and propagate it along the membrane at speeds of 0.3–80 mm ms^{-1} without loss of amplitude.

The voltage-gated channels generate the action potential as follows (figure 6.10). At resting potential, typically –70 mV to –60 mV, the voltage-gated channels for sodium and for potassium open with a low probability. When an analogue input depolarizes the membrane, the open probability increases and a small proportion of sodium channels opens immediately. Driven by their maximum force, sodium ions surge in and depolarize the membrane further, creating a positive feedback loop (figure 6.10). Almost all of the voltage-gated potassium channels remain closed because they respond to depolarization more slowly. A longer activation time constant is programmed into their finite state transitions to keep them closed while the sodium channels are starting to open. This delayed opening increases

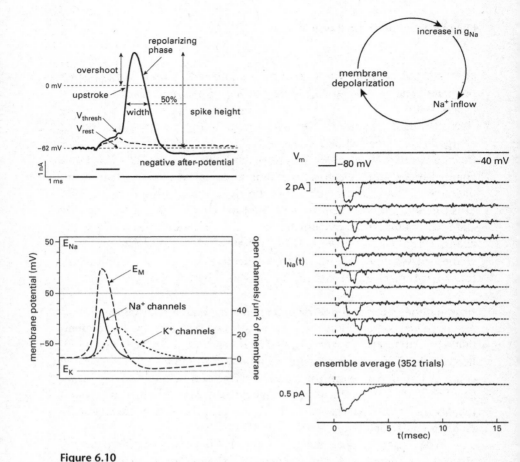

Figure 6.10
Voltage-gated sodium channels and voltage-gated potassium channels meet the need for speed by producing an action potential. Upper left: Action potential waveform. Spike initiated when suprathreshold current depolarizes membrane potential from resting potential, V_{rest} to threshold, V_{thresh}. Fast upstroke overshooting to peak height and repolarizing phase complete rapidly to produce spike with narrow width (measured at 50% spike height). Slower negative after-potential follows. **Upper right:** Positive feedback loop that accelerates spike upstroke and drives overshoot to maximum amplitude. Increase in voltage-gated sodium conductance, g_{Na}, increases inflow, depolarizes membrane and increases voltage-gated sodium conductance. **Lower left:** Time course of spike (E_M, left axis) and voltage-gated sodium and potassium conductance, plotted as density of open channels (right axis). The rapid increase in the number of open sodium channels that drives the upstroke is short-lived because sodium channels quickly inactivate. The voltage-gated potassium channels open more slowly to repolarize, and generate the negative after-potential. **Lower right:** Recordings of the activity of two voltage-gated sodium channels show that, following a step depolarization, each opens with a randomly varying latency for a randomly varying time. Averaging 352 individual responses demonstrates that a large array of channels averages out noise to produce a reliable sodium current. Upper left redrawn from Bean (2007). Upper right and lower left from Shepherd (1994) with permission. Lower left, data from J. B. Patlek, plotted after Hille (2001), with permission.

the efficiency with which sodium channels charge the membrane capacitance by preventing the charge being carried in by sodium from being negated by charge carried out by potassium. Blocking this futile cycle allows the action potential to develop and, by reducing the number of ions crossing the membrane, saves pump energy.

At a critical level of depolarization, the *threshold* potential (figure 6.10), sodium's positive feedback takes off. All available sodium channels open (figure 6.10), more sodium ions surge in, and, unopposed by the more sluggish potassium channels, their current depolarizes the membrane toward equilibrium potential (E_{Na} = 50 mV) in less than 1 ms. As the membrane potential approaches this peak, large numbers of voltage-gated potassium channels are starting to open (see figure 6.10). Potassium ions experience their maximum force and surge out, driving the membrane potential back down, toward rest. At the same time, the open sodium channels change conformation and lock shut. This *inactivation*, programmed into a sodium channel's state changes, stops incoming sodium ions from negating the charge being carried by outgoing potassium, thereby increasing efficiency. The voltage-gated potassium channels drive the membrane potential to resting potential within 0.5 ms and, being no longer depolarized, start to close. But because potassium channels change their state more slowly, many remain open; the membrane potential dips below rest and approaches E_K, creating a negative afterpotential (figure 6.10).

While potassium channels are repolarizing the membrane, the voltage-gated sodium channels remain inactive. To reset to its initial state (closed but responsive to depolarization), a sodium channel must experience the strong negativity of potentials close to rest. This state change is programmed to have a time constant of ~3 ms. The resulting delay, plus the residue of open potassium channels, makes it impossible to trigger another action potential during a *refractory period* of 2 ms.[4] Although being refractory places a ceiling on action potential frequency, it ensures that an action potential cannot trigger a resurgent sodium current during its repolarizing phase. This prevents a single action potential from starting a continuous train of spikes.

In summary, an action potential is the product of three electrical feedback loops, all formed by voltage-gated channels. Sodium's positive feedback loop depolarizes the membrane to the action potential's peak (figure 6.10), and potassium's delayed negative feedback repolarizes to rest. Speed and efficiency are enhanced by a third negative feedback loop, mediated allosterically by sodium channel inactivation. Because channels gate each other electrically, the action potential is brief. This increases timing precision and hence the number of bits carried by an action potential (chapter 3). Being electrical, an action potential travels rapidly along a neuron's membrane at speeds up to 100 mm in a millisecond (chapter 7) yet retains its information because it is faithfully regenerated by feedback. But how can the information carried by such an electrical signal drive a chemical circuit? The answer is a voltage-gated channel with a chemical output.

How a voltage-gated calcium channel links electrical to chemical

A voltage-gated calcium channel admits an ion that readily binds a protein and changes its conformation. As noted in a chemical synapse, calcium entering via channels opened by presynaptic depolarization binds the protein synaptotagmin, which then changes conformation and triggers vesicle release. A calcium ion is especially effective at changing a protein's conformation because, being divalent, it pulls negatively charged parts of a protein closer together.

Calcium is especially effective as a chemical messenger because cells pump it out to keep the internal concentration low, 30–200 nM. This creates a steep concentration gradient, equivalent to a battery of 130 mV that, aided by the –70 mV resting potential, drives calcium in through a channel at a rate of ~10^7 ions per second. With so little internal calcium, the proteins within nanometers of the channel experience a 100-fold increase in calcium concentration within 100 µs. This nanodomain calcium signal has a wide bandwidth because it decays as rapidly as it rises. The puff of calcium injected by a channel vanishes within 500 µs by diffusing rapidly into a large sink, the well-buffered bulk of the cell's cytoplasm. Viewed from the channel's nanodomain, this rapid removal mechanism comes for free. The calcium puff is mopped up by buffering proteins, distant pumps, and exchangers.

In summary, the simplest electrical circuits demonstrate how the brain satisfies the need for speed over distance. Whereas chemical signaling can send information in a millisecond, but only over 1 µm, passive electrical signaling can send it a millimeter in the same time—1,000-fold faster. Active electrical signaling (action potentials) can send it still faster, by another 100-fold, over much longer distances. Electrical circuits can be constructed to use the same operators as chemical circuits (figure 6.2; cf. figure 6.9). But operating more rapidly over longer distances requires more power. An electrical circuit consumes orders of magnitude more energy than a chemical circuit and, because electrical signaling uses wires, costs more space.

Given the costs, one expects efficient design. Since ion channels are allosteric proteins and operate stochastically, they present the same issues of

S/N, bandwidth, and redundancy that were identified for chemical circuits. We should also expect the same need to match output to input—symmorphosis. Moreover, when short-range chemical circuits have filtered signals into parallel streams with different S/N and bandwidth and information content, the electrical circuits that relay this information rapidly over distance need to match these inputs with appropriate outputs. This requires a diversity of ion channels with subtly different sensitivities and speeds—that is, access to the large "part list" contained in the genome.[5]

Constraints on information processing by circuits of ion channels

Biophysical constraints

Three biophysical factors limit the performance of electrical circuits formed by ion channels: (1) the high electrical resistance of single channels, (2) membrane capacitance, and (3) channel noise from thermal fluctuations in single proteins.

First, channel resistance. Despite having a high transport number for a protein molecule, a single channel nevertheless has a high resistance, R_{Ch} ~ 10^{11} Ω. The reason is that selectivity requires the ions to pass in single file. Driven by a typical range of voltages, 10 mV–100 mV, a channel passes 0.1–1 pA. In a neuron with typical input resistance, 10^8 Ω, such currents are sufficient to change membrane voltage by 10 μV to 100 μV. That's not much. For example, the voltage change needed to reliably trigger an action potential is about 1 mV–10 mV, that is, 10 to 1,000-fold larger. Moreover, voltage decays exponentially with distance, so a single channel's signal soon disappears in the membrane voltage noise. A larger voltage signal will travel further and support a workable S/N at its destination, and this is easily achieved (equations 6.9 and 6.11; figure 6.9)—by opening more channels.

Second, membrane capacitance. As noted, capacitance limits a signal's rate of change. One channel, passing 0.5 pA, charges the membrane slowly, and this limits temporal frequency and bandwidth. For example, one channel charges the 314 μm^2 membrane of a spherical neuron, 10 μm in diameter, with a time constant of 88 ms, giving a bandwidth of 12 Hz. This limit too can be raised—by opening more channels.

Third, channel noise. Channels, like other proteins, change conformational state stochastically because they are subject to thermodynamic fluctuations. Therefore, a channel opens and closes stochastically with probabilities that depend upon its input (figure 6.7). This stochastic opening adds noise. The ratio of signal to noise can be improved—by opening more channels.

Channels operating in an electrical parallel array, as in figure 6.8, obey the same rule as molecules in a chemical array (figure 6.5). The S/N of an array of M parallel channels increases as $\sqrt{M}$, and as M increases, efficiency falls. Consequently, an efficient electrical circuit will match its number of channels to three factors: fixed cost, costs of other signals in the circuit, and input S/N (figure 6.5). In summary, one adjustment, opening more channels, improves four measures of performance: signal amplitude, signal bandwidth, S/N, and information capacity (equation 5.6). So, what constrains the numbers of channels that a circuit can employ to improve its performance?

What limits the number of channels in a circuit?

A circuit could maximize its performance by maximizing the number of channels it uses. Some parts of protein circuits (e.g., ligand-gated channels on a postsynaptic membrane) achieve this locally by packing channels in the cell membrane as a crystalline array ($\sim 2.5 \times 10^3$ channels per μm^2). This produces tremendous local currents which charge the membrane with extreme rapidity, a design used by the electric eel to discharge its electric organ. However, such a power drain could not be sustained globally across an entire neuron.

The number of channels is limited by membrane space for pumps. A pump molecule has approximately the same footprint as a channel, but, operating at 200 cycles s^{-1}, it extrudes only 600 sodium ions s^{-1}. To match the throughput of one open sodium channel (6×10^6 sodium ions s^{-1}) requires 10,000 pump molecules, which occupy 4 μm^2 of membrane. Thus the density of *open* channels that a neuron can sustain is reduced to one channel per 4 μm^2, 10,000-fold less than their maximum packing density. This translates into a 10,000-fold lower bandwidth and a 100-fold reduction in S/N. Being proportional to bandwidth and $\log_2$ (S/N), the sustainable information rate is cut by almost five orders of magnitude. Placing the circuit's battery chargers (pumps) alongside the circuit's transistors (ion channels) limits a neuron's ability to process information, but cell biology offers few alternatives.[6]

Were a neuron to fully pack its membrane with channels and their obligatory pumps, could it power them? The essential ATP is generated within the neuron by mitochondria. These occupy space, so the maximum sustainable ATP production is proportional to cytoplasmic volume and mitochondrial density. Typically 4×10^5 ATP s^{-1} can be generated per μm^3 (based on a specific metabolic rate for cortical neurons of 40 μmoles ATP/g/min; Attwell & Laughlin, 2001), which means that generating the power for one open sodium channel requires about 5 μm^3 of cytoplasm. Thus, when operating at the pump limit of one open channel per 4 μm^2 of membrane, 5 μm^3 of cytoplasm is required to provide the pumps' ATP, giving a surface area to volume ratio of 1:1.25. Therefore, a spherical neuron must be greater than 7.5 μm in diameter to operate at the pump limit, but a smaller sphere has a larger surface area:volume ratio, so it is limited by the ability of mitochondria to generate ATP. Many neuronal cell bodies have diameters greater than 7.5 μm, but to connect efficiently they branch (chapter 13), and this increases surface area:volume. Thus, a pyramidal neuron, with a surface area:volume ratio of about 3:1, cannot reach the pump limit to open channel density. Forced to operate with fewer open channels, it must reduce the rate, temporal precision, and accuracy of its electrical signals. Housing the system that burns fuel to supply energy also limits a neuron's processing power, but again, that's cell biology.

In short, the molecular power transistor (ion channel), its molecular battery charger (ion pump), and its intracellular power station (mitochondrion) prevent the brain from reaping a major benefit of irreducibly small molecular components, high-density computing. Thus, unlike conventional engineering design, neural design must maximize performance at low-power density. Given that opening more channels inevitably costs space—membrane area for pumps and cytoplasmic volume for mitochondria—it is all the more critical to open the minimum number of channels required to meet functional specifications. To paraphrase a now familiar principle, a low-energy-density brain should send information with the lowest rate of channel opening.

Providing speed and accuracy with low energy density circuits
Given that low energy density limits the minimum time constant and maximum S/N by limiting the number of open channels, how can a brain respond quickly and accurately? A solution adopted by most brains is to open many channels infrequently in concentrated groups—that is, use powerful signals that are sparsely distributed in space and time, as happens with action potentials and synapses.[7] This design leads to an apparent paradox. These concentrated electrical signals are costly and consume most of the brain's energy, so they are part of the problem, but, given the need to send accurate signals far and fast, they are also part of the solution.

Although concentrated bursts promote temporal precision by increasing S/N and reducing the membrane time constant, their spatial and temporal sparsity enforces low mean rates. For example, the power density of cortical gray matter limits the mean firing rate, averaged across all classes of cortical neuron, to less than 10 Hz (Attwell & Laughlin, 2001; Lennie, 2003; Sengupta et al., 2010; Howarth et al., 2012). How the brain manages to process information effectively within this limit is a major theme in neural design.

Can arguments based on energy density be extended to establish an upper limit to a brain's processing power—as bits per volume per second? Possibly, but this would only consider the expensive electrical signals. Over short distances and longer times, chemical processing is orders of magnitude cheaper, thus the principle *compute with chemistry*. Chemical and electrical circuits can process information with similar operators, but at scales and costs that differ by orders of magnitude. Therefore the design task for a neuron is to integrate across these scales to achieve the best result in space, time, and energy. This is the subject of chapter 7.

7 Design of Neurons

Chapter 6 explained that much of the brain's computing occurs by chemistry at the scale of single protein molecules and protein circuits. Computing by chemistry offers good S/N at irreducibly low cost in space and energy. Moreover, where the reaction vessel shrinks, the principle of mass action can operate on high concentrations with small numbers of molecules. High concentrations allow low binding affinities to achieve useful signaling rates. Small volumes also shorten distances—over which diffusion is rapid. Also, because concentrations of diffusing molecules decay steeply in space and time, many computations can be accomplished wirelessly—simply by placing detectors at different distances from a source and letting Brownian motion do the math.

Computing with proteins allows a nearly infinite parts catalog—because a protein can be customized by changing a single amino acid—and that is effected simply by swapping a single base pair in the DNA. Thus, natural selection can shape every component precisely for a specific task—for example, to match a particular binding affinity and a particular cooperativity to a particular signal (figures 6.2 and 6.3). The ease of adjusting protein structure has generated immense diversity: overall, the mammalian brain transcribes 5,000 to 8,000 genes and uses alternative splicing to produce 50,000 to 80,000 distinct proteins.[1]

Chemical computing works brilliantly across a spatial scale of nanometers to micrometers and a temporal scale of 100 μs to seconds (e.g., rod phototransduction; chapter 8). Yet to serve behavior, computations must retain the same timescale but travel up to 1 millionfold farther. To achieve speed over distance requires recoding the chemical signals to electrical signals. Recoding begins with an allosteric trigger, such as ligand-binding or G protein activation, but allostery must eventually open an ion channel in the membrane to establish an electrical signal. This is one key task for a neuron: use allostery to send an electric signal somewhere fast.

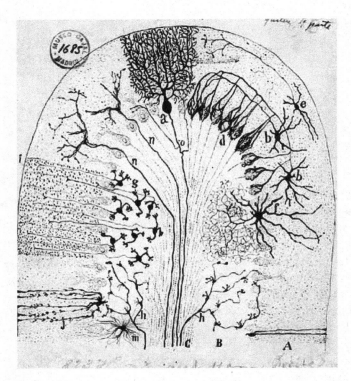

Figure 7.1
Neurons and glial cells of cerebellar cortex. Neuron types shown here (a, b, e, f, g) all express the standard polarized design: inputs to multiple dendrites converge to cell body and output to a single axon. Neuron types: a, Purkinje; b, basket; e, stellate; f, Golgi; g, granule. Input axons: h, mossy fiber; n, climbing fiber. Two types of glia (j, m) are shown at lower left. Each Bergman glial cell (j) wraps the dendritic arbor of a single Purkinje neuron (a). Drawing by S. Ramón y Cajal. Reprinted with permission from Sotelo (2003).

"Somewhere fast" has two parts. First, a chemical signal released by a *presynaptic* neuron targets a *postsynaptic* neuron on a short branch (*dendrite*). The chemical transmitter binding to a protein receptor allosterically opens its ion channel. This initiates an electrical signal that spreads passively along the dendrite toward a central locus (cell body or specialized cable segment) for integration with signals from other dendrites. Second, the integrated electrical signal recodes to an all-or-none pulse that spreads actively down a single cable (*axon*) toward presynaptic terminals that contact other neurons (figure 7.1).

Dendrites are 10–1,000 micrometers long, depending on neuron type, and over such distances "fast" means up to 50 micrometers per millisecond. Axons are 1–1,000 millimeters long, and over such distances, "fast" means at least 1 millimeter per millisecond. Thus, dendrites conduct passive electrical signals about 50-fold faster than chemical diffusion, and axons conduct active electrical signals at least 20-fold faster than dendrites.

A neuron steps up from the nanometer scale of protein circuits to the micrometer scale of a synapse (1,000-fold), then to the millimeter scale of a dendritic tree (1,000-fold), and then to the meter scale of the longest mammalian axons (1,000-fold), ultimately integrating processes that span a 10^9 range of spatial scale. This greatly increases the cost of space, materials, and energy. A protein molecule allosterically encoding 1 bit occupies about 50 nm^3; whereas the smallest neuron cell body encoding 1 bit occupies 10^9 greater volume; and the largest neuron cell body encoding 1 bit occupies 10^{12} greater volume and correspondingly more materials. The energy cost of encoding 1 bit rises from about 25 k_BT to about 10^9 k_BT.[2]

Such numbers explain why neurons need to be efficient. The microvessels that deliver oxygen and metabolic supplies distribute densely, forcing neurons to occupy their interstices (figure 7.2). Were neurons to be energetically less efficient, they would need more mitochondria to produce more ATP—and that would require a denser capillary network at the expense of efficient neuron layout (chapter 13). The same constraint applies to space and materials. For example, the diameter of a cerebellar Purkinje cell body is 10-fold greater than that of a cerebellar granule neuron, but its volume is greater by 1,000-fold (figure 7.1). Therefore, we must explain how the design of each neural component: synapse, dendrite, cell body, and axon match each other and conserve space and energy.

Synapse

Synapses enable neurons to process information in neural circuits by transferring and transforming signals at specific connections. The simplest synapses are electrical, made from proteins that form an array of channels that connect two neurons. Where it serves as a simple resistor, an electrical synapse is as inexpensive and noiseless as a connection can be.[3] This is why electrical synapses are widely used to weakly couple neurons, for example, to compute the mean signal over a patch of retina to reduce redundancy (chapter 11), to synchronize rhythmical activity among the cortical interneurons, and to synchronize motoneurons that drive the same muscle. But coupling with a resistor does not equip a circuit to compute much. More

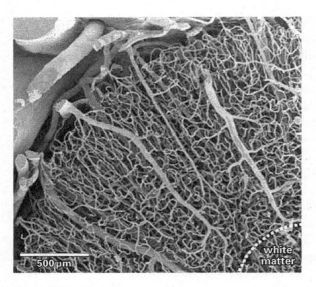

Figure 7.2
Blood vessels distribute densely in gray matter with even mesh. The 500-μm scale bar corresponds roughly to the dimension of largest local circuits. Therefore, neurons and glia must fit into the interstices of the capillary network. Were energy cost to rise, due either to lower neuronal efficiency or enhanced neuronal performance, vessel density would rise at the expense of efficient neuron layout. Cerebral cortex from superior temporal gyrus of monkey. Reprinted with permission from Weber et al. (2008).

transformations are required, and signals must be amplified to produce fast responses that are resistant to noise. These requirements are met by chemical synapses, so called because a presynaptic neuron transmits chemically by sending a pulse of neurotransmitter to receptors on a postsynaptic neuron.

Origin of graded chemical signal

A chemical pulse originates when a vesicle docked to a presynaptic *active zone* fuses with the plasma membrane and releases transmitter molecules through a *fusion pore* into the *synaptic cleft* (figure 7.3). The vesicle contains about 4,000 molecules of transmitter concentrated by a transporter protein in the vesicle membrane to roughly 100 mM. Discharge through the pore requires about 100 μs, during which molecules are diffusing away; yet their concentration at the postsynaptic receptor proteins clustered 20 nm across the cleft rises briefly to about 10 mM (figure 7.3). This suffices for

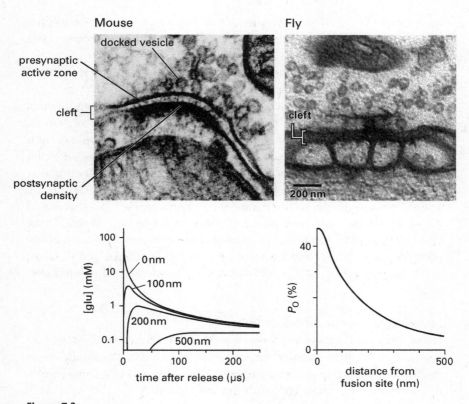

Figure 7.3
Fusion of one synaptic vesicle briefly raises the concentration of transmitter within the synaptic cleft to 10 mM. Upper left: Synaptic vesicles, about 40 nm, from mouse cerebellar cortex docked to presynaptic membrane across the synaptic cleft from the postsynaptic density. This density houses a complex of proteins that support, among other functions, long-term potentiation (see chapter 14). Courtesy of K. H. Harris. **Upper right:** Synaptic vesicles, about 30 nm, from *Drosophila* medulla. Note that cleft width and vesicle size are similar for mammal and fly. Fly vesicle contains similar number of transmitter molecules as mammal (Borycz et al., 2005). Courtesy of Zhi-yuan Lu, Patricia Rivlin & Fly EM Team, Janelia, HHMI. **Lower left:** Concentration decays steeply in space and time. **Lower right:** Concentration at 20 nm from fusion site suffices to bind and open roughly half of the postsynaptic ion channels. Half-saturation by one vesicle allows multivesicular release to enhance the response. Decay to lower concentration with time and distance can be exploited by other types of receptor molecules with higher binding affinities (figure 11.10). P_o, open probability of postsynaptic ion channels. Graphs are modified and reprinted with permission from Xu-Friedman & Regehr (2004).

transmitter to bind cooperatively to low affinity receptors, thereby opening their channels and initiating the electrical signal, a miniature postsynaptic current (*MPSC*).

This design requires matching closely the number of molecules and their concentration within a vesicle to its emptying time, diffusion distance, and receptor binding constant. Were the vesicle to contain fewer molecules or a lower initial concentration, the final concentration at the postsynaptic receptor would be too low for its binding affinity. The same would occur if the vesicle emptied more slowly or if the diffusion distance across the cleft were greater. Any of these factors could be compensated for by a higher affinity at the receptor, but that would sacrifice bandwidth (chapter 6). These factors could also be compensated by a narrower cleft, but that would increase cleft electrical resistance and reduce postsynaptic current. Thus, cleft width appears to optimally balance transmitter concentration at the postsynaptic receptors and electrical resistance (Savtchenko and Rusakov, 2007; Graydon et al., 2014). So here is symmorphosis at the nanometer scale.

Molecular mechanism of vesicle fusion

For vesicle fusion (*exocytosis*) to work at all requires multiple allosteric processes. And for it to transfer the information encoded chemically to an electrical signal, while preserving temporal precision and S/N, these allosteric processes must couple efficiently as now explained.

To preserve temporal precision, vesicle fusion must occur promptly as a triggered event. This requires *docking* it in advance to a specialized *active* zone and then *priming* the vesicle with multiple *SNAREs*, each a complex of four protein molecules. A SNARE, upon binding the vesicle tightly to the presynaptic membrane, adopts a high free energy conformation that is metastable (figure 5.4). Consequently, a small signal can push a SNARE over the hump on its energy landscape and trigger fusion.

The trigger is a surge of calcium ions reaching the docked vesicle through voltage-gated channels clustered at the active zone. When channels open in response to a presynaptic depolarizing electrical signal, several hundred calcium ions enter to raise the local concentration by 50-fold in less than 500 µs. Several calcium ions are bound by the protein *synaptotagmin* attached to the vesicle, which then binds to the SNARE and pushes it over the energy hump (Südhof, 2013). As the SNARE plummets to a lower energy conformation, the freed energy causes violent tugs on the vesicle. The combined force from three SNAREs suffices to fuse the vesicle to the presynaptic membrane and wrench open a pore with consequence already noted. The entire process, from presynaptic electrical signal to postsynaptic receptor activation, occurs fast enough to preserve temporal precision and to be completed within 600 μs.

Speed and temporal precision emerge from several design principles. The molecular components are irreducibly small and locate close together—within nanometers. This allows fast chemistry with irreducibly few molecules to achieve the high concentrations needed to transmit fast signals (high bandwidth). Chemistry achieves speed and gain by storing energy and then releasing it with concatenated switches: (1) voltage switch opens a calcium channel, releasing energy stored in calcium's electrochemical gradient (figure 7.4); (2) synaptotagmin binds calcium at low affinity, releasing energy stored within the SNAREs; (3) SNAREs fuse a vesicle, releasing energy stored by concentrating transmitter.

Timing is sharpened by cooperativity that steepens the response curves (figure 6.3): several voltage-gated calcium channels cooperate to establish a sufficient calcium concentration, several calcium ions cooperate to cause synaptotagmin to bind a SNARE, and several SNAREs cooperate to fuse a vesicle. Thus, via switches directing stored energy, the chemical signal recodes to electrical with good S/N and temporal precision.

In order to transmit high frequencies, a steeply rising chemical signal must also fall steeply. Thus, each stage must terminate quickly: (1) calcium channels close instantaneously as the membrane repolarizes; (2) calcium concentration collapses locally within tens of microseconds as calcium is bound rapidly by high-affinity buffering proteins; (3) synaptotagmin switches off sharply because of its steep dependence on calcium; (4) transmitter concentration decays within less than 1 ms by fast binding to transporter proteins on synaptic and glial membranes and by diffusing from the cleft.

In short, rapid release and rapid termination produce a chemical signal in the cleft that peaks within about 0.6 ms and lasts less than 1.5 ms, thereby transmitting information with irreducible delay and a bandwidth of about 1 kHz. This suffices to transmit most frequencies coded by the neuron's electrical signals because the membrane time constant is constrained by energy cost (chapter 6; Attwell & Gibb, 2005). Thus, the mechanism of chemical signaling at the synapse matches the bandwidth of the presynaptic neuron.

Vesicle release is stochastic

An action potential reaching a presynaptic terminal may cause a single vesicle to be released, or it may fail. Release is stochastic with a probability that

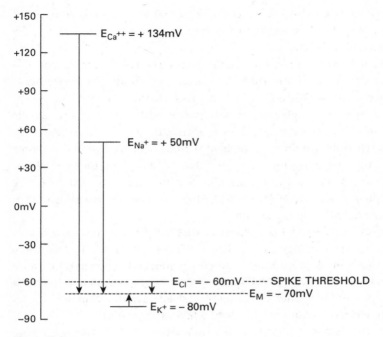

Figure 7.4
Driving forces on key ions at the neuron's resting potential. Calcium and sodium ions are driven strongly inward whereas the forces on potassium and chloride ions are weak until the membrane is depolarized. Then their driving forces increase with the degree of depolarization and tend to oppose it. Since a neuron must depolarize to release a graded chemical signal or to fire an all-or-none spike, inputs whose allosteric effects open calcium or sodium channels are excitatory, and inputs whose allosteric effects open chloride or potassium channels are inhibitory.

can vary between 0.1 and 0.9. The uncertainty of release introduces noise, thereby reducing information, but it can have advantages (Harris et al., 2012). For example, it reduces the likelihood that two vesicles will redundantly carry the same signal. It also offers a mechanism to adjust the effectiveness of a synapse—its *weight*—by tuning release probability. This provides mechanisms for homeostasis and plasticity (chapter 14). The costs and benefits of stochastic release are discussed further in later chapters.

Recovery and cost of presynaptic chemical signal
Vesicles fusing by exocytosis expand the presynaptic membrane, thus increasing its capacitance and time constant. To prevent this, the terminal retrieves each fused vesicle by *endocytosis*, folding inward the added patch

of membrane and pinching it off. This sounds simple, but, of course, it requires allosteric action by several types of proteins—to reverse what was accomplished by the SNAREs. Obviously, a synaptic terminal must maintain a strict balance between exo- and endocytosis.

Although a synaptic vesicle appears morphologically simple, it is really a complex molecular machine comprising about 800 protein molecules of about 40 protein species (Fernández-Chacón & Südhof, 1999). Beyond the structural membrane proteins and transporter proteins to fill it with transmitter, there are specific proteins for other tasks: to link the vesicle to neighbors near the presynaptic membrane for rapid recruitment to docking sites (Hallermann & Silver, 2013; Hallermann et al., 2010), to provide two of the proteins in each SNARE plus the synaptotagmins to trigger them, and to mark the vesicle for endocytotic retrieval. By specifically retrieving vesicle membrane, rather than nonvesicle membrane, the specific vesicle proteins are retrieved together, allowing the vesicle to be refilled and readied for rerelease within a minute.

Using allostery of chemically coupled proteins, the vesicle release mechanism is efficient in space, materials, and energy: to extrude the modest numbers of calcium ions, ~12,000 ATP; to energize the SNAREs, <100 ATP; to retrieve the vesicle, <500 ATP; and to fill the vesicle, ~11,000 ATP (Attwell & Laughlin, 2001).[4] Therefore, the total cost of a presynaptic chemical quantum is ~23,000 ATP. The postsynaptic electrical response to this signal costs roughly 10-fold more, as we now explain.

Postsynaptic electrical response

The transmitter molecules reaching receptor proteins across the cleft bind stochastically, and when two or more molecules bind cooperatively to the same protein, it changes conformation to open a channel (figures 6.6 and 6.7). The contents of one vesicle, a *quantum*, open about half of the available channels (figure 7.3). Therefore, enlarging the quantum by increasing transmitter concentration within a vesicle, or releasing several quanta (*multivesicular release*) can produce a graded increase in the fraction of open channels (see below, figure 7.17). Thus, the information packet presented to a postsynaptic receptor cluster is graded, as is the opening of channels that capture the amplitude and timing of the upstream event. Still in chemical mode, it is cheap.[5]

But now ions flow through the open channel with a direction and amplitude that depend on their electrochemical driving force (figure 7.4). When the transmitter is glutamate and the postsynaptic receptor is a ligand-gated cation channel (chapter 6), sodium ions, and in certain cases calcium ions,[6] are strongly driven inward to depolarize the membrane. The channel is also permeable to potassium, but its weaker driving force moves fewer ions outward. The net ionic current converts the graded chemical signal to a graded electrical signal. The cost is 10-fold more energy, but this is essential to send over distance at acceptable time delays. Recall that, whereas a graded chemical signal diffuses cheaply, about 1 μm in a millisecond, a graded electrical signal needs batteries but travels 50-fold farther in the same time.

The greater energy cost demands measures to reduce noise. This follows the principle *send only what is needed*, and that means sending the least possible noise. The noise sources include: stochastic release of vesicles, timing of vesicle fusion, size of a transmitter quantum,[7] times of receptor binding/ unbinding, and channels opening/closing (Ribrault et al., 2011). So the neuron takes various measures to mitigate them.

The number of molecules released in the quantal pulse from a vesicle varies across neuron types. The number depends strongly on vesicle diameter, d, since volume goes as d^3, and on the final intravesicle concentration, roughly 100 mM. Each neuron type selects a vesicle between 30 and 50 nm in diameter (figure 7.3). This allows a roughly fivefold range in number of molecules. Functional vesicles as small as 20 nm in diameter have been produced experimentally by genetic manipulation, but they contain fewer transmitter molecules than a 30-nm vesicle—insufficient to establish an effective postsynaptic concentration. Small vesicles with more transporter molecules might conceivably establish higher internal concentrations, but actually, overexpression of transporter proteins gives larger vesicles with similar mean concentration. Thus, the 30-nm vesicle seems to be a lower bound from fly to mammal (figure 7.3).

Increasing the number, M, of postsynaptic receptors improves postsynaptic S/N by $\sqrt{M}$ and reduces the time constant by charging the capacitance more briskly (chapter 6). A small synapse clusters about 20 receptors, improving S/N compared to one receptor by about 4.5-fold. A large synapse may expand the receptor cluster up to about 10-fold and thereby improve S/N by about 14-fold (figure 7.5).[8] But this threefold benefit comes with a 10-fold greater cost, so it is reserved for special purposes, such as auditory synapses that transmit with high temporal precision (chapter 10). In any case, the graded electrical signal now serves the neuron's next task: integrate and send an output.

Different protein receptors process on different timescales

A neuron must register events on different timescales and does so using postsynaptic receptors with different kinetics. Consider as an example one

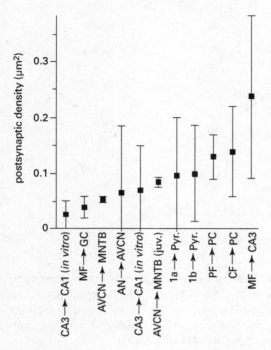

Figure 7.5
Postsynaptic receptor clusters from different synapse types span a 10-fold range of size. This reflects customized S/N for each type, according to needed benefit and cost. Area of postsynaptic density corresponds to size of the receptor cluster. CA3 → CA1= hippocampus; MF → GC = cerebellar mossy fiber to granule cell; AVCN → MNTB = anteroventral cochlear nucleus to medial nucleus of trapezoid body; AN → AVCN = auditory nerve to anteroventral cochlear nucleus; 1a, 1b → Pyr = pyriform cortex; CF → PC = climbing fiber to Purkinje cell; PF → PC = parallel fiber to Purkinje cell; MF → CA3 = hippocampal mossy fiber to hippocampal pyramidal cell. Reprinted with modification and permission from Xu-Friedman and Regehr (2004).

broad family of ligand-gated cation channels, the glutamate receptors (Attwell & Gibb, 2005). Two types, the NMDA receptor and the AMPA receptor,[9] bind glutamate with similar ON rates, but the NMDA receptor's OFF rate is 400 times slower. This slower OFF gives the NMDA receptor a more sustained response that covers a longer timescale (figure 7.6); it also causes a 400-fold higher sensitivity to glutamate: whereas an AMPA receptor requires glutamate concentrations of nearly 1 mM to cooperatively bind two glutamate molecules, an NMDA receptor is doubly bound at about 1 μM.

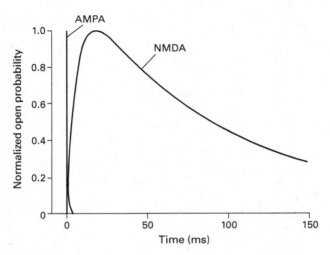

Figure 7.6
Neurons use different types of receptor to encode a range of temporal frequencies.
Graph shows time course of activation for AMPA and NMDA receptors by a 0.3-ms
pulse of 1 mM glutamate. To more finely optimize binding for a range of high frequencies, different AMPA types are employed along with different regulatory molecules. Reprinted with permission from Attwell & Gibb (2005).

The AMPA receptor covers a neuron's shortest timescale (figure 7.6). Its
high rates for glutamate ON and OFF match fast rate constants for channel
opening and closing, so an AMPA receptor completes its electrical response
with a time constant of about 0.8 ms. This speed preserves the temporal
information transmitted by fast vesicle release. It also matches the lower
limit to a neuron's membrane time constant imposed by the energy cost of
a low membrane resistance (time constant = capacitance × resistance). In
other words, an AMPA receptor equips an excitatory synapse to transmit
the full bandwidth of neuronal response.

For AMPA receptors to cover this fast timescale and wide bandwidth,
glutamate at the postsynaptic receptor site must decline promptly. Glutamate diffuses about 1 μm per millisecond, so by restricting AMPA receptors
to a small postsynaptic patch, less than 1 μm in diameter, its concentration
falls by e-fold within a millisecond (figure 7.3). Glutamate accumulation
during successive releases is prevented by active uptake. Transporter proteins on the surrounding neuron and glial membranes bind glutamate rapidly and move it into the cell, powered by the influx of three sodium ions.
The transporter works slowly, requiring a full minute to retrieve the 4,000

glutamate molecules released by one vesicle. Therefore, to mop them up on the required millisecond timescale requires transporters to be concentrated around the synapse: to bind 4,000 glutamate molecules within a millisecond requires 10,000 transporters that bind with high affinity (Attwell & Gibb, 2005).

To efficiently match components, AMPA receptors locate near the vesicle release site, within about 20 nm, where they see a fast, high peak and steep decay of glutamate concentration (figure 7.3). Thus, the receptor's molecular structure is matched by vesicle content and distance from the release site (figure 7.3). NMDA receptors in certain cases locate farther from the release site where they see slower rise and fall of glutamate concentration, allowing them to integrate contributions from successive release events. Their sustained responses can also be integrated with fast AMPA receptors (Clark & Cull-Candy, 2002).

The NMDA receptor's sustained response allows it a role in detecting *coincidence detection*. This exploits the NMDA receptor's voltage sensitivity. A receptor, binding glutamate when the membrane is at resting potential (–65 mV), reacts weakly because its channel is partly blocked by a positively charged magnesium ion. When the membrane depolarizes—for example, due to AMPA-mediated currents from synapses on the same dendrite—magnesium is forced out, allowing sodium and calcium to enter. Thus, an NMDA receptor detects the coincidence of a presynaptic input (glutamate) AND a postsynaptic response (depolarization) within a time window of about 100 ms set by the slow unbinding of glutamate.

The NMDA receptor's ability to detect and signal coincidence equips a neuron for pattern recognition and learning (chapter 14). An active receptor emphasizes the coincidence by amplifying and extending a synapse's excitatory input; moreover, it marks the synapses whose signals coincide. Only synapses that recently delivered glutamate have NMDA receptors primed for action. When these receptors are unblocked by depolarization, they admit chemical messengers (calcium ions) that initiate structural change. Because an action potential also depolarizes synapses, the NMDA receptor enables a neuron to take a first step in learning; it can identify and modify those synapses whose inputs coincide with a definitive output.

The duration of an NMDA receptor's time window is critical for learning. Shorter would increase false negatives—the receptor would miss correlations between events that take longer to unfold. Longer would increase false positives—more unrelated events would occur in the same time window. The NMDA receptor's OFF rate creates the 100-ms time window that seems about right for many of life's more immediate events.

However, this mechanism creates a problem (Attwell & Gibb, 2005). The slow OFF rate retains some glutamate bound until nearly every last molecule has been removed from the synaptic cleft. This requires a transporter to harness the energy of *three* sodium ions: two sodiums can pull extracellular glutamate down to 180 nM, but this leaves 13% of NMDA receptors still bound. Thus, for NMDA's time window to close within 100 ms requires a transporter with appropriate stoichiometry—at a cost of 50% more energy.

Intervals still longer than the NMDA receptor's time window are covered by a *metabotropic* glutamate receptor, *mGluR*. This receptor belongs to the same molecular family as the β-adrenergic receptor (chapter 5), and like that receptor, it activates a G protein to deliver an amplified signal. Responses driven by mGluR can be tuned to cover a range of time intervals, from about 0.1 s to tens of seconds. Moreover, mGluR's second messengers can, like calcium from the NMDA receptor, institute longer lasting structural changes.

Processing multivesicular release

We have explained how receptors process on different timescales by varying their output to a singular input, a puff of glutamate. Receptors can also detect variations in input timescale, produced by different temporal patterns of vesicle release, by varying the kinetics of receptor activation, deactivation, and desensitization. For example an AMPA receptor desensitizes in response to prolonged glutamate, causing the response to a sustained burst of action potentials to decline over time. This fast desensitization favors signals that change on a short timescale, thereby tuning AMPA's bandwidth to higher frequencies and eliminating redundancy. Conversely, an mGluR activates slowly and does not desensitize, thereby favoring inputs that change on longer timescales.

In summary, the glutamate receptor families enable a neuron to process on different timescales by producing synaptic responses of different durations. Receptors are constructed from different parts, that is, from different combinations of a receptor's protein subunits, to provide the key differences in kinetics, sensitivity, and output: an AMPA receptor that unbinds glutamate at a high rate to create a narrow window and desensitizes to favor high frequencies; an NMDA receptor that is voltage sensitive and delivers calcium ions; and an mGluR that acts more slowly via second messengers. But simply engineering receptors is insufficient. A receptor's kinetics and sensitivity must be matched by the stoichiometry and affinity of transporter molecules, by their density around a synapse, and by the dimensions of the synapse itself (Attwell & Gibb, 2005). This conclusion, that to be

effective requires design in depth, is more amply illustrated by photoreceptors (chapter 8).

Efficiency from synaptic inhibition

Neurons employ various forms of synaptic inhibition. All increase efficiency by improving timing precision (Sengupta et al., 2013) and reducing
redundancy. These effects all help concentrate information, so that the
space and energy used to send expensive electrical signals are well spent.
Inhibition also serves to delete information unneeded by a downstream
user. These effects are discussed with specific examples in chapters 9 through
12. Here we explain the mechanisms and why they are cheap.

One type of synaptic inhibition is achieved when transmitter binding
opens a membrane channel for chloride ions. The open chloride channel
reduces the depolarization produced by an inward cationic current and
hence reduces the probability of triggering an output—a vesicle release or
spike. The inhibition is achieved in two ways. First, when E_{Cl} is negative to
the membrane potential (figure 7.4), chloride enters and neutralizes the
charge carried by entering cations. Second, irrespective of whether chloride
flows in or out, the open chloride channel lowers membrane resistance,
thus shunting the depolarization. This effect dominates when chloride currents are small, which is generally so because E_{Cl} is generally near E_M (figure
7.4). Small currents require less restorative ion pumping, which makes
chloride's *shunting inhibition* energy efficient.

Shunting inhibition is also achieved by opening a potassium channel. As
for chloride, the potassium current is small because E_K is near E_M (figure 7.4).
The inhibitory potassium channels are not gated by chemical transmitter
but rather by G proteins, membrane voltage, or calcium. Thus, they add
substantially to the parts list for energy efficient inhibition.

The transmitters for ligand-gated chloride channels are GABA, glycine,
and histamine (figure 6.7). The GABA receptors (*GABA$_A$*) comprise a diverse
family of molecules. The receptor assembles as a pentamer from several
classes of subunit (alpha, beta, gamma, etc.), each of which has subtypes.
This permits customized properties, such as different binding constants,
different speeds of opening, and different rates of desensitization. The
GABA$_A$ receptor's ligand binding is modulated allosterically at several sites
on the molecule by brain chemicals, such as steroids and by various exog-
enous chemicals, such as alcohol, barbiturates, and benzodiazepine "tran-
quilizers." This suggests, by analogy with endogenous modulators of other
types of receptor (endogenous opiates and endocannabinoids), that there
should be endobenzodiazepines. One such molecule has been reported, a

secreted protein, *diazepam binding inhibitor*, that potentiates the GABA$_A$ receptor (Christian et al., 2013).

Dendrites expand a neuron's information capacity

To gather more information, a neuron must supply more membrane for synaptic contacts while minimizing the length of wire devoted to connecting (chapter 13). The design solution is to grow dendrites, which also offer compact compartments for electrical and chemical computing and for integrating signals. A dendrite is structured like an electrical cable—with a conducting core (cytoplasm) and outer insulation (the membrane's lipid bilayer). Voltages decay exponentially on a cable (figure 7.7) with a length constant[10] that depends upon resistance per unit length of insulating membrane, r_M, and conducting core, r_{cyt}:

$$length\ constant = \sqrt{(r_M/r_{cyt})}. \tag{7.1}$$

A dendrite does not transmit far: r_M is too low because potassium channels stay open to maintain resting potential, and r_{cyt} is too high because hydrated ions in cytoplasm conduct poorly. Therefore, a dendrite's length constant is generally less than 1 mm, and dendrites preserve signal amplitude by staying shorter than their length constant.

A dendrite may increase its length constant by growing thicker, thus reducing r_{cyt}, but the improvement goes only as the square root of diameter (Koch, 1999). With length constant increasing as $\sqrt{d}$ and volume increasing as $d^2 \times length$, the total cost of space and materials increases as $(length)^5$. Such a steeply diminishing return requires designs that keep dendrites short (chapter 13). Temporal resolution (bandwidth) also requires short dendrites because longer ones increase capacitance. This delays signals and spreads them out, thus attenuating high frequencies (figure 9.9). In short, signals traveling passively on a dendrite longer than 1 mm would be too weak and too slow to carry much information. Yet these cable properties that limit dendritic length can be exploited to process information as it is gathered.

Dendrites process directly

Dendritic biophysics provides cheap and robust analogue processing (Koch, 1999). For example, by placing input synapses that carry less information distally on the dendrite, they can be given less weight in the final output, whereas signals that carry more information can be given more weight by placing them nearer the cell body (figures 11.15 and 11.16). More generally, inward currents from excitatory synapses can be combined with outward

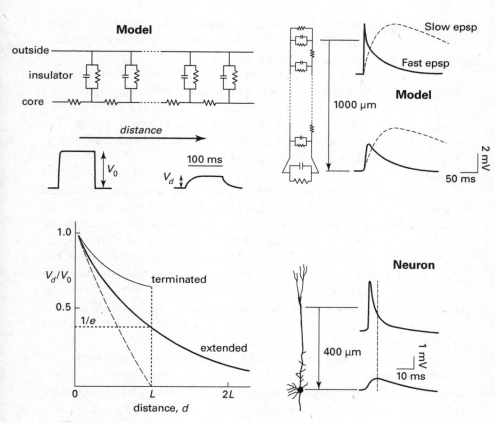

Figure 7.7
Passive transmission by dendrite. Upper left: Passive cable modeled as series of thin
segments, each contributing to resistance and capacitance of insulator (dendrite's
membrane), and to resistance of conducting core (dendrite's cytoplasm). **Middle left:**
Cable smooths square pulse as it transmits over distance, d, and attenuates steady-
state voltage amplitude from initial value, V_0, to V_d. **Lower left:** Amplitude transmitted, V_d/V_0, decreases exponentially when cable is extended many length constants,
L (middle curve, steady-state response of *semi-infinite cable*). Transmission improves
when cable is terminated and end is sealed (top curve). Transmission worsens when
end is open and short circuits core's signal (bottom curve). **Upper right:** Model of
passive dendritic cable transmitting to cell body (larger resistor and capacitor at base
of cable). Fast EPSP severely attenuated, slow EPSP less so. **Lower right:** Neuron transmitting fast EPSP to cell body via dendrite. Fast EPSP is smoothed, thereby delaying
time to peak (dashed vertical line), and attenuated. Cable transmission curves after
Rall (1959). Modeled transmission of fast EPSP and slow EPSP replotted from Spruston et al. (1998). Cortical pyramidal cell morphology and recordings of EPSP adapted
from Stuart & Spruston (1998) with permission.

currents from inhibitory synapses and potassium channels. Thus, a dendrite serves as an analogue electrical circuit that adds, subtracts, divides, multiplies, and takes logarithms (figure 6.2).

Such operations implemented directly by an RC circuit combine many synaptic inputs to produce an output within milliseconds. This rapid many-to-one integration, which serves behavioral requirements for prompt decision, would be difficult to implement with a chemical circuit. The many-to-one ability is further exploited by joining several dendrites to the cell body or an integrating segment to form a final common output (figure 7.1).

To increase their processing abilities, dendrites *complicate their design* (Branco & Häusser, 2010). For one example, a dendritic twig at the distal tip of an elaborate dendritic tree makes a coincidence detector by strategically expressing voltage-sensitive sodium channels (Harnett et al., 2013). Glutamate released at excitatory synapses binds AMPA and NMDA receptors, but only when AMPA currents from several synapses coincide does the dendritic twig depolarize sufficiently to unblock the NMDA receptors. Through them, calcium and sodium ions enter to amplify the coincidence, and voltage-gated sodium channels register it by producing a robust pulse, an action potential.

This pulse does not propagate far because the sodium channels are confined to the twig. However, it generates sufficient current to drive a detectable signal into the larger dendritic tree. Here, strategically positioned potassium channels shunt responses from particular twigs and branches, thereby selectively blocking some inputs or controlling their gain. Thus, more complicated dendrites provide two layers of processing, local within a single dendrite and global within the larger dendritic tree.

A dendrite can add another layer of processing by receiving a synapse on the globular head of a dendritic spine, 0.5–1.5 μm in diameter (figure 7.8; Yuste, 2013; Sala & Segal, 2014). Evoked current and chemical messengers pass from spine head to dendrite through a thin neck, 0.05–0.25 μm in diameter and 0.5–2 μm long. The neck resists the flow of intracellular current and chemicals, thereby creating a computing compartment in the head that can be regulated by varying neck diameter and length. This design neatly resolves two conflicting demands.

A synapse must merge its current with many other currents in an extended RC network (dendrite and tree). In doing so, it loses individuality because the amplitude of its *EPSP (excitatory postsynaptic potential)* depends largely on the state of the network, as determined by the multitude of synaptic and voltage-gated conductances (Yuste, 2013). Yet, a synapse must

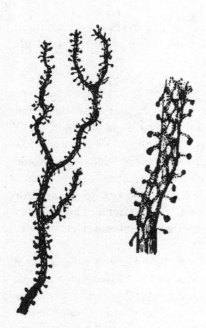

Figure 7.8
Purkinje cell dendrites studded with spines of varied morphology. Reprinted from
Ramón y Cajal (1909).

also *adapt, match, learn, and forget*. For this, it must retain individuality to
monitor and adjust its own EPSP (chapter 14). A specific example is
explained for the retina's horizontal cell (chapter 11, figure 11.5).

The spine neck's variable resistance allows it to adjust the amplitude of
the EPSP in the spine head.[11] This robust private measure of output can
then drive electrical and chemical circuits in the head that modify the output to adapt, match, learn and forget (chapter 14). The head's small volume
allows chemical processing to be fast, reliable and efficient. A spine also
increases wiring efficiency by extending a dendrite's reach at minimum
cost in space and materials (chapter 13, figure 13.3).

In short, dendrites collect and process analogue signals before final
recoding to faster, regenerative pulses for transmission down the axon. But
dendrites can also process in reverse. A dendritic tree may express voltage-
gated sodium and calcium channels that allow spikes triggered in the
axon to propagate back into the tree. This sends information about the
neuron's output back to the dendrites where it serves various purposes,
for example, to strengthen the synapses that generated the neuron's

output—for learning (chapter 14). Now we consider the design to minimize information loss in converting dendritic analogue to a pulse code at the axon, for faster transmission over distance.

The axon converts to a pulse code

Having gathered and processed analogue signals from multiple dendrites, the neuron needs to send the information over long distance to its synaptic terminals via a single specialized cable (figure 7.1). Passive conduction of a graded signal is too slow, and its exponential decay would lose information to noise (figure 7.7). Consequently, for distances greater than one length constant (~1 mm), an axon recodes to all-or-none pulses—action potentials (figure 6.10). These travel faster and avoid decay by regenerating as they go. Information encoded as spikes travels far with little loss.

However, the step of recoding from analogue to pulses loses a lot of information—as much as 90% (Sengupta et al., 2014). Whereas an analogue signal tracks changes continuously by allowing several response levels (figure 5.3), pulses allow the membrane potential to change intermittently between just two levels (figure 3.5). Thus, a neuron using analogue can transmit more than 2,000 bits s^{-1}, whereas using spikes, it can manage fewer than 500 bits s^{-1}. Furthermore, a 100-mV spike costs far more energy (chapter 6), so recoding to pulses massively reduces energy efficiency (bits per ATP).

A neuron tries to minimize these losses by converting to spikes at a specialized initiation site.[12] This *initial segment* locates at some distance from the cell body's large capacitance to shorten the membrane time constant (see below, figure 7.11). The initial segment also packs membrane channels extra densely to further reduce the time constant and increase S/N. With higher S/N and shorter time constant, the site triggers spikes faster and more reliably, both of which increase bits per spike, thereby reducing loss of information and improving efficiency.

To increase bits per spike right at the initial segment is critical because transmission down the axon is expensive. A transmitted spike charges the entire axon's membrane capacitance by about 100 mV by admitting sodium ions, and to pump them out costs 6.3×10^3 ATP per square micrometer of membrane (chapter 6). A pyramidal neuron in cerebral cortex uses an irreducibly thin axon (~0.3 μm in diameter) to send spikes through its intracortical circuit. Even so, the cost per spike is 6×10^6 ATP per millimeter, and over its length of 4 cm, the cost is 2.4×10^8 ATP (Attwell & Laughlin, 2001).

Considering the full population of pyramidal neurons, their spikes account for more than 20% of the energy used to process information in cortical circuits (Sengupta et al., 2010; Howarth et al., 2012).

In short, the conversion of synaptic and dendritic analogue signals to axonal spikes creates an expensive bottleneck. Bits are limited and costs increased, which makes it imperative to use each spike efficiently according to principles of neural design. In particular, an axon should *send only what is needed* by reducing redundancy and noise and by limiting transmission to what downstream neurons need to know (chapter 11). Thus, the carving out of essential features for transmission is a critical process in which the final step, the conversion of analogue to spikes by voltage-gated channels, plays a decisive role (Aguera y Arcas et al., 2003).

Supply and recycling

The neuron cell body must continually deliver fresh organelles, such as mitochondria, vesicles, and rafts of receptor proteins, outward to distant dendritic and axonal arbors. These nether regions must return recyclable materials and inform the cell nucleus of their needs.[13] Traffic in both directions needs to be faster than diffusion—at rates up to about 10 mm per hour. For efficiency's sake, bidirectional traffic goes along the same track, a protein monorail.

The *microtubule*, constructed from subunits as a cylindrical polymer, is irreducibly fine, about 30 nm in cross section. Two molecular motors, *kinesin* and *dynein*, step along the tubule, ferrying cargo, in opposite directions. The motors operate with lever arms that, jutting orthogonally from the tubule, require clearance; therefore, microtubule spacing can be no closer than about 50 nm. Attach some cargo, and the circumferential zone about the tubule needs to be about 30 nm. These molecular structures set a lower bound to the caliber of a neuronal process. The finest axons at about 100 nm are just thick enough to accommodate a single microtubule plus its molecular motors and cargo. Axons are encouraged to shrink because space and energy costs rise as d^2, but eventually they hit this lower bound because of their irreducible need for fast transport.

Transport along the microtubule monorail is relatively cheap. Each 8-nm step of kinesin down the tubule costs 1 ATP, so to move a cargo 1 mm costs about 10^5 ATP. Although this might seem expensive, the molecular stepping motor is slow, so the cost per second is only about 100 ATP, far lower than the cost of electrical signaling by an ion channel (chapter 6).

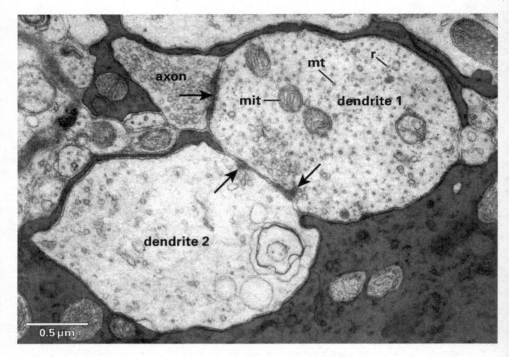

Figure 7.9
Local computing by chemical axodendrodendritic synapses. Axon terminal contacts dendrite 1 that contacts adjacent dendrite 2 that feeds back to dendrite 1. Note the array of evenly spaced microtubules (mt) in cross section and glial wrappings (shaded). Tiny dots are ribosomes (r) that support dendritic protein synthesis. Reprinted with permission from Famiglietti (1970).

Variations on the standard design

The standard polarized design requires every synaptic input voltage to travel a fair distance along a dendrite (up to a millimeter) over a significant time (2–20 ms) and then still longer distances along an axon, costing additional time and energy. Computations that avoid this long, slow loop would save considerable resources. Thus, neural designs include various features for computing locally, both at the synaptic and neuron levels.

Synapses designed for local computing

Certain designs allow direct computing between dendrites. A dendrite may form a chemical synapse onto a neighboring dendrite. The presynaptic

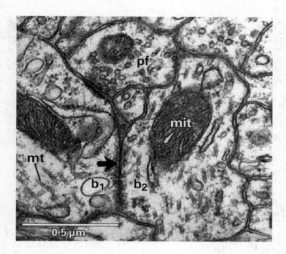

Figure 7.10
Local computing via electrical synapse between two basket cell dendrites. When chemical synapse from parallel fiber (pf) depolarizes b_1, current flows across the electrical synapse (arrow) to instantaneously and inexpensively depolarize b_2. Rat. Reprinted with permission from Sotelo & Llinás (1972).© 1972 Rockefeller University Press.

dendrite releases transmitter onto the postsynaptic dendrite, which may reciprocate with a return chemical synapse to its neighbor (figure 7.9). Such *dendrodendritic* chemical synapses are employed by circuits in many brain regions, including spinal cord, thalamus, olfactory bulb, superior colliculus, retina, and cerebellar cortex (Shepherd, 2004). They are efficient because (1) computations remain in analogue mode, which is direct and cheap; (2) expensive long-distance electrical signaling is avoided; and (3) different branches of the same neuron can compute independently, thereby expanding by up to 100-fold the computational possibilities of a single neuron (Grimes et al., 2010)

Dendrodendritic synapses can also be electrical, via a *gap junction*. This connection passes current directly between dendrites through a transcellular channel (*connexon*). It is fast—essentially instantaneous—because it dispenses with all the steps needed by a chemical synapse that take a millisecond or longer. Moreover, it does not amplify, so it is energetically cheap, and, as evident in figure 7.10, it requires no space at all. This type of synapse, illustrated here for cerebellar cortex, is ubiquitous. Specific computational functions will be treated in chapter 11.

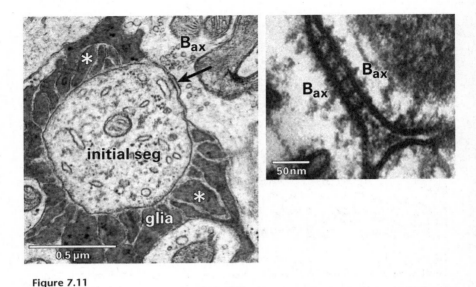

Figure 7.11
Two types of inhibitory synapse at the Purkinje cell axon initial segment. Left: Basket
cell axon terminal (Bax) forms a chemical synapse (GABA) onto Purkinje cell initial
segment. Glial fingers form a sheath around the axon. These fingers contain dense
arrays of potassium transporters that rapidly bind potassium released during the action potential. Rat. Reprinted with permission from Palay and Chan-Palay (1974).
Right: Basket cell axon tendrils form *septate-like junctions* with high resistance cross-
bridges. Their capsule surrounds the glial fingers and apparently elevates the extracellular resistance. Basket cell spikes that synchronously invade the tendrils depolarize the extracellular space, thus reducing the intracellular depolarization to modulate
spike timing. Reprinted with permission from Sotelo & Llinás (1972).

Local computing is also accomplished by axoaxonic synapses. One
design delivers a chemical synapse to an axon's initial segment (figure
7.11). This site, as noted, critically regulates spike frequency and timing, so
a chemical synapse can act powerfully without consulting the neuron's
slower and costlier integrative apparatus. Another design uses a special
form of electrical inhibition—fast and cheap. The cerebellar basket cell
axon uses both designs, as now explained.

The basket cell axon, having richly enveloped the Purkinje cell body
(figure 7.1), sends fine tendrils to surround the initial segment (see below,
figure 7.15). The tendrils penetrate the glial wrapping and deliver a chemical synaptic contact that releases GABA (figure 7.11). Additionally, the tendrils join together using high resistance cross-bridges to form a capsule
surrounding the glial fingers (figure 7.11). The capsule apparently elevates

extracellular resistance so that basket cell spikes, locally synchronized by the dendrodendritic electrical synapses (figure 7.10), invade the fine tendrils and depolarize the extracellular space. This instantaneously reduces the initial segment's intracellular depolarization and modulates spike timing (Korn & Axelrad, 1980). Polarizing the extracellular space to control a neuron's output is direct, economical, and relatively noise free—so it is used elsewhere (chapters 9 and 11).

Other types of local computing use an axoaxonic synapse, directed not to the initial segment, but rather to another synaptic terminal. Matched to function, these can be either electrical or chemical, and sometimes both are combined at the same junction. Electrical junctions between synaptic terminals of the same type improve S/N (chapter 11, figure 11.4). They also increase temporal precision because, as one synapse depolarizes, its coupled neighbor draws off some current, advancing its own depolarization and retarding the first (Pereda, 2014). This also increases synchrony between coupled terminals, a valuable property in many circuits achieved directly at negligible cost in space and energy.

A chemical axoaxonic synapse, depending on transmitter and receptor type, can be excitatory or inhibitory. For example, a retinal axon terminal releases glutamate onto AMPA receptors at an axon terminal of a thalamic interneuron (figure 12.4). Another synapse may release GABA onto a $GABA_A$ receptor, which gates a chloride channel. This connection is usually inhibitory because a particular protein pump (KCC2) sets E_{Cl} near the resting membrane potential (figure 7.4). However, certain terminals express a different pump (NKCC1) that sets E_{Cl} positive to the resting potential. In this case a synapse matching GABA to $GABA_A$ will *depolarize* the postsynaptic terminal and be excitatory. This allows a circuit to diametrically reverse its function, not by altering the anatomical structure, nor the transmitter, nor its receptor. Instead it simply swaps chloride pumps.

Neurons designed for local computation

Neurons differing from the standard polarized design are numerous, so here we note from retina two radical alternatives. One type radiates dendrites symmetrically from its cell body to collect chemical synaptic inputs. This *starburst* neuron lacks an axon but forms chemical outputs at the distal dendritic tips (figure 11.24). The design is such that a visual stimulus moving centrifugally, from cell body toward the dendritic tips, releases GABA onto a ganglion cell and thereby blocks its spiking to that direction of motion (figure 11.24).

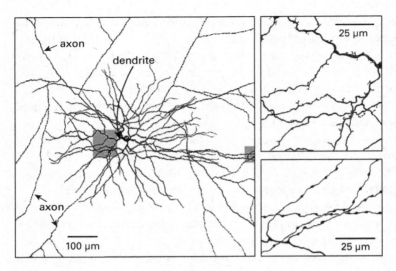

Figure 7.12
Polyaxonal amacrine neuron reverses the standard polarized design. Left: Instead of collecting input on dendrites and funneling to a single axon, it radiates multiple axons from distal dendritic tips. Gray boxes indicate regions shown at higher magnification. **Upper right:** Dendrites express spines for receiving inputs. **Lower right:** Axons express varicosities for sending outputs. Reprinted with permission from Davenport et al. (2007).

This neuron breaks another rule: whereas most neurons release only one chemical transmitter, the starburst neuron releases *acetylcholine* in addition to GABA. These transmitters are packaged by different transporters into different vesicles that cluster at different presynaptic sites and contact different dendrites. Thus, a starburst process, which computes locally and connects locally, can, by releasing different transmitters onto different neurons, evoke opposite responses to the same stimulus. One computation produces two outcomes for the same price.

Another radical design radiates dendrites symmetrically about the cell body to collect local information; then, each dendrite radiates an axon from its distal tip to broadcast the information over millimeters (figure 7.12). This *polyaxonal amacrine* neuron is polarized, but in reverse. The cell body, rather than converging information for a single axon, diverges via multiple axons in all directions (figure 7.12).

We conclude that the core rule for designing a neuron is to build it for a particular task. This achieves the needed performance for least cost (chapters 9, 12, and 13).

Glial cells in design of neurons

Glial cells comprise a substantial fraction of the brain's volume (Halassa & Haydon, 2010). In white matter (tracts) *astrocyte* cell bodies and processes use more than 30% of the space. Myelin sheaths occupy an additional 25%, and *oligodendrocyte* cell bodies that provide the myelin wrapping use an additional 13%. Thus, in tracts the space allotted to glia comes to about 65% (figure 13.21; Perge et al., 2009). In gray matter (circuits) the fraction for astrocyte processes varies locally by design, but overall is about 10%, plus some added allowance for cell bodies (Mischenko et al., 2010; Schüz & Palm, 1989). So on average, the total space for glia in gray matter is roughly 15%.

Glia expend considerable energy. For example, the mitochondrial volume fraction of astrocyte processes in white matter is more than 3%, more than twice that of the myelinated axons. In the optic nerve astrocytes contain more than 70% of the mitochondria (see below, figure 7.21; Perge et al., 2009). In gray matter less than 5% of the mitochondria are in glia, but gray matter processes information in dense neural circuits, so its overall metabolic rate per volume is threefold higher (Attwell & Laughlin, 2001; Harris & Attwell, 2012). Given glia's substantial costs in space and energy, what are the benefits to neural design?

White matter: Benefits of myelin and astrocytes

A naked axon conducts action potentials efficiently, but conduction velocity rises only as √d. Therefore, where speed is required, the naked axon must become inordinately thick. This cost is accepted for a command neuron that triggers the escape response of an invertebrate; most famously, the squid giant axon is about 1 mm in diameter. This works if there are only a few giant axons, but they could not be used routinely because they would take far too much brain space. Yet vertebrates move fast and need many fast axons.

When speed requires an axon thicker than about 0.5 μm, the design solution is for an oligodendrocyte process to wrap a segment of the axon in a multilayered, jelly roll of plasma membranes. This is myelin. Its multiple layers effectively reduce the axon's membrane capacitance and increase its resistance. This increases space constant and reduces time constant. These improvements allow the advancing foot of the action potential to spread further and faster. Thanks to myelin wrapping, action potential velocity increases in direct proportion to axon diameter at about 6,000 mm/s per micron diameter.

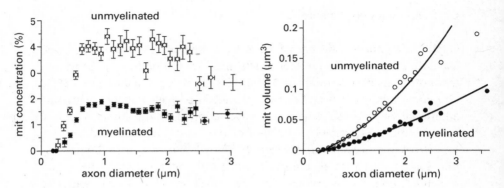

Figure 7.13

Myelination saves energy as well as conduction time. Left: In myelinated axons thicker than about 0.7 μm mitochondria occupy a constant proportion of cytoplasmic volume, about 1.5%. In unmyelinated segments of the same axons mitochondria occupy about 4% of cytoplasmic volume. **Right:** Mitochondrial volume per unit axon length rises linearly with diameter for fine axons (d < 0.7 μm) and quadratically for thicker ones. This figure compares ganglion cell axons within the retina, where they are unmyelinated, to their continuations in the optic nerve where they are myelinated. Thus, both plots represent the same neuron types. Reprinted with permission from Perge et al. (2009).

The myelin-wrapped segment extends for about 0.5–1 mm, so as the voltage pulse flashes passively across this distance, it soon encounters a naked spot of neural membrane (*node of Ranvier*) which concentrates sodium channels. These are of a specific type, Nav1.6, that open rapidly and synchronously to regenerate the action potential—which then continues its passive course to the next node. The sodium channels pack so densely at the node (up to $2,000/\mu m^2$) that the potassium channels needed to repolarize are displaced laterally.

Nodal spacing increases directly with axon diameter. This works because thicker axons produce larger nodal currents and increase the number of myelin wraps, further increasing the space constant. In systems where spike arrival time is critical, nodal spacing can be tweaked to compensate for different conduction distances (Cheng & Carr, 2007; Carr & Boudreau, 1993). One might imagine that concentrating sodium channels at a few sites rather than distributing them over the whole axon would save energy, and this proves to be so (figure 7.13). This saving, though substantial, does not begin to explain the threefold difference in energy cost of white matter compared to gray matter and, thus, its far sparser supply of blood vessels. That is explained by the absence of synaptic currents (Harris & Attwell, 2012).

Astrocytes in white matter are critical to the design that generates an efficient action potential. Following a spike, the axon repolarizes by releasing a pulse of potassium into the nodal extracellular space. Because neighboring axons are all firing, the concentration gradient needed for the pulse to diffuse from the node is diminished. Therefore, potassium must be removed rapidly by sodium-potassium pumps (chapter 6). But where to place them?

Space at the nodal axon membrane is so fully occupied by sodium channels that the potassium channels are displaced laterally. Thus, the nodal membrane cannot accommodate large numbers of pump molecules. So the design places many fast-binding pumps on the astrocyte membranes and some slower ones along the axon (Ransom et al., 2000). Thus, astrocytes in white matter are key to rapid removal of extracellular potassium, and this may explain their large proportion of space and energy capacity.

Gray matter: astrocyte compartments for transmitter diffusion

Certain computations benefit when neighboring synapses operate independently of each other. Other computations benefit when neighboring synapses share their transmitter. The degree of independence versus sharing is set partly by the degree of astrocyte wrapping. Certain types of synapse are individually wrapped, allowing each pulse of transmitter to bind the postsynaptic receptors and then to be removed by transporter proteins on the astrocyte membranes (see below, figure 7.17). Other types are poorly wrapped, allowing longer persistence and spread of transmitter to neighboring synapses (see below, figure 7.16).

Still other types of synapse provide multiple synaptic contacts from closely spaced release sites that are all wrapped together by a glial capsule (figure 12.4). The glial membranes densely express transporter proteins for the particular transmitter released within the capsule (Josephson and Morest, 2003). This allows pulses from one release site to spill over to neighboring receptor patches with consequences to be discussed below. Such *glomerular synapses* are used in various locations, including spinal cord, cochlear nucleus, thalamus (figure 12.4), and cerebellum (see below, figure 7.16).

Other tasks for glia

Astrocytes display myriad other properties. For example, they respond to neuronal activity and neurotransmitters via G-protein-coupled receptors. Moreover, they release gliotransmitters, such as glutamate, D-serine and ATP, which act on neurons. Astrocyte-derived ATP can modulate synaptic

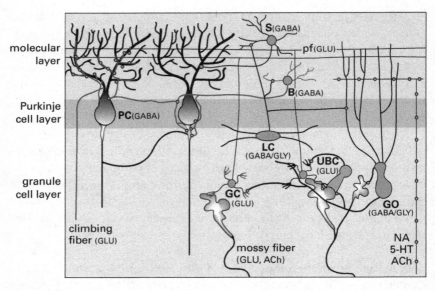

Figure 7.14
Wiring diagram of cerebellar cortex. Granule cell (GC) integrates excitatory contacts (from distant mossy fibers and local unipolar brush cells (UBC)) with inhibitory contacts (from Golgi neurons (GO) with excitatory contacts from parallel fibers). Purkinje cell (PC) integrates excitatory contacts from a single climbing fiber and 175,000 parallel fibers (pf) with inhibitory contacts from stellate (S) and basket (B) cells that receive inhibitory input from the Lugaro cell (LC). Purkinje cell sends recurrent GABAergic contacts to axon initial segment of neighbors. The basket cell extends fine processes to encapsulate the Purkinje axon initial segment, an arrangement associated with electrical inhibition. GLU, glutamate; Ach, acetylcholine; GLY, glycine; NA, noradrenaline; 5-HT, serotonin. Redrawn from Sotelo (2008).

transmission, either directly or through its metabolic product, adenosine (Schmitt et al., 2012). Astrocytes are also interposed between blood vessels and neurons and thus play important roles in regulating metabolic responses (Howarth, 2014). This does not exhaust the list of properties and contributions of glial cells to neural function. However, understanding remains too incomplete to fully grasp how the various features contribute to efficient design.

Each neuron's design serves a larger circuit

Cerebellum illustrates the extent to which neurons and glia are adapted for specific functions within a larger circuit. A schematic diagram of cerebellar cortex identifies the circuit's four types of excitatory neuron, and four types of inhibitory neuron (figure 7.14). Although a complete account of their cooperative effort remains a goal, enough is already known to see how some features of their functional architecture serve efficient processing.

The cerebellar circuit performs two operations. First, it remaps information coded at high mean rates by a modest number of mossy fibers to much lower mean rates carried by a much larger number of granule cells. This occurs in the inner synaptic layer. Second, it formulates an output by a quite small number of Purkinje cells in the outer layer. The granule cells project their individually sparse representation into the outer layer via a massive array of axons, running parallel to save wire (chapter 13). Each Purkinje neuron integrates single synaptic inputs from 175,000 parallel fibers to send an output pattern via its axon. To efficiently implement these two operations—remap and send an output—requires different functional architectures.

Functional architecture of inner synaptic layer

The inner synaptic layer densely packs astronomical numbers of granule cells in small clusters (figure 7.15). The cell bodies are irreducibly small, 6–7 μm in diameter. The cell body is filled almost completely by the nucleus, leaving a mere crescent of cytoplasm essential for protein synthesis. The dendrites are limited to four short processes, about 12 μm long, terminating in specialized claws that collect all the synaptic input (figures 7.1 and 7.14).

The granule cell transmits to Purkinje cells with an irreducibly thin axon. Its diameter can be less than 0.2 μm, which allows just enough space for 1–2 microtubules plus their motors and cargo, and an internal resistance just low enough to prevent noise. Any narrower and the current entering through one sodium channel would see an internal resistance so high as to bring the membrane to threshold. This would allow a lone channel opened solely by thermal buffeting to generate a spontaneous spike, thereby introducing noise (Faisal et al., 2005). This design—thin axon—supports only a very low mean spike rate because, with a high surface area/volume ratio, it can contain relatively few mitochondria.

Granule cell design allows no input synapses, except at the four specialized claws. This provides high membrane resistance, which reduces the cost of maintaining resting potential. Even so, the bill for resting potential is considerable (see below, figure 7.21). As a neuron shrinks, the ratio of surface area to volume increases as 1/diameter, so the ratio for a granule cell is nearly 10-fold greater than for a Purkinje cell. Such a relatively large membrane area presents an expanse for leakage, and since the granule cell is the

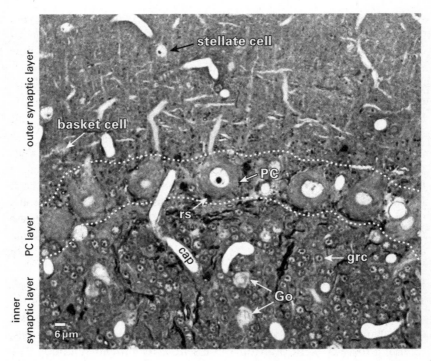

Figure 7.15
Largest cerebellar neuron occupies more than a 1,000-fold greater volume than smallest neuron. Thin section (~1 μm) through monkey cerebellar cortex. Purkinje cell body (PC) and nucleus are far larger than those of granule cell (grc). The latter cluster to leave space for mossy fiber terminals to form glomeruli with grc dendritic claws and space for Golgi cells (Go). Note rich network of capillaries (cap). Fine, scattered dots are mitochondria. Courtesy of E. Mugnaini.

brain's most numerous neuron, this small cost grows large (see also chapter 13).

Much of the inner synaptic layer is occupied by the large axon terminals of mossy fibers (figures 7.1 and 7.16). A terminal interlaces with multiple (~15) dendritic claws, each from a different but neighboring granule cell, and forms a complex knot (*glomerulus*), nearly as large as a granule cell body (figures 7.1 and 7.16). The mossy fiber axon fires at an unusually high mean rate (up to 200 Hz) and is therefore among the brain's thickest (figure 4.6).

To match the axon's high rate, a terminal expresses 150 active zones, 10 per postsynaptic granule cell (figure 7.16). These sites are capable of driving

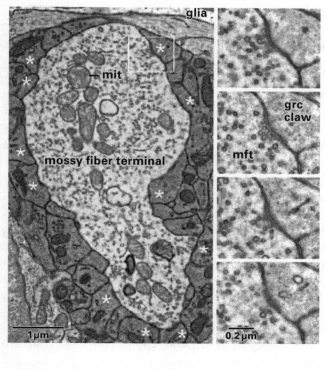

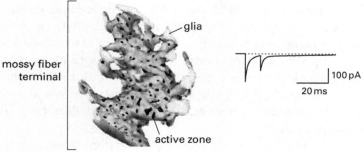

Figure 7.16

A large terminal may contact many small dendrites without intervening glia. This design pools transmitter from many synapses. It densensitizes postsynaptic receptors to reduce spike rate while improving temporal precision. **Upper left:** Central mossy fiber terminal contacts onto granule cell dendritic claws at release sites marked by *. Postsynaptic processes pack closely with no intervening glia. The whole structure (mft + grc claws) is partially encapsulated by glial membranes (above, shaded). **Upper right:** Serial sections through one release site, showing docked vesicles. **Lower left:** Three-dimensional reconstruction of mossy fiber terminal. Terminal provides hundreds of active zones. Each active zone has approximately seven neighbors closer than 1 μm. Glia cover about 20% of outer surface. **Lower right:** Granule cell strongly excited by first spike in mossy fiber but less strongly by next spike 10 ms later. Reprinted with permission from Xu-Friedman & Reghr (2003).

a granule cell synchronously, with one vesicle per presynaptic spike, up to frequencies of 700 Hz (Saviane & Silver, 2006). This direct mapping of spike input to vesicle output preserves the bandwidth and dynamic range of the mossy fiber axon. However, to sustain signaling, the terminal must replenish, dock, and prime vesicles at the axon's mean rate. The enlarged terminal provides the volume to accommodate a releasable pool of about 5×10^4 vesicles, about 300 per active zone. Given that synchronous release maintains bandwidth and signal range, how is the terminal designed to maintain the other determinant of information rate, S/N?

To maintain S/N, the enlarged terminal contacts many small dendrites without intervening glia (figure 7.16). This design allows transmitter released at one site to diffuse to neighboring sites. Thus, a postsynaptic receptor cluster on one granule cell dendrite receives a pulse of transmitter from its own release site, plus pulses from at least seven other sites less than 1 µm distant. This *spillover* of transmitter from several sites reduces noise produced by probabilistic release, and being adjacent, diffusion affects only slightly response duration. To allow spillover, the synaptic complex is largely devoid of glia expressing transporters. Consequently, transmitter released across the terminal at high rates tends to persist, and this too is put to good use.

The persistent spillover densensitizes postsynaptic receptor clusters, which acts as a negative feedback to reduce the amplitude of the granule cell's excitatory postsynaptic currents (EPSCs; figure 7.16). Thus, a granule cell can integrate numerous temporally correlated inputs to improve temporal precision, yet since each input delivers a small postsynaptic current, the mean spike rate is drastically reduced. In other words, a glomerulus with large terminal, multiple postsynaptic clusters, and scant glia, is well designed to remap information from densely coding mossy fiber axons to sparsely coding granule cells.

The roughly 50-fold step down of mean spike rate from mossy fiber to granule cell is efficient partly because bits/spike increases for lower rates (figure 3.5). Efficiency is further increased by improving a spike's temporal precision (chapters 5 and 6) to generate a *sparse code*, in which each granule cell is mostly silent and only fires brief, well-timed bursts at frequencies greater than 100 Hz (Ruigrok et al., 2011). Timing precision and brevity of the burst are enhanced by another contributor to the glomerulus: the Golgi neuron, whose cell bodies distribute sparsely within the inner synaptic layer and contribute GABAergic inhibitory contacts to the glomerulus (figure 7.14). This contribution to sparsifying the signal is relatively cheap because the individual cell is of modest size, low rate, and modest numbers. Moreover, inhibition is far cheaper than excitation (see below, figure 7.20).

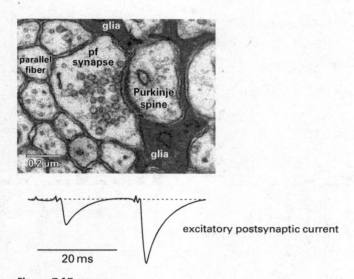

Figure 7.17
Parallel fiber synapse to Purkinje cell spine is ensheathed by glia and is facilitated at high frequencies. Upper: Glial wrapping is nearly 70%, which greatly reduces spillover between neighboring active zones. **Lower:** Excitatory postsynaptic response to brief burst of spikes at parallel fiber synapse (two spikes at 50 Hz). The second response is larger, probably because two vesicles were released simultaneously. The enhanced pulse of transmitter diffuses further from the release site to reach extrasynaptic NMDA receptors (Nahir & Jahr, 2013). Reprinted with permission from Xu-Friedman & Reghr (2001).

Now consider the circuit's second operation—integrate granule cell messages to form a Purkinje cell output. The granule cell axon ascends to the outer synaptic layer and branches as a T to run parallel with its neighbors and perpendicular to the fan-like Purkinje cell dendritic arbors (figures 7.1 and 7.14). Extending for 2 mm, it contacts one dendritic spine on many neurons, including Purkinje cell, stellate, and basket cells. The presynaptic active zone docks substantial numbers of vesicles (figure 7.17) so that the release of one vesicle does not deplete the ready pool. This design allows a second spike to admit sufficient calcium to release several vesicles simultaneously and produce a larger postsynaptic response (figure 7.17). The glutamate from single and multiple releases spills over to NMDA receptors just beyond the postsynaptic density, thereby extending the Purkinje neuron's time window for coincidence detection (figure 7.5).

The parallel fiber contacts upon spines tend to be well wrapped by glia, which reduces spillover between neighboring synapses (figure 7.17). Thus,

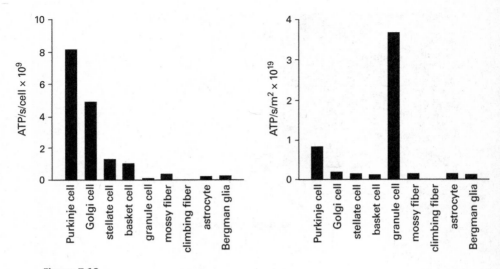

Figure 7.18
Energy costs by cell type. Left: Purkinje cell is the most expensive neuron, and granule cell is cheapest. **Right:** Granule cell array is the most expensive, and Purkinje cell array is far cheaper. Glial cells are cheap individually and as arrays. Reprinted with permission from Howarth et al. (2012).

a granule cell's output synapse is structured to reliably deliver a precisely timed message—privately (Nahir & Jahr, 2013). All 150,000 parallel fiber synapses onto an individual Purkinje cell tend to be wrapped by the same glial cell (*Bergman glia*), whose form mimics that of the Purkinje cell's extensive dendritic tree (figure 7.1).

The different tasks of inner and outer cerebellar layers and their consequent different designs illustrate why there can be no generic neuron. In the inner layer, high-rate synapses improve S/N by pooling excitatory responses and sharpen timing precision with feedback inhibition—to allow a burst of information-rich spikes (figure 7.16). In the second case, spikes deliver this information by a synaptic design that facilitates to a burst (figure 7.17). The first design reduces glial wrapping to enhance synaptic spillover; the second design does the opposite. Now we can ask: what are the costs of these two designs?

Costs of different neuron designs

Energy costs by cell type
When the various energy costs are totaled, the individual Purkinje cell proves to be the most expensive neuron, and granule cell proves to be the

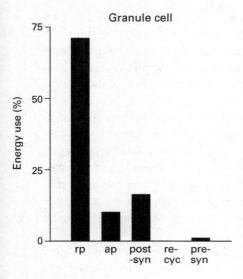
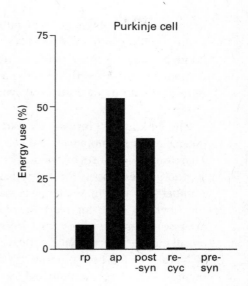

Figure 7.19
Energy costs by cell function: Granule cell versus Purkinje cell. rp, resting potential; ap, action potential; postsyn, postsynaptic receptor currents; recyc, transmitter
recycling (ATP for uptake by glial transporters, metabolic processing, and vesicular
transporters); presyn, presynaptic calcium entry and vesicle cycling. Reprinted with
permission from Howarth et al. (2012).

cheapest (figure 7.18). This should be no surprise since the Purkinje neuron
is far larger, receives many more synapses, and fires at a greater than 10-fold
higher mean rate. On the other hand, it seems mildly surprising that, the
granule cell array is approximately fourfold more expensive than the Purkinje cell array. This reflects the granule cells' outnumbering the Purkinje
cells by 274 to 1. The local inhibitory neurons are cheap individually and as
arrays, except for the Golgi neuron. The latter is costly as a single cell
because it receives a high rate of excitation from mossy fibers and is considerably larger than the granule cell (figure 7.14). Its major cost (75%) goes for
postsynaptic receptor currents (Howarth et al., 2012). However, it is cheap
overall because the array is sparse. Cerebellar glial cells are cheap individually and as arrays.

Energy costs by cellular function and computational stage
Each part of a neuron has its own cost, and the proportions vary according
to the cell's design (figure 7.19). Thus, the granule cell's thin axon (which,
ascending to the outer synaptic layer, branches as a T to become the irreducibly fine parallel fiber) sends each action potential cheaply. And, because

of sparse coding, its mean spike rate is low. Therefore, less than 10% of a granule cell's energy goes for spikes (Howarth et al., 2012). On the other hand, because the fine axon has a high surface area/volume, much energy is needed to maintain the resting potential against leaks. Postsynaptic currents at the input are costly, but synaptic release along the parallel fiber is cheap.

The Purkinje cell reverses the pattern. Its thick axon sends each spike at greater expense; moreover, the cell fires at high mean rates (figure 4.6). Therefore, the cell uses most energy for action potentials. The Purkinje cell's second greatest cost is for postsynaptic excitatory currents due to its vast number (>10^5) of glutamatergic contacts from parallel fibers and the climbing fiber. Its costs for recycling vesicles and transmitter are negligible because, except for a very few recurrent contacts, a Purkinje cell's outputs are all outside the cerebellar cortex.

Note that for both neuron types there are presynaptic costs. These include extruding accumulated calcium from synaptic terminals via a sodium/calcium exchanger, retrieving vesicles by endocytosis, refilling vesicles via a transporter, and retrieving and resynthesizing transmitter. For both granule and Purkinje neurons these presynaptic costs are negligible (figure 7.19), and the same is true for all the inhibitory interneurons (Howarth et al., 2012). Some of these processes are cheap because they use chemistry (neural exo- and endocytosis, metabolic processing of transmitters). Calcium extrusion is cheap because, although it uses energy for active transport, the calcium current to release a vesicle is miniscule compared to the current needed to open the calcium channel and compared to the postsynaptic current that the vesicle evokes. This reemphasizes the economy of chemical processes and the high cost of electrical amplification.

Cerebellar cortex contains more types of inhibitory neurons than excitatory ones (figure 7.14). However, the inhibitory neurons distribute sparsely and their synaptic currents are far cheaper. Therefore, excitation costs nearly four-fold more than inhibition (figure 7.20). Moreover, since the inhibitory processes serve to reduce redundancy and restrain expensive excitation, they seem a particularly good investment.

Cerebellar neurons parcel out their computations so neatly that the computational costs can be evaluated (figure 7.21). The inner synaptic layer is tasked to step down mossy fiber firing rates and concentrate information with a sparse code in granule cells. This costs slightly more than half of the total energy. The sparse code must then propagate to all the neurons in the outer synaptic layer, and this costs nearly one third of the total. Finally,

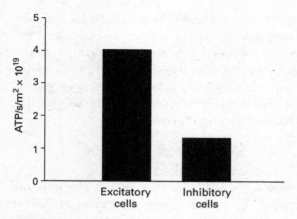

Figure 7.20
Excitatory neurons in cerebellar cortex cost nearly fourfold more than inhibitory neurons. Reprinted from Howarth et al. (2012).

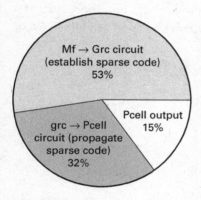

Figure 7.21
Input layer to cerebellar cortex consumes most energy. Intermediate layer consumes less, and output is cheapest. Mf, mossy fiber; Grc, granule cell; Pcell, Purkinje cell. Relabeled from Howarth et al. (2012).

the result of the outer-layer computation must be sent as Purkinje cell output which, due to the step down in neuron numbers, costs least (15%). These calculations correspond rather neatly to the distribution of *cytochrome oxidase*, which serves the final step of mitochondrial energy production: dense patches within the inner synaptic layer, corresponding to the synaptic glomeruli versus broad but weak distribution in the outer synaptic layer and strong in the Purkinje neurons (figure 13.20).

Conclusion

This chapter has focused on a few of the many ways that a neuron integrates information encoded at the input as chemical signals to transmit an output electrically at speed over distance. A core point is that "the neuron" is a shape-shifter. It can assume any form within limits ultimately set by physics, chemistry, and cell biology in order to function according to the basic principles of economy in neural design. The chapter did not explain how a neuron matches its internal chemical signals to achieve high efficiency, or how it couples them with equally high efficiency to its electrical outputs. That is the topic of chapter 8.

8 How Photoreceptors Optimize the Capture of Visual Information

Chapter 5 set out fundamental reasons why the brain computes with intracellular proteins. Protein circuits encode information with the least possible space and least possible energy; moreover, they optimally match their inputs and outputs, and they adapt quickly and cheaply as conditions change. Chapter 6 explained that protein circuits can be fast over short distance, but for speed over long distance they must couple to costly electrical circuits. That chapter explained designs for various circuit components as read from a catalog of neural parts: a linear amplifier, a nonlinear gain control, and so on. Chapter 7 explained how a neuron's various parts communicate with each other efficiently but not how they manage to accomplish something useful. It is the difference between explaining the principle of a flying buttress and a keystone arch—versus the cathedral at Chartres.

This chapter explains how a protein chemical circuit captures and chemically amplifies a photon's energy and then couples to a protein electrical circuit for transmission over distance. For this process of phototransduction, animals have evolved two types of chemical circuit (figure 8.1; Yau & Hardie, 2009). Both start with the same *chromophore*, a small molecule (*11-cis retinal*) that couples to the same type of protein (*opsin*). Cis-retinal, absorbing a photon's energy, unbends to *trans-retinal (photoisomerizes)*, thereby donating its energy to the opsin (R) and changing its conformation. The activated opsin (R^*) binds and activates a G protein (G^*), as described for the β-adrenergic receptor (figure 5.6). Beyond this point, the two circuits have strikingly different designs. One leads to closing a cation channel that is open in the dark, while the other opens a cation channel that is closed in the dark. The first type serves mammalian vision; the second serves fly vision (figure 8.1).

The mammalian design seems counterintuitive. Channels open in the dark cause a steady current that requires continual pumping to maintain an ionic gradient. Moreover, because the current is depolarizing, it opens a

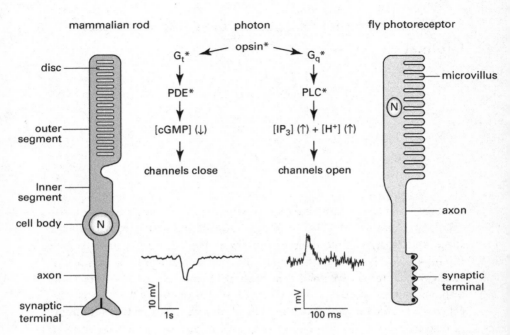

Figure 8.1

Mammalian rod and fly photoreceptor amplify the energy of a single photon using different protein circuits. In both transduction schemes a photon isomerizes an opsin to activate a G protein. Thereafter the schemes diverge: the rod closes cation channels to hyperpolarize sharply (~3 mV, peaking ~125 ms); the fly photoreceptor opens cation channels to depolarize sharply (~ 1 mV, peaking ~ 20 ms). Both responses can be resolved against background noise, but the fly response is faster. N, nucleus; G_t^*, activated G protein transducin; G_q^*, activated G_q protein; PDE*, activated enzyme phosphodiesterase; PLC*, activated enzyme, phospholipase C; [cGMP], concentration of the messenger molecule cyclic guanosine monophosphate; [IP$_3$], concentration of the messenger molecule inositol triphosphate; [H$^+$], concentration of protons. Rod recording is from mouse, reprinted from Cangiano et al. (2012); fly recording is from *Drosophila*, adapted from Niven et al. (2007).

voltage-gated calcium channel in the synaptic terminal that causes continual release of synaptic vesicles at a substantial rate, about 100 vesicles per second (Rao-Mirotznik et al., 1998; van Rossum & Smith, 1998). Only when a photon capture closes channels does the current pause, hyperpolarizing the cell and thus interrupting vesicle release. This double expense (tonic current, tonic vesicle release) seems profligate, especially in the absence of any photons. Indeed, this design is often cited as an instance where the brain is highly inefficient.

The fly design seems more intuitive. There is substantial fixed cost to maintaining the resting potential in darkness, associated in part with tonic transmitter release. This fixed cost equals 20% of the signaling cost in daylight, but beyond that, the design is pay-per-view: photon capture initiates a depolarizing current that increases transmitter release. This follows standard design and would seem to be cheaper. One might well wonder, did mammals somehow get stuck with an awkward, expensive system by some evolutionary accident?

We doubt this explanation for two reasons. First, the mammalian retina retains the insect-type system and uses it for certain key functions, for example, in retinal neurons that project to central neurons that entrain the circadian clock (chapter 4; Xue et al., 2011). Second, both rod and cone use less energy than a fly photoreceptor (Fain et al., 2010). To understand why, we consider first the mammalian rod under the simplest condition—dim light.

Phototransduction: Mammal

To capture an image in starlight
Starlight spreads photons so sparsely that a mammalian rod with a collecting area of about 1 μm^2 encounters a photon about once in 10 minutes, so a rod must capture and amplify single photons. But integrating single photon signals over minutes would not provide an animal with useful vision. To move fast, an animal must see fast, so the rod must integrate for briefer intervals, successive "snapshots" of about 200 ms. These catch one photon within a retinal patch of about 100 × 100 μm, that is, 10,000 rods. These limits to localizing photons in time and space allow only a sparse "pointillist" image (figure 8.2).[1]

Chances of two photons simultaneously striking the same opsin molecule are infinitesimal even in daylight. A double-hit would require the light intensity of a powerful laser, such as used in the "two-photon" microscope. Because the image is intrinsically sparse, the chemical amplifier should try

$10^{-5}R*$/rod/integration time

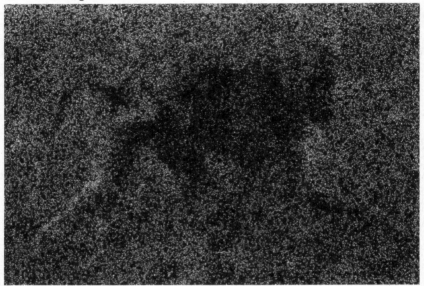

Figure 8.2
Baboon in starlight. A mammal needs to capture this photon-sparse image within 200 ms. Because photons are captured according to a Poisson distribution, S/N of the image goes as square root of the number of photons. Therefore, if many photons were missed, the image would disappear in a dim haze; if photons were mimicked by random noise events, the haze would brighten, but again the image would disappear. These performance requirements—speed, reliability of photon capture, and S/N—set key design features of the rhodopsin molecule and the transduction cascade, also of rod structure, the rod array, and the neural circuit. Reprinted with permission from Sterling (2004a). Original image from Botswana; see Tkacik et al. (2011).

to encode every photon but add the least possible noise. This requires precise tuning of opsin's folding by the amino acid sequence. Were the opsin too stiff, it would dissipate the energy from a photon hit with a shudder but fail to change conformation (false negative). Were the opsin too flabby, it would change conformation to a Brownian blow and report a photon when there is none (false positive). The compromise is a protein soft enough to support photoisomerization with nearly 70% efficiency, yet stiff enough to resist thermal isomerization at body temperature less than once per 200 years (Burns & Pugh, 2013).

This degree of thermal stability just barely suffices. To capture a rare photon in a patch of 10,000 rods requires an aggregate of 10^{12} opsins. So, even though opsin stoutly resists thermal blows, one molecule among this astronomical number inevitably receives an extra-hard jolt to set it off. This triggers the whole transduction cascade, causing a rod to falsely report a photon. The thermal event rate per rod is about $0.0024R^*/200$ ms, so summed over all rods in the patch, it is about $24R^*/200$ ms. Objects in starlight are about fivefold brighter than this. Consequently, an image such as that shown in figure 8.2 emerges above a noisy haze—but just barely. In short, opsin's thermal stability sets an absolute threshold for vision (Naarendorp et al., 2010).

To catch the rare photon, a rod packs opsins densely in quasi-crystalline, planar arrays on discs of intracellular membrane that orient perpendicular to the light path (figure 8.3; Liang et al., 2003). Each disc is a double membrane with opsins on both outer faces that provide about 80,000 opsins (Liang et al., 2003). These fill only about 50% of the surface, allowing 10% for other transduction proteins and free space for proteins to diffuse, collide, and transfer information. To improve the chance of catching the photon, the rod uses many discs, stacking them closely, nearly 40 per micrometer, along the outer segment (figure 8.3). The mouse rod with about 800 discs baits its trap with 6×10^7 opsins. These raise its probability of photon capture to about 0.4, and given opsin's probability of 0.67 for isomerization, the probability that a photon reaching a rod will cause an R^* exceeds 0.25.

A rod could improve photon capture by adding more discs. Indeed rods of certain creatures of the deep sea, where photons are profoundly sparse, do so. But thermal events increase proportionally with the number of discs (n); whereas the number of captured photons increases according to the law of diminishing returns because each disc added to the bottom of the stack is shielded by the discs above. So in mammals the upper limit is about 900 discs. Indeed, across mammals differing greatly in eye size, mouse to cow, outer segment length (number of discs) is conserved to within a factor of about 2 (Leibovic and Moreno-Diaz, 1992; reviewed in Sterling, 2004a).

The outer segment diameter, slightly more than 1 μm, is also conserved, thus requiring 10,000 rods to tile a patch that receives one photon per 200 ms. An amphibian rod, being fivefold thicker, tiles this territory with only 400 rods, so the mammalian design costs more than 25-fold more cell membrane and ion channels. The payoff is speed: reducing volume accelerates the mammalian rod's single photon response by 25-fold (Lamb & Pugh, 2006; Reingruber et al., 2013)—for reasons that we now explain.

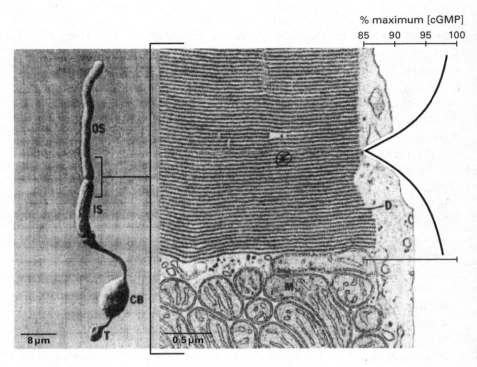

Figure 8.3
How a mammalian rod captures a photon. Left: Isolated mammalian rod (rabbit).
Outer segment (OS) contains a stack of about 900 membrane discs bearing densely
packed molecules of rhodopsin. Inner segment (IS) contains mitochondria that fuel
the sodium-potassium pumps that maintain the dark current. Cell body (CB) contains instructions and machinery for synthesizing proteins; axon terminal (T) transmits the single photon signal to a second-order neuron. Bracket indicates level of
longitudinal section to the right. **Middle:** Discs (D), each presenting two membrane
surfaces, stack densely to increase probability that a passing photon will be captured. ⊗ indicates site of one photon capture. Space between the disc stack and the
plasma membrane allows longitudinal diffusion of cyclic guanosine monophosphate
(cGMP) from site of photoisomerization. Mitochondria (M) pack densely in the inner segment. **Right:** Curve shows cGMP sink at peak current. cGMP decrement at the
initiating disc is modest, down less than 15% from the dark concentration. cGMP
deficit spreads longitudinally away from the initiating disc as cGMP flows toward the
sink. This distributes channel closings along much of the outer segment, preventing
local saturation and allowing the hydrolysis of cGMP molecules at one disc to amplify optimally. This curve (truncated) is reprinted with permission from Gross et al.
(2012). Images are from Townes-Anderson et al. (1985, 1988), reprinted from Sterling
& Demb (2004) with permission.

How the transduction cascade optimizes the single photon response

To generate a fast hyperpolarizing current following photon capture, cation channels must close promptly. This requires the rod to rapidly reduce its intracellular concentration of the small molecule, cyclic guanosine mono-phosphate (cGMP), that opens the channel. The enzyme for hydrolyzing cGMP, *phosphodiesterase* (PDE) is among the most efficient enzymes known, its catalytic rate being limited by the time for diffusion of substrate to the catalytic site (Liebman et al., 1987; Leskov et al., 2000; Reingruber et al., 2013). Since the time to reduce cGMP concentration reaches this lower bound set by physics, natural selection cannot accelerate catalysis. Instead it shrinks the rod diameter (d), thus shrinking intracellular volume as $1/d^2$. This shrinks correspondingly the number of cGMP molecules that must be hydrolyzed to reduce the intracellular concentration (figure 8.4).

Now enters another key feature of protein circuits: cooperativity. The membrane cation channel binds three cGMP molecules to open. There-fore, as cGMP concentration falls, the channel closure rises more steeply (figure 8.4). Greater speed could be achieved by still higher cooperativ-ity, but a steeper curve (figure 6.3) would reduce the dynamic range over which cGMP concentration could modulate channel current. Coopera-tivity matches other design features that set the spatiotemporal concen-tration of cGMP, including its diffusion coefficient, spontaneous rate of hydrolysis, and rate of resynthesis. Threefold cooperativity matches these parameters to optimally exploit the evoked fall in concentration (Gross et al., 2012).

The single photon response rises sharply to mark the instant of capture, and decays more slowly as cGMP rises and opens channels (figure 8.4). Since S/N goes as the square root of the number of open channels, a current produced by a few channels with large conductance (~2 pS) would fluctuate substantially. So the rod membrane expresses many channels (~10^6) with small unitary conductance (0.1 pS) and low binding affinity (K_d ~ 20 μM). By using many channels of low affinity, most (≥95%) can remain closed at a low dark concentration of cGMP (~4 μM). This reduces the steady synthe-sis of cGMP and the still costlier dark current yet still leaves open about 10^4 channels to respond to a photon-induced fall in cGMP. This number of channels, each open for a short time, supports an S/N at 36°C of 6.8 ± 2.8 (figure 8.1; Cangiano et al., 2012).

The reaction compartment between two discs spans only 30 nm (figure 8.3), and the band of plasma membrane encircling it bears at most 10 open channels. So, how could 100 channels close? This feat requires the zone of low cGMP to extend longitudinally to reach more open channels (figure

8.3). Concentration decays exponentially in space and time, so were the zone generated as a submillisecond event, like transmitter release (figure 7.3), it would reach few channels. But cGMP hydrolysis persists long enough for the zone of low concentration to move far enough (several micrometers) to close the requisite number of channels. The optimum time course for the zone of low cGMP to close the necessary channels with the fewest hydrolyzed cGMP has been calculated—and it matches the time course of the measured response (figure 8.4).

The single photon response achieves its optimal shape via three well-coordinated processes: activation, deactivation, and recovery. Following opsin's fast isomerization (within milliseconds), R* collides with transducin (G), activating it to G* that, in turn, activates the phosphodiesterase (PDE) to G*-PDE* to initiate cGMP hydrolysis. R* persists long enough to activate about 20 G* (Lamb & Pugh, 2006) and is then thoroughly deactivated in two stages with an overall time constant of roughly 40 ms. Once this is accomplished, the response time course depends on the accumulation and persistence of G*-PDE* molecules. G* → PDE* is catalyzed by a protein complex on the membrane that regulates G protein signaling ($RGS_{complex}$), so it is fast.

The G*-PDE* complex persists far longer than R*, so it accumulates. When the current peaks at about 125 ms, about 70% of all G*-PDE*s produced are active simultaneously. This deepens the cGMP sink (figure 8.3), which otherwise would tend to dissipate as it formed. G*-PDE* molecules continue to hydrolyze cGMP as the response decays, and deactivates gradually with a time constant of approximately 200 ms.

R* deactivates by multiple phosphorylations from *rhodopsin kinase* and then by capping with the protein *arrestin* (chapter 5). The kinase is restrained in darkness by a calcium binding protein (*recoverin*), but as channels close, intracellular calcium falls, triggering conformational change in recoverin that disinhibits the kinase. Calcium decreases because its entry via the channels is blocked and the existing calcium is vigorously removed by a sodium/calcium exchanger. Thus, early channel closings trigger the first deactivation step, which is completed within 40 ms. G*-PDE* deactivates when a GTP bound to G* hydrolyzes—catalyzed by the $RGS_{complex}$. Since this complex governs both activation and deactivation, it critically shapes the single photon response.

As cGMP hydrolysis declines over 200 ms, the zone of low concentration continues to spread longitudinally. An additional process is needed to replace the hydrolyzed cGMP and restore its initial dark concentration. cGMP is synthesized continuously in darkness by an enzyme, *guanylate*

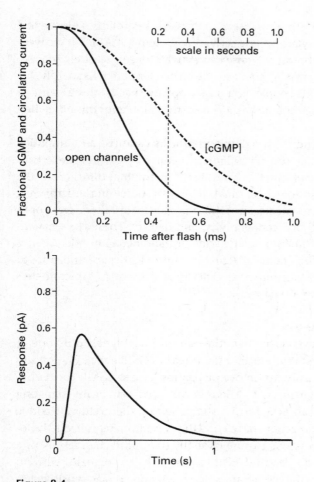

Figure 8.4
As the cGMP concentration falls, channels close more sharply to cause a steeply rising current. Upper: These curves were calculated omitting the deactivation reactions. Channel curve was calculated for threefold cooperativity. **Lower:** Steep current marks the instant of photon capture. Modified and reprinted with permission from Lamb & Pugh (2006) and Gross et al. (2012).

cyclase, to replace its continuous loss from low-level hydrolysis. This enzyme is controlled by a guanylate cyclase activating protein (GCAP). But recovery following the single photon response cannot wait for this leisurely process, so the disc bears a faster GCAP that activates when calcium falls sharply due to channel closure. This couples to a faster cyclase that creates a *source* of cGMP to spread longitudinally, canceling the sink and thus speeding recovery.

In summary, the energy of a single photon is captured and amplified chemically by a protein circuit in a highly structured physical context: two-dimensional diffusion of proteins on a disc; longitudinal diffusion of the final product's sink and source; optimal shaping of source and sink to reach channels on the plasma membrane whose properties are themselves optimized for binding affinity, cooperativity, and density. These close matchings of structure and function at the scale of nanometers to micrometers enhance efficiency. The chemical signal, optimized in space and time to physical limits, is now transmitted electrically across scores of micrometers to reach and then cross a synapse.

How the rod transmits 0 or 1

The rod in starlight over successive integration times signals only one of two values. In darkness it produces a high steady cGMP concentration signaling: **000** . . . Eventually a photoisomerization causes cGMP concentration to plummet, signaling **1**. Then cGMP recovers, again signaling **000** . . . This representation is further sharpened by the cyclic nucleotide gated channel's threefold cooperativity, so that **1** is ultimately encoded allosterically as the synchronous closure of about 100 cGMP channels. Now a chemical **1** recodes to an electrical **1** with a sharply hyperpolarizing current. The latter is filtered by the membrane's RC circuit and voltage-sensitive ion channels at the inner segment, producing a 3-mV hyperpolarizing voltage pulse with S/N of about 7 (figure 8.5; Cangiano, 2012). This pulse spreads rapidly down the axon to the synaptic terminal.

The rod synaptic terminal, tonically depolarized in darkness, represents **000** . . . as a steady release of vesicles from a single active zone, averaging about 20 vesicles per 200-ms integration time (Rao-Mirotznik et al., 1998; van Rossum & Smith, 1998). These deliver pulses of glutamate sufficient to close most of the postsynaptic cation channels on the dendritic tip. The synapse represents **1** when its 3-mV hyperpolarization causes a brief pause in release that allows postsynaptic channels to open and depolarize the second-order neuron (figure 8.5; Taylor & Smith, 2004). This scheme works for the average single photon response that exceeds the continuous vesicle

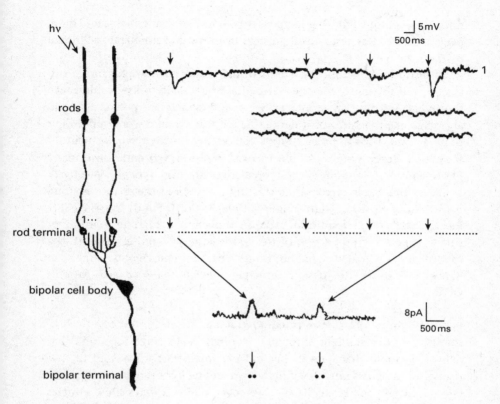

Figure 8.5
How a rod transmits *0* or *1*. Left: Early stages of rod circuit. n rods converge on a
bipolar neuron, each contacting a dendritic twig. Upper traces show the rod voltage
response. Absorbed photon (*hv*) causes a discrete hyperpolarizing response of mean
amplitude ~3mV and S/N ~7. The single-photon response varies in amplitude and
time course. Traces from two other rods exemplify the noise that will be summed
postsynaptically from ~20 rods along with the signal at the bipolar cell. **Upper:** Trace
indicates release of synaptic vesicles from rod synaptic terminal. Dark rate is ~100
vesicles/s—signaling 0000. Hyperpolarization evoked by a photon briefly suppresses
vesicle release—signaling 1. **Lower:** Trace shows rod bipolar current response. Tonic
release closes cation channels in the bipolar dendritic tip; the pause in release opens
these channels, allowing an inward current that evokes a brief burst of vesicle release
at the bipolar terminal. The two smaller single-photon responses are absent from the
bipolar cell's current response and its vesicle release. These responses would be indistinguishable from the noise of 20 rods summed at the bipolar cell. Therefore, the
bipolar cell employs a nonlinear mechanism that amplifies the larger responses more
than the smaller ones. Anatomical diagram from Rao et al. (1994); voltage traces
from Cangiano et al. (2012), modified and reprinted with permission; current trace
from Berntson et al. (2004), modified and reprinted with permission.

noise by sevenfold. But what happens to responses that are smaller and more ragged, that is, less easily distinguished from the continuous noise (figure 8.5)?

To discard smaller photon events would render the starlight image still sparser, but to transmit small events that are actually noise would make the image still noisier. Moreover, when rod signals are pooled to build an image like that shown in figure 8.2, a linear synapse would pool their noise as well. For example, 20 rods converging on their bipolar neuron[2] through a linear synapse would increase noise as $\sqrt{20}$ and efface many photon events. To address these challenges the rod synapse selectively amplifies the larger events, thus creating a threshold that removes noise before pooling (van Rossum & Smith, 1998; Taylor & Smith, 2004; Field & Rieke, 2002[3]). The nonlinear synapse transmits a high proportion of true R* and a low proportion of the continuous dark noise. In short, the rod integrates a graded signal but thresholds it to remove noise; and in so doing, it satisfies the design principle *combine analogue and pulsatile processing.*

Rod in brighter light transmits a graded signal

Intensities from starlight through moonlight span three log units that deliver no more than one R* per rod per integration time, and in that regime a rod signals only *0* or *1*. But dawn and twilight provide multiple R* per rod, thereby increasing its S/N. Moreover, higher R* rates allow a briefer "snapshot" that captures faster motion. Daylight provides still higher rates of R*, which a rod continues to encode by reducing its chemical gain. Operating well into daylight with a quantum efficiency nearly twice that of cones, and occupying about 90% of the receptor mosaic, rods capture 10- to 20-fold more photons than cones.[4] The rod's greater S/N and speed in bright light both increase its information rate (equation 5.7), but then comes a puzzle: how could a synapse designed for *0* or *1* transmit these richer signals?

The terminal's dynamic range of release rates does not increase in brighter light. Quite the opposite: light tends to hyperpolarize the terminal, reducing its mean rate from a maximum of 20 vesicles per integration time. Also, the integration time shortens, from roughly 200 ms to 100 ms, further reducing mean vesicle rate per integration time. So, although the rod active zone can modulate stochastic release to transmit a graded signal, its mean rate <10 vesicles per integration time would transmit with an S/N < 3, that is, $<\sqrt{10}$. This could suffice for dawn and twilight but not for the rod's S/N in daylight.

To escape this conundrum, the rod terminal couples to neighboring cone terminals via electrical synapses (figure 7.10). A cone terminal contains about 20-fold more active zones than the rod and thus modulates a higher mean vesicle rate (figure 11.9). This allows the cone terminal to transmit a higher S/N and use cone bipolar circuits specialized for this purpose (figure 11.12). Thus, the rod synapse triumphs by versatility: (1) a chemical mode operates at high gain as a nonlinear filter to transmit *0* or *1* nearly optimally; (2) a chemical mode operates at lower gain to transmit a coarsely graded signal; and (3) an electrical mode operates without gain or noise to transmit a finely graded signal into the cone synapse.

When the rod finally saturates in bright daylight, a few minutes predicts many hours of the same. Therefore, the rod shuts down its transduction machinery. Key proteins move off the discs and diffuse in bulk down to the inner segment (Calvert et al., 2006) where they park at low cost while the cones take over. A cone costs somewhat more to operate than a rod (see below), but distributes sparsely, comprising only 5% of the photoreceptor array. This seems to be another payoff for using the "backwards design" that depolarizes tonically and hyperpolarizes to send a signal. It allows the large array needed for dim light to saturate and reduce its major costs, so that a far smaller array of modestly greater unit cost can provide far better performance (Fain et al., 2010).

How cone design provides a finer, faster image

The briefest comparison of figure 8.6 to figure 8.2 emphasizes how fine an image can be captured in daylight. So it is natural to ask, what features of cone design allow this?

The cone opsin molecule, like the rod's, transduces a single photon, but the response is fivefold faster: about 25 ms to peak for cone versus about 125 ms for rod (figure 8.7). This allows the cone a briefer snapshot that captures temporal frequencies up to 100 Hz (Lamb, 2013). The single photon response is far smaller than the rod's (figure 8.7), and in brighter light it becomes still smaller. Consequently, a cone with dark current and maximum response like a rod's can sum many more R* during its integration time and thereby achieve a more finely graded signal. For example, capturing 10,000 photons per 100 ms allows S/N of about 100, that is, $\sqrt{10{,}000}$ (Rose, 1974). These properties require modifications to the opsin molecule and all the other proteins in the transduction circuit, including the cGMP-gated channel.

Modification starts with opsin. Once the chromophore has isomerized to R* and reported a photon, the opsin must reset by releasing the trans

Figure 8.6
Baboon in daylight. Photons arriving at far higher rates than starlight (figure 8.2) allow far better S/N with finer localization in space and time. For example, 10,000 photons/100 ms set an upper bound S/N ~100 by integrating over 1 μm^2 for 100 ms. Because each cone in a dense array sends a private output, the brain can resolve spatial images up to 60 cycles per degree and temporal differences up to 100 Hz (chapter 11). These opportunities for high performance (S/N, acuity, and speed) are boosted by rods but best exploited by a different photoreceptor design: the cone. Reprinted with permission from Sterling (2004a). Original image from Botswana; see Tkacik et al. (2011).

chromophore and binding a fresh *cis*. The trans diffuses to a nearby cell in the pigment epithelium where it is reset metabolically to *cis* and then returned to the photoreceptor layer to bind an empty opsin. A rod fails in very bright light because its opsin binds the chromophore so tightly to resist thermal bumps that it cannot release fast enough to keep pace with high rates[5] of R*.

Cone opsin supports the high release rates in bright light by binding its chromophore weakly. This makes cone opsin more vulnerable to thermal bumps, which is why a cone's rate of thermal events exceeds the rod's by 1,000-fold (Fu et al., 2008). Moreover, the cone's faster transduction circuit gives more noise from spontaneous hydrolysis of cGMP. Also, the faster cGMP channel gives more noise due to state transitions in gating (Angueyra & Rieke, 2013). All sources together give the cone a dark noise equivalent to

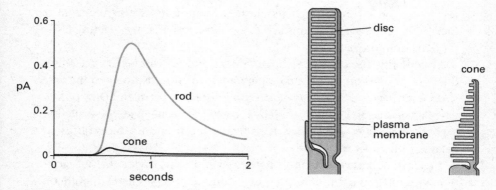

Figure 8.7
Cone single photon response is smaller than the rod's but faster. Left: Cone responds to photons of lower energy (longer wavelength, ~570 nm), whereas rod responds to photons of higher energy (shorter wavelength, ~500 nm). Reprinted with modification and permission from Lamb (2013). **Right:** Cone outer segment lacks internal discs and contains more plasma membrane per volume. The cone [cGMP] sink is established at inner surface of plasma membrane, closing local membrane channels rapidly, but with poor access to more distant channels. This speeds the cone single photon response but makes it smaller.

600R*/cone integration time, about 250,000-fold greater than the rod's thermal rate per rod integration time (Burns & Pugh, 2013; Angueyra & Rieke, 2013).

Clearly the cone transduction cascade is ill-suited for intermediate light levels (dawn/dusk) because the low rate of photon capture is swamped by the high rate of dark noise. But the cone is surrounded by 50 rods with far lower dark noise, so their electrical synapses set the cone terminal's S/N (Borghuis et al., 2009). In daylight the cone's rate of photon capture exceeds its dark noise and allows the cone terminal to be driven by fast, finely graded signals from its own outer segment. Several features of the cone protein circuit and subcellular architecture support this.

The cone's high speed and low gain arise from several interrelated features. R* lifetime is brief (cone ~3 ms vs. rod ~40 ms). This is due to faster phosphorylation by a faster kinase. G*-PDE* lifetime is also briefer (10 ms vs. 200 ms). This is due to a faster $RGS9_{complex}$ and more of it. Calcium turnover is faster (3 ms vs. 50 ms), as is regeneration of cGMP due to a faster GCAP driving a faster cyclase. Thus, all stages of the cone protein circuit accelerate to produce a sharper sink and a sharper source for cGMP (Burns & Pugh, 2013). Finally, the cone uses a light-gated channel that is modified to more rapidly run through its allosteric state transitions (Angueyra & Rieke, 2013). In short, the cone achieves a faster response by selecting faster proteins from the parts list.

Structurally, the cone lacks floating discs and instead locates the transduction proteins on the plasma membrane, which folds finely to increase surface area (figure 8.7). This reduces the distances between PDE*, cGMP-gated channels, and cyclase, thereby focusing the zone of cGMP depletion on a small number of channels. This tighter localization reduces diffusion delays at the expense of gain.

In short, because two neuron types, rod and cone, divide the full range of environmental light intensities, each can sculpt its cellular structure on the scale of microns to serve more efficiently. On the scale of nanometers, each can optimize key proteins for activation, deactivation, and recovery, to work at just the right speed and gain. They match ratios and lifetimes to provide just the right amount of product for best S/N, and couple to channels of just the right properties. Each optimized design, rod or cone, carries particular disadvantages, but these are mitigated by the larger scheme: a retina that uses two specialized designs to get the best of both.

Whereas the rod chemical synapse transmits only at low S/N, the cone synapse transmits at higher S/N. Whereas the cone outer segment has lower S/N, the rod outer segment has higher S/N. So cross-coupling allows them to match their advantages for each condition and avoid their disadvantages. This duality carries forward on the scale of neural circuits that cross the retina to the ganglion cells which serve as a final common pathway for starlight and all brighter levels (see figure 11.20). Now we can ask, what are the costs of maintaining a dual system, rod and cone, especially since each is designed "backwards" to close channels in response to light.

Economics of mammalian phototransduction

There are two energy costs to transducing a photon. The first is powering the chemical amplifiers that rapidly hydrolyze and resynthesize cGMP. The second is powering the ionic currents that generate electrical signals and transmit them at synapses.

The chemical amplifiers are cheaper. Following photon capture in a rod, 20 ATP are used to activate 20 G* and 3 ATP to inactivate R* via a kinase. Following its activation, PDE* hydrolyzes cGMP at a high rate, and that requires regeneration by guanylate cyclase, costing 2 ATP per cGMP. In darkness the chemical amplifier is driven steadily by noise, and the static guanylate cyclase consumes only 6×10^5 ATP s^{-1}. However, in light, the dynamic guanylate cyclase consumes up to 10-fold more ATP s^{-1}. At its

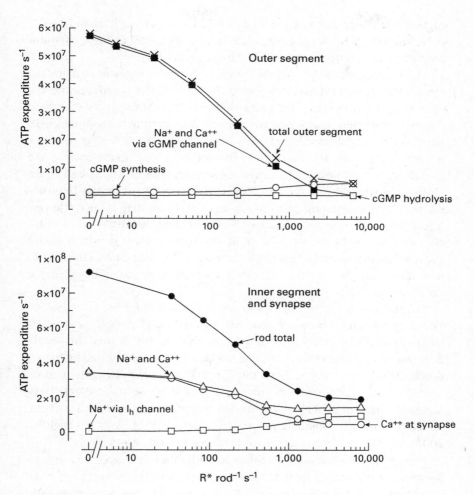

Figure 8.8

Chemical amplification in a mouse rod uses far less energy than electrical signaling.
Upper: Outer segment chemical processes (activation, deactivation, and recovery) are
cheap and increase with light level whereas restoring ions that pass through outer
segment channels is expensive and decreases with light level, given as R^* rod^{-1} s^{-1}.
Lower: The contribution of inner segment ion channels to total energy consumption. The cost of presynaptic calcium current declines with increasing light level,
but the cost of I_h current rises. Thus, the inner segment's electrical circuits consume
a significant proportion of the total rod's total energy, particularly at higher light
levels. Reprinted with permission from Okawa et al. (2008).

highest rate the cyclase costs at least 60-fold more ATP per R* than R* + G*, reflecting PDE*'s higher gain (figure 8.8). Yet, compared to these chemical amplifiers, the electrical one costs far more (figure 8.8).

Most energy is used by sodium pumps in the inner segment to maintain electrical currents in the outer segment (figure 8.8). The pumps remove the sodium that enters via cGMP-gated channels. They also supply energy to remove the calcium that enters this way, by maintaining the sodium gradient that drives sodium/calcium exchangers. In the dark this all costs about 6×10^7 ATP s^{-1}, 100-fold more than the chemical amplifier. However, the electrical cost declines as light closes channels and reaches zero when all are closed and the rod is saturated. Nonetheless, over most of the intensity range electrical signaling costs greater than 10 times that of chemical (see figure 8.8). This is vividly apparent from the dense distribution of mitochondria and cytochrome oxidase at the inner segments which house the rods' sodium-potassium pumps (figures 8.3, 13.19), and it well exemplifies the principle of neural design *compute with chemistry* (because it's cheaper).

The steady current through presynaptic calcium channels elicits steady vesicle release and this adds to the cost (figure 8.8). As the rod hyperpolarizes in light, the presynaptic calcium channels close, but most of this saving is negated by a negative feedback mechanism; I_h channels opened by hyperpolarization, admitting sodium ions at the rod inner segment. Consequently, the cost of electrical currents in the inner segment is relatively constant across all light levels (figure 8.8). Recycling and refilling a synaptic vesicle is a mainly chemical process so it costs less (Attwell & Laughlin, 2001), 10^6 ATP per second at the tonic release rate of 100 s^{-1}.

In darkness the rod array accounts for half of all the oxygen consumed by retina. However, as light reduces the outer segment current the rod's share falls to 10%, freeing resources for the retina's cone circuits to encode information at much higher rates (chapter 11). Thus, savings that accrue from rod saturation are immediately invested in better vision.

The individual cone costs somewhat more than a rod. First, whereas the rod's circulating current shuts down daily for hours, the cone's current circulates ceaselessly. Second, the cone's considerable dark noise involves tonic activity of all components of the transduction cascade (activation, deactivation, and recovery). Third, the cone's greater speed means turning over all the transduction components at a higher rate. Finally, the cone signal encodes far more bits per second, so it needs a larger synaptic terminal with 20-fold more active zones that release nearly 20-fold more vesicles (see chapter 11). The larger terminal needs a thicker axon with about

20-fold more microtubules that occupy an approximately 20-fold greater cross-sectional area (figure 11.9). Despite its greater unit costs, the cone array is small enough (10% of the rod array) that the overall cost is comparable.[6]

In summary, the mammalian rod matches protein components and structure across many levels to improve efficiency. Here, we considered: (1) outer segment length and volume; (2) numbers of key transduction proteins on a disc, their molecular ratios, and lifetimes; (3) functions of the disc; (4) number of cGMP-gated channels; (5) channel binding affinity, cooperativity, and conductance; and (6) space in the photoreceptor mosaic. These features all cooperate to set transduction efficiency, gain, S/N, and bandwidth for a particular environmental state (starlight). The design achieves most of the amplification and filtering by allosteric protein chemistry, which is locally fast and cheap, leaving the final stage for expensive electrical currents to transmit fast over distance.

The cone does the same for a different state (daylight) by drawing suitable proteins from the parts list and modifying the structures in which they are housed. Then, to cover the intermediate light levels (around dawn and dusk), the two cell types cooperate.

Thus, for rod and cone various biophysical factors jointly set the spatio-temporal integration of photons that structures the information packet to be relayed forward. This, in turn, determines many features of the down-stream neural circuits as will be discussed in chapters 11 and 12. Now we turn to fly phototransduction, where a different set of proteins are adapted to meet a fly's need for speed. Although these circuits work "forwards" by opening channels on demand, they turn out to be less energy efficient than the "backwards" design of rods and cones.

Phototransduction: Fly

A fly photoreceptor encounters the same physical constraints as mammal. First, dim light provides sparse photons—which must be captured efficiently by densely packing an opsin that resists thermal bumps. The fly photoreceptor contains about 3×10^7 opsins, each binding its chromophore tightly for thermal stability. Although it has half as many opsins as a mouse rod, the fly eye superimposes the receptive fields of six photoreceptors (figure 9.1) so that each sampling point (image pixel) is covered by a similar number of opsins ($\sim 10^8$), with similar rates of thermal events.

Second, because bright light provides photons at high rates, their capture requires opsin molecules that replace their spent chromophore promptly. A fly opsin does not release its *trans* chromophore but clasps it tightly, holding it such that a second photon converts *trans* back to *cis*. This *photoreisomerization* allows a fly opsin to clasp its chromophore tightly to resist thermal knocks in dim light (like a rod) and to regenerate it to *cis* at high rates in bright light (like a cone). Thus, a fly makes do with a single receptor type.

Third, a fly photoreceptor encounters the same range of contrasts in natural scenes over the same wide range of mean light levels ($>10^6$). In dim light a fly, like a mammal, uses single-photon sensitivity plus broad spatial and temporal integration to capture a coarse image (figure 8.2) whereas in bright light it uses finer spatial and temporal integration to capture spatial detail and low contrasts (figure 8.6).

Fourth, image motion reduces the number of photons reaching a photoreceptor from a particular feature in the field and, because a photoreceptor integrates over time, motion also blurs the image A blowfly moves fast, turning at thousands of degrees s^{-1}, and, although it partially stabilizes the image with rapid, compensatory head movements (van Hateren & Schilstra, 1999), motion across the receptor mosaic still reaches hundreds of degrees s^{-1}. Such fast moving images are severely blurred in mammals because of the slower mechanism for phototransduction. Consequently, when our brief saccadic eye movements impel our gaze across a scene at hundreds of degrees s^{-1}, our brain suppresses vision (Burr, 2004).

But for a fly, high image speeds are the norm, so a fly photoreceptor must use every opportunity to reduce its integration time. In dim light the fly's single photon response is 5 times faster than a rod's (figure 8.1). In bright light a blowfly photoreceptor's electrical response to a brief flash starts within 3 ms and completes within 12 ms. This exceptional speed enables a blowfly to resolve flicker up to 300 Hz whereas we stop at 60 Hz. So, how does a fly photoreceptor respond so quickly, and at what cost?

Speed is achieved with a radically different design

Let us recall why the rod single photon response is slow. The smallest reaction compartment, the cytoplasmic space between two discs, is relatively large, about 2×10^{-17} L. Therefore to significantly change the concentration of the second messenger cGMP requires hydrolysis of many molecules. Although the hydrolytic enzyme, PDE, is fast, concentration changes slowly. Moreover, the change spreads slowly within a still larger reaction compartment, the outer segment, to close more distant channels. Recovery is slow for the same reasons. The rod's larger compartment allows recruitment of many channels, thereby increasing S/N, but speed is sacrificed for

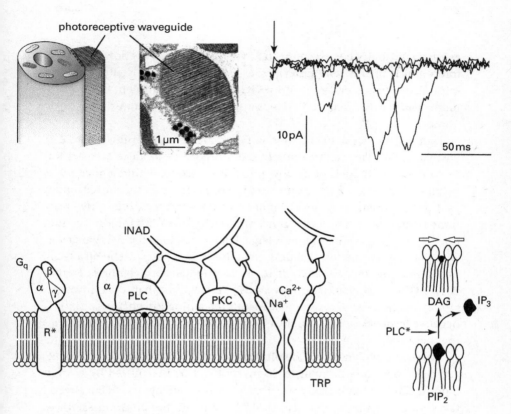

Figure 8.9
A microvillus contains the complete transduction circuit and generates discrete responses to single photons. Upper left: A fly photoreceptor forms its photoreceptive waveguide from microvilli. Cross section of the waveguide showing the microvilli densely packed in a regular array. **Upper right:** Single photon responses produced by microvilli. Four recordings of photoreceptor membrane current, each showing the brief pulse of inward current produced when one of the photoreceptor's 30,000 microvilli transduced a photon. Arrow indicates the 1-ms stimulus. Note the variations in response latency, amplitude, and duration. **Lower left:** Organization of the transduction circuit at the microvillus membrane. The circuit's proteins R* (activated rhodopsin); G_q comprising α, β, and γ subunits; phospholipase C (PLC); protein kinase C (PKC); and the TRP ion channel are held together by the microvillus membrane and the scaffolding protein INAD. Black dot indicates a molecule of PIP_2 (phosphatidyl inositol biphosphate) in the inner leaflet of the microvillus membrane. **Lower right:** The transduction circuit's mechanical response to a photon. Activated PLC cleaves PIP_2 into diacylglycerol (DAG) and inositol triphosphate (IP_3). DAG remains membrane bound and IP_3 floats free. Neighboring phospholipids move to close the gap, so the membrane contracts. Photoreceptor drawing adapted from Hardie & Rhagu (2001) with permission. EM of fly microvilli and recordings of single photon responses courtesy of Roger Hardie. Diagrams of transduction circuit and its mechanical response based on Hardie & Postma (2008) and Hardie & Franze (2012).

gain and reliability because diffusion over these distances is slow. A fly photoreceptor is intrinsically faster because, to maximize concentrations and minimize diffusion delays, it shrinks its reaction compartment to the minimum, a *microvillus* (figure 8.9) that is just large enough to transduce a single photon.

Transduction in a microvillus is faster for several reasons (Hardie & Postma, 2008). First, all of the molecules needed to produce an electrical response to an R* are kept in close proximity. Second, within a microvillus volume of about 2×10^{-18} L, one extra molecule or ion raises concentration by 1 µM. Consequently, small numbers of ions and molecules drive reactions at high rates. Third, the concentration gradient driving the key ion, calcium, into the compartment is large. Fourth, the ion channel that generates the compartment's electrical response is not opened chemically by ligand binding, but *mechanically* by membrane tension (Hardie & Franze, 2012). Fifth, this channel drives a positive feedback that, once initiated, rapidly opens most of the compartment's remaining ion channels. Sixth, a negative feedback terminates the electrical response promptly by inactivating these channels.

Now some explanation. Chemical amplification starts like a rod. A fly R* activates about 5 G proteins in 20 ms and is then inactivated by arrestin binding.[7] This is fewer than activated by a rod R* (about 20 G*), but the rod takes longer, meaning that the fly's first stage in amplification sacrifices gain and S/N for speed. At the next stage of amplification, the fly's G* activates one molecule of a hydrolytic enzyme that breaks down its substrate at a high rate—as in a rod. However, the fly's enzyme, *phospholipase C (PLC)*, does not attack small molecules diffusing in a cytoplasmic compartment; rather, it cuts a larger molecule confined to the inner leaflet of the microvillus membrane, the phospholipid *PIP_2*. The cut releases a bulky component, *inositol triphosphate (IP_3)*, into the microvillus lumen along with protons. PLC* is held against the inner membrane leaflet by the cytoskeletal protein, *INAD (inactivation no afterpotential D)* so that it is near abundant substrate— PIP_2 forms 2% of the inner lipid leaflet. This allows a PLC* molecule to work quickly, releasing about 100 IP_3 molecules in 50 ms.

Now comes a more radical difference. The fly photoreceptor's ion channel, *TRP (transient receptor potential)*, is not opened chemically by binding a diffusible messenger, but mechanically by membrane tension (Hardie & Franze, 2012). When PLC* releases IP_3, the membrane shrinks and stretches; and because INAD binds both PLC and TRP, most shrinkage occurs around a channel (figure 8.9). Were this all, nothing would happen because the TRP channel is insensitive to stretch. However, protons that were released simultaneously with IP_3 rapidly sensitize TRP channels. At a critical membrane tension the TRP channel snaps open, admitting sodium and calcium ions at 10 times the rate of a rod cGMP channel.

Driven by a high concentration gradient across the membrane (10^6-fold), calcium ions surge into the tiny microvillus, rapidly raising the internal concentration from about 0.15 µM to above 1 µM. This sudden increase in calcium further sensitizes TRP channels, so more open, which further increases calcium, and so on until about 15 of the microvillus's 20 channels have been opened (Hardie & Postma, 2008). Thus, as with an action potential, runaway positive feedback initiates a rapid surge of inward current. Moreover, to produce a brief pulse, the current is quickly shut off by inactivating channels.

As the current entering through TRP channels is peaking, calcium inside the microvillus is rising above 50 µM, heading for 500 µM. At these higher concentrations, calcium closes and inactivates TRP channels via a low-affinity mechanism that depends on phosphorylation by *protein kinase C* (PKC). Because each TRP channel also binds a PKC molecule via INAD, all open TRPs are closed within 15 ms. PKC also inactivates another near neighbor, PLC*, to prevent it from unnecessarily depleting the stock of PIP_2. Closure of the microvillus's last TRP channel completes the fly's response to R*, a mini–action potential that rises clearly above the noise for about 20 ms to generate a single photon response which is one fifth the duration of a mouse rod's. All that remains is to reset the microvillus to transduce another photon. To remove TRP inactivation, sodium/calcium exchanger molecules in the microvillus membrane return the lumen's calcium concentration to its original nanomolar level. This takes about 100 ms, and during this *dead time* the microvillus is refractory: it cannot respond to R*.

The fly's design achieves high speed and high gain by approaching physical limits. The microvillus is irreducibly small. Extending 1–2 µm,[8] it can be no shorter because it forms the waveguide that traps photons. A shorter structure with the same refractive index would leak photons.[9] The microvillus inner diameter, about 50 nm, suffices to house the cytoskeletal scaffold needed to support a cylindrical tent of membrane. To minimize noise and delays from diffusion, most interacting molecules are held adjacent by the scaffold molecule INAD. Protein molecules that must diffuse, G* and PIP_2, are confined to the membrane so that they can save time by diffusing in two dimensions, rather than three. For speed and S/N, the output of the chemical amplifier couples to ion channels mechanically through changes in membrane volume. This volume change is an efficient integrator of

PLC's chemical output because it responds instantaneously to the release of each IP_3. With such mechanical integration, signal loss and temporal delay due to diffusion are negligible.

But the fly's microvillar design has disadvantages. Positive feedback applied to a small compartment is risky. As in a very thin axon, stochastic opening of a single channel can exceed threshold and trigger a false positive (chapter 7). The microvillar design prevents this dark noise by incorporating an AND gate. The TRP channel does not open to a proton, nor does it open to tension—but only to their coincidence, a proton AND tension. Thus, when one TRP channel opens spontaneously in response to thermal blows, it cannot engage other channels for positive feedback because they are not primed by protons. In short, the inclusion of molecular logic allows positive feedback in a small compartment for speed and high gain while preventing one source of noise. However, the fly's microvillar design introduces other noise sources.

Noise in microvillar transduction

Although R* registers a photon with nanosecond precision, the microvillus's electrical response varies stochastically in latency, rise time, amplitude, and duration (figure 8.9). Latency varies between 30 and 100 ms, depending on the time taken for about 5 G* to activate about 5 PLC*, and for these PLC* to cleave enough PIP_2 to open the first TRP channel. Amplitude varies due to fluctuations in the small numbers of G*, PLC*, and TRP used for amplification. Response duration varies due to fluctuations in the positive and negative feedbacks that shape the electrical response. A rod signal is less noisy because it is produced more slowly with many more molecules, and does not use positive feedback to boost gain.

Random variations in timing reduce the information carried by a single photon response. Even so, fly single photon responses are randomly dispersed over a time interval that is less than half the duration of a rod's response, giving the fly better temporal resolution. At higher light levels, when photons are transduced at higher rates, currents from several microvilli coincide to produce a continuous analogue signal. Now random fluctuations in both amplitude and timing reduce S/N, and the latency fluctuations reduce bandwidth by doubling the duration of the flash response. Consequently, the noise in a microvillus's response to a photon continues to reduce information at higher light levels.

Fly and mammal use same coding strategies to encode a finer, faster image

As photons become abundant, the image's S/N improves, making finer detail visible (figure 8.6 vs. 8.2). Fine detail is susceptible to motion blur at

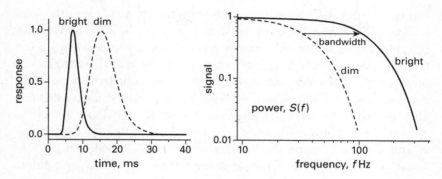

Figure 8.10

A faster response codes higher frequency signals. Left: Response to a bright flash delivered in daylight is completed 3 times faster than the response to a dim flash delivered in darkness. 1-ms flash delivered at time $t = 0$, response normalized to peak amplitude. **Right:** Faster response in bright light enables photoreceptor to code higher temporal frequencies, f, than slower response in dim light. Arrow shows the increase in bandwidth. Plots are of signal power versus frequency, obtained by squaring the Fourier transforms of the flash responses and normalizing peak power to 1. Bandwidth is defined as frequency at which $1/\sqrt{2}$ of the input power is transmitted (i.e., 0.5 amplitude transmission) but photoreceptor is still responding at $f = 3 \times$ bandwidth and transmitting approximately 10% of power. Plotted from data of Tatler et al. (2000).

a slow "shutter speed," so now a photoreceptor responds more briskly. The fly photoreceptor, like the cone, accelerates its response to optimally balance motion blur versus photon noise and thus maximize information uptake (figure 9.11; van Hateren, 1992b). The fly's flash response reduces in duration threefold to produce the threefold increase in bandwidth that maximizes information rate in daylight (figure 8.10).

Both fly and cone photoreceptors use an identical coding strategy: each adapts its gain to the local mean intensity, I, to code *contrast*. Contrast is $\Delta I / I$, where ΔI is the difference between the intensity at the receptor and the local mean. Photoreceptors code contrast using the hyperbolic I/O function of protein and electrical circuits (figures 8.11, 6.3, and 6.9). In the function's logarithmic midregion, changing the light intensity by a fixed proportion produces a constant change in membrane potential, ΔR. Thus, a given contrast, $\Delta I / I$, is coded by a constant response, ΔR, independent of the mean intensity I.

Coding contrast simplifies visual processing in two ways. First, dividing by the mean reduces a 10^6 range of light intensities to proportions that are better accommodated within a neuron's limited response range. Second, coding contrast helps achieve one of vision's goals, to identify the same object under different conditions of illumination. Because most natural objects generate an optical signal by reflecting and/or transmitting a particular fraction of the light that falls upon them, an object that is differently illuminated sends a different signal to the eye. Dividing by a measure that depends on local illumination, the local mean intensity, factors this out. The resulting contrast signal depends not on illumination, but on an object's physical properties, its reflectance and transmittance.

Thus, contrast coding starts the process of *generalization*, whereby the visual system assigns signals that differ because of viewing conditions (e.g., a face viewed in different light or from different angles) to the same object.

The logarithmic region for coding contrast spans about 2 log units of background intensity. It is the same for both fly and vertebrate and suits the intensity range that both encounter in an evenly illuminated natural scene (figure 8.11; Normann & Werblin, 1974). But during the day the earth's turning changes the background light level by 5 log units, far exceeding this fixed function's range. In addition, merely stepping from shade into sunlight or flicking the eyes between bright highlights and shady recesses takes the signal out of range. Fly and mammal solve the problem with similar strategies, one for slow changes in background and another for faster ones.

As noted for rod, the fly photoreceptor adapts to slow increases in background by reducing the numbers of transduction proteins in the reaction compartment.[10] Over many minutes, G proteins and TRP channels leave the microvillus to reduce gain, and arrestin moves in to terminate responses more quickly (Hardie & Postma, 2008). The reduction in response duration increases bandwidth, thereby increasing information rate by 28% (Burton, 2002). Cone and fly photoreceptors adapt to faster, briefer changes in background using the same gain control signal, a rapid change in calcium ion influx that tracks the activity of their channels.

The cone adapts by averaging the calcium signal in the outer segment, a significant volume, and feeding it back to several control points in the transduction circuit to accelerate deactivation and recovery (Burns & Pugh, 2013). This reduces the amplitude and duration of changes in [cGMP] that modulate channels, so reducing gain and increasing bandwidth in bright light. Lowering the gain shifts the I/O function to a higher intensity range while retaining the same two log unit range for contrast coding. The fly photoreceptor shifts similarly also using calcium to control gain[11] (figure 8.11).

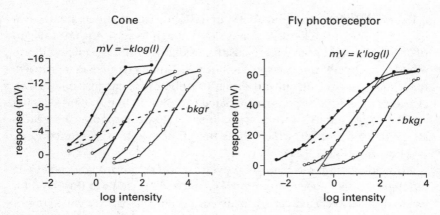

Figure 8.11
Cone and fly photoreceptors encode contrast using the same analogue primitive, the midregion of a hyperbolic input/output (I/O) function. Voltage response is proportional to log intensity. When flashes are presented in darkness (solid circles), sensitivity is high, and the logarithmic region (indicated by the straight line) covers the lowest intensity range. Adapting a photoreceptor to a steady background lowers sensitivity, and the I/O curves (open circles) shift to keep the response to the background (dashed line) in the logarithmic range. The brightest background shifts the curve by 3–4 log units, enough for full sunlight. Turtle cone, replotted from Normann & Perlman (1979). Blowfly photoreceptor, replotted from Matić & Laughlin (1981), with permission. In fly the slope of the curve determined with dimmest flashes is depressed because a fully dark-adapted photoreceptor light-adapts during the rising phase of its response (Matić & Laughlin, 1981).

Every time a microvillus transduces a photon, it admits a surge of calcium ions, and it could use these to reduce its gain. However, a microvillus transduces photons stochastically, and, being just one of 30,000 microvilli, its rate and S/N are low. Consequently, a microvillus receives too little information about the mean to control its own gain. Instead, a fraction of each inward calcium surge diffuses from a microvillus into the cell body. This larger reservoir maximizes S/N by averaging the calcium concentration across all microvilli, over an appropriate time window, and then updates the gain of every microvillus.

Calcium diffuses from the cell body into the microvillus lumen. Because the reservoir is vast and the microvillus is tiny and contains just a few sodium/calcium exchangers, the lumen's calcium concentration cannot fall below that of the larger reservoir. As a result, the reservoir sets a baseline calcium concentration in the lumen that represents the combined photon rate of all microvilli. As this baseline rises with light intensity from about

0.15 µM in the dark to about 10 µM in daylight, TRP channels are desensitized. This form of desensitization is slow enough to allow the fast calcium rise triggered by an R* to engage positive feedback within the microvillus to produce a single photon response. But with some channels desensitized, fewer are opened, until in full daylight a photon opens just one or two channels for a 3-ms response. Compared to a photon in dim light opening about 15 trp channels for an approximately 30-ms response, the gain is about 100 times lower, which shifts the photoreceptor's I/O function by two log units (figure 8.11).

This adaptation mechanism provides another advantage. Increasing luminal calcium concentration moves a microvillus closer to threshold for its minispike. Averaged across many microvilli, this tighter clustering of briefer responses increases peak response amplitude and temporal precision, thereby increasing S/N, bandwidth and hence information rate (figure 8.12). With a bandwidth in excess of 100 Hz, a fly photoreceptor codes over 1,000 bits s^{-1}, but information coded at these high rates is expensive.

Economics of fly phototransduction

Space

The fly design, which packs all transduction components into each microvillus and gathers all microvilli to form a waveguide, requires most of the supporting components to locate elsewhere. Consequently, the power supply (sodium–potassium pumps that charge the ionic batteries), the metabolic furnace that produces ATP (mitochondria), and the regulators of the electrical gain and bandwidth (voltage-gated potassium channels) are all placed alongside the microvilli, in the cell body (figure 8.1). The latter also contains machinery for protein synthesis and other housekeeping tasks.

To transmit its information-rich analogue signals, the fly photoreceptor requires synapses with many active zones (chapter 9). The challenge resembles that of the cone terminal, but fly information rates are far higher, and this requires many more active zones and more space. Also, vesicle release by a fly photoreceptor terminal is triggered, as elsewhere, by calcium entering via voltage-gated channels (chapter 7). Consequently, the photoreceptor synapse should be isolated from the transduction apparatus where calcium ions are used for gain control.

To meet these needs, the synaptic terminal locates at the end of an axon, well away from the transducer/waveguide and cell body. Like mammalian photoreceptors, the fly photoreceptor is elongated, with the transducer/waveguide alongside the cell body at one end, and the synaptic terminal at

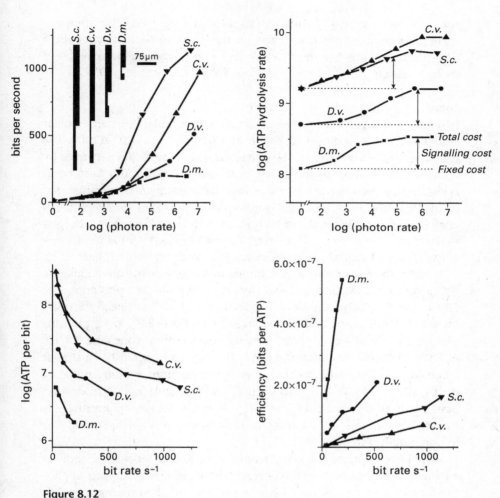

Figure 8.12

Bit rate, energy cost, and energy efficiency in fly photoreceptors. Photoreceptors of different species perform the same task but have different lengths. *Drosophila melanogaster (D.m)*, *Drosophila virilis (D.v)*, *Calliphora vicina (C.v)*, *Sarcophaga carnaria (S.c)*. **Upper left**: Bit rates encoded as analogue signals increase with photon rate and are higher in longer photoreceptors. **Inset**: photoreceptor lengths. **Upper right**: Energy consumption rises with photon rate from a baseline in the dark, the fixed cost. Costs are higher in longer receptors, but the proportions of signaling to fixed costs are similar for all receptors (vertical arrows). **Lower left**: Energy cost per bit falls with bit rate as the fixed cost per bit falls and gain control increases bandwidth and reduces energy cost per photon. Thus, efficiency rises. **Lower right**: Shorter receptors with lower information capacities are more efficient. Replotted from data of Niven et al. (2007).

the other end (figure 8.1). Although its microvilli are irreducibly compact, there must be many of them to achieve the necessary S/N. They require a long photoreceptor. As a result, a blowfly's photoreceptors make up 4% of its body mass.

Energy

In darkness, individual photoreceptors in mammal and fly consume similar amounts of energy (figure 8.12; Fain et al., 2010; about 1×10^8 ATP s^{-1}). The early chemical stages require ATP to regenerate the molecules destroyed by hydrolysis, and PIP$_2$ costs somewhat more than cGMP. But the greater cost is electrical: powering the ion pumps that maintain the concentration gradients that drive ions through channels. Inward fluxes of sodium and calcium are high for mammal and fly because both types keep channels open in darkness. Their current depolarizes the membrane potential to voltages where synapses transmit single photon signals with high gain (figure 8.12).

In light, the costs for fly and mammal move in opposite directions, the fly opening channels and mammal closing them. Thus, at a photon rate of 500 s^{-1} the fly photoreceptor consumes 1.5×10^8 ATP s^{-1}, nearly 5 times more than a rod (figure 8.12, *D. melanogaster*, cf. figure 8.8). At 10^5 photons s^{-1}, near the top of a cone's range, the fly photoreceptor consumes 3×10^8 ATP s^{-1}, 10 times more than the cone. The latter, because it has a similar bandwidth and photon rate, should code information at a rate similar to the fly, about 200 bits s^{-1}.[12] Thus the mammal photoreceptor's "backwards" design is 10 times more energy efficient. However, its biggest advantage is that with two classes of photoreceptor, the majority are rods that shut down in bright light (Fain et al., 2010).

A fly photoreceptor has every reason to be energy efficient. A blowfly resting in sunlight uses 8% of its energy to power electrical currents in photoreceptors. A fly saves some energy by regenerating its chromophore with a photon. The photon comes for free, whereas the mammal regenerating biochemically requires several ATP (Okawa et al., 2008). Photoregeneration, being direct and quick, keeps pace with photoisomerization. Even in bright sunlight, half of a fly's chromophores are *cis* and poised to transduce, whereas in a cone only 1% are *cis*, and in a rod, practically none. More significantly, by controlling gain and bandwidth to match *photon rate* (the rate at which a photoreceptor transduces photons), a fly photoreceptor seamlessly converts from rod-like to cone-like function. This "neatening up" halves the number of photoreceptors a fly needs in its compound eye, so saving space, materials and energy. However, these advantages are offset by the inefficiencies introduced by the fly design.

Three factors reduce a fly photoreceptor's efficiency. First, transduction has intrinsically low quantum efficiency, because cylindrical microvilli pack rhodopsin less efficiently than the rod's flat discs and the cone's folded membranes. Second, signals amplified by positive feedback are noisier. Therefore, to achieve a given sensitivity and S/N, a fly photoreceptor must be larger. A larger neuron draws more current, and this increases energy cost. Third, and most significant, the fly's one-type-fits-all design is inherently inefficient.

Whereas a rod synapse can transmit in dim light with a single active zone, a fly photoreceptor cannot. In bright light, it must transmit at high rates like a cone, and that requires many active zones. Indeed, the fly photoreceptor uses 45 active zones, twice that of a foveal cone (chapter 11). Yet, like a rod, the fly's large synaptic terminal must be tonically active in the dark to transmit a single photon signal with high gain. As stated, this requirement is met by an inward current that depolarizes the photoreceptor by 10–15 mV, and this greatly elevates the dark cost. Thus, as engineers know (chapter 1), a component given two tasks seldom does both with optimal efficiency.

Unable to eliminate the inefficiency that comes with a one-fits-all design, the fly does the next best thing: it attempts to maximize the efficiency with which it performs its allotted task. A photoreceptor's function is to code and transmit the information it extracts from photons. Consequently, structure and mechanism are adapted to reduce the cost of coding information at the rates needed for adequate vision.

Analogue information rate and efficiency

To code efficiently, a photoreceptor avoids the expense and inefficiency of spikes by coding directly in analogue (figure 8.13). A fly photoreceptor's information rate rises with light level according to equation 5.7, as S/N increases as square-root photon rate and calcium's gain control widens the bandwidth (figure 8.13). Information rate reaches its maximum, a photoreceptor's *information capacity*, in full sunlight. Capacity varies among species and is highest in the longer photoreceptors of larger, faster flies (figure 8.12). The higher capacities are needed to code the faster moving images experienced by larger flies and, at 1,000 bits s^{-1}, are twice that of spikes (de Ruyter van Steveninck & Laughlin, 1996).

Irrespective of its ultimate capacity, a photoreceptor's efficiency increases with rate (figure 8.12). At low photon rates, and hence low bit rates, the signaling cost is small compared with the fixed cost of maintaining the resting potential in the dark (figure 8.12). In this regime the efficiency rises

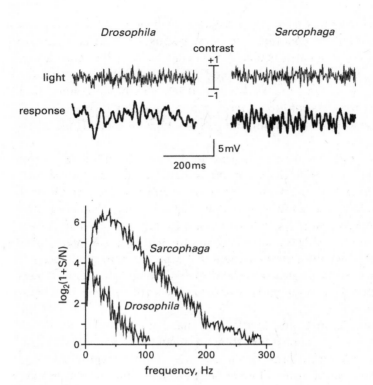

Figure 8.13
Fly photoreceptor increases information capacity by increasing bandwidth and S/N.
Upper left: Response of fruit fly *Drosophila* photoreceptor to rapidly changing contrast of bright light is slower than that of fleshfly *Sarcophaga* photoreceptor. *Drosophila*, therefore, fails to signal at higher frequencies, as shown below. **Lower:** A measure of S/N, $\log_2(1 + S/N)$, plotted against frequency for both photoreceptors. *Drosphila*'s inability to respond to rapid changes restricts its signal to frequencies below 100 Hz. *Sarcophaga*'s quicker response generates signal out to 300 Hz; i.e., it has three times *Drosophila*'s bandwidth. *Sarcophaga*'s longer photoreceptor provides higher S/N at most frequencies. Plotting $\log_2(1 + S/N)$ gives bits per Hz. Thus graph shows how with an increased bandwidth (more Hz) and a better S/N (more bits per Hz) *Sarcophaga* has capacity to code more information (figure 8.12). Adapted from Niven et al. (2007) with permission.

with rate as the dominant fixed cost is divided by increasing numbers of bits. Parallel arrays of proteins and synapses show this same behavior (figure 6.5).

At higher rates, signaling cost dominates (figure 8.12). Nonetheless, efficiency continues to rise as the photoreceptor adapts to higher photon rates. Calcium feedback increases the information per photon, by reducing latency dispersion and increasing bandwidth, and it also decreases the cost per photon, by reducing response amplitude and duration.[13] Thus, gain control makes a photoreceptor most efficient where efficiency is most valuable, at its highest bit rate (figure 8.12). Why then are the photoreceptors that achieve higher maximum rates, that is, have higher information capacities, less efficient than those that achieve lower maximum rates?

Capacity and efficiency

To increase the information capacity of its analogue signal, a neuron must increase its bandwidth and/or S/N (equation 5.7). A fly photoreceptor increases both (figure 8.13). In doing so it loses efficiency. The fixed energy cost rises as (capacity)$^{1.5}$ and the total energy cost of operating at full capacity rises as (capacity)$^{1.7}$ (Niven et al., 2007), These increases are unavoidable; they are dictated by biophysical and biochemical constraints on bandwidth and S/N.

Once the bandwidth of light-gated current has been increased by reducing the duration and latency of single photon responses, the next obstacle to achieving a higher information rate is the membrane time constant. A photoreceptor reduces membrane time constant by increasing the density of open potassium channels (equation 6.10). Membrane bandwidth increases in proportion to open channel density, as does information rate (equation 5.7). The energy cost also increases in proportion to open channels; consequently, when bandwidth is increased by reducing membrane time constant, efficiency is little changed. It follows that energy efficiency is lost primarily by increasing S/N.

To increase S/N, a photoreceptor must increase its photon rate. When operating at full capacity, in sunlight, there is no shortage of photons entering the photoreceptor, so to increase rate a photoreceptor's microvilli must transduce more of the available photons. Because dead time limits a microvillus's photon rate to about 50 s^{-1}, this can only be done by adding more microvilli (Howard et al., 1987; Anderson & Laughlin, 2000; Song et al., 2012). These constraints explain why the high S/N *Sarcophoga* photoreceptor is 4 times longer than the low S/N *Drosophila* photoreceptor (figures 8.12, 8.13). A longer photoreceptor with more microvilli has a greater membrane area, which increases fixed cost and signaling cost. These costs rise in proportion to maximum photon rate and hence $\sqrt{(S/N)}$. However, information increases as $\log_2(1 + S/N)$ (equation 5.7), so efficiency falls. A higher rate photoreceptor also needs more synapses in a larger terminal (chapters 9, 10, and 11), and their higher fixed and signaling costs decrease efficiency still further.

In summary, increasing S/N requires a larger photoreceptor with more microvilli and synapses. Costs increase as the square of S/N but information increases as the log(S/N). Thus, increasing a fly photoreceptor's information capacity inevitably reduces efficiency. This fact has a profound influence on neural design.

An increase in cost with capacity punishes excess capacity

With fixed and signaling costs rising out of proportion to capacity, a bit of information costs more in a high-capacity cell. Fixed cost elevates cost per bit at low rates, and signaling cost elevates cost per bit at high rates (figure 8.12). It follows that an efficient design reduces the cost of all transmitted bits by eliminating excess capacity. Thus the capacity of an efficient photoreceptor is matched to the information supplied by the eye's optics.

Matching a photoreceptor's information capacity to optical supply

The information presented to a photoreceptor depends on the optical quality of the retinal image and the speed of image movement. An image sharply focused by high-quality optics delivers more information than an image blurred by low-quality optics. Increasing image speed increases the rate at which a photoreceptor receives this information by presenting more parts of the image per second.

Differences in image quality and speed explain why a *Drosophila* photoreceptor has fivefold lower information capacity than a blowfly photoreceptor, 200 bits s^{-1} compared with 1,000 bits s^{-1} (figure 8.14). *Drosophila*'s small, low-quality eye has 5 times less spatial resolving power than a blowfly's, and the poorly focused image moves more slowly because *Drosophila* is less agile in flight. Consequently, a *Drosophila* photoreceptor receives information at a lower rate. By lowering capacity to match *Drosophila* increases efficiency by sixfold (figure 8.14). The saving is large because S/N is reduced (figure 8.13). But, given the large savings made by reducing S/N, why does a larger fly increase capacity by increasing S/N? Capacity increases in proportion to bandwidth at constant S/N, with no loss of efficiency.

S/N codes parts of a moving image that bandwidth cannot. As an image moves over a photoreceptor, each spatial frequency in the image is

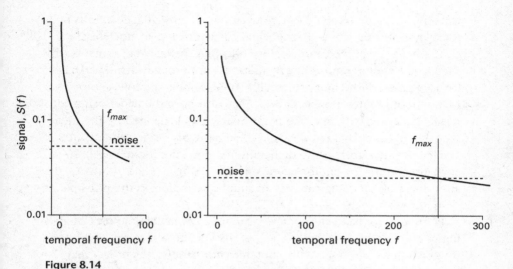

Figure 8.14

**A photoreceptor must increase both bandwidth and S/N to retrieve spatial informa-
tion from a faster moving image. Left:** The spectra of signal and noise in a photoreceptor coding a slowly moving image. Signal amplitude per Hz of temporal frequency, $S(f)$ is plotted against temporal frequency, f. This temporal spectrum, $S(f)$, is
generated by the image of a natural scene moving across the photoreceptor. According to natural image statistics signal falls off steeply with increasing spatial frequency. Consequently $S(f)$ falls likewise, according to the scaling relationship *temporal
frequency = spatial frequency × image speed*. Photoreceptor signal $S(f)$ falls below photon
noise (flat spectrum) at a temporal frequency f_{max} = 50 Hz. Thus the photoreceptor retrieves spatial information up to a spatial frequency limit of f_{max} / *image speed*. **Right:**
Image speed is increased fivefold. To code the same spatial information the temporal
bandwidth increases fivefold to 250 Hz. However faster movement reduces signal at
each temporal frequency, $S(f)$, by spreading the spatial signal power over a fivefold
wider range of temporal frequencies. Consequently noise is reduced to retrieve signal
as indicated, and this requires an increase in S/N.

converted into a temporal frequency at the photoreceptor according to a
simple formula: *temporal frequency = spatial frequency × image speed*. The signal delivered to a photoreceptor decays sharply with increasing temporal
frequency for two reasons. The spatial frequency spectrum of a natural
scene goes as $1/f^2$, and the roll-off is steepened by optical blur. Consequently, when a photoreceptor extends temporal bandwidth to code faster
moving images, it soon reaches a temporal frequency where the signal dips
below the noise (figure 8.14). Higher temporal frequencies are lost in noise
and so, therefore, are the higher spatial frequencies they represent (van

Hateren, 1992a,b). If the fly is to recover these finer spatial details, its photoreceptors must increase their S/N and pay with a loss in efficiency.

The necessity of increasing S/N to code finer spatial detail explains why the longest photoreceptors in a fly retina are in the zone where spatial acuity is highest. With more microvilli, these longer photoreceptors have higher photon rates and better S/N. They also have the wider bandwidth needed to code the higher temporal frequencies generated by the movement of higher spatial frequencies. These necessary improvements in S/N and bandwidth translate into higher information rates (Burton et al., 2001). Such fine-tuning of photoreceptors across a single retina speaks to the advantages of matching capacity to supply, according to the principle of symmorphosis.

Photoreceptor bandwidth is also matched to behavioral needs. Comparing photoreceptors in a wide variety of insects, those that are flightless or fly slowly have *slow eyes*, with photoreceptor bandwidths of less than 30 Hz, while faster flying insects have *fast eyes*, with bandwidths from 30 Hz to 120 Hz.[14] Because a significant cost of bandwidth is opening more potassium channels to reduce the membrane time constant, the photoreceptors of slow and fast eyes use different types of potassium channels (Laughlin & Weckström, 1993). These potassium channels have been selected from the parts list to code signals more efficiently.

How potassium channels increase efficiency

A fast photoreceptor uses noninactivating potassium channels whose voltage sensitivity and dynamics match the gain and bandwidth of the membrane to the signals being coded (Weckström et al., 1991). Voltage sensitivity is adjusted so that the membrane resistance decreases progressively with light-level, as the photoreceptor depolarizes from –65 mV in the dark to about –30 mV in full daylight. Membrane resistance determines membrane's gain, in millivolts per nanoampere, so resistance and gain are highest in the dark, to produce large single photon responses with least current and energy, and lowest in the daylight, to save energy and prevent saturation. Reducing resistance also reduces the membrane time constant ($\tau_M = RC$). Consequently the membrane's bandwidth increases in step with the increasing bandwidth of the light-induced current, thus protecting higher frequency signals from attenuation.

To further protect high-frequency signals, a fast cell's voltage-gated potassium channels activate and deactivate relatively slowly (figure 8.15). This slow response spares high frequencies because the channels cannot keep pace with rapid changes in voltage. It also confines attenuation to low frequencies, thereby reducing redundancy (chapter 9). By not inactivating, the channels maintain the membrane resistance at the same value during steady depolarization, thereby holding the membrane's gain and bandwidth at the values appropriate for the intensity that produced that depolarization.

These noninactivating potassium channels are expensive to operate—channels that are continually open constantly consume energy. Nonetheless, the channels are used efficiently. They are largely closed in the dark, and this reduces fixed cost. The channels then open progressively with increasing light level, allocating energy to lowering gain and widening bandwidth according to need.

A slow photoreceptor uses a different type of potassium channel. This opens quickly in response to a sudden depolarization, to damp excessive responses, but then closes by inactivating (figure 8.15). Closure brings three benefits. The membrane has a longer time constant, and this improves S/N by removing higher frequency photon noise. Indeed, some slow photoreceptors tune their membrane to be a *matched filter* that optimizes S/N by having the same frequency response as the light-induced current signal (Laughlin, 1996). Second, contrast signals are transmitted with a higher gain (Niven et al., 2003b). Third, energy is saved (Laughlin & Weckström, 1993).

The savings made by using the right type of channel are substantial. *Drosophila's* slow photoreceptor uses inactivating potassium channels, and when these are disabled by mutation, the photoreceptor compensates by inserting noninactivating potassium channels in its membrane. The substitution of an inappropriate channel halves the photoreceptor's information rate and doubles the energy cost per bit (Niven et al., 2003a,b). Some insect photoreceptors substitute channels purposefully. A locust switches from noninactivating channels during the day, when vision is faster, to inactivating at night, when vision is slower, using the neuromodulator serotonin (figure 8.15; Cuttle et al., 1995).

The fine-tuning of potassium channels continues a familiar theme. Almost every component of a fly photoreceptor is adapted to code information efficiently. But information is a general measure of representational capacity (chapter 5). When an eye is specialized for a particular task, such as detect a mate, photoreceptors select voltage-gated channels that adapt them to this task.

Photoreceptor as mate detector

A drone bee's one purpose in life is to mate with a queen. A queen flies across the sky, and drones take off and race to be the first to intercept her.

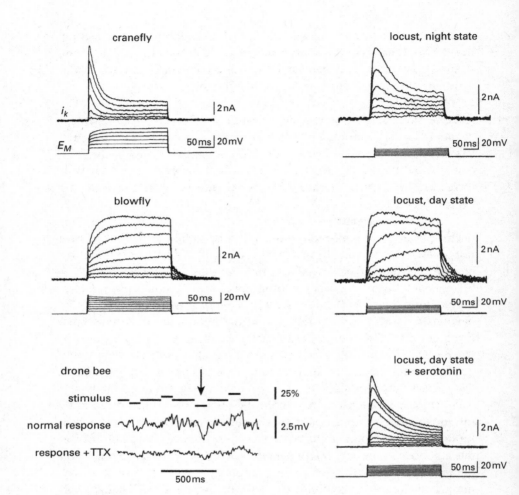

Figure 8.15

Insect photoreceptors select voltage-gated channels from parts list to increase efficiency and improve specific behaviors. Upper left: Cranefly photoreceptor. Potassium current i_K, (upper traces) rises quickly as stepwise increases in membrane potential, E_M (lower traces) opens voltage-gated potassium channels. Current then decays as channels are inactivated by sustained depolarization. Channel activation and inactivation increase with depolarization. **Middle left:** Blowfly photoreceptor, type R1–R6. Potassium channels have different properties; channels activate more slowly with almost no inactivation. Blowfly is more aerobatic than cranefly, and must resolve faster moving images. Blowfly channels sustain potassium conductance to increase membrane bandwidth; cranefly channels inactivate to save energy. **Upper right:** Locust photoreceptor in night state, vision is slow, so channels behave like slow cranefly. **Middle right:** Locust photoreceptor in day state, vision faster, channels behave like fast blowfly. **Lower right**: Locust in day state but photoreceptor treated with serotonin, a neuromodulator that mediates circadian changes in visual system. Currents are night state. **Lower left**: Drone bee photoreceptor uses voltage-gated sodium channels to amplify response produced by queen bee flying overhead. Queen produces small decrements in light intensity (as at arrow, upper trace). Photoreceptor responds by hyperpolarizing (middle trace). Blocking sodium channels with tetrodotoxin (TTX, lower trace) reduces amplification, thereby reducing responses to decrements and increments, and noise level. Range bars in plots of potassium currents; i_K 2 nA; E_M 20 mV; time base 50 ms. Cranefly and blowfly data replotted from Laughlin & Weckström (1993). Locust data replotted from Cuttle et al. (1995). Drone bee adapted from Vallet et al. (1992).

For this task a drone has evolved a pair of exceptionally large eyes. To improve S/N, the eye has wide lenses and long photoreceptors. The opsin is tuned to short wavelengths to enhance the contrast of a dark target, a queen, against the blue sky.[15] A drone photoreceptor also amplifies signals using the voltage-gated sodium channel that other neurons use to generate spikes, but its membrane is designed to prevent spikes; these would interrupt the flow of analogue information. Sodium channels and potassium channels balance to allow sufficient positive feedback to accelerate and amplify the signal, but too little for a spike.

The sodium channels amplify both depolarizing and hyperpolarizing signals (figure 8.15), in the latter case by closing to reduce inward current. This *subthreshold amplification* by sodium channels doubles the amplitude of small blips in membrane potential caused by a queen crossing a photoreceptor's field of view (figure 8.15), so that she is more easily detected by neurons in the drone's brain. The subthreshold amplifier's limited dynamic range matches the levels of depolarization generated by bright skylight. Under dimmer conditions, when photoreceptors are less depolarized, drones fail to take off in pursuit of the queen (Vallet & Coles, 1993). This design, where sodium channels amplify small voltage signals of either polarity, is also used by neurons in mammalian cerebral cortex to accentuate their synaptic inputs.

Summary and conclusions

This chapter asked initially: (1) how are chemical and electrical circuits designed to accomplish something useful; (2) what are the dominant costs for such hybrid circuits; and (3) why do mammalian receptors work "backward"? These questions seem now to be answered.

Photoreceptor efficiency improves measurably by following the broad principles of neural design. For economy, they *process by chemistry*, using molecules, complexes and compartments that are *irreducibly small* and *adapted* to their task. For example, by selecting different versions of signaling proteins from the parts list, a rod achieves low dark noise, high gain, and slow response whereas a cone achieves low gain and rapid response. For efficiency photoreceptors *match structures and components* at many different levels. This includes matching numbers, molecular ratios, and affinities of the key transduction proteins on a rod disc and matching the sensitivity of voltage-gated potassium channels to bandwidth in a fly photoreceptor.

Photoreceptors *process in analogue*, both to *compute directly with analogue primitives* (e.g., the log transform used for contrast coding) and to achieve the high bit rates needed to code images that are rich in detail. However, they must process and transmit signals electrically for speed over distance, and this is the dominant cost. With high gain in dim light and transmission at high information rates in bright light, high costs create intense pressure to be efficient. This leads photoreceptors to *adapt, match, and send at lowest acceptable rates*.

Most broadly, one may conclude that mammals were not trapped with an inefficient design. Quite to the contrary, the fly design is less efficient, first because it has not *specialized*, either for transmitting low S/N signals with high gain (like rods in dim light) or for transmitting high S/N signals with low gain (like cones in bright light). Second, because the fly design boosts speed and gain via positive feedback in a small compartment, which increases noise.

The good news for flies is that their design achieves the speed and bandwidth that the fly requires but that the mammalian design did not deliver. Now the fly photoreceptor's task is to maintain the speed and bandwidth of synaptic transmission at least cost. For this, it uses a specialized synaptic interface, the *lamina*, as discussed next (chapter 9).

9 The Fly Lamina: An Efficient Interface for High-Speed Vision

The first layer of the blowfly's visual system, the lamina, is a large and costly structure, occupying 10% of the brain's volume, using 20% of the brain's neurons, and consuming 2% of the blowfly's resting energy production. Yet the lamina is just an interface that receives signals from the compound eye's photoreceptors, processes them a little, and transmits them onwards in the brain. Why invest so much in such a seemingly simple task? The answer lies in the economics of one of a blowfly's life essentials, visual information.

The aerobatic blowfly depends on visual information because without a clear view from its cockpit window it will surely crash. The fly moves and turns fast, so to adequately resolve its surroundings, its compound eyes must provide for sensitive, high-speed vision. For contrast sensitivity, each of the eyes' 6,000 image pixels is coded by eight photoreceptors. For high-speed vision the eye's photoreceptors code frequencies out to about 300 Hz to achieve information rates of 1,000 bits s^{-1}—3 times faster than a human cone (chapter 8). High performance costs energy, materials, and space (chapters 6 and 8). Consequently, the compound eyes' photoreceptors account for 8% of the blowfly's resting energy consumption and constitute 4% of body weight. A sensitive, high-speed image acquisition system does not come cheap.

To profit from its expensive eye, the blowfly must translate the information captured by photoreceptors into useful behavior, and for this it invests 1% of its body mass, 50% of its brain's volume, and 80% of its brain's neurons[1] in visual circuits (figure 4.12). However, when these circuits fail to receive some photoreceptor information, vision deteriorates. The view from the cockpit window is hazier, the fly must throttle back, and, proceeding more cautiously, it could lose its race with a mate or a rival or a predator. Thus, for the fly to profit from its investment in vision by detecting the features that guide behavior, the information coded by photoreceptors must be transmitted to circuits in the brain.

This is the lamina's task, to transmit information from photoreceptors to circuits in the medulla, for feature extraction (chapter 4). For this, the lamina uses neurons specialized to transmit at high bit rates, the large monopolar cells (LMCs). But to transmit at high rates, the lamina's synapses and circuits must solve a serious practical problem. Photoreceptors code information in a fragile format. Natural scenes are low contrast, the average is 0.4 (Laughlin, 1981), and a photoreceptor has a low contrast gain, 3–4 mV per unit contrast (Anderson & Laughlin, 2000). Consequently, most of the information is in analogue signals of less than 3 mV and is, therefore, vulnerable to noise and attenuation during transmission.

To preserve the information coded by photoreceptors' weak signals, the lamina obeys information theory (equations 5.6 and 5.7)—it maintains bandwidth and S/N. It maintains spatial bandwidth by mapping the photoreceptor array retinotopically onto an equivalent array of output neurons; it maintains temporal bandwidth by transmitting analogue signals across fast synapses and keeping wires short; and it maintains S/N with large numbers of high-gain synapses that amplify the signal and average out noise. But biophysics and cell biology make bandwidth and S/N expensive commodities, which explains why the lamina is large and costly to operate. Costly processes should be sparingly applied and efficiently implemented, which explains why the lamina preserves and transmits information efficiently, according to principles of neural design.

Wiring preserves information and increases efficiency

Precise connections preserve spatial information

The regular array of image pixels formed by the compound eye is mapped 1:1 onto a corresponding array of neural modules, *lamina cartridges* (figure 9.1). A cartridge takes its inputs from the eight photoreceptors that code the same pixel and sends its outputs to the corresponding module at the next level of processing, the *medulla cartridge*. This precise *retinotopic projection* from pixel to cartridge preserves spatial information in two ways. First, giving each pixel its own set of neurons maintains spatial resolution, as happens in our fovea where each cone projects to its own pair of midget ganglion cells via its own pair of midget bipolar cells. Second, a retinotopic projection maintains the spatial continuity of objects in the world and this simplifies spatial processing. It also minimizes wire by reducing the length and complexity of the neural connections that are used to compute the

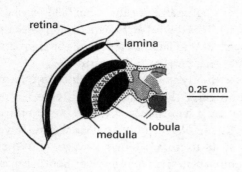

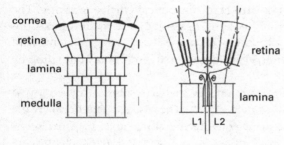

Figure 9.1
Layout and wiring of fly lamina. Upper: Section through head of housefly *Musca*, showing extensive lamina close to retina and rest of optic lobe, medulla then lobula. **Lower left:** Retinotopic projection to lamina cartridges and then to medulla cartridges. **Lower right:** Neural superposition wiring pattern showing 3 of the 7 co-aligned photoreceptor axons projecting to a single lamina cartridge. Lamina cartridge sends output to medulla cartridge with large monopolar cells; 2 shown, L1 and L2. Upper redrawn from Strausfeld (1976).

spatial and temporal relationships that define objects. For these reasons, many visual systems project information retinotopically (e.g., chapters 4, 11, and 12).

Using a wiring pattern to gather more information

An ingenious combination of optics and wiring, the *neural superposition* eye, enables a blowfly to see better by gathering more photons per pixel (figure 9.1). Many other insects, such as bees, crickets, and dragonflies, use the simpler *apposition* eye in which each of the eye's many facets is a lens that focuses light onto a single waveguide, approximately 2 μm in diameter. The waveguide is constructed by a column of eight photoreceptors and because they all share the waveguide's light, all have the same field of view. Thus they all code one image pixel. This optical arrangement simplifies the retinotopic projection of image pixels to lamina cartridges; the eight photoreceptors' axons form a bundle that projects to one cartridge.

The optics and wiring of the fly's neural superposition eye is complicated, to gather more light (figure 9.1). As in an apposition eye, there are eight photoreceptors under each lens, but to gather more photons, they make separate waveguides. The photoreceptors R1–R6 make one waveguide each, and R7 sits on top of R8 to make a single waveguide. With seven waveguides, the photoreceptors beneath a lens receive about 7 times more photons,[2] and this increases S/N by about $\sqrt{7} = 2.65$, without having larger lenses. However, there is a downside to this optical efficiency saving. The waveguides under one lens are "looking" in seven different directions. Consequently, were the fly eye to follow the apposition wiring rule and send these axons to the same lamina cartridge, spatial resolution would drop by two thirds.

The neural superposition eye restores spatial resolution by adjusting both optics and wiring (Kirschfeld, 1967). The visual angle between the waveguides under a lens is made equal to the visual angle between lenses (figure 9.1). Now seven waveguides under seven different lenses are co-aligned. Wiring combines these independent samples of the same signal by directing the axons of the co-aligned photoreceptors to the same lamina cartridge. This "neural superposition" of signals involves axons crossing over to form complicated "minichiasms" between the retina and lamina, but is worth the effort. The wiring pattern restores spatial acuity and, by increasing the S/N by about $\sqrt{7}$ gains more information without enlarging the eye—a useful improvement in efficiency.

Neural superposition depends on the accurate wiring of 6,000 minichiasms between retina and lamina. Any mistake will destroy spatial information by superimposing signals from different points in space. To avoid this loss, the developmental growth rules that direct axons across minichiasms are at least 99.8% accurate (Horridge & Meinertzhagen, 1970). Their precision makes an important general point—neural circuits are not obliged to cope with inaccurate wiring. When valuable information is at stake, developmental mechanisms deliver. This suggests that where the brain does wire imprecisely, it is a matter of efficiency in that greater precision is not worth the cost (Lightner, 2011).

Information is represented more efficiently by minimizing wire

Just as the accurate projections via minichiasms are faithfully replicated across the eye, so are their target structures, the lamina cartridges (figure 9.2). Every cartridge contains 16 neural components, the axon terminals of photoreceptors R1–R6 and 10 interneurons, all encapsulated by 3 types of glial cell. Each neuron and glial cell has a characteristic shape and sits in a characteristic position. This replication of cartridges is worthwhile because a cartridge's structure is optimized to save wire (Rivera-Alba et al., 2011).

Efficient wiring places the neurons that make the largest numbers of synapses closest together (chapters 2 and 13). The cartridge's wiring diagram (connectome) shows that R1–R6 photoreceptors provide most of the cartridge's synapses (about 70%; figure 9.3; Rivera-Alba et al., 2011). Of these, by far the most numerous (60% of all synapses) are output tetrads (figure 9.3) at which a presynaptic site on a photoreceptor terminal drives four postsynaptic elements—one each from the principal output neurons, L1 and L2, one from an amacrine cell, and one from either the output neuron L3 (in the distal cartridge) or a glial cell (in the proximal cartridge). To make these most numerous connections as short as possible, the R1–R6 photoreceptor terminals form a cylindrical palisade around L1 and L2 (figures 9.2 and 9.3).

The amacrine cell dendrites form a distribution hub that participates in over 90% of the cartridge's synapses, so they are placed midway between the cartridge's central axis and its outer margin. The dendrites climb up the outside of the palisade next to the terminals of R1–R6, close to both their most frequent inputs, the photoreceptor tetrads, and their most frequent outputs, the three epithelial glial cells that sheath the cartridge (figure 9.3). The next most frequent amacrine output is to the basket cell, T1, which climbs the palisade alongside the amacrine.

Neurons that form fewer synapses are placed outside the palisade. The LMC, L3, which forms 40% fewer tetradic contacts with photoreceptors than L1 and L2, is placed outside the photoreceptor palisade so as not to obstruct the more numerous dendrites of L1 and L2, but remain close to R1–R6. So intense is the pressure to reduce wire that even the neurons that make fewer than 20 synaptic connections are near their optimal positions. Thus replicating cartridges in minute detail upholds a principle of neural design, *minimize wire*.

The intricacy and crystalline regularity of the lamina's cartridges (figure 9.2), and of the medulla's too, greatly impressed Cajal. He likened the insect visual system to a small and exquisitely crafted precision instrument, a hunting case watch and the vertebrate retina to a rude wall clock (Ramón y Cajal, 1917).[3] Chapter 11 will show that Cajal's comparison was superficial. The mammalian retina appears less orderly because, for efficiency, every point on the retinal image is sampled by more than 20 neurons, and their

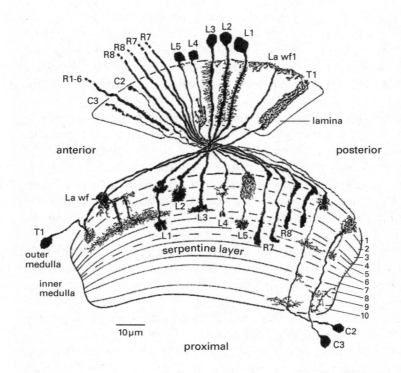

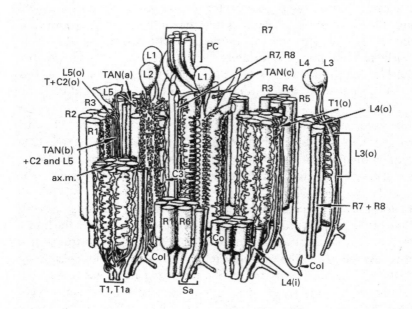

Figure 9.2
The neurons that comprise a lamina cartridge. Upper: Each shown in a different cartridge, then projecting across chiasm to medulla. **Lower:** Positions of neurons within cartridges. Note regularity of cartridge array. Each contains same set of neurons, identically positioned to save wire. Cartridges cut away to reveal internal structure. Note cartridge built around cylindrical palisade of six photoreceptor axon terminals and LMCs L1 and L2 within palisade with numerous dendrites. These receive synapses from terminals. Further details in text. Upper, *Drosophila* lamina neurons from Fischbach & Dittrich (1989) with permission. Lower, housefly cartridges from Strausfeld (1971) with permission.

overlapping dendritic arbors obscure their intricate connectivity patterns that also save wire.

Providing the capacity to transmit photoreceptor information

To preserve information, a photoreceptor's output synapses must transmit over 1,000 bits s^{-1}, so they transmit analogue signals (figure 9.4) with a wide bandwidth and a high S/N. For analogue synaptic transmission vesicles release histamine molecules that bind and open postsynaptic chloride channels (figure 6.7). Vesicles are released continuously, even in darkness, at a rate that is modulated by the photoreceptor signal. Release rate increases when a photoreceptor depolarizes in response to brightening and decreases when it hyperpolarizes to dimming. Consequently, synapses transmit an inverted version of the photoreceptor input (figure 9.4). Negative contrasts, 0.0 to –1.0, are coded as graded depolarizations, and positive contrasts, 0.0–2.0, as graded hyperpolarizations (figure 9.4). This dynamic range covers over 95% of natural contrast signals.

To maintain the temporal bandwidth of the photoreceptor input, a synapse responds 5 times faster than the photoreceptor (figure 9.5). To maintain S/N in the face of synaptic noise, each photoreceptor transmits with 220 parallel synapses, which means that with six photoreceptors R1–R6, a pixel's contrast is conveyed by an array of 1,320 parallel synapses. As with a parallel array of M noisy signaling molecules (figure 6.5), the S/N for transmission is improved by $\sqrt{M}$, which for $\sqrt{1,320} \approx 36$.

The S/N of the parallel synaptic array meets the fly's specifications for resolution and rate. With a combined mean vesicle release rate of 250,000 s^{-1}, the synaptic noise level is equivalent to a photoreceptor input signal of 70 µV and an image contrast of 1.2%. Because this contrast approximately equals the behavioral threshold for contrast detection, the cartridge fulfills

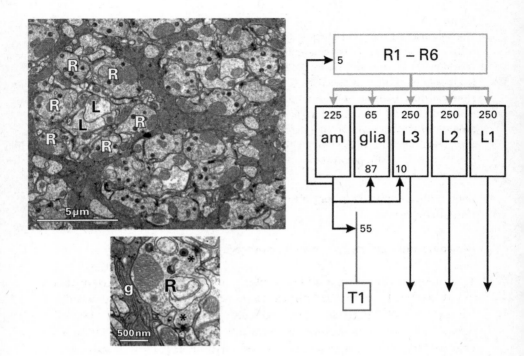

Figure 9.3
Lamina neurons, synapses and circuits. Upper left: A cross section through house-fly cartridge. Note palisade of six photoreceptors terminals, R, surrounding pair of monopolar cells, L1 & L2, and dense glial sheath enveloping whole cartridge. **Lower left**: Photoreceptor terminal, R, showing two tetradic output synapses (*) each with synaptic vesicles and release site with presynaptic density. Density faces four post-synaptic dendrites, only two seen in this section, L1 and L2. Note proximity of glia, g. **Right**: Wiring diagram of circuits in lamina cartridge of fly *Drosophila*. Numbers in-dicate numbers of synapses. For clarity strongest connections (10 or more) are shown and correlated inputs from six photoreceptor terminals, R1–R6 are lumped together. Single feedback connection onto photoreceptor terminals also shown to emphasize its weakness. Note that text gives numbers of synapses in blowfly lamina. In smaller *Drosophila* lamina numbers are about one quarter blowfly's. Upper and lower left, electron micrographs courtesy of Ian Meinertzhagen. Right, from Laughlin (2010); synaptic numbers updated from improved wiring diagram (River-Alba et al., 2011).

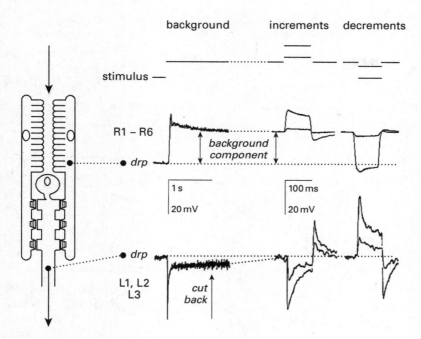

Figure 9.4
Analogue responses to light stimulus are transformed as they pass from photore-ceptors R1–R6 to LMC via an array of parallel synapses. Response to background is discounted. Response to increments and decrements about background, as produced by objects, is enhanced. **Left:** Schematic showing flow of signal from stimulus to LMC. **Right:** Responses of photoreceptor, type R1–R6, and responses of postsynaptic LMC (type L1, L2, or L3) to identical set of stimuli (top traces). Photoreceptor maintains steady response to bright background, about 20 mV above dark resting potential (drp). LMC response cuts back as predictive coding removes best estimate of background. Removal of background component permits amplification of responses to increments and decrements (compare amplitudes of photoreceptor and LMC responses). Amplified responses cut back during 100 ms increments and decrements, as predictive coder updates estimate of background. From Laughlin (2010) with permission.

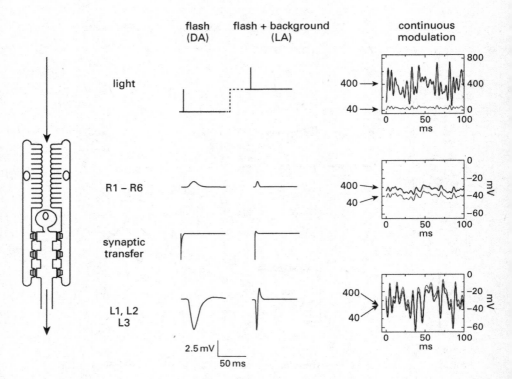

Figure 9.5

Photoreceptor output synapses process analogue signals as they are transferred to a large monopolar cell, L1, L2, or L3. Left: Schematic showing flow of signal from stimulus to LMC. **Middle:** Responses to brief flash delivered when photoreceptor dark adapted, flash (DA), and when light adapted by steady background, flash + background (LA). Light adaptation decreases duration of photoreceptor flash response. LMC flash response decreases in duration and changes waveform. Impulse response for synaptic transfer, deduced from pre- and postsynaptic flash responses shows that speed of LMC response is set by photoreceptor, change in waveform is produced during synaptic transfer. **Right:** Moving image of varied contrast generates continuous modulation of light intensity as it passes across pixel coded by LMC (top panel). Increasing illumination 10-fold, 40 units to 400, increases background and modulation 10-fold. Photoreceptor response (middle panel) generates contrast signal, modulations normalized to same amplitude, superimposed on different responses to background. During synaptic transfer to LMC, background signal is removed and contrast signal amplified to fill LMC response range. From Laughlin (2010) with permission.

the blowfly's requirement for high-definition vision. Combining S/N and bandwidth (equation 5.6 and 5.7), the array's information capacity is 2,100 bits s^{-1} (de Ruyter van Steveninck & Laughlin, 1996). This suffices to transmit over 80% of the information coded by the R1–R6 photoreceptors. The obvious question: "Why not transmit the remaining 20%?" we address next.

We close this section by noting that the equivalent neurons of the vertebrate retina, bipolar cells (chapter 11), also receive signals at high-vesicle-rate photoreceptor synapses and transmit them onwards in analogue. However, they divide the photoreceptor input into two parallel streams; ON bipolar cells that depolarize in response to positive contrasts and OFF bipolar cells that depolarize in response to negative contrasts. Chapter 11 will explain how this division is made and why.

The high cost of a high-capacity synaptic array

The information rates of parallel arrays are subject to the law of diminishing returns (figure 6.5), and this inflates the cost of information transmitted by photoreceptor output synapses (figure 9.6). A single output synapse transfers 55 bits s^{-1}, with an efficiency of 0.25 bits per vesicle, at an energy cost of 5×10^4 ATPs per bit, but 1,320 parallel synapses transmit only 40 times more bits, 2,100 bits s^{-1}, at 50 times the cost per bit, 2×10^6 ATP (figure 9.6). Increasing the capacity by another 400 bits s^{-1} to transmit the remaining 20% would require 50% more synapses. These would take 50% more space (see below) and add 1% to the blowfly's resting energy consumption. Perhaps this is why 20% of the photoreceptor information is discarded—it is simply too expensive to transmit.

As in other synapses (chapter 7), energy consumption is dominated by the ion pumps that, by recharging ionic batteries, sustain postsynaptic currents (Laughlin et al., 1998). The presynaptic energy costs are significant but difficult to estimate. Approximately 10% of the ATP is used to recycle vesicles and neurotransmitter, and at least 10% goes to sustain the presynaptic calcium fluxes that control vesicle release.

Now to the materials that synapses use, and the space they occupy. The cartridge's largest structure, the 40-μm-long, 2-μm diameter photoreceptor axon terminal, houses the machinery needed to release and recycle about 5×10^4 vesicles s^{-1}—namely, a pool of synaptic vesicles, vesicle release sites, recycling machinery, pumps, and mitochondria (chapter 7). The terminal cannot be much smaller—it is refilling vesicles equal to its own volume every 10 minutes. The postsynaptic neurons, L1 and L2, house the

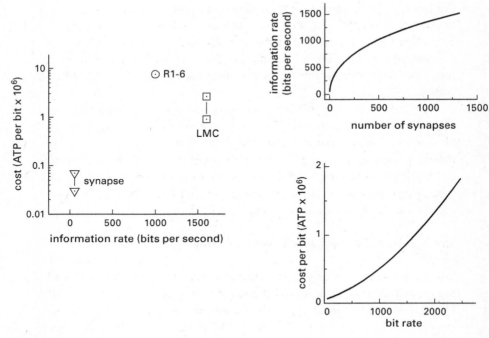

Figure 9.6
Energy cost per bit rises with bit rate in analogue neurons, and in parallel arrays of
analogue synapses. Left: Cost per bit versus bit rate for a photoreceptor, an LMC, and
a single synapse connecting photoreceptor to LMC, operating alone. **Upper right:**
Information rate rises with number of such synapses in array. Note law of diminishing returns, as seen in parallel array of signaling molecules (figure 6. 5). **Lower right:**
Because of diminishing returns, array's energy cost per bit rises with bit rate on a
steepening upward curve. Plotted from results of Laughlin et al. (1998).

histamine-gated chloride channels and the pumps and mitochondria
needed to power their currents (chapter 6).

L1, L2, and the six photoreceptor terminals occupy two thirds of the
cartridge's volume and make two thirds of the cartridge' synapses, suggesting that, as found elsewhere (chapter 7), neurons are sized to match the
number and activity of their synapses. Indeed, tetradic output synapses are
equally spaced over the surfaces of photoreceptor terminals, at one per 1.6
μm^2, irrespective of individual differences in terminal size (Meinertzhagen,
1993), suggesting that this is just enough surface area to support the needs
of the high-release-rate synapse. L1 and L2 also receive tetradic inputs with
equal density. In short, synapses are packed as densely as possible.

Consequently, any increase in numbers of synapses increases a neuron's surface area and hence its use of space, materials, and energy. This constraint helps explains why synapses are densely packed and why numbers are reduced by using synapses efficiently.

Synaptic structure increases synaptic efficiency

The efficiency of a photoreceptor output synapse is quadrupled by directing a vesicle's neurotransmitter to four output elements at a tetrad (figure 9.3). With this arrangement, one tetrad releasing 200 vesicles s^{-1} is equivalent in S/N and postsynaptic drive to four conventional monadic synapses, releasing a total of 800 vesicles s^{-1}. With 4 times fewer output synapses the terminals are 4 times shorter and, because terminals run the full length of a cartridge, the lamina's volume shrinks accordingly (figure 9.2). Although tetrads reduce the need for space and materials fourfold, they have little impact on energy efficiency because consumption is predominantly postsynaptic. A rod or cone's active zone also drives two or more postsynaptic neurons for the same reason, to economize on synaptic space and materials (chapter 11).

Why use an inefficient parallel array?

This perplexing question is raised by information theory. A single photoreceptor output synapse has a capacity of 55 bits s^{-1} (figure 9.6) so according to Shannon (chapter 5) a coding scheme that uses just 39 synapses can achieve the output terminals' capacity of 2,100 bits s^{-1}. However, 1,320 synapses are used. Why not implement the optimum coding and improve efficiency ×34?

Achieving the theoretical optimum makes heavy demands on neural circuit design. There can be no redundancy (chapter 5); every one of the 39 synapses must transmit a unique signal that does not correlate with the other synapses'. To achieve this the cartridge's circuitry must sum analogue signals from R1–R6 and then split the result into 39 independent components, each driving just one synapse. Such a split would require some complicated signal processing,[4] and for high-speed, high-resolution vision the processors must maintain the system's bandwidth and S/N. To complicate matters, local processing within a terminal is not an option because chemical and electrical signals from nearby synapses would interfere. A set of fast and accurate neural circuits would be needed, and these would likely cost more than the parallel array.

Here is an important lesson in neural design: what is most efficient in theory is not always most efficient in practice because neural implementation stands in the way. Interestingly the vertebrate retina does improve its coding efficiency by splitting photoreceptor signals into two independent components, ON and OFF (Von der Twer & MacLeod, 2001), but does so *postsynaptically* with bipolar cell glutamate receptors (chapter 11). Given that practical considerations force the lamina cartridge to use the noise reduction method of last resort, a parallel array of many synapses, an array should be used as efficiently as possible (chapter 6).

Matching synaptic capacity improves efficiency

In theory, an array's efficiency is improved by matching the number of synapses to the S/N and information content of the signals being transmitted (figure 6.5). Flies of different sizes demonstrate this match. A blowfly's photoreceptor codes information at 5 times the rate of *Drosophila*'s (1,000 bits s^{-1}; cf. 200 bits s^{-1}), and makes 5.5 times as many tetradic synapses (blowfly 220; *Drosophila* 40; Nicol & Meinertzhagen, 1982; Rivera-Alba et al., 2011). The S/N of a blowfly photoreceptor is 2.3 times that of *Drosophila*, and so is the S/N of its synaptic output array ($\sqrt{5.5} = 2.3$; Laughlin, 1994). In mammalian retina, cone photoreceptors adjust the numbers of their output synapses to make a similar match for the same economic reason; to use space, materials, and energy more efficiently (figure 11.9).

Sending only what is needed improves efficiency

The information capacity of a synapse or neuron is limited by its bandwidth and S/N, and, for efficiency, these resources must not be wasted on noise and redundancy; *send only what is needed*. Noise is reduced before synaptic transmission by electrically coupling photoreceptor axons as they enter the cartridge. Chapter 11 will consider primate foveal cones and explain how this presynaptic coupling works and why it is efficient. Reducing redundancy is more complicated but more profitable, and a cartridge uses efficient methods.

Removing redundancy with predictive coding
Predictive coding removes a large source of redundancy in natural images. Signals in nearby pixels, and hence signals in nearby lamina cartridges, are correlated because they are quite possibly representing the same object, similarly illuminated. Furthermore, the optical point spread function

increases local correlation by distributing light from a single point to several pixels. Because there are spatial correlations, the signal expected at one pixel can be predicted from the signals in surrounding pixels, and this prediction is, by definition, redundant (chapter 5). A predictive coder makes this prediction and then removes it from the incoming signal. As a result, signal amplitude is reduced by ~ 3/4 without loss of information. Because smaller signals can be transmitted at lower cost the need for space, materials, and energy drops accordingly.

Predictive coding is easily implemented in a retinotopic array of lamina cartridges (Srinivasan et al., 1982; figure 9.7). To code a single pixel, a cartridge takes the signal from the six photoreceptors sampling its pixel. It also forms a prediction of the signal it expects to receive from its pixel by taking a weighted sum of signals from surrounding pixels. These surround signals are readily obtained from neighboring cartridges in the retinotopic array and weighted and summed to form the prediction. This is subtracted from the signal delivered by the six photoreceptors and the difference, signal minus redundancy, is transmitted onwards to the medulla by an LMC.

Predictive coding uses a concentric receptive field, in which the surround, the predictor, produces a response of opposite polarity to the center, the signal from the coded pixel (figure 9.7). Many visual neurons use this form of lateral inhibition to code more efficiently by reducing redundancy (chapters 11 and 12).[5] Thus, predictive coding, an image compression algorithm invented by engineers almost 60 years ago to code TV signals efficiently, is implemented in animals by a basic sensory interaction, lateral inhibition, that has been used for at least 400,000,000 years.[6] We will return to the lamina cartridge's predictive coding mechanisms after describing how it removes temporal correlations, and deals with the different patterns of correlation found in natural images—forest, sky, savannah, and so on.

How predictive coding removes temporal correlation

The output of a single pixel changes over time as the eye moves across a scene and objects move within a scene. These temporal signals are also correlated. Movement converts spatial correlation into temporal correlation and, analogous to the optical point spread function, the photoreceptor flash response spreads signals over time, leaving traces of the past in the present.

A lamina cartridge removes these temporal correlations by temporal predictive coding (Srinivasan et al., 1982; figure 9.7). The cartridge uses the signal being received now, at time t, to predict how it will influence the

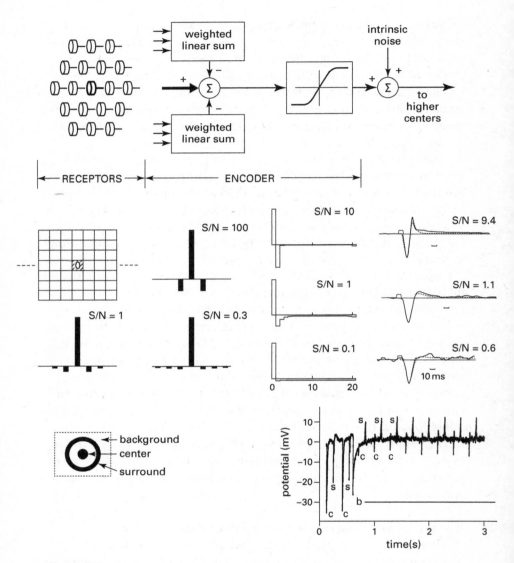

Figure 9.7
Spatial and temporal predictive coding by a large monopolar cell (LMC). Upper:
scheme for spatial predictive coding. Encoder subtracts weighted sum of signals from
surrounding receptors (the prediction) and subtracts from the signal at central receptor. This is lateral inhibition. Information now concentrated in smaller signal,
which is amplified with higher gain to fill response range, giving more protection
from intrinsic noise. **Middle left:** Predictive surround contracts, weights increase at
higher S/N. Weights plotted along central row of pixels in square array. **Middle right:**

Temporal predictive encoder subtracts weighted sum of earlier signals from present signal using flash response with off transient of opposite polarity. **Left column:** Theory. Off response duration decreases, weights increase at higher S/N. **Right column:** Theory versus experiment. LMC flash responses (solid traces) at different light levels and measured signal:noise follow theory (dashed traces). Pulse on central trace shows duration of flash. **Lower:** LMC rapidly adapts receptive field to light level to implement predictive coding. c, responses to flashes in receptive field center; s, responses to bright annulus in receptive field surround; b indicates presence of bright widefield background. Schematic on right shows extent of stimuli delivered to centre, surround and background. In darkness (no background), c and s same polarity; s is generated by light reaching center. Background activates subtraction of surround from center for predictive coding; s now has opposite polarity. Note surround subtraction activates within 100 ms of onset of background. Lower from Laughlin and Osorio (1989) with permission. Remainder from Srinivasan et al. (1982) with permission.

signals that follow, at $t + \Delta t_1, \Delta t_2, \ldots$ etc., and then subtracts this prediction away at $t + \Delta t_1, \Delta t_2, \ldots$ etc., as the new signals arrive.

This procedure sounds complicated, but it is easily implemented by the time course of a neuron's response to an instantaneous input, its *impulse or flash response* (figure 9.7). The flash response of neurons L1, L2, or L3 has two phases (figure 9.7). The fast initial response codes the signal received here and now, at time t. The smaller, more slowly decaying response is the prediction of what *should* follow and, because it is of opposite polarity, this prediction subtracts from the signals that *do* follow. This temporal predictive coding is also performed by bipolar cells of the vertebrate retina (chapter 11), for they too work more efficiently when redundancy is reduced. Note that the cartridge's predictive coder, the biphasic impulse response, eliminates the most obvious form of redundancy, a photoreceptor's steady response to the background signal (*cut back* in figure 9.4).

Matching predictive coding to image statistics

Consider a blowfly pursuing a delicious scent on a bright summer's day. Dashing through a wood, through a hedge, and out into a pasture dotted with cows, the blowfly encounters three visual scenes in rapid succession, each with a different pattern of spatial correlations. Fortunately, predictive coding does not have to adapt its predictors to scene changes because, when photoreceptor signals are reliable (i.e., on bright summer days), the nearest neighbors provide the best estimate (figure 9.7). Thus, one spatial predictor and one temporal predictor fit most natural scenes (Srinivasan et al., 1982).

However, there is an image statistic that must be adapted to, photon noise which reduces correlations by randomizing input signals. The effects of photon noise increase as the light level falls so, to avoid the S/N disaster of subtracting an unreliable estimate from a noisy signal, predictive coding adapts to the changing noise level. Like an election pollster faced with a population of swinging voters, coding's predictors increase the reliability of their estimates by taking more samples. The spatial predictor widens to include more pixels in the surround, and the temporal predictor prolongs the inverted phase of the flash response (figure 9.7). Weightings are reduced so that when faced with a totally predictable image, a large area of uniform brightness, the prediction equals the incoming signal. Mechanisms in the lamina cartridge make these adjustments precisely, within tens of milliseconds (figure 9.7), to make good predictions over a 10^6-range of background (i.e., mean) light levels. Many other visual systems, our own included, adapt rapidly to the lowering of intensity by extending receptive fields and impulse response for the same reason, to continue to remove redundancy and *send only what is needed*, information.

How the cartridge implements predictive coding

The basic plan for a predictive coder is straightforward—build spatial and temporal predictors and subtract their outputs from incoming photoreceptor signals (figure 9.7). However, the costs of predicting and subtracting depend on the mechanisms used and where they applied. By applying economical mechanisms at strategic locations, the lamina cartridge increases its *overall* efficiency more than sixfold. Mechanisms are economical because they avoid using noisy and expensive chemical synapses, and they are applied presynaptically to reduce expenditure on vesicle release and postsynaptic current.

The advantage of subtracting presynaptically
A presynaptic mechanism is advantageous because, by acting on vesicle release, it eliminates redundancy from the vesicle stream. Redundancy constitutes 75% of the input, so its removal reduces the number of vesicles required for transmission by 75%.[7] The number of tetrad synapses needed to transmit information is reduced by the same factor, thereby cutting the costs of materials, space, and energy by 75%. No information is lost by subtracting redundancy, so efficiency increases fourfold.

Presynaptic subtraction saves yet more space, materials, and energy by avoiding the wasteful alternative—postsynaptic subtraction. To subtract postsynaptically, the synaptic chloride current that drives LMCs must be opposed by a current flowing in the opposite direction, and this current would double energy consumption by doubling ion flux. Indeed, opposing one current with another is similar to slowing a car by putting the *left* foot on the brake pedal while keeping the right foot on the gas. By acting presynaptically, subtraction regulates the gas without stepping on the brakes.

Synapses would be needed to deliver the opposing current to LMCs, these would consume extra space and materials, and to maintain a high S/N, many would be used. The photoreceptors use 1,320 tetrads to maintain a high S/N in LMCs, but the subtracted signal (the prediction) is usually smaller, so 700 synapses might suffice. To receive these extra synapses, the neurons L1 and L2 would have to be 50% longer, increasing the cartridge's volume by the same percentage. Totaling up the extra costs of subtracting postsynaptically, space and materials increase sixfold and energy increases twofold. This makes subtracting presynaptically 6 times more efficient in space and materials and twice as energy efficient. Olfactory receptor neurons and mechanoreceptors use presynaptic mechanisms to regulate their vesicle release to make similar efficiency savings (Nawroth et al., 2007).

Why nonsynaptic mechanisms are used for presynaptic subtraction

The wiring diagram for lamina circuits (figure 9.3) indicates that nonsynaptic mechanisms play the primary role in subtracting redundancy from the presynaptic terminal. Only 10% of the terminal's synapses are inputs, and only input synapses can subtract, so their contribution is either too slow or too noisy. The prediction is mainly subtracted by two nonsynaptic mechanisms—a molecular feedback circuit within the terminal membrane and a slower electrical circuit that changes the potential of extracellular space. These mechanisms have three advantages: they avoid adding synaptic vesicle noise; they make efficient use of existing resources; and they not only subtract the prediction; they formulate the prediction. In other words, they neaten up.

The feedback circuit within the photoreceptor terminal

Feedback within the photoreceptor terminal generates the biphasic impulse response used for temporal predictive coding (figure 9.5). Feedback is fast; it acts within 1.5 ms (figure 9.5), so electrical circuits are being used, but the ion channels have not been identified.[8] Speed is essential because the prediction must keep up with the high information rates of photoreceptors in bright light. However, when light levels fall, information rates decrease, noise increases, and the feedback mechanism adapts by slowing down and progressively weakening its effect (figure 9.5). This adaptation to input matches the predictor to decreasing S/N (figure 9.7).[9]

Changing the potential of extracellular space

When light depolarizes the photoreceptor terminal, the cartridge's extracellular space depolarizes more slowly, by a lesser amount (figure 9.8; Laughlin, 1974; Shaw, 1975). Conversely, when light dims and the photoreceptor hyperpolarizes, the extracellular response does likewise. These extracellular responses subtract directly from the intracellular signal at the photoreceptor terminal because the voltage-gated channels that regulate vesicle release experience the difference between the intracellular and extracellular electrical potentials. And, because the extracellular potential changes more slowly than the intracellular signal, the extracellular potential is subtracting a part of the temporal prediction. Indeed, it eliminates a large proportion of standing background signal (figure 9.8; Weckström & Laughlin, 2010).

The extracellular potential is set up by glial cells, which envelop the cartridge (figure 9.8) and interlock with tight junctions to block the extracellular flow of ions (and hence current) into and out of the cartridge. The currents that enter or leave the cartridge in neurons (e.g., receptor terminals and LMCs) must complete their circuits by returning to their origins in extracellular space, and in doing so they set up an extracellular potential as they cross the glial resistance barrier.

Consider the current that enters photoreceptors by light-gated channels, flows down to the axon terminal, and depolarizes the terminal membrane as it flows out into the cartridge's extracellular space (figure 9.8, lower). To complete its circuit, this current must return to its point of origin, the photoreceptor layer's extracellular space, and on this return leg it crosses the glial resistance barrier and depolarizes the extracellular space. The currents that drive electrical signals along the axons of the output neurons L1, L2, and L3 to the medulla also return to their point of origin, the extracellular space adjacent to postsynaptic chloride channels, and as these return currents cross the glial resistance barrier, they increase the extracellular depolarization produced by photoreceptor current.

An extracellular mechanism that depends on return current is both economical and noise free because it directs the flow of existing currents across a passive resistance without directly engaging chemical synapses. The mechanism is also effective; it removes much of the standing background signal (figure 9.8). However, we still do not know how the changes in extracellular potential are slowed down for temporal predictive coding, by what is effectively a capacitor in parallel with the resistance barrier.

The extracellular mechanism also performs spatial predictive coding. The glial cells that separate cartridges (figure 9.8) allow some current to flow laterally between cartridges. Consequently, a part of the extracellular signal in one cartridge is the sum of extracellular signals from neighboring cartridges, weighted by the resistances of the glia that separate cartridges. These weighting resistors adapt to light level, as required for predictive coding. Resistance increases in bright light, both to strengthen the temporal prediction by increasing the extracellular potential produced by current injected into the cartridge, and to narrow and deepen the spatial prediction by restricting spatial spread of current. The large numbers of synapses made by photoreceptors and amacrines onto glia (figure 9.3) are well placed to makes these adjustments by changing glial membrane resistance.[10]

Circuits that compute by changing the potential of extracellular space are found elsewhere. Mammal retina apparently uses an extracellular field potential to implement spatial predictive coding (chapter 11). Vertebrate hair cells and invertebrate mechanoreceptors actively polarize the extracellular space by using pumps to drive transducer currents. Certain axons can generate extracellular current that, when enclosed by a glial capsule cause rapid electrical inhibition at the initial segment of a projection neuron (figure 7.11).

Transmitting information to the medulla

Once synapses have transferred information from photoreceptors to the LMCs L1, L2, and L3, the information must be transmitted to output synapses in the medulla, 0.5–1.0 mm away. To maintain high bit rates, L1 and L2 transmit analogue signals along axons that are passive cables, designed to conserve information by preserving S/N and bandwidth (van Hateren & Laughlin, 1990).

The S/N at an axon's output synapses depends on three factors, the amplitudes of the signal and noise transmitted to the output synapses, and the amplitude of the noise added by output synapses. To preserve S/N, the axon is designed to reduce the attenuation of transmitted signal and increase the attenuation of transmitted noise. The axon membrane has an unusually high specific resistance, $5 \times 10^4 \, \Omega \, cm^2$, which is equivalent to one open potassium channel per 6 μm of axon. According to cable theory (chapter 7) this reduces signal attenuation by reducing leakage during transmission. With this well-insulating membrane, the 3-μm diameter axon transmits low frequencies virtually without loss (figure 9.9).

Higher frequencies are attenuated by the membrane's capacitance (chapter 7), and to reduce this effect, that is, to conserve bandwidth, the signal is

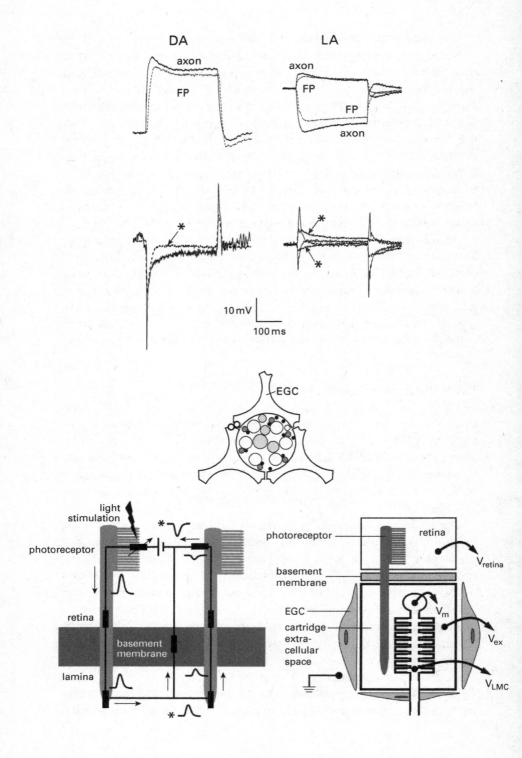

Figure 9.8

Extracellular field potential in lamina cartridge subtracts from presynaptic signal at photoreceptor axon terminal. Upper: Field potential (FP) responds to light like axon but slower and by lesser amount. DA, light pulse delivered in darkness; LA, increment and decrement about constant background. **Middle:** Subtraction of FP from axon response gives membrane potential at axon's synapses (*). Response is transient and resembles larger postsynaptic response of LMC. **Lower center:** Diagrammatic cross section of lamina cartridge showing tight sheath made by three epithelial glial cells (egc). **Lower right:** Barriers, compartments, and potentials. Glia separate cartridge's extracellular space from lamina's extracellular space, which being directly connected to body cavity is ground. Glia in basement membrane isolate retinal extracellular space from lamina extracellular space and from ground. Potentials relative to ground, V_{retina} = retinal extracellular space, V_{ex}, cartridge extracellular space; $V_{M(Vm)}$, membrane potential of LMC cell body; V_{LMC}, membrane potential at LMC synaptic zone. **Lower left:** Circuit that generates FP. Pulse of light-gated current entering photoreceptor in retina depolarizes axon terminal membrane, crosses into lamina extracellular space and returns to retina, most across glial resistance barrier at basement membrane, but small amount through photoreceptor in same cartridge if not so depolarized. Return current depolarizes cartridge's extracellular space. From Weckström & Laughlin (2010) with permission.

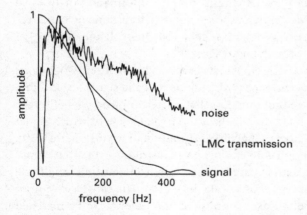

Figure 9.9

LMC axon is a matched filter that selectively attenuates noise as it transmits passively from lamina to medulla. Analogue signal amplitude at medulla terminal, normalized to lamina input, is plotted against signal frequency to give frequency response for passive transmission (LMC transmission). Transmission curve matches amplitude spectrum of input signal, and filters out high frequency input noise. A perfect match between frequency response and signal spectrum optimizes output S/N in the presence of broad band noise but LMC's passive axon cannot meet this specification, a cable's frequency response is too shallow. Nonetheless it attempts to match. From Van Hateren & Laughlin (1990) with permission.

forced down the axon by the array of 1,320 photoreceptor output synapses. Their high synaptic conductance rapidly charges and discharges the axon's capacitance, so that signals up to 100 Hz are attenuated by less than 25%, and transmission extends to 400 Hz (figure 9.9).

To selectively attenuate noise before it arrives at an axon's output synapse, the frequency response for transmission follows the input signal spectrum. Thus, by acting as a *matched filter*, the axon strikes the optimum balance between attenuating signal and eliminating high-frequency synaptic noise (figure 9.9).

Although LMC axons are designed to be efficient cables, signals are inevitably attenuated by passive transmission, so some information will be lost to intrinsic noise. To minimize this loss the axon's input signal must be as large as possible. Amplification is required, and to be most efficient, it too is optimized.

Optimizing amplification

With redundancy removed by presynaptic predictive coding, the photoreceptor synapses amplify signals to fill the LMC response range. Amplification is matched to input statistics to maximize the information that can be coded within the confines of the response range. This is done by maximizing signal entropy and this optimum is achieved when all analogue signal levels are used with equal frequency (figure 5.2).

The frequency with which signal levels occur depends on two factors: first, the frequency of occurrence of the inputs that the signal levels represent, and, second, the coding function that relates inputs to signal levels. One particular coding function equalizes the frequency with which signal levels occur, the cumulative probability distribution of inputs (figure 9.10). The cumulative distribution converts any given distribution of input amplitudes into equally frequent signal levels because it maps input amplitude (its x-axis) onto a linear probability scale (its y-axis). The cartridge's output neurons, LMCs, are coding pixel contrast, so the function to be matched relates contrast to response amplitude. This function is optimized: it follows the cumulative probability distribution of the contrasts that are encountered when coding natural scenes (figure 9.10). Retinal bipolar cells use the same strategy (figure 9.10).

In the fly lamina cartridge, the optimum coding function depends on the gain of the photoreceptor synapses that drive LMCs. The synapses take small photoreceptor signals that code contrast linearly with a low slope and amplify them nonlinearly to produce robust signals that code contrast

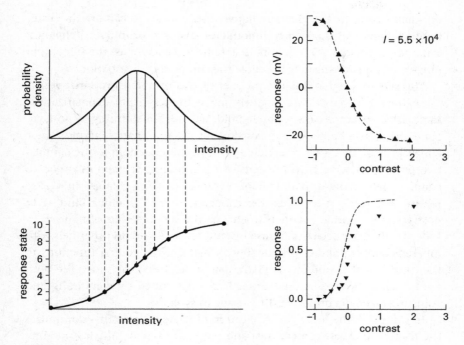

Figure 9.10
Using an input's cumulative probability distribution to code an output optimizes coding efficiency. In theory coding is optimized when all output states are used with equal probability (figure 5.2). **Left:** Given an input probability distribution (e.g., upper curve) coding according to the cumulative probability (lower curve) achieves this optimum by converting equal areas under the input probability distribution into equal increments of response. Noise divides response into discriminable response states, all used equally often. Thus coding is optimized by matching an input statistic. **Upper right:** Blowfly LMC codes optimally. Amplitudes of responses to given contrasts (triangles) follow cumulative probability in natural scenes (dashed curve). **Lower right:** Bipolar cell in vertebrate retina (tiger salamander) codes likewise. Left, from Laughlin (1981) with permission. Upper right, from Laughlin et al. (1987) with permission. Lower right, replotted from data of Burkhardt et al. (2006), after converting their measure of log contrast to contrast.

optimally, according to the cumulative distribution. To capture the distribution's sigmoidal shape, a commonplace analogue primitive, a chemical synapse's sigmoidal I/O function, is stretched to encompass the full output range and adjusted in slope to follow the cumulative distribution.

The synaptic sigmoid is the product of two mechanisms. First, vesicle release rate, which increases exponentially with presynaptic depolarization, increases slope at the base of the sigmoid. Second, self-shunting by postsynaptic channels (chapter 6) reduces slope at the top of the sigmoid. These effects provide two ways of matching the synaptic sigmoid to the cumulative probability. One is to use sensitive presynaptic calcium channels to couple small changes in presynaptic voltage to large changes in vesicle release. The other is to adjust the cooperativity of histamine binding to postsynaptic chloride channels to change the sigmoid's slope (figure 6.7). Indeed, different species of fly use histamine-gated chloride channels with different cooperativities, presumably to match their coding functions to their individual requirements (Skingsley et al., 1995). In short, the optimum coding function is constructed and fine-tuned by combining and adjusting analogue primitives that reside in synapses.

In summary, the judicious application of analogue primitives optimizes the relationship between contrast and response by matching it to natural image statistics. Like the removal of redundancy, this is a necessary step in optimizing an LMCs efficiency. To complete the optimization, another step is necessary. The dynamics of the LMC response, as described by the waveform of its flash response, must be tuned to the statistics of signal and noise.

Tuning response dynamics to optimize efficiency

The dynamics of vision varies with illumination to improve resolution. At low light levels photon noise dominates and must be reduced by averaging over a slow response. At high light levels photon noise is less severe, brief changes can be resolved, and these are captured by a fast response (chapter 8). In other words, S/N (slow response) is traded for bandwidth (fast response). This explains why an LMC's response changes with illumination, longer in starlight, shorter in daylight (figure 9.11), but it does not explain how precisely how slow or fast these responses must be if they are to optimize efficiency.

LMCs provide the answer; their flash response waveforms are optimized within constraints imposed by signal, noise, and limited response range (van Hateren, 1992b). Two adjustments are made (figure 9.11). Response speed is adjusted to optimize the trade-off between bandwidth and S/N

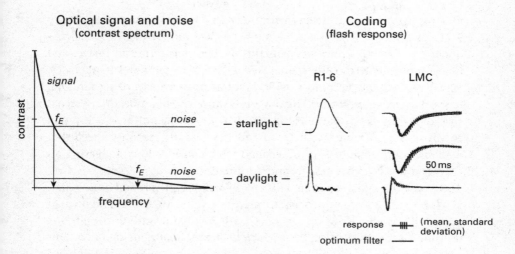

Figure 9.11
LMC response dynamics adapt to input statistics to maximize transmitted information. Left: Relationship between temporal spectra of natural signals and photon noise. Ratio between signal and noise is lower in starlight than daylight so frequency at which signal equals noise, f_E, is lower. Frequencies above f_E are inefficient, noise > signal. **Right**: Flash responses of receptor R1–R6 and LMC in starlight and daylight. Middle trace is LMC response at intermediate light level. In starlight LMC responds slowly to suppress transmission of frequencies above low f_E. In daylight LMC responds quickly to suppress above high f_E. Receptor does likewise, showing it regulates LMC response speed. In daylight LMC responds with off transient that implements predictive coding. Bars on LMC waveforms are means and standard deviations of responses recorded from several LMCs at given time intervals, mean is center of bar. Continuous curves are theoretical predictions of waveforms that maximize information transmitted at given light level, within constraints imposed by natural image statistics, including photon noise, intrinsic noise introduced during transmission and LMC's limited response range. LMC optimum waveforms and data from Van Hateren (1992c), with permission. Receptor recordings, Laughlin (unpublished).

and, as noted with predictive coding, response waveform adapts to reduce redundancy, thereby optimizing the use of response range (figure 9.7). The completed response waveform maximizes the transmission of information. To understand how, we start with response speed.

The optimum response speed depends on the power spectra of signal and noise (figure 9.11). An LMC codes moving images—note that the response to stationary stimulus is transient (figure 9.4). Movement changes contrast within its receptive field, and the LMC responds by changing membrane potential (figure 9.5, right). The power spectrum of this time varying signal declines steeply with increasing temporal frequency (figure 9.11), as follows. A natural scene's power spectrum falls as 1/(spatial frequency)^2; the eye's optics steepens this decline as it forms an image, and image movement converts spatial frequencies into temporal frequencies. Signal is inevitably accompanied by photon noise which, being random, has a flat power spectrum. The key observation is that a declining signal spectrum intersects a flat noise spectrum at the frequency at which signal equals noise, f_E (figure 9.11). This frequency sets the response speed that maximizes information. To understand why, consider how increasing an LMC's bandwidth changes S/N and information.

Because the signal spectrum falls rapidly with frequency and noise is flat, S/N is highest at low frequencies. Consequently an LMC's S/N is high when bandwidth is low. However, an LMC with low bandwidth and high S/N transmits little information because *information = bandwidth* $\log_2(S/N)$. This equation favors increasing bandwidth at the expense of S/N—but only up to point. Beyond f_E, the point at which signal equals noise, frequencies are inefficient because they carry more noise than signal. So for efficiency the frequencies beyond f_E must be suppressed, and bandwidth is adjusted accordingly.

This matching of bandwidth to the power spectra of signal and noise explains LMC response speed. Response speed determines bandwidth (figure 8.10) and bandwidth is adjusted the frequency at which signal equals noise, f_E, to maximize transmitted information. Thus an LMC's optimum response is slow in starlight when a low S/N places f_E at a low frequency and fast in daylight when a high S/N places f_E at a high frequency (figure 9.11), and to maintain optimality response speed adapts to all light levels in between (van Hateren, 1992b).

The second change in LMC response dynamic, a change in waveform (figure 9.11), has been noted. An OFF response of opposite polarity grows in amplitude and narrows in duration with increasing illumination, to reduce redundancy by predictive coding (figure 9.7). The removal of redundancy from an LMC's limited response range allows the informative parts of signals to be transmitted with higher gain, thereby reducing the loss of information to intrinsic noise added during transmission. Note that the matching of bandwidth to f_E has a similar effect. With less noise from uninformative frequencies above f_E, the informative frequencies below f_E can be amplified with higher gain.

The mechanisms that optimize transmission by changing LMC response speed and waveform have also been noted. Photoreceptors determine LMC response speed and they adapt to illumination to optimize bandwidth, responding slowly at low at light levels and quickly at high (figure 8.10). Two presynaptic mechanisms act on photoreceptor output synapses to change response waveform. Feedback within the presynaptic terminal and changes in extracellular potential (figure 9.8) generate and adapt the OFF response. All of these mechanisms act in concert to produce an optimal response, and for this the adaptation of each must be matched to the others'. The mechanisms reside in different cells and different parts of cells, in a photoreceptor's microvilli, in the membrane of its axon terminal, and in resistance barriers constructed and regulated by glia. Coordinating them to precisely optimize an LMC's responses is not a trivial task.

In summary, to transmit optimally an LMC continuously adapts its response waveform, converting from a slow integrator that sums over time to reduce impact of photon noise in starlight to a brisk differentiator that emphasizes rapid changes in daylight (figure 9.11). To optimize the transmission of information, the response adapts to match the bandwidth of transmitted signal to the statistics of signal and noise in a natural image, and eliminates redundancy. Thus an LMC follows one principle, *adapt and match*, to satisfy another, *send only what is needed*.

Summary

The lamina accepts high-rate photoreceptor signals and processes them to transmit most of the information to the medulla for further processing for high speed vision. The lamina is efficient. Part of its efficiency comes from not transmitting 20% of photoreceptors' information. The remaining 80% can then be sent at a lower rate, reducing cost per bit. The remaining increases in efficiency are made without loss of information, by adhering to principles of neural design.

High bit rates are maintained economically by processing and transmitting in analogue. Information capacity is matched to input rates (*symmorphosis*) to minimize the use of resources, especially synapses. Coding is matched to input statistics to use resources to their full capacity, and matches are made efficiently by fine tuning analogue primitives. Circuits minimize noise and redundancy to *send only what is needed*, using an advantageous motif, removing redundancy presynaptically by mechanisms that do not depend on noisy vesicle release. Another motif, tetradic synapses at which one active zone drives the postsynaptic dendrites of four cells, cuts the need for space and materials by 75%.

No part is left idle. Extracellular space is electrically polarized to remove redundancy at presynaptic terminals, using resistance barriers set up and controlled by glia. The axonal cables that transmit analogue signals to the medulla are tuned as matched filters that maximize the transmission of signal and minimize the transmission of noise. Last, almost every neuron and synapse is positioned to fill space efficiently and *minimize wire*.

The rigorous design of so many components, mechanisms and processes makes substantial savings. A lamina that failed to implement the measures listed above would be a much less efficient interface. It would be 6 times larger and consume at least 4 times more energy to achieve the same bit rates. These extra costs would make a fly's brain 50% larger, and increase a resting fly's energy consumption by 10%.

Although an efficient interface, the lamina's design is not a universal. Its efficiency depends on analogue signals travelling the short distances within a fly's head. When signals have to travel longer distances in larger structures, efficient transmission requires different designs. These are considered in the next chapter.

10 Design of Neural Circuits: Recoding Analogue Signals to Pulsatile

Chapters 8 and 9 explained that a sensory neuron collects information via a specialized transducer that modulates ion channels. Small currents through these channels sum to produce an analogue membrane voltage that encodes information efficiently and with a potential for high capacity—hundreds of bits per second. Moreover, these signals can transfer across a synapse at comparably high information rates. However, analogue voltages, spreading passively, decay, and to relay rapidly changing signals beyond about 1 mm requires recoding to regenerative pulses—action potentials—in pulsatile mode.

Mammalian sensors, being substantially further than a millimeter from the brain, all require action potentials to relay their information. Therefore, they must all recode from analogue to pulsatile (A to P), and to minimize irretrievable loss, they must maximize the ratio: bits out/bits in. They should also use resources efficiently (bits/ATP and bits/neural volume). All sensors must accomplish this by A-to-P recoding, but they differ in where and how. Certain sensors recode directly to action potentials whereas others require prior synaptic processing (figure 10.1).

Olfactory and many skin sensors recode directly to spikes. Sensors of sound and head motion both use one synaptic stage, recoding to synaptic vesicles and thence to spikes in a second-order neuron. Photosensors use two synaptic stages: first, they recode to synaptic vesicles that modulate a graded voltage in a second-order neuron, staying largely in analogue mode; then they recode to spikes in a third-order neuron. These connectivity differences have been known for a century (Ramón y Cajal, 1909) and thus lie at the base of Data Mountain. The reasons for these diverse arrangements are set out for the first time in this chapter.

smell   touch   hearing   vision

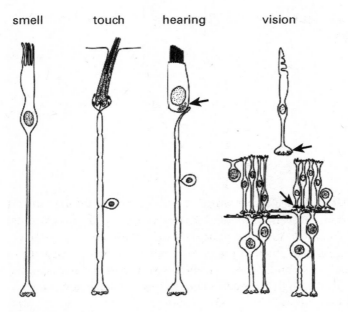

Figure 10.1
Analogue sensors recode to spikes at different stages. Smell and various touch sensors recode directly to spikes; sound sensors use one synaptic stage (arrowed), and photo sensors use two synaptic stages (arrows) before spiking. For exceptions to this broad rule, see Baden et al. (2013).

The coding challenge

To recode analogue voltages carrying more than 100 bits per second to spikes requires high firing rates. For example, to recode 100 bits per second, assuming no noise and no temporal correlation between spikes, would require about 30 spikes per second. However, real axons *do* have noise, plus temporal correlations that increase with spike rate. For example, an optic axon firing even at a modest mean rate (~10 Hz) fills only about 30% of its theoretical channel capacity (Koch et al., 2004, 2006). Moreover, this fraction declines as spike rate rises (Koch et al., 2006). Therefore, to encode 100 bits per second would require the spike rate to substantially exceed 100 Hz. Although neurons can fire transiently at much higher frequencies, those frequencies are uneconomical and largely unsustainable.

The stage selected for recoding depends on the magnitude of the initial information rate. Recall that higher spike rates need larger diameter axons

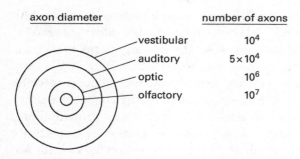

axon diameter	number of axons
vestibular	10^4
auditory	5×10^4
optic	10^6
olfactory	10^7

Figure 10.2
Sensor axon caliber trades off with axon number. Axon diameter varies across types by 10-fold, so cross-sectional area varies by nearly 100-fold. Array size (axon number) varies reciprocally by 1,000-fold. Shown here are mean diameters and axon numbers for human (Perge et al., 2012).

and thus use disproportionately more space and energy because they rise as diameter squared (chapter 3). Recall that vestibular axons, which fire continuously at about 100 Hz, are extremely thick (figure 4.6). This design works because vestibular axons are relatively few. However, optic axons are 100-fold more numerous, so if they had the same caliber as a vestibular axon, our optic nerve would be 10-fold thicker, one centimeter instead of one millimeter—and the *blind spot* where the optic nerve exits the retina would be 100-fold greater in area, 75 mm^2 instead of 0.75 mm^2 (B. Peterson and D. Dacey, *M. nemestrina*, unpublished data). Consequently, sensory neurons must either pay a high unit price, like vestibular axons, or use lower mean spike rates (figure 10.2).

Low-rate sensors code directly

An olfactory sensory neuron collects information at low rates. A sensor expresses only a single type of receptor protein, and there are about 1,000 types, so each sensor patrols a relatively small fraction of the full odorant spectrum. Odorant particles travel slowly, spreading out as they go, and therefore an olfactory source is blurry in space and time. To localize an odorant roughly in time and intensity requires a sensor to capture relatively few particles, each corroborating the others, and capturing more would add little information. Therefore, the sensor's delicate cilia express receptor molecules sparsely (figure 10.3).[1] Moreover, when a neuron has signaled the binding of a few odorant molecules, it adapts. Thus, the messages are rare, slow, and brief.

The olfactory sensor neuron transmits its low information rate (few bits per second) with a low spike rate (figure 10.3), and this allows it to recode directly from A to P. The low rate allows the olfactory axons to be thin, which is good because they are numerous (figure 4.2).

Mechanosensory neurons in skin also narrow their stimulus space to restrict the information rate. Each neuron expresses stretch-sensitive channels in its axon terminal membrane—which embeds a specialized capsule at a particular depth in the skin. Skin + capsule together create a mechanical filter that passes only certain amplitudes and frequencies of deformation. For example, the onionskin capsule of *Pacini's corpuscle*, located beneath the epidermis, transmits rapid indentations (up to about 300 Hz) from the skin surface to the stretch-sensitive channels—mechanical stimuli that we sense as textures and vibration (Loewenstein & Mendelson, 1965; Werner & Mountcastle, 1965; figure 10.3). Channels close in the absence of stretch and therefore produce only the occasional, brief depolarization—which can directly recode to spikes.[2]

In short, both olfactory and rapidly adapting cutaneous systems restrict their information rates by rigorously narrowing the stimulus space. The mammalian nose and insect antenna do this by molecular filtering; the mammalian skin and insect cuticle do it by mechanical filtering.[3] Low information rates at the input allow the output mechanism to recode directly to spikes (figure 10.3). How much information is lost at the final transition from A to P is unknown for these sensors. Yet, once recoded to spikes, information transfers across many levels of central synapses to where it finally guides behavior with no further loss (Werner & Mountcastle, 1965). The means to manage lossless transfer will be discussed (chapter 12).

High-rate sensors need a synapse

Auditory sensors (*hair cells*) also detect changes in pressure; however, compared to skin sensors, the amplitudes are more than 1,000-fold weaker, and the frequencies are up to 100-fold higher. Consequently, a hair cell needs to enormously amplify the miniscule, rapid variations in air pressure to modulate its cation channels and thus its membrane voltage (Dallos, 2008; Hudspeth, 2005; Jia et al., 2007; Ashmore, 2008). The amplifiers, both mechanical and electromechanical, are also filters that restrict a cell's input bandwidth so that the array of hair cells can code sound frequency. Nevertheless, a hair cell encodes a significant range of frequencies (Taberner & Liberman, 2005) and responds with a time constant of less than 0.5 ms.

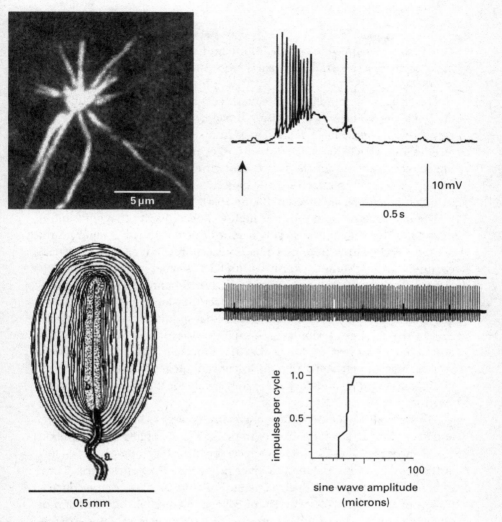

Figure 10.3

Olfactory and skin mechanosensors restrict their information rates, thus allowing them to recode directly to spikes. Upper: Olfactory sensor uses a single species of molecular filter distributed sparsely on the cilia. The neuron responds slowly, fires at low rates, and adapts. Shown here are the cilia of a sensor neuron in the intact olfactory epithelium. Recording shows response to an odor pulse (1 μM octanoic acid for 1 s). Arrow marks stimulus onset. Mouse courtesy of Minghong Ma (unpublished). **Lower:** Each mechanosensor uses a specific type of mechanical filter. Shown here is Pacini's corpuscle which responds to a punctate stimulus (2-mm diameter probe) to the palmar surface of the hand. Sensor is silent without stimulation, but shallow deformations of the skin (19 μm) at high frequency (150 Hz) evoke a spike to nearly every cycle. The I/O curve is steep, indicating sharp tuning to intensity. Drawing is reprinted from Ramón y Cajal (1909); recording and graph are modified and reprinted with permission from Talbot et al. (1968).

This enables a hair cell to encode information at rates that exceed the coding capacity of a spike train, so neural processing is required at the synapse[4] (figure 10.4).

This synapse accomplishes three tasks: (1) recodes the rapidly fluctuating analogue voltage to a precisely timed pattern of vesicle release, (2) carves away considerable redundancy from the release pattern, and (3) recodes each vesicle to a spike (Rutherford et al., 2012). Note that the carving of redundancy from the vesicle pattern also carves it from the spike pattern. The key is to match specific molecular and cellular components in a unified functional architecture (figure 10.4).

The hair cell's calcium sensor is highly cooperative: to trigger fusion, it must bind five calcium ions. This requires calcium to rise steeply near a docked vesicle, either from one adjacent calcium channel or from several channels opening simultaneously within tens of nanometers. Near the hair cell's resting potential, about –70 mV, its Cav1.3 channels have a low probability of opening (Zampini et al., 2010), yet they are present in sufficient numbers and proximity to docked vesicles that spontaneous channel openings in the absence of sound cause single vesicles to fuse at a substantial rate, about 20–40 s^{-1} (Graydon et al., 2011; Kim et al., 2013). This requires a special organelle (synaptic ribbon) to concentrate vesicles from the cytoplasm, and to prime and dock them to be ready for the next calcium surge (Matthews & Fuchs, 2010).

The hair cell's synaptic vesicles are relatively large (~45 nm in diameter) and dock along a ring at the base of the spherical ribbon. Facing the ribbon across the synaptic cleft, glutamate receptors distribute in a gradient—low at the center and high along the outer ring where the vesicles fuse. Thus, each fusion delivers a relatively large puff of glutamate to a dense concentration of receptors. This matches a high spatiotemporal concentration of glutamate to the low binding affinity of a particular receptor isoform (GluR2–3) to give good S/N, and a fast OFF rate that allows the postsynaptic response to follow high frequencies (figure 10.4). Thus, the pre- and postsynaptic structures serve to concentrate the information to be sent by each expensive spike.

The postsynaptic knob is small (~1 μm) for low capacitance (rapid charging) and high input resistance (sharp depolarization to small current). The knob couples directly to an axon whose fast sodium channels (Nav1.6) are suited for high firing rates. Thus, the overall design allows nearly every release event to trigger a spike. This avoids spatiotemporal integration, which would add delay and jitter. But the cost is substantial: continual spiking at high rates—noise in the absence of sound.

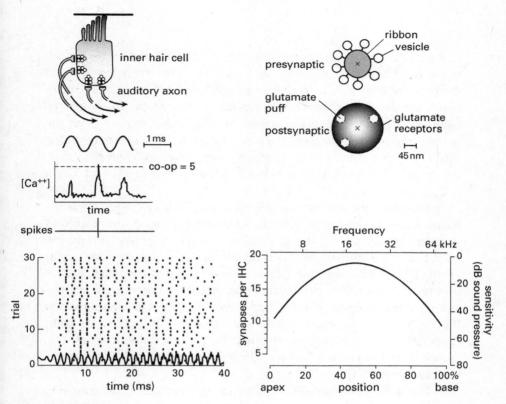

Figure 10.4

Auditory hair cell, transducing high frequencies, captures too much information to recode directly to spikes. Instead, it recodes to vesicles. **Upper left:** Each active zone drives one dedicated spiking axon (four of about 20 are shown). **Middle left:** Temporal precision is preserved by concatenating two cooperative mechanisms for vesicle fusion. This also reduces redundant spiking. Calcium concentration rises at the peak of each stimulus cycle (1 kHz), but only the middle cycle opens sufficient calcium channels for the concentration to reach threshold for binding five calcium ions to the vesicle's calcium sensor. **Lower left:** Spike responses in postsynaptic axon to pure tone (676 Hz) to which it is selectively tuned (chick). Spikes all align with the peak of every cycle, but many trials fail. Reading across 25 cycles of one trial (~35 ms), one has little uncertainty about the correlation of spike to stimulus timing, so adding more spikes where there are failures would be redundant. **Upper right:** Synaptic ribbon tethers to the presynaptic membrane (x) and itself tethers relatively large vesicles (45-nm diameter), bringing them into contact with the membrane along a ring. Postsynaptic glutamate receptors with low affinity (fast) cluster postsynaptically as a gradient that peaks at the ring and declines toward the center. Thus, the receptors are distributed to catch the fast peaks in glutamate concentration. **Lower right:** Number of active zones peaks near middle of cochlea at middle of frequency range and peak of sensitivity. Recording is modified and reprinted with permission from Moser et al. (2006); distribution of glutamate receptors, vesicle size, and distribution of active zones are modified and reprinted with permission from Meyer et al. (2009).

Spikes that follow single, thermally evoked vesicle fusions show considerable delay and temporal jitter (300–1,500 µs). However, when a sound drives membrane voltage to the steep region of Cav1.3's current/voltage curve, simultaneous channel openings occur more often. The larger calcium surges tend to fuse several vesicles simultaneously, causing larger glutamate puffs and thus larger postsynaptic currents. These reduce delay and jitter (500 ± 50 µs) and efficiently recode sound to spikes. Three simultaneous quanta suffice to reliably recode a fusion event to a spike event (Grant et al., 2010).

In summary, one or a few neighboring calcium channels open to cause a fusion event. A fusion event with one vesicle reliably causes one postsynaptic spike, but that spike jitters and therefore correlates weakly with the calcium channel opening(s). Multiquantal fusion also causes one spike, but its greater postsynaptic strength reduces spike jitter and so correlates strongly with the channel opening(s) that caused it. Precise timing of fusion involves two levels of cooperativity: (1) at the nanometer scale, high cooperativity at the vesicle's calcium sensor to release a vesicle; (2) at the 100-nm scale, cooperativity between calcium channels to release multiple vesicles. This second level appears to be key to reducing spike redundancy—as we now explain (figure 10.4).

The cooperative opening of calcium channels that locks vesicle fusion to the voltage peak also makes spiking unreliable because the chances of success depend on the likelihood of the channels opening jointly. This is the product of their individual probabilities, so if four channels were needed, each with $P_{open} = 0.15$, the chance of joint opening would be about 5×10^{-4} (Zampini et al., 2010). Consequently, a voltage repeating over many cycles at, say, 3 kHz would sometimes cause fusion, but usually not. When a spike does occur, it is precisely timed; however, many cycles are skipped. Skipping cycles reduces redundancy and improves coding efficiency (figure 10.4). One problem solved.

But another problem remains. As described so far, a hair cell's recoding from vesicles to spikes is 1:1—one large fusion event triggers one spike in one postsynaptic axon. Yet, the cell's analogue signal contains far more information than one axon can send by spikes. To send more, the hair cell employs multiple active zones, up to about 20–30, that each contact a separate axon. The active zones differ, both in numbers of docked vesicles and calcium channels. This allows channel cooperativity to create a particular release pattern for each active zone and thus a particular firing pattern for each postsynaptic axon. Thus, a hair cell's full analogue message is custom filtered and converted to pulses by about 20–30 axons whose total rate is approximately 800–1200 Hz.[5]

These are *expensive* axons since their high mean rates require thick caliber and high energy (figures 10.2 and 4.6). This illustrates that a neuron can encode so much information in analogue mode that sending it by pulses requires a veritable cable of thick, power-hungry axons. Auditory systems hold down the total cost by restricting the number of axons (about 30,000 in human). But the overall high cost is unavoidable for transmitting information sensed at high frequencies beyond a millimeter.

The middle region of the cochlea is more sensitive than either end. This region in mouse serves frequencies near 16 kHz—the frequency emitted by squirming pups. Sensitivity peaks in human near 3.2 kHz— the frequency of a baby's cry. These hair cells express the most active zones and thus send the most axons to the brain (Meyer et al., 2009). Thus, the auditory hair cell recodes its rich analogue signal to multiple pulses that carry different amounts of information to the brain. There they contact different subsets of neurons using different types of synapse. The advantages of such a parallel design will be explained in chapter 11.

When high-rate sensors do not rectify

The auditory hair cell rectifies its input. Each pressure cycle bends a stereocilium that tensions a protein that yanks channels open like a trapdoor. Unbending releases the tension, so the channels snap shut. Current flows only once per cycle rather than at each half cycle. This is efficient because the two half cycles are perfectly correlated, so to code both would be redundant.

The vestibular hair cell does *not* rectify. Instead it linearly encodes both increases and decreases in pressure. This design works for sensing position and velocity of the head—a large inertial mass that changes slowly (0–20 Hz). These frequencies are 3–4 orders of magnitude lower than sound frequencies, so there are no redundant cycles to be skipped. Moreover, the change to be sensed may occupy far less than one cycle—the head may turn slightly and pause indefinitely before resuming its initial position. So both directions are independently informative, and neither should be deleted by rectification.

The vestibular hair cell at rest is partially depolarized by a few open channels. Head motion in one direction opens more channels, and motion in the other direction closes them (figure 10.5). The postsynaptic axon does not fire to a single or compound fusion. Rather, it integrates many release events from multiple active zones in one hair cell and even converges release events from several adjacent hair cells (Goldberg et al., 2011). Thus, the vestibular axon is not about encoding single, well-timed events over many cycles but rather about encoding a good S/N over less than one cycle. Whereas an auditory hair cell must *diverge* its message, directing one active zone to each of 20 high-rate axons, a vestibular hair cell must *converge* multiple active zones onto one axon (figure 10.5).[6]

This design, which encodes low temporal frequencies by modulating a high tonic spike rate, has advantages for the vestibular system's mechanisms for fast central processing. One certainly expects a payoff because the design is so costly per unit. Yet, the system is affordable because vestibular axons are few (figure 10.2). But what if a high-rate sensor without rectification also needs to be numerous? Suppose that a sensor's usefulness involves an extended, dense array? The vestibular design would fail due to soaring costs. This is the challenge for cone photoreceptors in the retina.

Why does a cone photoreceptor need *two* synaptic stages?

An olfactory neuron captures individual particles, amplifying each capture via a G protein cascade (chapter 8). From the air's universe of molecules the olfactory sensor selects a small subset with a discrete molecular shape/charge that bind to its single type of odorant receptor expressed from the DNA catalog of 1,000 types. Thus, an olfactory sensor uses stringent molecular filters to select which particles to bind. A cone also captures individual particles, but its protein detector (*cone opsin*) is broadly tuned. True, cones can choose between two or three receptor proteins with somewhat different spectral tuning. However, while these differences are critical for discriminating color, they do not significantly narrow a cone's rate of particle capture.

Odorant particles travel slowly and spread out as they go. Therefore, an olfactory source is blurry in space and time. An olfactory sensor, having identified an odorant roughly in time and intensity, based on relatively few corroborative particle captures, would gain little by capturing more. That would violate the rule match sensor to signal quality (chapter 8). Light particles travel directly at, well, the speed of light and go directly from object to cone—via the optics that match image quality to quality of the sensor array (chapter 8). Thus, the more photons that a cone captures during a brief interval, the more finely it localizes a point in space and time (Sterling et al., 1992).

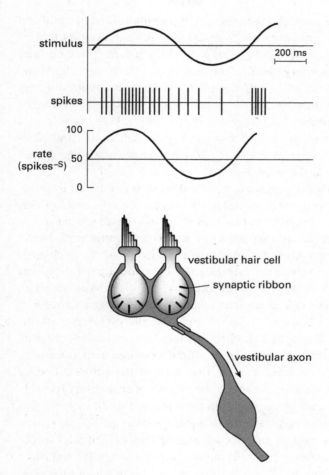

Figure 10.5
Vestibular hair cells, transducing low frequencies, can sum their analogue signals before recoding to spikes. Upper: Head rotates slowly (1 Hz). Spikes from second-order vestibular axon are modulated linearly through the full cycle around 50 spikes per second. **Lower:** Adjacent hair cells each converge multiple active zones onto single afferent fiber. Modified from Eatock et al. (2008).

In short, because photons are numerous and well localized in space and time, there is an opportunity for fine imaging. A cone exploits this by investing heavily in detector molecules, packing them densely at far greater numbers than an olfactory neuron. This design transduces particles at tremendous rates to build a finely graded photovoltage that imbues the cone with too much information for direct recoding to spikes. It is even too much to be accomplished by a single synaptic stage. Lacking chemical and mechanical filters to reduce the information stream (figures 10.3 and 10.4), the cone is forced to filter its analogue photovoltage *neurally* before finally recoding. This is what the retina is "for."

The first stage is quite general, removing noise and redundant information that no downstream element will need. This involves discarding information about mean intensity and transmitting only signals greater or less than the mean—that is, *contrast* (chapter 9). The second neural filter sharply reduces the event rate—throttling down to a "sparse code." To implement a sparse code, the retina can no longer signal objects brighter and dimmer than the mean by modulating a high tonic rate (like vestibular neurons). Therefore, it rectifies—creates separate channels to send signals greater or less than the mean (ON and OFF). The second stage also performs "custom" filtering—selecting special aspects of the captured information for routing to particular downstream targets at rates sufficient for their particular needs.

Why two synaptic stages? Imagine that each of the cone's 20 active zones were to directly excite an axon in the manner of an auditory hair cell. These axons could send messages edited for noise and mean intensity but would relay the full image, unrectified, leaving further editing to central mechanisms. That would require all optic axons to fire at high rates and be thick (figure 10.2). Auditory hair cells number approximately 10^3, but cones number approximately 10^6. To send a fine spatial image where each pixel needs 20 thick axons could not work.

Regarding the structure and function of the retina's two-stage "neural editing," a great deal is known—sufficient, we feel, to justify a separate chapter. The possible appeal, for readers blind to its intrinsic fascination, is that these neural circuits exemplify design principles applicable to the rest of the brain. For a fuller explanation of why photoreceptors need two synaptic stages, proceed to chapter 11.

The eye in daylight is like the camera on a planetary rover. Both capture rich images, and both are constrained to send them via a channel of low information capacity.[1] The rover solves this by transmitting slowly, using minutes per image, and storing them at base for leisurely assembly and analysis. But the eye's stream of raw data cannot be stored—there is far too much. Moreover, an animal needs the images to guide action in real time. The eye functions under a time constraint: it must capture, process, and send an image within about 100 ms.[2] Consequently, ganglion cell and behavioral performance both improve with temporal summation up to about 100 ms but not beyond (Geisler, 1989).[3]

For an image processor this presents a significant computational challenge. The good news is that on the chemical/electrical scale 100 ms is a decent amount of time. Although an auditory hair cell's temporal precision cannot tolerate any neural integration beyond a millisecond (figure 10.4), the retina can integrate for 100-fold longer. The brain exploits this opportunity by establishing complex neural circuits within the eye. When these have processed images in situ, the output channel can transmit at the specified rate at an acceptable cost—thus, a retina.

Given that the retina lies on the optical path, it must stay thin; therefore, processing should minimize the volume of neurons and wires. Processing must also minimize energy because to supply fuel and oxygen requires blood vessels. Although photoreceptors are themselves energetically demanding, they are supplied by vessels outside the optical path. But the neural circuits require blood vessels within the optical path—which scatter and absorb light heading for the receptors (figure 11.1). For these reasons the thickest synaptic layer in mammalian retina never exceeds about 30 μm.[4]

The retina's architecture has been thoroughly mapped. Most of its roughly 80 neuron types and their array structures are known on a scale of

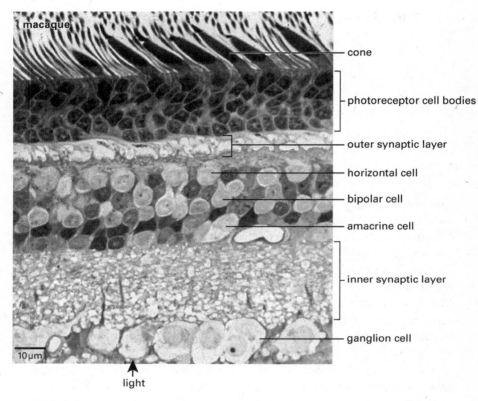

cone

photoreceptor cell bodies

outer synaptic layer

horizontal cell

bipolar cell

amacrine cell

inner synaptic layer

ganglion cell

light

Figure 11.1
Vertical slice through monkey retina. The arrow indicates the optical path. The neural retina contains three cell layers and two synaptic layers. By staying thin, it preserves good optics. Light micrograph courtesy of Noga Vardi.

millimeters. Many synaptic circuits are known on a scale of micrometers, and many receptor proteins, vesicle distributions, and ion channels are known on a scale of nanometers. The maps span a scale of 1 millionfold—the United States down to a house lot. The retina has accumulated its own Data Mountain, yet it is no scree of unconnected facts. We know the fundamental plan (Masland, 2012; Wässle, 2004; Sterling, 2004a; Light et al., 2012; Wässle et al., 2009), the key mechanisms, and the overall efficiency (Borghuis, et al., 2009; et al., 2006; Ala-Laurila et al., 2011). By now integrating this knowledge with the constraints (time, space, energy), the goal of reverse engineering can be approached. One might then explain *why* the retina is designed just so.

The recoding challenge

A cone in daylight captures about 10,000 quanta s^{-1} to encode a finely graded voltage (chapter 8). To recode directly would require 10,000 spikes s^{-1}—exceeding the brain's highest mean spike rate by 100-fold. Obviously the retina must recode to a lower rate. But *how* low? And what is the governing computational scheme? Also, how does the retina's implementation follow principles of neural design?[5]

Consider this example (figure 11.2). A spot of daylight 1% brighter than the background delivers about 10^9 photons to a patch of 4,000 cones in 100 ms. The photons (spot + background) isomerize about 10^7 cone opsin molecules, thus reducing quanta by 100-fold. The cone synaptic terminals release about 10^5 vesicles, reducing quanta by another 100-fold. The bipolar cell terminals then achieve a radical transformation: they collapse tonic release nearly to zero; however, to the spot's onset they release quanta in small bursts. About 10 vesicle quanta suffice to reliably trigger one ganglion cell spike (Freed, 2005). In short, a pattern reaching the photoreceptors as 10^7 events is compressed by retinal circuits for the most sensitive ganglion cell to a *single* event—one spike.

This suffices for the spot to be perceived (Borghuis et al., 2009). For example, based on that single spike, a monkey can report the appearance of a spot and thereby earn a sip of juice. This implies that at perceptual threshold, a single spike can impart meaning. In this respect the eye is like the skin, where perceptual threshold is also set by a single pulse sent by a single axon (Werner & Mountcastle, 1965; Barlow, 1972; chapter 10).

Across the stages from cone input to ganglion cell output considerable information is discarded, and this reduces sensitivity. As the quantal rate steps down by 100-fold at the cones, sensitivity falls by 10-fold (figure 11.2). This is the square-root law, which determines signal-to-noise ratio when it is based on random processes, such as photon arrival (chapter 8). Vesicle release is also random, yet the 100-fold decrease in rate at the cone terminal reduces sensitivity by only about fourfold, and the more than 100-fold decrease in rate bipolar cell terminal reduces sensitivity by only about 2.5-fold. Since successive losses are multiplicative, overall neural loss across the retina is about 10-fold (Borghuis et al., 2009; Ala-Laurila et al., 2011).

How can neural stages outperform the square-root law? By discarding what is uninformative, thus reserving signaling capacity for what is most informative. The cone terminal begins this process by removing noise and redundancy. The synaptic architecture for these two filtering operations is

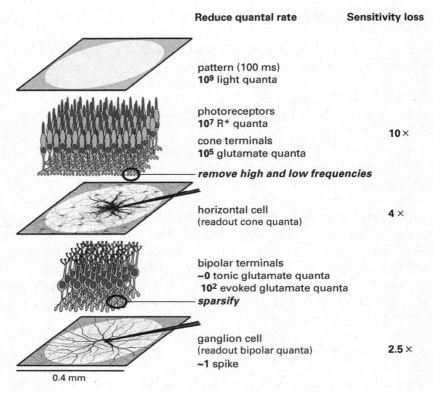

Figure 11.2
Quantal rates step down from photon input to ganglion cell output. The initial stage
loses sensitivity by 10-fold, following the square-root law; later stages preserve sensitivity because of neural processing. Electrode recordings from single horizontal and
ganglion cells read out the cone and bipolar quanta. Guinea pig. Modified and reprinted with permission from Borghuis et al. (2009).

evident in a slice through a cone terminal (figure 11.3). The filtering strategies and implementation are explained in the next section.

Reducing noise

The strategy for reducing noise is simple. Light intensities at adjacent points
in an image tend to be correlated. For example, when you view a dark patch
on a bird's wing, all the receptors collecting from that patch receive similar
intensities and thus register similar photovoltages. But they are never identical because photons arrive stochastically. Their fluctuations (photon
noise) occur independently in each receptor, so membrane photovoltages

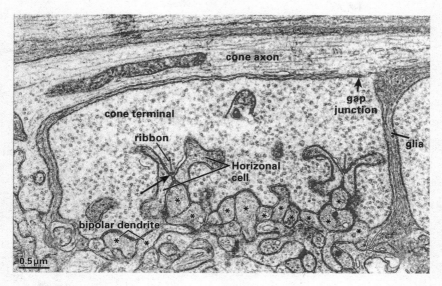

Figure 11.3
Section through cone synaptic terminal from monkey fovea. Gap junction electrically couples this terminal to neighbor. Arrow indicates site of vesicle release at base of synaptic ribbon onto horizontal cell processes and cone bipolar dendrites (*). Gap junctions reduce cone noise; horizontal cell negative feedback to calcium channels at release site reduces redundant signals; bipolar dendrites near cone release sites encode fast signals, whereas dendrites far from the release sites encode slower ones (see figure 11.10). Glial membranes, bearing high concentrations of glutamate transporters, separate the cones and reduce spillover between them (Szmajda & DeVries, 2011; Burris et al., 2002). Electron micrograph courtesy of Yoshihiko Tsukamoto. Modified and reprinted with permission from Sterling (2004a).

contain both a correlated component (signal) and an uncorrelated component (noise). Both components flow between neighboring receptors across the gap junctions (figure 11.3). The correlated signals shared by cones add linearly, but the uncorrelated noise adds as the square root. Consequently, when n receptors pool noisy signals, S/N improves as √n (figure 11.4).

Another effect of gap junction coupling is to attenuate high spatial frequencies (fine spatial detail) and pass low ones. This is accomplished directly in analogue by resistively coupling the photoreceptor terminals and integrating voltages through their membrane capacitance (figure 11.4). Thus, the retina's first neural circuit follows the principle *compute directly with analogue primitives*. This filtering operation is implemented identically in an electronic circuit—by passing a current through a resistor and a

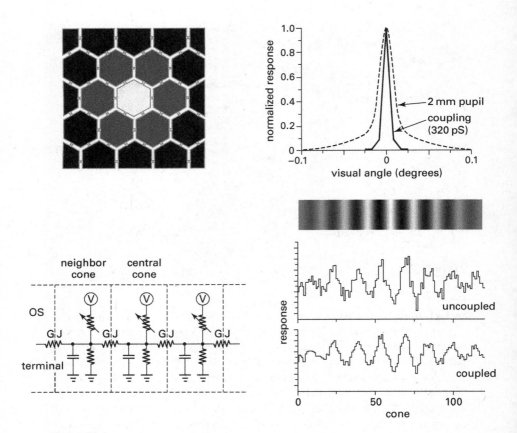

Figure 11.4
Cone electrical coupling reduces noise from phototransduction. Upper left: Foveal
cones pack triangularly and couple their synaptic terminals via gap junctions. **Lower
left**: Schematic circuit for electrical coupling between two neighbors. Photon flux at
the cone outer segment (OS) controls light-modulated conductance (variable resistors)
which, in series with voltage source (V) and membrane conductance (2×10^9 S), generates the voltage response in the terminal. **Upper right**: Current spread to neighbors blurs
the cone image, but the blur due to measured coupling (320 pS) is narrower than the
blur due to optical factors in bright light. **Lower right**: Sinusoidal grating with contrast decreasing from center. This grating flashed on the cone array produces noisy
fluctuations that obscure responses at low contrast; but coupling reduces this noise,
allowing the terminals to send more signal. Modified and reprinted with permission
from DeVries et al. (2002).

capacitor. In the retina, however, the circuit operates at the molecular level, so the currents and capacitances can be far smaller (chapter 6).

Coupling at the photoreceptor level has several advantages. First, discarding noise saves resources. In particular, it allows the signal to be sent with fewer vesicles, thus holding to the steep limb of the square-root law. This in turn reserves postsynaptic currents for encoding signal. Second, surrounding rods also couple to cones and contribute *their* photovoltages, further improving the cone's presynaptic S/N (Sterling et al., 1988; Borghuis et al., 2009). Third, downstream stages introduce nonlinearities, such as the two stages of cooperativity for vesicle release described in chapter 10, and rectification, which differently affect signals above and below the mean. So coupling cones obeys another important rule of circuit design: *reduce noise before it becomes distorted by nonlinear processing*. This rule is followed by all photoreceptor arrays, both insect and vertebrate (chapter 9).

This noise filter operates efficiently. The gap junction is essentially two-dimensional—simply a patch of close apposition between two cell membranes—so it requires zero extra space (figure 11.3). Also, its coupling via *transcellular* channels keeps the ion flows within cytoplasm, therefore electrical coupling requires little extra energy. Finally, the gap junction, while attenuating noise *within* a neuron, adds very little noise of its own. In short, this filtering mechanism costs no space and little energy or noise. It is even tunable to stay optimal as S/N changes with light intensity (Li et al., 2013).

A Ponzi scheme?

At this point a reader might worry that this mechanism sounds too good. Would not pooling signals across the photoreceptor mosaic reduce its ability to represent fine spatial detail? Only if spatial acuity were set by the photoreceptor lattice, but in general it is not. The main reason for a fine receptor mosaic is not spatial acuity but rather to reduce receptor volume in order to improve transduction speed (chapter 8). Spatial acuity is generally set at the retina's *output* layer, that is, at the ganglion cell arrays. Since a ganglion cell typically pools signals from about 10^2–10^3 cones, local pooling at the cone terminal will not affect spatial acuity at a ganglion cell mosaic.

"Midget" ganglion cells in the primate fovea are exceptions to this general rule. They do collect from single cones, so strong coupling between foveal cones would indeed reduce acuity. However, the average coupling conductance between foveal cones is small enough that the resulting "neural blur" is narrower than the eye's optical blur. Thus the neural blur little affects spatial acuity; yet it suffices to improve S/N by nearly 80% (figure 11.4; DeVries et al., 2002).

Would not coupling cones of different spectral tuning blur their spectral differences, which are key to color vision? Yes, but this effect is reduced by clustering cones of like spectral type. The uncorrelated noises are attenuated, but the correlated spectral responses are not. The strategy works especially well for cones sensitive to middle (M) and long (L) wavelengths because their random distribution creates many patches of like type. The strategy fails for cones sensitive to short (S) wavelengths, which, being rare (5–10%), tend to be surrounded by M and L cones. If S cones coupled strongly to M and L, spectral differences would indeed be severely attenuated. The circuit avoids this by not coupling the S cones (Hsu et al., 2000; Hornstein et al., 2004; Li & DeVries, 2004).

In summary, gap junctions perform no magic—they offer neither perpetual profit nor an inconceivably large return for the investment. They simply allow linear trade-offs—a small spatial or spectral blur is traded for better S/N. The exact degree of coupling appears to maximize total information from the array (Garrigan et al., 2010). This "truth-in-lending statement" should reassure readers that photoreceptor coupling is no Ponzi scheme but simply an intelligent design.

Compress files by subtracting the mean

The other component to be discarded from the photosignal is the *mean*. This might also seem to be a trick—like Garrison Keillor's mythical town "where all the children are above average." It is no trick, but simply the predictive coding scheme already described for the fly photoreceptors (chapter 9). A cone sees a scene as a succession of changes between bright and dark, each a deviation from the mean intensity. These brief dimmings and brightenings are informative whereas the steady mean is not, so they are what the retina should transmit.

The strategy is simple: measure the mean precisely by summing raw intensity responses across a broad patch of cones. This predicts the intensity at the patch's center. Deviations from this prediction—in either direction—represent the *contrast* signal—the component most worthy of transmission. The contrast signal is then isolated by subtracting the prediction from the actual center signal (figure 9.7). To measure the mean specialized neurons (*horizontal cells*) spread thick processes as a planar arbor just beneath the layer of cone terminals (figure 11.5).

Adjacent horizontal cell processes overlap extensively and couple electrically by gap junctions to form a low-resistance network—good for

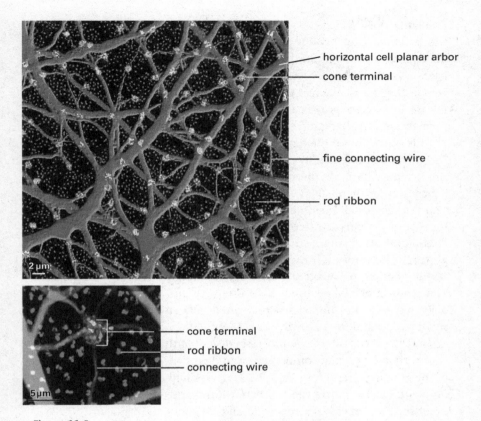

Figure 11.5
Functional architecture for filter that removes low spatial and temporal frequencies from cone signal. Upper: Horizontal cells form a planar meshwork of thick branches beneath the layer of cone terminals (figure 11.1) and connect to every cone. This connection uses irreducibly fine wires (~0.1 μm) that invaginate the cone terminal and expand parallel to each ribbon. Every cone ribbon is flanked by two expansions that lie within 20 nm of every docked vesicle (figure 11.3). Note that this type of horizontal cell connects exclusively to cone terminals, entirely avoiding rod terminals. **Lower:** Final connection to cone terminal uses short, thin wire that reserves space for other essential connections. Shown are type A cells from rabbit. Reprinted with permission from Pan & Massey (2007).

averaging noisy inputs as described for the cone terminals. These thick trunks send irreducibly fine processes (~ 0.1 μm) to invaginate every overlying cone terminal. Within the invagination, the process expands along the base of a ribbon so as to detect and process the glutamate pulse from every vesicle that fuses (figure 11.3). The expansion lies within 20 nm of the release sites and expresses a low-affinity, AMPA-type isoform of the glutamate receptor (Haverkamp et al., 2001b; DeVries, 2014).

This expansion of the horizontal cell within the cone terminal is clearly designed to capture rapidly changing signals. First, it locates near the release site to reduce diffusion distance and give it a fast pulse of glutamate at high concentration (figure 11.3). Second, it expresses fast glutamate receptors. Third, it uses small membrane area to reduce capacitance for rapid charging. Fourth, it connects to the horizontal cell with a short process to avoid attenuating fast changes. All these properties—short diffusion distance, fast glutamate receptors, low capacitance, and short connection—together represent a design to average the cone signal's rapidly changing components. The connection, being irreducibly fine (0.1 μm) attenuates the voltage delivered to the low-impedance horizontal cell network, giving each cone a small weight and thus allowing broad summation.

Horizontal cells also have processes deeper in the synaptic layer that see slowly changing signal components and detect them with slower isoforms of the glutamate receptor (figure 11.3; Haverkamp et al., 2001b). Thus, the horizontal cell captures the full bandwidth carried by every vesicle from every cone and averages across thousands of cones.

During the 100 ms allowed for summation, the low-resistance horizontal cell network sees about 10^5 glutamate pulses, sensing both rapidly changing and slowly changing signal components. Their summation improves S/N across the full bandwidth by more than 300-fold.[6] So the horizontal cell membrane voltage with high S/N ratio accurately predicts local intensity (Borghuis et al., 2009). It then subtracts this prediction from the cone terminal. Ideally, this computation should cost minimal extra space, energy, or noise. This seems to preclude a standard chemical synapse to shunt cone photovoltage. Instead, the horizontal cell network provides a current source to feed electrical signals back over the same fine wires to the expansions located within 20 nm of the cone's voltage-gated calcium channels. Feedback implements this subtraction by changing the response of these channels, either electrically, as demonstrated in the fly (chapter 9), or via protons that modulate the channel (Klaassen et al., 2012; Hirasawa & Kaneko, 2003; Thoreson & Mangel, 2012; Davenport et al., 2008). Either way or both, the circuit uses the same wires for input and output.

One problem remains: how to weight the averages for the filters that remove noise and redundancy. S/N improvement by pooling partially correlated signals from the patch of cones near the center declines with the strength of correlation. Correlations for a given cone are strongest with its nearest neighbors and decay exponentially in space and time. Thus, averaging for both center (by cone coupling) and surround (by horizontal cells) should be weighted for the decay in correlation strength. The optimal weightings for both center and surround filters are roughly Gaussian, so when summed at the cone terminal, the overall optimal weighting is a two-dimensional difference-of-Gaussians. For a digital computer it is a somewhat formidable expression:

$$\Gamma_{\sigma,K\sigma}(x,y) = I * \left(\frac{1}{2\pi\sigma^2} e^{-(x^2+y^2)/(2\sigma^2)} - \frac{1}{2\pi K^2\sigma^2} e^{-(x^2+y^2)/(2K^2\sigma^2)} \right)$$

But the retina computes this filter directly in analogue—by convolving the eye's optical spread function with the output of the cone/horizontal cell circuit (figure 11.6). To optimize the weightings for both center and surround requires subtle matching of the optics to the coupling strength between cones and between two types of horizontal cell, one narrow-field (H_B) and the other wide-field (H_A) (figure 11.6).[7] The optimal weighting shifts with background intensity, as noted for fly (chapter 9), so this circuit adapts by changing coupling strengths and feedback gains. For example, coupling strength can be altered by phosphorylating connexin 36 via a kinase activated by retinal dopamine neurons that track changing intensity and circadian time (Li et al., 2013).

Having introduced the first set of retinal interneurons, we should note that they exemplify an important design rule: *minimize wire caliber*. Horizontal cells use coarse neurites to conduct passively over longish distances (H_A) and fine ones to reach cone terminals over short distances (H_B). Both types are only as thick as they need to be for their particular task. The cells also obey the rule *minimize wire length*—by arborizing strictly in two dimensions beneath the cone array and connecting to it via the shortest possible wires (figure 11.5). This means placing the neurons as close to their targets as possible, given the other claims on the space. Indeed, horizontal cell bodies are as close as they can be to the layer of cone terminals and not interfere with the connecting wires (figure 11.1).

Quantizing the cone's analogue signal

The problem for the cone terminal is how to represent the filtered analogue membrane voltage as a varying stream of pulses (transmitter quanta).

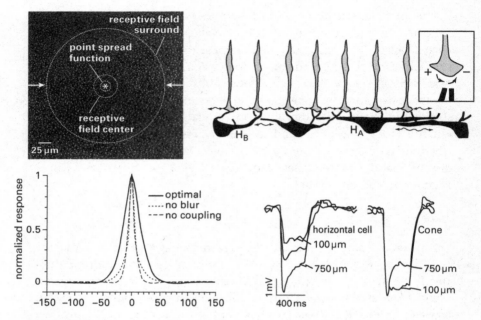

Figure 11.6

A local circuit computes the cone terminal's difference-of-Gaussians receptive field (Smith, 1995). **Upper left:** A bright point at the cornea blurs on the optical path to stimulate about 10 cones (*point spread function*). The transduced signal spreads further via cone coupling to create a receptive field center for one cone (*) that includes about 50 neighbors. Horizontal cells sum cone signals more broadly and feed back negatively to create a receptive field surround that includes about 1,200 cones. **Upper right:** The neural circuit. Arrows between the terminals denote coupling; the surround is shaped by inhibitory feedback (inset). **Lower left:** Sensitivity profile across the cone receptive field (between arrows above). Optimal weighting for the center requires combining both optical blur with cone–cone coupling. Optimal weighting for the surround requires combining a narrow, deep contribution from the narrow-field H_B cell with a broad, shallow contribution from the wide-field H_A cell. **Lower right:** Horizontal cell response increases with spot size, consistent with its broad collecting area, but negative feedback from horizontal cells causes the reverse in a cone: large response to spot filling the receptive field center and small response to spot filling the surround. Modified and reprinted with permission from Sterling (2004a); based on studies by Smith (1995), Leeper & Charlton (1985).

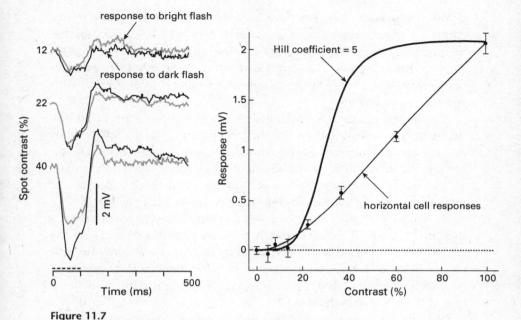

Figure 11.7

Horizontal cell responds to dark and bright equally and linearly at low contrasts.
Left: Horizontal cell responses to flashes dimmer than the mean (dark trace) and
brighter than the mean level (pale trace). Responses are of opposite polarity, but
equal up to 20% contrast. Higher contrasts saturate the response to bright whereas
responses to dark continue increasing. Each trace averages about 140 trials. Modified
and reprinted with permission from Borghuis et al. (2009). **Right**: Horizontal cell response to dark is linear across wide range of contrasts because cooperativity of cone
vesicle release is low (Hill coefficient ~2). If cooperativity were high (Hill coefficient
~5), as for the auditory hair cell (figure 11.8), the response would saturate at lower
contrasts. Graph courtesy of Bart Borghuis (2014).

Several requirements must be satisfied. Dark and bright should modulate
quanta oppositely but equally. This is evident in the horizontal cell's
response to increments and decrements of the cone vesicle stream (figure
11.7, left). Second, linearity should extend over a significant dynamic range
(figure 11.7, right).

To meet these requirements, the cone terminal sets its mean voltage
around –50 mV by balancing the photocurrent in background light against
horizontal cell antagonism. This voltage half-activates the calcium channels (Cav1.4; Mercer et al., 2011), so a depolarizing dark stimulus sharply
increases the rate of channel openings, and a hyperpolarizing bright stimulus sharply decreases the rate. The balance is actually somewhat

asymmetrical, such that responses to dark continue to rise linearly with contrast whereas responses to bright saturate (figure 11.7, left). The mechanistic explanation is that larger bright contrasts drive vesicle release toward zero, so that is the floor. But larger dark contrasts drive vesicle release to higher rates, limited only by the calcium current that continues to rise with stronger depolarization.

The mechanism raises a deeper question: why not adjust the balance to maintain symmetrical responses to dark and bright across the full dynamic range? Natural scenes are known to contain more dark contrasts than bright ones, so efficient design should allot more resources to transmitting dark. This asymmetry, arising at the cone output, carries through later stages of retinal circuitry and will be revisited later (see below, figure 11.19).

In summary, a visual scene is first quantized at the cone terminal as a flickering pattern of calcium channels, then as a flickering pattern of entering calcium pulses, and finally as a flickering pattern of vesicle release. Sensitivity, linearity, and the optimal balance for dark and bright are promoted by several molecular features governing vesicle release and by the architecture of the active zone (figure 11.8).

First, the cone vesicle's calcium sensor exhibits weak cooperativity, requiring only two calcium ions to bind for release. Second, it binds calcium with high affinity, so a low concentration suffices. Third, it *un*binds slowly, thus allowing temporal integration of the calcium influx. Consequently, the calcium concentration effective for release is 10-fold lower than for any other known synapse (Duncan et al., 2010). These properties render the release rate sensitive to small voltage changes,[8] exactly what is needed to finely quantize the cone's analogue voltage and to match the terminal's output range to the distribution of natural contrasts (chapter 9).

Fourth, each active zone uses small vesicles and docks them along a line. Both features promote independence between the glutamate pulses, thus optimal S/N improvement by their linear temporal summation across a wide dynamic range. Here is the reasoning. Small vesicles release small glutamate pulses which decay steeply in space and time. Thus, two vesicles released simultaneously will not sum much unless they are neighbors. Given tonic release of approximately five vesicles per ribbon per 100-ms interval, neighbors will rarely release simultaneously (figure 11.8).

Vesicle release, because it is initiated by the stochastic opening of one calcium channel, is itself stochastic (chapter 10). Therefore, the tonic rate of about five vesicles per 100 ms per active zone, if suppressed to zero by a bright stimulus, could yield a S/N ratio of about 2.2. This could discriminate three levels, dark versus grey versus white. The cone terminal improves this ratio by using multiple ribbons of identical structure, spacing them out to avoid major cross talk (figure 11.9). Typically a cone terminal uses 20 ribbons, yielding S/N of about 10.[9] Cones in peripheral retina expand their capacity to collect information (larger outer segments); correspondingly, they expand their capacity to distribute information (more active zones) and also their capacity to resupply the terminal (more axonal microtubules) (figure 11.9). This solidly illustrates the design principle *match capacity to information*.

In summary, contrasts encoded linearly in the cone's analogue signal are quantized linearly in the flickering pattern of voltage-gated calcium channels, then in the flickering pattern of inward calcium pulses, and finally in the flickering pattern of vesicle release. The mechanism requires meticulous design at the molecular and nanoscales to properly match (1) the calcium channel's current/voltage function to the cone's membrane voltage and bandwidth; (2) calcium channel proximity to vesicle docking sites; (3) binding affinity, OFF-rate, and cooperativity of the vesicle's calcium sensor; (4) vesicle volume and docking architecture; and (5) number and spacing of active zones within the terminal. These presynaptic features demand equally meticulous matching of postsynaptic features, as we now explain.

Discrete-to-analogue recoding at cone synapse creates parallel channels of different capacity

Each cone vesicle delivers glutamate into the synaptic cleft as a discrete (all-or-none) pulse. Upon reaching a bipolar dendrite, glutamate molecules bind to about 10 ligand-gated ion channels and cause a depolarizing current of about 5 pA. Such quantal currents from all sites across the dendritic tree sum to cause a graded voltage. Thus, the cone's analogue voltage that was first discretized by vesicles now reverts at the bipolar cell to an analogue voltage. This provides a critical opportunity to divide the cone's information packet and send smaller packets at lower bit rates. These parallel channels continue in various formats across much of the visual brain, so it seems worth identifying their humble origins in nanoscale chemistry and synaptic architecture.

Diffusional filtering
Bipolar cells implement *diffusional filtering* (Rao-Mirotznik et al., 1998). Certain types place their glutamate receptors near to vesicle release sites (20–400 nm). These receptors see a fast-rising, concentrated glutamate pulse. Other types locate their glutamate receptors far (400–1,800 nm) from

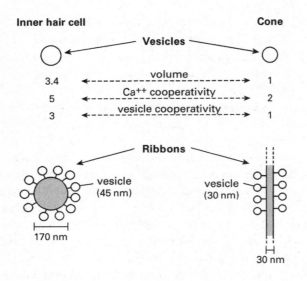

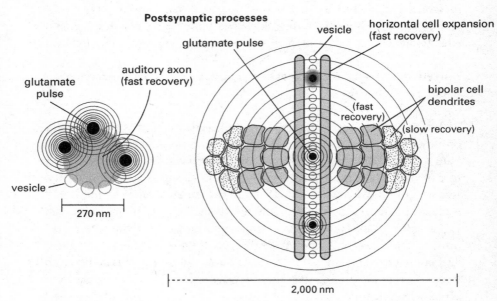

Figure 11.8
Active zone architectures match functions. Inner hair cell: Synaptic vesicle is large
with high calcium cooperativity. Consequently, it can be released only by a sharp,
fast depolarization that causes a large calcium current. Release in turn gives a large
pulse of glutamate. Docked vesicles are arranged on a spheroid or ellipsoid, so several
vesicles released simultaneously allow large glutamate pulses to sum across a single
postsynaptic disk served by fast glutamate receptors. This design promotes high S/N

and high temporal reliability that delivers enough information to justify a postsynaptic spike. **Cone:** Synaptic vesicle is small with low calcium cooperativity. Consequently, it can be released by a small depolarization that causes a small calcium current. Release gives a small pulse of glutamate. Docked vesicles are arranged along a line and have low release probability (~0.1 per 100 ms) and thus a low probability that two neighbors will release simultaneously. This arrangement promotes independence of postsynaptic events—which, while individually noisy, sum postsynaptically in space and time to increase S/N and temporal precision, thus concentrating information. Diffusion perpendicular to the ribbon spreads the transmitter in space and time. It is detected first by horizontal cell expansions that parallel the elongated ribbon to catch every pulse. Being nearest to release sites, the horizontal cell expansion sees high, fast pulses and detects them with fast receptors. Bipolar cell dendrites are next, the near dendrites using fast receptors, and the far dendrites using slower receptors. All structures are shown to scale, vesicle diameter being 30 nm; 2,000 nm indicates spacing of active zones.

release sites (Calkins et al., 1996). Those receptors see a slower-rising, more diluted glutamate pulse that has been spread out by diffusion (figures 11.8 and 11.10). The "near" receptors are fast isoforms—which desensitize quickly and recover quickly. The "far" receptors are slow isoforms—which desensitize like the fast isoforms but recover more slowly. These slow isoforms are thus well matched to encode the diffusionally blurred glutamate pulse (DeVries et al., 2006).

One might have imagined that distant receptors would express higher binding affinities to match the lower concentrations. But molecular properties, such as binding affinity and desensitization rate, are identical, the near and far glutamate receptors differing only in recovery time (DeVries et al., 2006). This property, combined with diffusional filtering, suffices for each postsynaptic site to select a particular bandwidth and S/N, which together define its information rate.[10] Thus, a key computation with huge consequences arises from a small structural difference between two protein isoforms (slow vs. fast recovering) and nanoscale differences in their locations.

It seems remarkable that *each* vesicle from a cone manages to distribute its information to 12 types of neuron (2 horizontal + 10 bipolar). Equally remarkable, each vesicle contributes a specific amount of information to each type. The vesicle's custom tailoring employs about 2,000 glutamate molecules (Rao-Mirotznik et al., 1998; DeVries et al., 2006). Were the number much smaller, say 500, diffusional filtering would fail because it would need higher receptor affinities with slower off-rates (like hormone receptors); furthermore, such sparse ligand would be too noisy. Were the number much larger, say 8,000, diffusional filtering would also fail because of

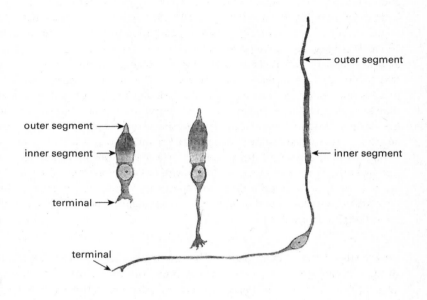

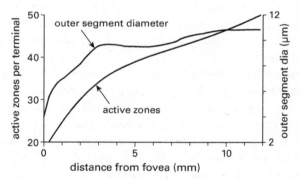

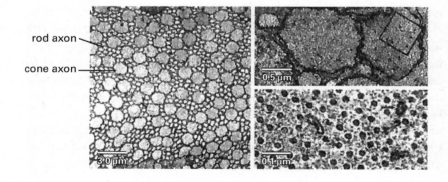

Figure 11.9
Cones match resources to information capacity. Upper: Peripheral cones (left, middle) with thick inner segments, thick axons and large synaptic terminals compared to a foveal cone (right). **Middle:** Outer segment diameter and number of active zones increase roughly in parallel. **Lower left:** Cross section through cone and rod axons near fovea. Rod terminal signals single photon events (0 or 1) with a single active zone and thus a thin axon. Cone terminal signals finely graded voltage—requiring many active zones and thus a thick axon. **Lower right:** Higher magnifications show axon cross section filled with evenly spaced microtubules that serve as monorails for molecular motors. Cone drawings are from von Greefe (1899), reprinted with permission from Sterling (2004a). Graph: active zones modified from Haverkamp et al. (2001a); outer segments modified from Packer et al. (1989).

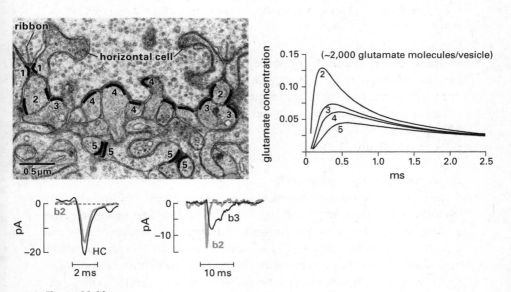

Figure 11.10
Diffusional filtering at the cone synapse. Upper: Horizontal cell processes invaginate the terminal to locate within 20 nm of vesicle release sites at base of ribbon (1). These processes extend parallel to the ribbon (perpendicular to this section) in order to catch every quantum; they express fast AMPA receptors. Horizontal cell processes also locate deeper in the synaptic layer (5) to read out slower modulations of glutamate using slower glutamate receptors (Haverkamp et al., 2001a). Thus, by monitoring cone glutamate release at different diffusion distances with different receptor types, the horizontal cell encodes the cone's full bandwidth. Bipolar dendrites also locate at different distances from the release sites (2–4) and use different glutamate receptors to segment the cone bandwidth. **Lower:** Electrical responses show fast and slow components of horizontal cell (HC) response and fast versus slow responses of two bipolar types (b2 and b3) to the same quantum. Micrograph reprinted with permission from Sterling 2013, contrast of postsynaptic receptor sites enhanced; responses modified and reprinted with permission from DeVries et al. (2006).

OFF-bipolar cells   ON-bipolar cells

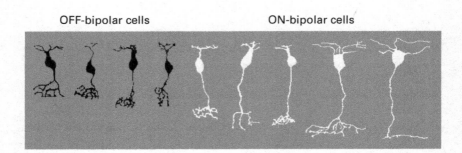

Figure 11.11
The cone subdivides its large information packet among multiple bipolar types. Cells
shown here are from mouse, but all vertebrate retinas follow this pattern. Latest
count, probably complete, finds five OFF and six ON types. Reproduced with permission from Wässle et al. (2009).

receptor saturation and the impossibility of removing so much glutamate
because transporters are already packed at maximum density (DeVries,
2011). Thus, the number of glutamate molecules in a cone vesicle matches
what is needed by the circuit's layout and its diverse molecular properties.
This explains why cone's vesicle diameter should be 30 nm and not the hair
cell's 45 nm (figure 11.8).

Duplicate the image and invert it

To increase its information capacity, a bipolar cell's dendrites sum currents
from the active zones of four to eight cones. The dendrites of a given type
all express a particular receptor isoform at a particular number of sites and
a particular diffusion distance. This creates bipolar types with distinctive
information rates: some types with many, high-rate contacts and other
types with fewer, low-rate contacts (Ratliff & DeVries, 2011).[11] The membrane voltage achieved by integrating about 500 quanta/100 ms is finely
graded. The bipolar cell's input message, concentrated presynaptically by
reducing transduction noise and redundancy, is about 10–30 bits per 100
ms (Ratliff & DeVries, 2011). This rate, 300 bits s^{-1}, would be expensive to
transmit via spikes because it requires thick axons (figure 4.6). It would be
acceptable for a few axons, but not for the optic nerve's one million axons.
Consequently, there is no alternative but to further subdivide into smaller
packets of information for transmission to ganglion cells.

A bipolar cell transmits only half of its input signal, dark contrasts or
bright ones. Certain types depolarize to dark and hyperpolarize to bright
(*OFF cells*) whereas other types invert the signal, depolarizing to bright and
hyperpolarizing to dark (*ON cells*) (figure 11.11). Simply inverting the input

Information:
lower rate → fewer outputs, thinner axons, upper strata
higher rate → more outputs, thicker axons, lower strata

ganglion cells select
different information rates

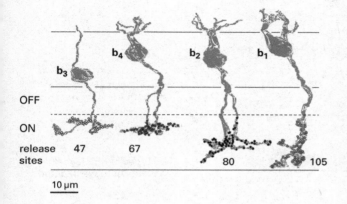
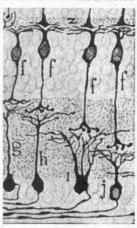

OFF

ON

release   47   67
sites   80   105

10 µm

Figure 11.12
**Bipolar types with higher information rates use more active zones and stratify to
connect with least wire. Left:** ON bipolar cell types with different information capacities use different axon calibers and different numbers of active zones. **Right:** Strati-
fied bipolar axons connect to different ganglion cell types with minimal wire. Left
from Cohen & Sterling (1990); right from Ramón y Cajal (1917). Both are modified
and reprinted with permission.

signal in this manner does not reduce the packet size because the dendritic
voltages linearly encode the full range. So at the outer synaptic layer the
ON cells and OFF cells encode identical information and are redundant. But
signal inversion is a key step—because it prepares both classes for the inner
synaptic layer which introduces a profound nonlinearity with numerous
profound advantages.

To execute this design at the outer synaptic layer requires no extra wire,
but only oppositely modulated cation channels. Dark contrasts excite via
ligand-gated receptors that *open* a cation channel (OFF cells) whereas bright
contrasts excite via a metabotropic receptor that *closes* a cation channel
(ON cells) (Xu et al., 2012; Morgans et al., 2010). OFF cells are of four types,
each diffusionally filtered to carry a particular-sized information packet.
ON cells are of five types, also carrying particular-sized information packets
as evidenced by their different axon calibers and different numbers of output active zones (figure 11.12). Ganglion cell dendritic arbors costratify
with particular types of bipolar terminal and thus assume their basic properties, slow ON, fast ON, and so on (figure 11.12).

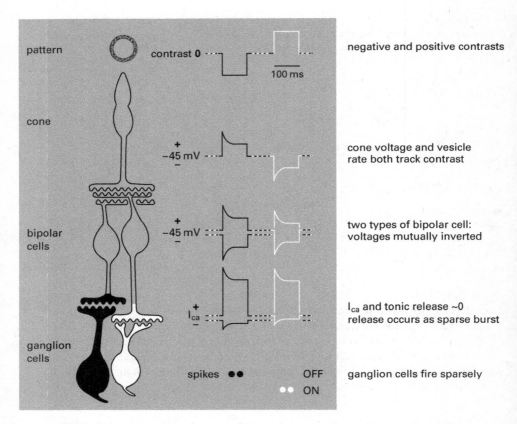

Figure 11.13
Bipolar cell outputs simultaneously sparsify and rectify. Dark and bright contrasts
activate calcium currents (I_{Ca}) reciprocally in OFF and ON bipolar terminals (dark and
bright, respectively). This reciprocally evokes brief bursts of glutamate quanta that
reciprocally excite spikes in OFF and ON ganglion cells. Modified and reprinted with
permission from Ratliff et al. (2010).

Forward through retina: A new code

The steady rain of cone quanta onto bipolar cell dendrites holds them near
−45 mV. At this voltage, calcium channels in the bipolar cell axon terminals
are mostly closed and the calcium current (I_{Ca}) is near zero. Consequently,
bipolar cell vesicle release is also near zero. Now, a dark contrast further
depolarizes the OFF terminals, turning on the calcium current and releasing
a burst of glutamate quanta (Freed & Liang, 2010). A bright contrast does
the same for the ON bipolar terminals (figure 11.13).

This design—inversion followed by rectification—simultaneously solves two problems. First, the image that was encoded by a high quantal rate is now encoded by far fewer quanta, a sparser code (Berry et al., 1997). With lower rate and sharper timing, each event carries more information. Second, the image that was simply duplicated at the outer synaptic layer has been rectified. Now each bipolar cell transmits only half the contrasts, dark or bright. Rectification, though doubling the number of transmission lines, halves the information per line, which reduces transmission costs by more than fourfold (chapters 3 and 9). Sparsification and rectification both arise at the bipolar terminal's voltage-gated calcium channel and then propagate across subsequent stages.

The circuit is tuned by familiar fine-scale features. For example, the vesicle's calcium sensor uses high cooperativity (about 4; Heidelberger & Thoreson, 2005; Matthews & Fuchs, 2010); this steepens the input/output curve compared to the cone and sharpens timing. Timing is also sharpened by negative feedback. The glutamate pulse that excites a ganglion cell simultaneously excites a reciprocal synapse that feeds GABA back onto the bipolar terminal, shunting it and curtailing release (figure 11.14). The specific isoform ($GABA_C$) is slower by 10-fold than the standard receptor ($GABA_A$), probably to optimize timing (Freed et al., 2003; Lukasiewicz & Shields, 1998).

Timing is further sharpened by multivesicular release at the bipolar cell's active zone (figure 11.14). The latter can be small because the mean release rate is low. This compact ribbon, like that of the auditory hair cell, can simultaneously fuse several vesicles to deliver a double or triple pulse of glutamate (Matthews & Sterling, 2008; Singer et al., 2004). Because the active zone docks only about 10 vesicles, multivesicular release significantly depletes it. This instantly reduces sensitivity, providing a lagless gain control that comes for free with the synaptic architecture (Oesch and Diamond, 2011). The various mechanisms establish a timing precision of about 14 ms, far short of the hair cell, but optimized to encode the cone's bandwidth (Freed, 2005).

Sparsification and rectification together reduce the input rate from about 500 quanta per 100 ms at the bipolar dendrites to an output rate of about 0.01 quanta per 100 ms at each active zone (Freed, 2005). This allows a bipolar terminal with 100 active zones (figure 11.12) to release one quantum per 100 ms which, measured postsynaptically, conveys about 0.2 bits (Freed, 2005). Therefore, the information rate for this bipolar cell axon is about 0.2 bits per 100 ms. Should this single quantum trigger a spike? Certainly not. A spike costs 100-fold more than the glutamate quantum, so to

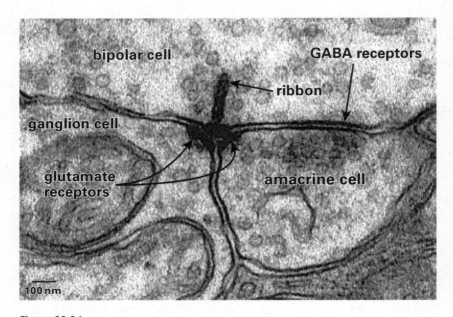

Figure 11.14
Feedback enhances timing precision at the bipolar terminal. Bipolar cell active zone with ribbon excites a ganglion cell dendrite and an amacrine process that feeds back onto slow GABA receptors at the bipolar terminal. This small ribbon has the capacity to release several vesicles simultaneously to a strong stimulus but, docking only about 10 vesicles, it can be depleted by strong stimuli. Thus ribbon structure directly implements contrast gain control. Electron micrograph courtesy of David Calkins.

be cost-effective, a spike should convey a much larger packet of information.

This observation identifies the ganglion cell's core computational task: concentrate information from about 0.2 bits per bipolar quantum to about 2 bits per spike. Two bits suffice for one spike to reliably report a weak flash on the receptive field (figure 11.2). To concentrate information by 10-fold requires several stages: (1) quantal currents caused by individual glutamate pulses convert to an analogue voltage: ganglion cell membrane potential; (2) the analogue signal then converts to a pulsatile output: spikes. Consider now some of the key design features that concentrate information by improving S/N and timing precision.

Ganglion cells optimally sum correlated bipolar cell inputs

To improve S/N from synapses that are individually unreliable, a ganglion cell collects from many whose responses are partially correlated in space

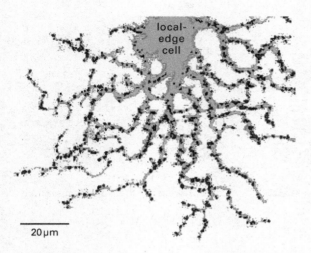

20 μm

Figure 11.15
How a ganglion cell creates a Gaussian weighting. Bipolar active zones distribute evenly across a thin stratum of the inner synaptic layer (figure 11.12), and ganglion cell dendrites collect them evenly, as shown here for a local-edge cell. Gaussian weighting arises from the denser branching near the soma that becomes sparser toward the periphery. Redrawn from Xu et al. (2008).

and time. Synapse density is constant on the membrane, so the numbers increase with dendritic length (figure 11.15). Because quanta are sparse, they occur widely across the arbor, allowing their currents to sum linearly (Freed, 2000, 2005; Liang & Freed, 2010).

To sum optimally, as noted for cone coupling, quanta should be weighted according to their strengths of correlation across the receptive field. These correlations arise from correlations in the structure of natural scenes. Thus, a ganglion cell optimally summing its input quanta optimally improves its S/N. For natural scenes optimal weighting across the receptive field is roughly Gaussian (figure 11.6). Thus, a ganglion cell should weight strongly the quanta near its receptive field center and weight less strongly quanta near the edge (Tsukamoto et al., 1990).

This optimal weighting arises as an analogue primitive computation. The ganglion cell distributes its membrane as a two-dimensional Gaussian by branching densely near the center and more sparsely near the edge. Bipolar axons distribute their active zones evenly across a thin stratum (~2 μm) of the inner synaptic layer, so ganglion cell arbors that costratify with these synapses receive them at constant density (~40 contacts/100 μm² of dendritic membrane). The Gaussian weighting is computed for free

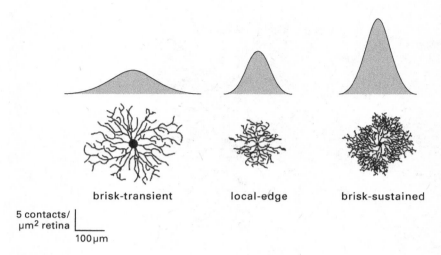

brisk-transient   local-edge   brisk-sustained

5 contacts/
µm² retina
100 µm

Figure 11.16
Ganglion cells directly compute a two-dimensional Gaussian filter. This filter weights
bipolar inputs to optimally improve S/N. Each of these cell types distributes its dendritic membrane in two dimensions across a thin stratum of the inner synaptic layer, where it costratifies with synaptic terminals of a particular bipolar type (figure
11.12). The bipolar terminals distribute evenly across the stratum, but the dendritic
membrane distributes as a Gaussian. Brisk-transient dendrites distribute as a broad,
shallow Gaussian, whereas local-edge and brisk-sustained dendrites distribute as a
narrow, steep Gaussian. A spot covering their receptive field centers activates all synapses which are far more numerous for brisk cells, about 6,000, than for local-edge
cells, about 1,400. Modified and reprinted with permission from Xu et al. (2008).

by the ganglion cell's branching pattern, as shown in figure 11.16 (Kier et
al., 1995; Xu et al., 2008).

Could a cell then continue to improve its S/N by extending its dendrites
ever farther to collect more synapses? No. Spatial correlations decline exponentially across natural scenes whereas S/N improves only as the square
root of added synapses. Thus, the strategy of extending dendrites soon produces diminishing returns. The broadest ganglion cell arbors are typically
about 0.5 mm in diameter and collect about 5,000 bipolar cell contacts.
This number is conserved across mammalian species of vastly different eye
size.[12] This suggests that the size of ganglion cell dendritic arbors reflects
the structure of natural scenes—which is similar for all terrestrial species.

Ganglion cells optimally sum excitatory and inhibitory inputs
Among a ganglion cell's synaptic inputs, 30% to 80% (depending on type)
are inhibitory—from amacrine cells (Freed & Sterling, 1988; Cohen &

Sterling, 1992). For a ganglion cell to code efficiently, the inhibitory currents must balance the excitatory currents. Moreover, since the cell's spatio-temporal filter adjusts with adaptation state (figures 9.7 and 9.11), the balance of inhibitory and excitatory currents should shift correspondingly to maintain optimality across conditions. Indeed, S/N in the brisk-transient ganglion cell is optimized by the balance that exists at normal resting potential. And, when challenged with a different contrast or stimulus size, the circuit dynamically adjusts both the inhibitory and excitatory conductances such that the amplitudes of the inhibitory and excitatory currents continue to maximize S/N (Homann & Freed, 2012).

Overlap in ganglion cell arrays: S/N versus redundancy

To capture spatial images requires an array—like pixels in a digital camera. The array's acuity is set by the neuron spacing because of Nyquist's rule (chapter 4). Therefore, each type of ganglion cell has a characteristic spacing to serve the acuity required for its particular function. As ganglion cells with fixed spacing extend their dendrites to improve S/N, nearest neighbors soon overlap (figure 11.17).

The overlap causes some of a ganglion cell's information to be redundant with respect to its neighbors. Thus, the scheme to concentrate information for each spike by improving a cell's S/N inevitably adds redundancy to their array. This is not all bad because redundancy in the presence of noise can improve signal detection (chapter 6). But ganglion cells must not squander their limited dynamic range on highly redundant signals. So how far should the dendrites extend? Since this question applies to most brain regions, an answer for retina may have general significance.

Optimal design should maximize the information sent by the *array*. This requires balancing the S/N improvement of individual neurons against the redundancy due to their overlap. An array is calculated to be most informative for natural images when receptive field centers are spaced at two standard deviations (σ) of their Gaussian sensitivity functions. That causes receptive field centers to overlap by about sixfold, but because a dendritic field is narrower than the receptive field center, dendrites need to overlap only by about threefold. This calculated optimal array corresponds precisely to what is observed (figure 11.18; DeVries & Baylor, 1996; Borghuis et al., 2008).

There are exceptions to this default design, but they are consistent with the basic rule. For example, dendritic fields of midget ganglion cells outside the fovea tile neatly without overlap (Calkins & Sterling, 1999; Dacey, 2004). Nevertheless, their receptive fields overlap to the same degree as for the larger parasol ganglion cells and obey the standard two-σ separation

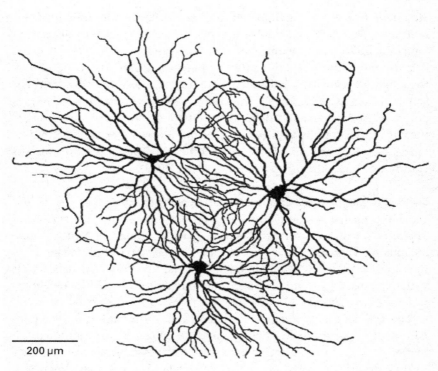

Figure 11.17
Neighboring ganglion cells in an array overlap their dendrites, achieving threefold coverage of all points in the field. Illustrated here are OFF brisk-transient cells, but the rule holds for brisk-sustained, local-edge, and so forth. Although exceptions exist, this is the default design. Reproduced with modifications and permission from Liu, Whitaker, & Massey (unpublished).

(Gauthier et al., 2009). In this case optical blur convolves with cone coupling to enlarge the bipolar cell receptive fields, allowing the ganglion cell receptive fields to overlap even though their dendritic fields simply tile. Thus, although the dendritic fields of midget ganglion cell contradict the rule of overlap, the receptive fields achieve optimal overlap.

Nonlinear mechanisms help concentrate information
Beyond linear summation, various nonlinear mechanisms further concentrate information in the ganglion cell's analogue signal. For example, voltage-gated sodium channels in the dendrites selectively amplify larger, faster excitatory postsynaptic potentials (Dhingra et al., 2005). These

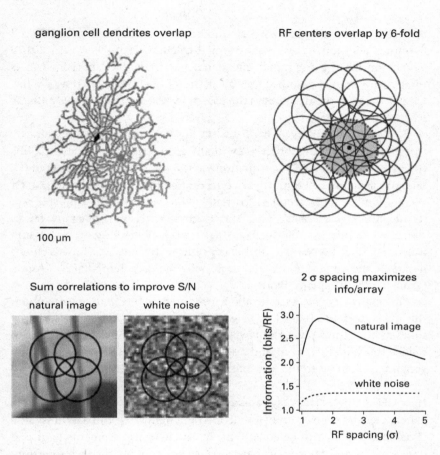

Figure 11.18
Receptive field overlap maximizes information from the array. Upper: Overlap of
ganglion cell dendritic fields causes sixfold overlap of the receptive field centers (RF).
Lower: Adjacent regions in natural images are correlated, so a larger field integrating
more correlated inputs improves S/N (figure 11.16). But this causes overlap and thus
redundancy. Total information from the array peaks for natural scenes when receptive fields are spaced at two standard deviations (σ) of their Gaussian centers. Adjacent regions in white noise lack correlations, so the optimal array simply *tiles* without overlap. Modified and reproduced with permission from Borghuis et al. (2008).

outweigh the smaller, noisy signals and improve timing precision. OFF cell dendrites additionally use voltage-gated calcium channels (Euler, 2010). When the multiquantal analogue signal has been thus sharpened, it is finally thresholded (another nonlinear step) to be requantized as a spike. Clearly the retinal output obeys the principle *combine analogue and pulsatile processing*.

By linearly summing tens of quanta, by filtering nonlinearly, and by thresholding, the ganglion cell can finally send a spike worth about 2 bits (figure 3.6). Compared to one quantum at the input that conveys about 0.2 bits, the ganglion cell output has concentrated information by 10-fold. Of course, some of its 2-bit spike is partially redundant with the 2-bit spikes of its neighbors. This seems puzzling, given that designs generally try to decrease redundancy. Yet, the next stage collects the redundant signals and uses a specialized synapse to further concentrate the message. So this redundancy at the ganglion cell output is accepted as a temporary measure. Chapter 12 will explain (figure 12.3).

Although one spike encodes about 10-fold more bits than one quantum, the spike costs 100-fold more to generate in the ganglion cell and many times more to transmit centrally via the optic nerve. This creates pressure to reduce total spikes per cell and also mean rates. One strategy is to match ganglion cell arrays to the distribution of information in natural scenes.

How OFF and ON arrays save spikes

Where Genesis states, "And God divided the light from the darkness," the division is assumed to be equal. The Taoist circle also segments light and dark equally, and our visual sensations do not contradict this impression. But it is not so. Natural scenes actually contain more dark regions than bright ones (Laughlin, 1981; Richards, 1982). And since there are more dark regions, they must also be smaller (figure 11.19).

Correspondingly, OFF arrays are denser than ON arrays and use narrower dendritic fields. In other words, OFF cells cover the same territory as ON cells but with a finer grain (Ratliff et al., 2010; Chichilnisky & Kalmar, 2002; Dacey & Petersen, 1992). Moreover, OFF dendritic arbors branch more than the ON arbors so, despite having smaller dendritic fields, their total dendritic lengths are similar. Because synaptic contacts follow dendritic length, individual OFF and ON arbors collect equal numbers of synapses—and thus equal amounts of information (Ratliff et al., 2010).

An ON ganglion cell, collecting the same amount of information as an OFF ganglion cell, sends similar numbers of spikes. However, the array of ON cells, with its twofold coarser grain, sends half as many spikes. Thus,

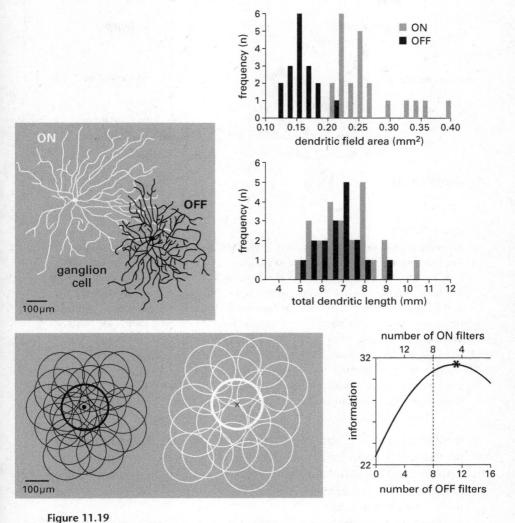

Figure 11.19
Ganglion cell dendritic arbors match the natural distribution of contrasts. Upper:
OFF dendritic arbors are smaller than ON arbors of same cell class (brisk-transient)
but branch more densely to achieve equal dendritic length and thus equal numbers
of synapses. **Lower:** OFF array is finer than ON array because dark contrasts distribute
more finely than bright ones; array information is maximized (*) when OFF cells
(filters) are twice as numerous. Modified and reprinted with permission from Ratliff
et al. (2010).

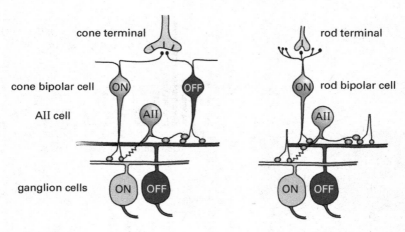

Figure 11.20
AII amacrine cell serves a cone circuit in daylight and a rod circuit in starlight. Left:
The ON cone bipolar axon terminal, depolarized by a bright contrast, couples electrically to the narrow-field AII cell, depolarizing it as well and thereby inhibiting, via
chemical synapses, the OFF bipolar terminal and OFF ganglion cell dendrite. This circuit enhances rectification of the OFF ganglion cell and, via "push–pull" amplification (excitation + disinhibition), extends the linear range for coding dark contrasts.
The rod terminal couples to this circuit via gap junctions (not shown). **Right:** The rod
bipolar cell amplifying single photon responses in starlight, depolarizes the AII cell,
which conveys the rod signal to the axon terminals of ON and OFF cone bipolar cells.
Thus, excitatory cone bipolar synapses upon ganglion cells serve as a final common
pathway for both cone and rod signals.

rectification into ON and OFF channels, beyond halving the information
per channel to allow lower rates (figure 11.13), also allows each channel to
better match the natural distribution of its inputs, and this conserves spikes.
Moreover, there is a further benefit whose explication requires a step back
to the bipolar neurons.

When an OFF bipolar cell has filled its information capacity, the ON
bipolar cell, collecting from the same cones, still has capacity to spare. To
fill it, the ON bipolar cell remains partially unrectified—maintaining some
degree of tonic vesicle release. Consequently, a depolarizing bright contrast
can sharply increase its release, and a dark contrast can decrease it. Therefore, the ON ganglion cell's spike train conveys some information regarding
dark contrasts. Moreover, the ON bipolar cell cleverly injects its modulation
by dark contrasts into the OFF circuit (figure 11.20).

The ON bipolar synaptic terminal couples electrically to the *AII amacrine
cell* that, in turn, provides inhibitory (glycinergic) synapses onto OFF

bipolar terminals and OFF ganglion cell dendrites. Consequently, bright contrasts depolarize the AII cell and thereby inhibit the OFF cells. This reinforces their rectification, further protecting them from encoding any bright information and preserving their full dynamic range for dark information. When dark contrasts hyperpolarize the ON bipolar cell, the AII cell hyperpolarizes, thereby disinhibiting the OFF bipolar synapses and the OFF ganglion cell dendrites. The result is "push–pull" amplification: excitation by opening the glutamate-driven cation channel, plus disinhibition by closing the glycine-driven anion channel. Combining the excitatory driving force ($E_{rev} \sim 0$ mV) and the inhibitory driving force ($E_{rev} \sim -70$ mV) extends the steep, linear range for encoding dark contrasts.

This modulated *crossover inhibition* from the ON to OFF channel improves efficiency at the output. The OFF ganglion cell is slightly hyperpolarized (roughly –5 mV), moving it farther from spike threshold and reducing spontaneous spikes. This reduces noise in the spike output, and thus mean spike rate, thereby increasing bits per spike (chapter 3). ON and OFF ganglion cell classes better match their signaling capacities to the natural distribution of contrasts. The benefits: fewer spikes, lower mean rate, and more bits per spike justify the cost of doubling the cell types (Renteria et al., 2006; Molnar et al., 2009; Liang & Freed, 2010; Manookin et al., 2008; Demb & Singer, 2012).

Neatening up

The AII cell, in mediating crossover inhibition from ON cone bipolar cells to OFF cone bipolar cells, uses a narrow dendritic arbor. In this respect, it resembles a narrow-field ganglion cell. This is no coincidence: crossover inhibition needs to be cospatial with direct excitation, so the collecting arbors for both the inhibitory and excitatory mechanisms should match (figure 11.20). Of course, a narrow arbor requires a dense array. Indeed, 20% of all amacrine layer neurons are AII cells. So it is expensive; yet the retina has found a way to squeeze out more profit.

The AII array, in addition to serving the cone bipolar circuits in daylight, serves a rod bipolar circuit in starlight (chapter 8). Rods depolarize a dedicated bipolar type that, via glutamatergic chemical synapses, depolarizes the AII cell. Now current passes via electrical synapses from the AII cell into the ON cone bipolar terminal, thereby releasing its glutamate onto the ON ganglion cell. The electrical AII → ON bipolar synapse is the same one that in daylight passed current in the opposite direction (figure 11.20). This ability to conduct in either direction allows the AII cell to serve either cone or rod circuit, depending on time of day.

The bidirectionality of this connection between AII and ON cone bipolar allows the rod and cone circuits to use the same set of synapses to excite the ganglion cells (figure 11.20). This avoids the cost of maintaining two sets and using each only half the time. And it effectively doubles the number of contacts that each circuit (cone or rod) can deliver to the available dendritic membrane (figure 11.15).

This design offers several efficiencies. First, the costly AII array can work two shifts—day and night. Second, the day circuit that sums narrowly can enlarge at night to sum broadly with little extra cost. This is accomplished by coupling AII neurons to each other via gap junctions that uncouple during the day to restrict signal spread and recouple at night to promote it. Third, excitatory synapses on the ganglion cell membrane can serve as a final common pathway for both shifts. This exemplifies the engineer's injunction to "neaten up" the design by using an expensive resource to its utmost without sacrificing performance (chapter 1).

More types saves more spikes

Since doubling ganglion cell types (ON and OFF) justifies the cost (Von der Twer & MacLeod, 2002), one can imagine that adding more types could do the same.[13] This seems straightforward because bipolar types have already segmented the bandwidth in the outer synaptic layer to reduce the amount of information that they must transmit (figure 11.10). Bipolar types deliver slowly changing signals to the inner synaptic layer via finer axons with fewer, slower synapses and deliver rapidly changing signals via thicker axons with more, faster synapses (Freed & Liang, 2010). Each bipolar type costratifies with a particular ganglion cell to transfer its portion of bandwidth (figure 11.12).

This establishes a functional ganglion cell type—which requires a structure to match. A type that encodes high temporal frequencies cannot sum temporally to improve its S/N; therefore, it must sum spatially, and this requires a wide dendritic field. Conversely, a type that encodes low frequencies *can* sum temporally, so it can use a narrow dendritic field.

In this design, bandwidth sets field size; field size sets cell spacing ($2\text{-}\sigma$ rule); and spacing sets array structure—sparser for cells coding high temporal frequencies and denser for cells coding low temporal frequencies.[14] Furthermore, segmenting the bandwidth with more cell types reduces mean spike rates. Since cost per bit increases with spike rate and higher bandwidths entail higher rates (Koch et al., 2006), the reasons to segment bandwidth are the same as those to segment contrast: resources are used more efficiently and better matched to information.[15]

For example, a broad-bandwidth, wide-field ganglion cell (brisk-transient) receives more than fourfold more contacts from fast bipolar cells than a low-bandwidth, narrow-field cell (local-edge) receives from slow bipolar cells. When these ganglion cell types view a nature video, fast features trigger bursts of quanta from fast bipolar cells and cause the brisk-transient cell to spike. Fast features trigger no quanta from slow bipolar cells, so the local-edge cell is silent. Yet a slow feature (an edge slowly going dim-then-bright) triggers a burst of quanta from slow bipolar cells onto the small ganglion cell, and this evokes a few spikes (figure 11.21).

If a narrow-field ganglion cell were to encode high temporal frequencies by collecting from fast bipolar cells, its limited spatial summation would reduce S/N and cause a noisier spike train. Expensive signaling capacity would be squandered. Conversely, if a broad-field ganglion cell were to encode low temporal frequencies, its spatial pooling (needed for high frequencies) would reduce certainty regarding the location and trajectory of the local edge; moreover, the spatially pooled slow signals would be redundant. Thus, efficiency improves by matching types to their frequency bands. But if that were the whole story, 10 ganglion cell types would suffice to transmit all the information delivered by the 10 bipolar types. Yet ganglion cell types are roughly twice as numerous.

What sets the *types* of types?

The exact number of ganglion cell types is unknown for any species, but in all mammalian retinas that have been studied carefully, it is at least 20. Although types are differently named across species, their receptive field properties, dendritic branching, and stratification are all rather well conserved. As noted, even field diameters are conserved: differences are less than twofold across eyes that differ in diameter up to 10-fold.[16] Cell types are conserved across different terrestrial habitats and across different modes of existence, such as predator and prey. For example, cat and mouse both have brisk-transient and local-edge cells, as do human, monkey, rabbit, and guinea pig (Crook et al., 2013).

But why are 20 types better than 10? And, since that is so, why not 40? Moreover, why are types so strongly conserved? If they were truly "feature detectors," would not species with different lifestyles need different types? Wouldn't mice need "hawk-detectors" and hawks need "mouse detectors"? In short, why don't the *types* of types vary more according to habitat and lifestyle? These questions are not philosophical; rather, they belong to the realm of reverse engineering—and thus are potentially answerable.

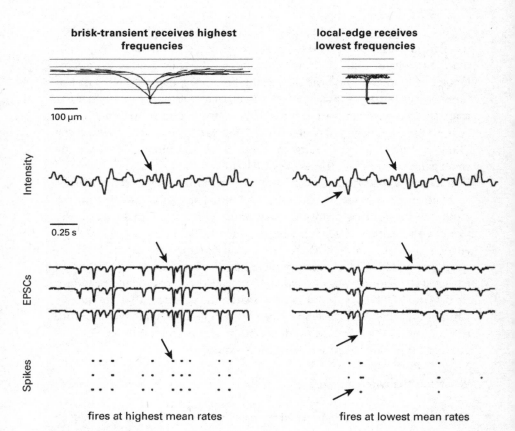

brisk-transient receives highest frequencies

local-edge receives lowest frequencies

100 μm

Intensity

0.25 s

EPSCs

Spikes

fires at highest mean rates

fires at lowest mean rates

Figure 11.21
Ganglion cell types transmit different bandwidths with different firing rates. Brisk-transient ganglion cell collects many high-rate bipolar contacts (about 5,000); whereas local-edge ganglion cell collects fewer low-rate bipolar contacts (500). When a natural scene video was presented repeatedly, a high-frequency feature (downward arrow) evoked bursts of synaptic quanta to the brisk-transient cell but not the local-edge cell. A low-frequency feature, an edge going dark then bright (upward arrow), evoked a burst of quanta to the local-edge cell. EPSC, excitatory postsynaptic current. Anatomical diagram modified and reprinted with permission from Rockhill et al. (2002); electrical recordings (guinea pig) are from Koch, Freed, & Sterling, unpublished.

Why types are conserved

One reason why ganglion cell types are conserved types across habitat and lifestyle is that physical properties, such as photon noise, vesicle noise, diffusion speed, synapse size, and so on are invariant. Therefore, a mouse eye can shrink the number of neurons compared to a cow eye, but it cannot greatly shrink either the neurons themselves or their synapses.

Another reason is that the statistical structure of natural scenes is invariant. All scenes contain mostly low temporal and low spatial frequencies in the same relative amounts; moreover, they are scale invariant and thus unchanging as a pattern moves closer or farther away, or as a pattern is magnified on the retina of a larger eye. The same is true for various motions that sweep, jump, track, or jiggle patterns across the retina: their distributions of temporal and spatial frequencies are all the same (figure 11.22).

Considering that retinal circuits are selected to encode optimally, once they reach the point of optimality—where costs just balance benefits—there is no pressure for change but instead pressure to preserve what works. Therefore, ganglion cell types are conserved. The many types of interneuron are also conserved because they help create ganglion cell types. Just as brisk-transient and local-edge ganglion cells are easily recognized across species, so are interneurons, such as the AII and *starburst amacrine cells* (MacNeil et al., 1999; Vaney, 1990).

Dendritic field sizes are conserved across species due to the invariance of various physical factors upon which the neurons rely. Scene statistics, photon noise, diffusion distances and speeds, synapse size, receptor binding properties, and channel properties are the same for mouse and human. Consequently, a local-edge cell, to fill its coding capacity, must spread its dendrites sufficiently for its overlying patch of cones to capture sufficient photons. And it must branch the dendrites sufficiently for the membrane to capture sufficient synapses. Similarly, the AII amacrine cell, which serves crossover inhibition between OFF and ON ganglion cells, must restrict its dendritic spread in order to compute the local differences between dark and bright contrasts[17] which, belonging to the scene statistics, are the same for all eyes.

For conservation of ganglion cell types, consider this example: the image cast on a mouse's retina by a distant hawk is small and slow-moving; so is the image cast by the mouse on the hawk's retina. The hawk has greater acuity because of optical factors and higher density of photoreceptors, but the mouse casts a smaller image. Thus, as the two species track each other, they will both use the local-edge type of ganglion cell. Species may have opposite goals, but when the images are similar, they use the same cell type.

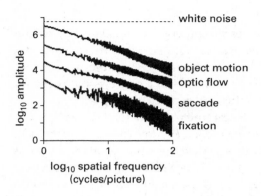

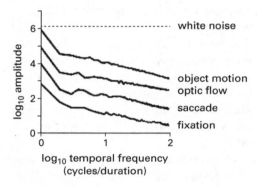

Figure 11.22
Natural scenes contain similar distributions of spatial and temporal frequencies across four types of motion. Both spatial and temporal frequencies decline with similar slopes. Spectra have been separated (shifted up) for clarity. Reprinted with modifications and permission from Koch et al. (2006).

Why some types should be less selective

A frog's auditory axon encodes frequencies characteristic of the species mating call with extremely high efficiency, filling about 90% of its information capacity. This requires a stringently tuned nonlinear filter to selectively amplify the signal (Rieke et al., 1995). This design serves brilliantly because the signal can be anticipated. But it could not serve ganglion cells' coding of natural scenes because they vary unpredictably.

A stringent filter applied to a natural scene would reject most information, so a ganglion cell would be mostly silent. Its channel would be filled by redundancy (*0 0 0...*) and would encode too few bits to pay its maintenance cost. Moreover, because one type would capture so little

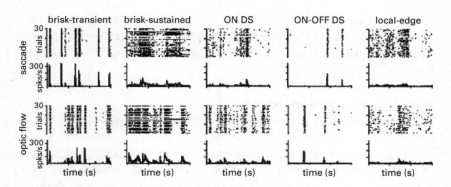

Figure 11.23
Each type of ganglion cell responds stereotypically to all scenes and all types of motion. Upper row: Video jumped across a natural scene to mimic saccades. Brisk cells fire at high mean rates; direction-selective (DS) and local-edge cells fire at low mean rates. **Lower row:** Video panned smoothly across a natural scene to mimic optic flow. Response patterns resemble those to saccadic motion. Reprinted with modifications and permission from Balasubramanian & Sterling (2009); data are from Koch et al. (2006).

information, many types would be needed, and none would pay their maintenance cost. Thus, the uniformity of natural scene statistics restrains the specificity of ganglion cell tuning and thereby reduces the number of types. This leads to a dual design: some nonstringent filters to capture the broad distribution of spatial and temporal frequencies in natural scenes, and some quasi-stringent filters to match stereotyped aspects present in most scenes that are needed by various downstream users.

The nonstringent filter fills only about 30% of a ganglion cell's information capacity, threefold less than the frog auditory axon. The remaining 70% of capacity is occupied by noise and redundancy. Small ganglion cells have more noise, and large ganglion cells have more redundancy (Koch et al., 2006). The quasi-stringent types fill capacity to the same extent as the nonstringent types. Thus, they are no more efficient than the nonstringent types, as would be expected because they share the same noisy inputs. However, the quasi-stringent types do manage to delete certain signals that their specific downstream targets will not need.

Since all scenes contain similar distributions of spatial and temporal frequencies, each type responds similarly to all scenes (figure 11.23). This is true equally for both nonstringent and quasi-stringent types. Initially this seems surprising since the traditional experiment searches for particular stimuli that a cell "likes best," leading to the idea of ganglion cells as "feature detectors." That idea is not wrong; it is just that the features are present in all scenes, so neurons wired to encode them must always "see" roughly the same thing!

Why some types are more selective

The nonstringent strategy employs five types to parcel the natural distribution of achromatic spatial and temporal frequencies, plus one more to parcel the spectral frequencies. The six types are: ON and OFF brisk-transient, ON and OFF brisk-sustained, local-edge, and blue–yellow.[18] The number of types employed by the complementary strategy, quasi-stringent filters, is uncertain, but it could easily exceed 15, as will be evident from the following examples.

Among the low-level users of retinal information are central pattern generators that coordinate motion of the eyes and head. All pattern generators require feedback that, as chapter 4 noted, is obtained most efficiently from sensors that send only what is needed to correct the output. To supply several of these low-level mechanisms, the retina constructs directional-selective ganglion cells—and not just one type, but *three* types that together cover the retina with 10 separate arrays.

First, consider eye movement controllers that maintain eye position as the head rotates, so that the image remains stationary on the retina. When the head rotates, signals from the vestibular semicircular canals command the eyes to smoothly counter-rotate and thus stabilize the retinal image. When eye motion lags head motion, the eye movement controllers need to know the slippage. The detector is the ON directionally selective (DS) ganglion cell. It signals slow global motion, and its axons target a low-level neuron cluster that uses the slippage signal to reduce the error (Simpson, 1984; Vaney et al., 2012).

The ON DS type responds exclusively to slow motion. That suffices because, even when the head moves fast, slippage is slow. It responds exclusively to ON. That suffices because global motion moves bright and dark regions together, so to measure slippage requires tracking only one. ON suffices and is cheaper because bright regions, containing less information per retinal area, trigger fewer spikes. The type responds to motion only in the direction of slippage, and that suffices because that is all the motor circuit needs to know.

Responses to irrelevant directions are carved away. To send them would waste retinal resources and incur additional costs downstream. The slippage detector should report only directions needed to correct for one pair of semicircular canals. But there are three pairs of semicircular canals, and thus three arrays of ON DS cells. Finally, the mechanism requires both transient and sustained information, which is cheaper to send on separate lines. So the downstream users that stabilize the image during head rotation are supplied most efficiently by six arrays of ON DS cell, as indicated in figure 11.24.

Inverse to the problem of stabilizing the whole retinal image when the head moves slowly is the problem of tracking a fast object with the head still. The eyes track smoothly until slippage triggers a fast jump for recapture. Slippage can be faster than in global motion, so the detector must sense faster motion. The object's leading edge can be bright or dark, so the detector must sense both. And it should detect slippage in each of the four directions of eye movement. This task falls to the ON-OFF DS ganglion cell.

This type responds to faster motion than the ON DS. And it responds to both bright and dark by spreading one dendritic arbor in the ON stratum and another in the OFF stratum (Vaney, 1994). Both arbors tile, but predictably, the ON arbor is coarse and the OFF arbor is fine. Thus, the ON arbor covers the same territory as the OFF arbor but with fewer synapses. ON-OFF DS cells send spikes in response to motion along only one of the four critical axes, so there are four separate arrays. Selectivity for four orthogonal directions is established by cofasiculating the ganglion cells with an array of specialized interneurons (*starburst amacrine*) that release GABA to motion centrifugal to its dendritic field. The ganglion cell connects asymmetrically and so receives inhibition from the sector opposite to the preferred direction (figure 11.24; Vaney et al., 2012).

The point of these selective filters is to transmit only the particular information needed by each downstream user. By discarding the rest, the ganglion cell can send fewer spikes at lower rates. Thus, a major task for amacrine circuits is to carve away all that is unneeded—in the sense of Michelangelo who famously may have said, *I saw the angel in the marble and carved to set him free*. Such carving for each of 20 ganglion cell types probably explains much of the amacrine cells' diversity (Vaney, 1990; MacNeil et al., 1999).

New types can be added

The nonstringent types relay information toward the cortex for further processing to reduce high-level uncertainties—for example, to identify a face (chapter 12). These types should represent all aspects of all scenes without judging what might be important—thus their low stringency. Types could

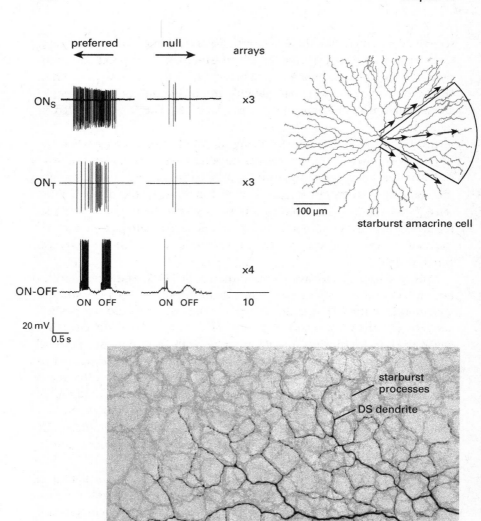

Figure 11.24
Three types of directionally selective ganglion cell express 10 complete arrays. ON
directionally selective (DS) are of two types, sustained (S) and transient (T); Hoshi et
al., 2011; Kanjhan & Sivyer, 2010). These form three arrays to match the semicircular
canals. ON-OFF DS forms four arrays to match the cardinal directions of eye movement. Starburst amacrine cell releases GABA to all centrifugal motion (away from its
cell body), but the sector shown here connects to ON-OFF DS cells that prefer motion
in the opposite direction. The ganglion cell dendrites cofasciculate with the network
of starburst processes but select only contacts that produce the appropriate selective
response (see figure 13.6).

be fewer if stringency were relaxed still further, but that would increase spike rates. Spike rates could be lowered by tightening stringency, but that would require more types. So the number of types that relay to higher levels is probably a compromise.

The quasi-stringent types that relay information to specific low-level users are engendered by the design rule *compute at the lowest possible level* (chapter 4). Therefore, the number of quasi-stringent types depends on the number of low-level users. The needs for both classes, nonstringent and quasi-stringent, are similar across terrestrial species because they see the same image statistics and share the same brain organization (chapter 4). So the *types* of types are conserved. There are roughly 20 types because that suffices: 10 are too few, and 40 are too many.

The reasoning resembles Darwin's, who claimed that the different beak structures of Galapagos finch species suffice to exploit the observed distributions of seed size and hardness. Biologists need this logic to make sense of their mountains of observation, yet they remain suspicious of its circularity. The strongest test is prospective: change the distributions of seed size/hardness and observe in real time whether beak structure follows. This experiment, conducted over decades on that very archipelago, confirmed the hypothesis (Weiner, 1994).

The test for retina is retrospective: new skills acquired by Old World primates needed greater visual acuity; and for that the retina added two additional arrays of bipolar cells and ganglion cells. Thus primates, including humans, have the same 20 or more types as other mammals, plus ON and OFF midget bipolar cells and midget ganglion cells that connect each cone to the brain by two private lines. These support a spatial resolution of about 60 cycles per degree, which is ≥10-fold that of most mammals. This fine acuity facilitates many human activities, such as threading a bone needle, reading nuances of facial expression, and nuances orthography.

These same private lines also preserve the cone's spectral tuning to short, middle, or long wavelengths; and this provides a simple mechanism for trichromatic color vision (Jacobs, 2009; Crook et al., 2011). Thus, the primate brain by investing in a double array of midget bipolar and midget ganglion cells, obtained in return the essential building blocks for both fine spatial vision and for trichromatic color vision. To exploit these basics has required the primate brain to invest half of the cerebral cortex (chapters 12 and 14), but this is the deal upon which we have built our civilizations. In short, when new neuron types offer great advantages, animals add them.

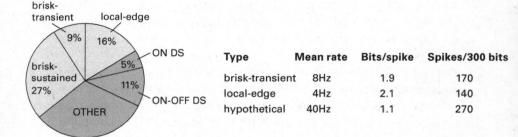

Type	Mean rate	Bits/spike	Spikes/300 bits
brisk-transient	8Hz	1.9	170
local-edge	4Hz	2.1	140
hypothetical	40Hz	1.1	270

Figure 11.25

How ganglion cell types apportion information. Total spikes are about equally divided between nonstringent types (pale) and quasi-stringent types. The brisk-transient cells send only about one tenth of the total whereas the local-edge cells send almost twice as many spikes at half the rate. Hypothetical ultrahigh-rate type, compared to local-edge, would halve bits/spike and nearly double the number of spikes to send fixed number of bits. DS, directionally selective. Modified and reproduced with permission from Koch et al. (2006).

How ganglion cell types spend the information budget

Total spikes divide about equally between nonstringent and quasi-stringent types (figure 11.25). Brisk-transient cells send about one tenth of the total spikes—at the highest mean rate and thus lowest efficiency. Local-edge cells send nearly twice as many spikes as brisk-transient cells—at the lowest mean rate and thus higher efficiency. To send a fixed number of bits, local-edge cells send about 10% more bits/spike and about 20% fewer spikes. To send the same number of bits, a hypothetical ultrahigh-rate type (resembling a mammalian auditory axon) would halve the bits/spike and nearly double the number of spikes.

Optic nerve: The last information bottleneck

The Mars camera, sending rich images under the constraints of space and energy, transmits every bit at low rates to a receiver on Earth that can process over indefinitely long times. The retina, sending comparable images under those same constraints, is further constrained by time. Most information from the retina has an early expiration date—100 ms. This forces the retina to set different bit rates to match the specific need of each downstream user (figure 11.26). Too low a rate would compromise function; too high would squander resources.

Resource usage depends critically on firing rate because that sets axon diameter—lower rates allow thinner axons (chapter 3). Consequently, the

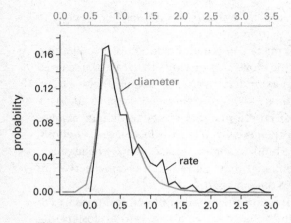
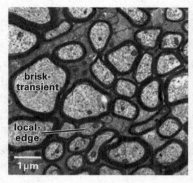

Figure 11.26
Low mean firing rates allow thin axons that quadratically reduce space and energy.
Left: Distribution of mean firing rates matches distribution of axon diameters. **Right:**
Optic nerve in cross section. Modified and reproduced with permission from Perge
et al. (2009).

distribution of firing rates sets the distribution of axon diameters, and this
determines the structure of the optic nerve (figure 11.25). Recalling that
space and energy costs rise as rate squared, one can appreciate the strong
pressure to reduce firing rates that skews the distribution toward finer
axons. Using this design, the human optic nerve compresses 10^6 optic
axons into a nerve less than 2 mm in diameter. However, if the mean rate
doubled, matching that of the high rate types, the optic nerve cross section
would quadruple.

What allows these rates to be as low as they are? All the mechanisms
discussed in this chapter. This includes reducing noise and redundancy
at the cone synapse, temporally filtering by diffusion of glutamate
quanta, dividing the contrast range and the temporal frequencies, carving away unneeded information, and (finally) matching rates to downstream needs. These mechanisms are integrated so as to ensure that every
spike, whose costs are evident in the very structure of the optic nerve,
encodes as many bits as possible and that the fewest possible spikes
are sent.

Conclusion

This chapter opened by asking how the eye, unlike the Mars camera, manages to send all its information in 100 ms. The basic answer, given in figure

11.25, is that the eye uses multiple, parallel pathways, relying on each to send a subset of the total information at low rates.

This neural design involves many computational mechanisms: filtering, inversion, sparsification, rectification, and carving. To create the skewed distributions of firing rate and axon diameter at the output (figure 11.26), the retina uses three cycles of analogue-to-discrete and discrete-to-analogue recodings.[19] This obeys several more principles of neural design: *compute with chemistry* because it is cheaper; *compute directly with analogue primitives*, also cheaper, but finally threshold to remove noise. These cheap stages reduce the rates and total spikes so that the retina and optic nerve can afford the cost of long-distance transmission.

The stringent ganglion cells report their modest but critical messages to lower centers for direct coupling to behavior but also to central clusters that relay to cerebral cortex. The nonstringent ganglion cells report richer versions of the scene to central clusters that relay to cerebral cortex for further analysis. These latter types segment the spatiotemporal bandwidth to transmit economically, and this raises a question: how does the cortex reassemble the segmented information to recognize objects and motion in real time? This topic is addressed in chapter 12.

Leaving retina, there are at least 20 distinct representations of the scene. But where and how do these representations finally serve behavior? Are they ultimately reintegrated, and if so, where? Most critically for our theme, how do these representations beyond the retina maintain efficiency in space, time, and energy?

That efficiency is preserved beyond the retina is known. Our behavioral threshold for discriminating a brief, low-contrast spot approaches that of a noiseless discriminator (Crowell et al., 1988; Savage & Banks, 1992). Although we are roughly 10-fold less sensitive, the difference is accounted for by noise at the retinal output (figure 11.2). Two points follow: (1) the many noise sources along the central visual pathways are so effectively managed that they do not degrade the S/N carried by a single spike from the retina; (2) perceptual decisions must be nearly noiseless since they match the efficiency of a noiseless discriminator. Moreover, discriminations based on integrating across modalities combine the relative reliabilities of the sources (Burge et al., 2008; Burge et al., 2010).

Principles for connecting the optic nerve

The optic nerve distributes to central neuron clusters mostly at low levels. These are sites in brainstem that govern functions that need little processing beyond what was achieved in retina (figure 12.1; Dacey, 2004). This follows the principle that sensing should guide behavior at the lowest possible level, allowing the brain to engage only the circuits needed for a particular task. For example, ganglion cell types designed to sense slow changes in light intensity terminate directly at the SCN to govern the central circadian clock (figure 4.1). This pathway uses less than 1% of all ganglion cells and retinal spikes.

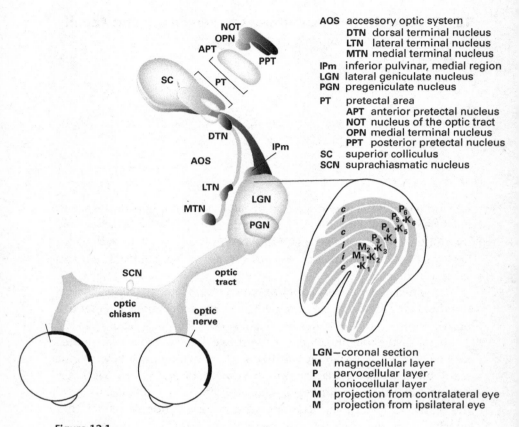

AOS   accessory optic system
    DTN   dorsal terminal nucleus
    LTN   lateral terminal nucleus
    MTN   medial terminal nucleus
IPm   inferior pulvinar, medial region
LGN   lateral geniculate nucleus
PGN   pregeniculate nucleus
PT   pretectal area
    APT   anterior pretectal nucleus
    NOT   nucleus of the optic tract
    OPN   medial terminal nucleus
    PPT   posterior pretectal nucleus
SC   superior colliculus
SCN   suprachiasmatic nucleus

LGN—coronal section
M   magnocellular layer
P   parvocellular layer
M   koniocellular layer
M   projection from contralateral eye
M   projection from ipsilateral eye

Figure 12.1
Retina sends different versions of the scene to 20 distinct central clusters. This
diagram shows primate pathways to neuron clusters in hypothalamus, dorsal and
ventral thalamus, and midbrain. The lateral geniculate nucleus (LGN) comprises 12
clusters that collect from ganglion cell with high, medium, and low information
rates (M, P, and K, respectively). Reprinted with modification and permission from
Dacey (2004).

Other examples include a ganglion cell type that senses brief intensity
changes and couples directly to central clusters that regulate pupil diameter
for optimal visual performance. Pupil constriction at high light intensities
improves vision by holding intensity on the photoreceptors near their optimum capacity (Borghuis et al., 2009; Laughlin, 1992). The narrower pupil
also improves the eye's optics to deliver the best image at the optimal intensity. These advantages are so straightforward that one cannot imagine a
circumstance where higher brain levels would need to comment on this

matching process. This exemplifies an important behavior (coordinated muscle contractions that continuously optimize vision) that is accomplished at the lowest central level. Examples that employ directionally selective ganglion cells to guide low level behaviors were discussed in chapter 11.

One target in the midbrain, the superior colliculus, undertakes considerable further processing of optic signals. It integrates visual, auditory, and somatosensory signals to produce various outputs: (1) an ascending pathway toward cerebral cortex that decides "where to look"; (2) a descending pathway to brain-stem motor mechanisms that move the eyes, ears, and head toward that site; and (3) another ascending pathway to correct for these self-initiated movements (chapter 4). To oversimplify: the superior colliculus helps a baseball player orient his head and eyes toward the pitcher. However, to discriminate a fastball from a curve and guide the bat requires much more detail—which is encoded by ganglion cell types with nonstringent filters and higher information rates.

The nonstringent ganglion cells supply a dozen neuron clusters within the thalamus, and these relay to primary visual cortex (*V1*). The latter initiates processing for visual perception and behavior, diverging signals widely across the cortex and downward to many subcortical structures. This chapter follows the pathways initiated by the nonstringent types through thalamus up to cortex and across it. We indicate how design principles identified at the scale of synapses and synaptic circuits play out at larger scales: neuron clusters, layers, columns, patches, and stripes. The same principles act on a still larger scale: clusters of cortical areas and the hierarchical arrangement of areas from back to front.

Design of a thalamic "relay"

The nonstringent ganglion cell types terminate in a complex of neuron clusters, termed collectively the *lateral geniculate nucleus* (LGN). Each type of optic axon supplies a particular cluster; consequently, parallel pathways arising in retina (brisk-transient, brisk-sustained, local-edge, blue-yellow, etc.) remain parallel within the LGN and exit without converging. The thalamic relay cells continue to reflect the informational properties of their retinal inputs: types with higher information rates use higher energy rates and, therefore, higher levels of cytochrome oxidase. This mitochondrial enzyme reflects local mean energy consumption and correlates with capillary density (figure 12.2; Weber et al., 2008).

In retina the order of cytochrome oxidase expression goes: brisk-transient → brisk-sustained → local-edge. This order continues through the LGN and

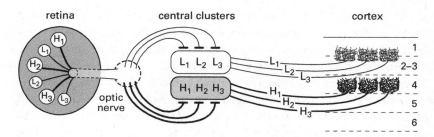

Figure 12.2
Ganglion cell types that intermingle in retina segregate centrally into compact clusters but maintain their nearest neighbor relationships. This preserves the retina's two-dimensional map within each central cluster and in the subsequent relay to cortex. Types sending at high information rates (H) with thick axons relay to middle layers of cortex; types sending at low information rates (L) with thin axons relay to upper layers. High-rate types at all three stages express more cytochrome oxidase than the corresponding low-rate types. Medium rate types are omitted for simplicity. These rules hold also for auditory and somatic systems (Jones, 2001).

also through V1 where expression is strongest at the input layer for non-stringent inputs (layer 4) and in patches of layers 2 and 3 that carry higher information rates (figure 12.2; Kageyama and Wong-Riley, 1984, 1985). These cytochrome-oxidase patches project to cytochrome-oxidase *thin stripes* in the second visual area (*V2*; Sincich & Horton, 2005; Federer et al., 2009; Federer et al. 2013). The broad point is that neurons from various distinct aggregations in V1 project separately to V2. Consequently, the energy savings achieved in retina by splitting into low-rate and high-rate channels continues as processing proceeds in parallel streams.

One could imagine a less orderly design—where various relay types, rather than segregating, would intermingle. However, that would force various types of optic axons to crisscross in order to find their target neurons. Thus, by segregating into parallel streams, optic axons avoid mutual interference and connect with less wire. Segregating also allows neurons within a type to share inputs without mutual interference and save additional wire (figure 12.3). For example, in human retina ganglion cells of a given type distribute across more than 1,000 mm^2, but in a central cluster their relay neurons gather compactly while still preserving their nearest neighbor relationships. Since the signals carried by neighboring ganglion cells are partially correlated, processing circuits can be shared economically. Therefore, their proximity in a central cluster serves efficient processing. In short, *neurons that fire together should locate together.*

Thalamic relay neurons preserve these nearest neighbor relationships in their output axons (figure 12.2). Consequently, as their output tract heads toward cortex, the map of visual space can be compressed, twisted, crumpled, or flattened—whatever is locally efficient. Then, upon reaching cortex it can re-expand with nearest neighbor relationships intact to connect with least wire. All these advantages apply equally to both eyes: until the stage where their messages are finally integrated (V1), separate processing is most efficient. So their thalamic relay clusters locate in alternate LGN layers (figure 12.1).

Why and how thalamic relays "repackage" their input signals

Most sensory modalities use nonstringent cell types to transmit high information rates to cortex (Jones, 2001). These types all use thalamic processing to enhance their efficiency—more bits per spike. So this design warrants some explanation. We now explain the mechanism for vision,[1] but a similar story could be told for somatosensory and auditory systems. The stringent ganglion cell types, such as local-edge and directionally selective cells, do project to thalamic clusters and continue in parallel up to cortex, but their thalamic and cortical processing remain murky.

Spatial acuity for each visual function is set by the spatial grain of a particular ganglion cell array (chapter 11). To preserve this acuity going forward requires providing each ganglion cell a private line to one relay cell (Weyand, 2007; Rathbun et al., 2010). This imbues the relay cell with the properties of its input type; for example the brisk-transient relay cell has higher spike rates and higher information rates (figure 12.2). Recall, however, that each retinal point is covered by the receptive field centers of six ganglion cells of the same type and that this redundancy is unavoidable because ganglion cells of fixed spacing extend their dendrites to collect more contacts and improve their S/N (figure 11.18).

The main reason, it seems, for a thalamic relay neuron to repackage retinal inputs is to reduce the redundancies caused by the sixfold overlap of centers. Repackaging allows information to reach cortex at the same bit rate but half the spike rate (Sincich et al., 2009). This saves space and energy by fourfold; moreover, these savings, by allowing postsynaptic neurons to operate at lower rates, earn interest going forward. This appears to be the LGN's main computational task, and that for other thalamic relays as well. Given such a substantial improvement by thalamic repackaging, we should explain how it works.

Redundancy in retina arises from the broad correlations of intensity across natural scenes, so it is reduced by summing signals broadly across

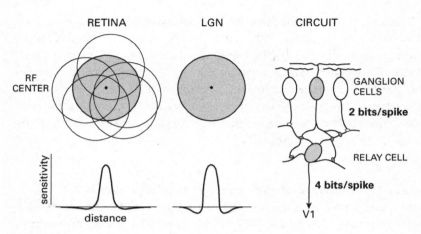

Figure 12.3
The quasi-secure thalamic synapse concentrates information for relay to cortex. A
ganglion cell receptive field (RF) center (shaded) is overlapped by five neighbors for
sixfold coverage. Its axon in lateral geniculate nucleus (LGN) connects strongly as a
private line, providing multiple contacts that cluster on proximal dendrites of the
relay cell (arrows) where they are ensheathed by glial membranes as a glomerulus
(figure 12.4). Neighboring axons connect weakly (fewer contacts, no glomerulus) to
that relay cell. Thus, a stimulus that evokes redundant spikes causes the shaded cell
to fire more reliably and thus to send more bits per spike. Connections to neighboring relay cells are omitted for clarity. Inhibitory surrounds, omitted for clarity from
the receptive field cartoons, are evident in the sensitivity plots.

receptive field *surrounds* and then subtracting the mean from the forward
pathway (figure 11.6). But redundancy at the thalamus arises from the overlap of retinal receptive field *centers*, so it is spatially narrow—tightly localized to nearest and next-nearest neighbors. To reduce this highly local
redundancy, the thalamus employs an entirely different design: the "quasi-
secure" synapse (figure 12.3).

The private line to a relay cell provides many active zones, so its input
spike frequently evokes an output spike (p ~ 0.5). Observers, initially
impressed with the strength of this connection, termed it a "secure" synapse. However, the key to its integrative function actually lies in its partial
*in*security—its frequent failures. When a spike on the private line evokes
too small an EPSC to cause a postsynaptic spike, the EPSC can sum with a
second one if the two arrive within about 30 ms of each other. This "supplementary" EPSC is driven by a spike from one of the overlapping neighbors
through a weaker synaptic connection (figure 12.3). So the quasi-secure

connection harnesses information delivered by the redundant spike to improve S/N of each private line, thus sending more bits per spike and fewer spikes.

What causes the supplementary spike is, of course, a pattern that overlaps both receptive field centers. Each neighbor's connection is roughly fivefold weaker than the quasi-secure input, so it cannot alone fire the relay cell. However, when it sums with the larger EPSC from the private line, it helps trigger a spike (Carandini et al., 2007). This design allows the relay cell to recapture redundant information from the overlapping neighbors without loss of spatial acuity.

A broad stimulus covering the receptive field centers of all the six ganglion cells that connect to the relay cell would be less effective than a narrow stimulus covering only the main ganglion cell input. Despite stimulating all six receptive field centers that contribute to the relay cell, the stimulus would reduce the number of spikes sent through the strongest connection (via lateral inhibition within the retina). Moreover, the surrounds of the overlapping neighbors, being broader than their centers, overlap more, so a broader stimulus would affect the surrounds more, thereby also reducing spikes from the overlapping neighbors.

In short, the quasi-secure synapse, while preserving the original information, uses the redundant input spikes to send the same message with half the output spikes. Spike reduction in an awake animal freely viewing a natural movie can be still greater, up to fourfold (Dastjerdi et al., 2003, 2007, 2011; Dong et al., 2005), implying a 16-fold reduction in space and energy (Perge et al. 2009).

This circuit, whereby the relay cell integrates a strong private input with redundant weaker ones, deepens the antagonism nearest to the center, giving the receptive field the appearance of a real sombrero (figure 12.3). This deepening of the near surround occurs via lateral inhibition onto the relay cell and also via excitation as the LGN relay neuron convolves multiple, overlapping ganglion cell surrounds. This is the same process that deepens the surround of ganglion cells as they convolve the broadly overlapping receptive fields of bipolar cells (Freed & Sterling, 1988; Smith & Sterling, 1990).

Implementation of the quasi-secure synapse

The local circuit for the quasi-secure synapse of a brisk-sustained relay cell involves a *synaptic glomerulus*—a knot of intertwined pre- and postsynaptic processes encapsulated by glial processes (figure 12.4; Sherman & Guillery, 2002; Sherman, 2004). The capsule surrounds the cluster of terminals

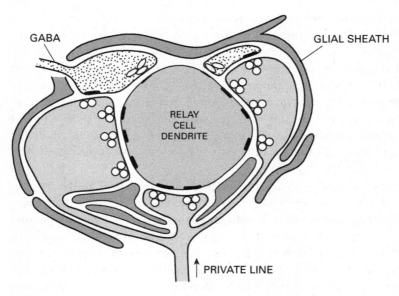

Figure 12.4
Synaptic glomerulus integrates quanta from multiple active zones of the private retinal axon. The glial sheath excludes glutamate transporters, thus prolonging glutamate spillover, so this glial architecture serves the design. Outside the sheath, glial membranes densely express glutamate transporters to mop up (Josephson & Morest, 2003). The relay cell expresses multiple glomeruli for its private line, and outside the glomeruli it collects additional contacts from adjacent retinal axons (figure 12.3). The retinal terminal also contacts GABA-ergic processes (stippled) that feed inhibition forward from retinal input to relay cell dendrite. This inhibitory circuit sharpens timing precision at the relay cell output. After Sherman & Guillery (2002).

provided by the private retinal axon and thus establishes an extracellular compartment for glutamate spillover, something like that described for the cone synapse (figure 11.3) and for the mossy fiber synapse (figure 7.16). The relay cell dendrite expresses mainly a fast AMPA receptor, plus some slower NMDA receptors (Ziburkus et al., 2000) that pass larger currents as the dendrite is depolarized. Together the two receptor types capture the full temporal bandwidth of the input and extend the range of linearity (chapter 7; Attwell & Gibb, 2005). The combination works as follows: fast initial depolarization progressively decreases the driving force on the entering cations. However, as the AMPA current declines with dendritic depolarization, the NMDA current rises to compensate.

The retinal terminals also contact presynaptic terminals that release GABA onto the relay cell dendrite (figure 12.4). This feedforward inhibition helps sharpen timing precision at the relay cell output. Thus, better S/N via multiple synapses and better timing precision via feedforward inhibition both increase information (bits per spike) at the output to cortex.

The primary retinal axon distributes about half of its active zones to its private line target and divides the rest among the five neighbors. Consequently, each relay cell collects about half of its retinal synapses from the private line and the rest from the five overlapping retinal neighbors. When the main retinal axon releases insufficient vesicles to trigger a spike from its main target, a temporally correlated spike from one of the overlapping neighbors can increment the main EPSC by about 20%.

Retinal axons reserve some of their outputs for a second type of relay neuron, smaller but with the identical receptive field. This type also uses a glomerular synapse but with essentially the reverse pharmacology: mainly slow glutamate receptors, including NMDA, plus a fast GABA receptor driven by an interneuron excited by the same retinal input. Their arrangement establishes a delay line: the retinal spike that delivers brief, fast excitation to the standard relay neuron also delivers slow, sustained excitation + fast, brief inhibition to the small relay neuron (Vigeland et al., 2013). So it fires but with a defined lag. The cortex sees an early spike from the standard relay cell, then a delayed spike from the "lagged" cell. This delay line apparently exists for all types of LGN relay neurons, and thus represents another of the LGN's important functions (Saul, 2008).

The quasi-secure synapse (many contacts from a private line arranged to fire the postsynaptic neuron with $p \sim 0.5$) is a common motif of neural circuit design. Many thalamic relay clusters, perhaps all, use this mechanism, as do other regions, including mossy-to-granule-cell synapses in cerebellar cortex and auditory synapses to various brainstem neurons (Grande & Wang, 2011). Cerebellar Purkinje neurons converge on deep cerebellar neurons with similar benefits (Heck et al., 2013). These are: high-rate inputs are throttled down (with help from inhibition), and partially redundant inputs (a design issue in all neuron arrays) are integrated to improve S/N and concentrate information. Wherever a neuron can reduce its mean rate, it can use a thinner axon (Deleuze et al., 2012), and the benefit goes as $1/(diameter)^2$.

We wish to acknowledge that the issues of thalamic circuitry are complex, involving diverse inputs, including cortical feedback, brainstem regulators of attention, rich pharmacology, and "channel-ology." For example, all thalamic relay neurons express a type T calcium channel that switches the neuron from tonic firing in wakefulness to burst mode in sleep. Recent evidence suggests a role for these channels in tonic firing as well. We omit these important topics to simply trace the excitatory pathway from retina to cortex, interpreting the functional architecture in terms of efficient signaling.

How synaptic resources are invested from retina to cortex

We have noted that an analogue voltage recoding to synaptic quanta in doubling its information rate more than quadruples the number of synapses. For example, a high-rate retinal ganglion cell (brisk-sustained) encoding twofold more bits per second than a low-rate ganglion cell collects 10-fold more bipolar cell contacts (6,000 vs. 600; figure 11.16). But what happens when the voltage signal is much larger and faster—an action potential traveling down the ganglion cell axon to its central arbor? How many synapses are then required to transfer the message, and what are the governing rules?

A ganglion cell axon reaching the LGN reduces the number of active zones used to transfer its message by 6- to 10-fold (figure 12.5). The high-rate ganglion cell that received 6,000 bipolar cell contacts transfers its primary message using 600 active zones; 300 to the private line relay cell + 300 apportioned among the five overlapping neighbors (Wilson et al., 1984; Sur et al., 1987; Roe et al., 1989; Raczkowski et al. 1988).[2] And the low-rate ganglion cell that received 600 bipolar cell contacts transfers its primary message using only 100 active zones (Raczkowski et al., 1988). How can 6- to 10-fold fewer synapses relay the same amount of information?

First, the information to be transferred is actually less. Recall that the ganglion cell's A-to-P converter, in thresholding the graded signal, discards nearly half the information (chapter 11; Borghuis et al., 2009; Dhingra & Smith, 2004). Second, synaptic transmission can be more reliable. Consider that a full sized spike (~100 mV) reaching an active zone at the retinogeniculate terminal evokes a larger, faster surge of calcium than generally occurs at the bipolar terminal; therefore, retinogeniculate release is more reliable and more sharply timed. In short, a pulse-driven active zone transmits more information than the analogue version, thus reducing the number of active zones required at the main relay.

The entire retinogeniculate terminal expresses 1,600 active zones, many more than the 600 used by the main relay. The extra 1,000 active zones are apportioned among the lagged relay neurons and assorted inhibitory interneurons. These types need to establish signal quality comparable to that of the main relay: inhibitory neurons should be no noisier or less reliable than the neurons that they modulate. Nor should lagged relay neurons send lower quality signals than standard relay cells—given that the signals

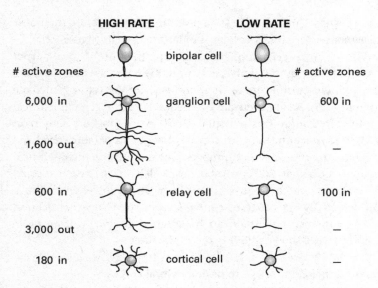

HIGH RATE   **LOW RATE**

active zones   bipolar cell   # active zones

6,000 in   ganglion cell   600 in

1,600 out   –

600 in   relay cell   100 in

3,000 out   –

180 in   cortical cell   –

Figure 12.5
Synaptic investment from retina to cortex. High-rate bipolar cells contribute 6,000
contacts to a high-rate ganglion cell (brisk-sustained) whereas low-rate bipolar cells
contribute only 600 contacts to a low-rate ganglion cell (local-edge). This 10-fold
difference matches a twofold difference in information rate. High-rate ganglion cells
contribute only 600 contacts to a relay cell, a 10-fold step-down. Low-rate ganglion
cells contribute only 100 contacts to a relay cell, a sixfold step-down. These step-
downs match a twofold step-down in information rate at the ganglion cell output
caused by spike thresholding. The high-rate relay cell expands its output synapses
at cortex by fivefold (600 in vs. 3,000 out). This corresponds to a twofold increase
in information per spike at the relay cell output caused by the quasi-secure synapse
(figure 12.3) and by the large divergence to many simple cells. One simple cell collects about 180 contacts from about 20 relay cells, thus approximately nine contacts
from one geniculocortical terminal (da Costa and Martin, 2011). Compared to the
private retinogeniculate connection, the connection from a single relay cell is highly
insecure (see figure 12.7).

recombine in cortex. Thus, one can see a point to reserving more than 60% of the total synapses for targets besides standard relay cells (figure 12.5).

Reaching cortex, relay cell axons greatly expand their numbers of output contacts. A standard relay cell delivers about 3,000 contacts, fivefold more than it received and twofold more than are used by the retinogeniculate terminal (figure 12.5). Again, there are two issues. First, the geniculocortical axon carries twofold more bits per spike than the retinogeniculate axon (figure 12.3), so to transmit that with active zones of comparable speed and reliability would require greater numbers. Second, the geniculocortical axon diverges to more than 200 cortical neurons, delivering to each approximately six contacts, which are driven by a large, sharp spike. Of course, a connection served by six separate contacts, compared to a connection served by 300 contacts sequestered in thalamic glomeruli, will be quite insecure—and that we shall explain is key to the design.

Six reasons for a thalamic "relay" to primary visual cortex

We can now summarize various reasons to gather optic signals at the thalamus before projecting them to V1. First, the thalamic station provides a control point where the rate of signaling can be gated according to level of brain and behavioral arousal. The gating mechanisms involve several brainstem nuclei that are, in turn, gated by the circadian clock—which are all nearby. Gating at the relay level allows fewer spikes to cortex, reducing energy and wire volume used by that pathway and other downstream circuits.

Second, each retinal neuron, to improve its S/N, overlaps its neighbors and thus carries substantial redundancy (chapter 11). By connecting strongly to its main retinal input and weakly to the overlapping neighbors, each relay cell uses the spatial redundancy to improve spike timing and thus increase bits per spike at the relay cell output. Thus, all the information arriving from retina to the LGN is relayed to cortex by half as many spikes.

Third, the thalamic relay expands certain cell types, creating "copies" to be used in different cortical areas for distinct computational purposes. For example, a brisk-transient pathway relays to V1 for "linear" processing and separately to V2 for "nonlinear" processing, in all expanding its numbers by about 10-fold (Yeh et al., 2003).[3] This expansion makes it even more important for the relay cell to concentrate information and thus halve the spike rate.

Fourth, the main relay types are all accompanied by secondary types with the opposite temporal phase relation to the initial signal ("lagged" vs.

standard). These lagged signals are sent at lower spike rates over finer axons and need fewer synapses at the cortical output. A possible reason to establish the lagged types at the LGN rather than at V1 is that they also use a glomerular synapse and presumably also benefit from the doubling of bits/spike. Moreover, it may be efficient to prepare all the essential subunits for "direct assembly" at the cortex.

Fifth, the LGN nuclear complex, by gathering all members of each type into a single cluster, can project each type as a bundle and so avoid the need to disentangle them at the cortex—again, saving wire. This also explains the separate layers for ON and OFF relay cells and the separate layers for each eye: all types are sorted and ready for direct assembly as subunits of cortical receptive fields.

Finally, various neurons in cortical layer 6 project down to the LGN and provide numerous excitatory synapses to the distal regions of relay cell dendrites. These cortical output neurons, diverse in dendritic morphology and axon caliber, constitute different types (Katz et al., 1984). Each type probably targets a specific type of relay neuron, cortical output neurons with thinner and thicker axons serving, respectively, low-rate and high-rate relay types. This provides the cortex with distinct circuits to gate its input from each type of relay cell. The clustering of different relay types in LGN probably improves their efficiency in collecting specific descending inputs from V1. The functions of this rich, highly specific, gating pathway have been the subject of much speculation and experiment—but remain uncertain.

All six reasons for a thalamic relay apply equally to other senses, for example, tactile and auditory systems, and also to relays for the motor system. All of these systems also express the quasi-secure synapse, and all sharply reduce their spike rates before projecting to their primary cortical areas.

How images are processed in V1

The purpose of cortical image processing is to identify patterns that predict and guide useful behavior. It involves combining signals from different regions of the retina and from the two eyes. Processing for patterns begins in V1.

V1 is the largest cortical area. It occupies in macaque monkey about 13% of the total cortex and about one quarter of cortex devoted to vision (Van Essen, 2004; Dougherty et al., 2003). V1 is constrained to restrict each category of pattern coding (space, motion, color, depth) to a single serial mechanism—there are no parallel pathways for "backup." The brain devotes

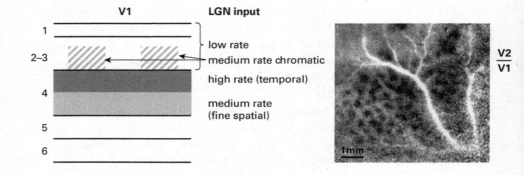

Figure 12.6
Energy capacities of input layers to V1 matches information rates. Left: Vertical section through cortex indicating cytochrome oxidase expression: strongest in upper layer 4 which receives high-rate relay cell input; medium in lower layer 4, which receives medium-rate relay cell input for fine spatial processing; and weakest in layers1–3, which receive low-rate relay cell input, including blue–yellow color contrast. Layers 2–3 express cytochrome oxidase in patches whose relay cell inputs code both red–green color contrast and achromatic contrast (double duty; Crook et al., 2011; Stockman and Brainard, 2009). **Right:** Tangential section through layers 2–3 showing dark patches activated by red–green isoluminant flicker. These functionally imaged patches correspond precisely to patches of elevated cytochrome oxidase (not shown). Reprinted with permission from Valverde Salzmann et al. (2012).

neither space nor energy to rarely used safety nets. Consequently, whatever integrations are effected in V1, they must rigorously preserve each distinct aspect of a scene that was relayed—or irretrievably lose it.[4]

The separate lines initiated in retina and continued in thalamus remain separate in V1, their terminals segregating in different strata (figure 12.6).[5] High-rate neurons for temporal processing terminate in upper layer 4, medium-rate neurons for fine, achromatic spatial processing terminate in lower layer 4, and low-rate neurons[6] terminate superficially, in layers 1–3. The inputs to cortex typically fire at higher rates than the outputs, so these strata expend more energy and express correspondingly higher levels of cytochrome oxidase. Patches in layers 2 and 3 also express more cytochrome oxidase (figure 12.6). These patches receive input from medium-rate relay neurons that encode red–green chromatic contrast, as now explained.[7, 8]

These red–green inputs are axonal branches from a subset of the medium-rate relay neurons whose full array terminates in deep layer 4. In the full array each neuron is excited by a single cone, so the array resolves spatial patterns down to the Nyquist limit, about 60 cycles/degree in humans. The receptive field centers, driven by either a red or green cone, are spectrally tuned. The center cone is antagonized by horizontal cells which collect unselectively from red and green cones (Crook et al., 2011). Consequently, some surrounds will comprise equal numbers of red and green cones, but others will comprise more red than green, or vice versa, and will either reinforce or antagonize the center's spectral response. The subset of neurons with center/surround spectral antagonism encode red–green chromatic contrast (Crook et al., 2013). This array resolves only about 40 cycles/degree and segregates in the layer 2–3 cytochrome oxidase patches.

We now follow the relay pathway for brisk-sustained signals. Since this pathway serves spatial acuity, it needs to preserve in cortex the acuity established in retina and preserved by LGN relay cells. Consequently, the initial pooling in V1 must be restricted to one dimension.

Indeed, the first-stage cortical neuron integrates inputs from a single row of relay cells with overlapping receptive fields. Thus, it can resolve a linear pattern of the same spatial frequency as the original sampling array (Green, 1970). But how far should this line extend—across tens of relay cells or thousands? Also, what about integration in the orthogonal dimension? And what about dark and bright contrasts that segregated in retina onto different relay lines—how are they finally reintegrated? Given that the earlier stages approach theoretical optimality, should we also expect the cortical neuron to resemble its four predecessors as an optimal encoder? What do theories of image processing suggest?

Theory of optimal coding predicts early V1 circuits

The theory starts with an observed property of natural scenes, namely that their spatial and temporal correlations produce power spectra that tend to decay as $1/(\text{frequency})^2$ (chapter 8). This means, for example, that trees (up close) and forest (at a distance) show the same shaped distribution of correlations. Given this statistical property, there exists an optimal way to represent any *particular* scene—by concentrating the activity that specifies that scene in the fewest units—a boon for any designer!

The optimal unit, actually a weighting function termed *Gabor filter*, optimally represents spatial position and spatial frequency. These two variables are reciprocal, so an element that is larger in space is narrower in spatial frequency, and vice versa. The linear filter that occupies the smallest possible volume in both domains is the Gabor (Daugman, 1985; Marcelja, 1980). Put another way, the Gabor filter strikes the optimal balance between two conflicting demands: integrating across space to establish the polarity and scale of intensity changes, and pinpointing within space to establish location.

Gabor filters tested on natural images prove most efficient with a length-to-width ratio of about 2 and a spatial frequency bandwidth of about 1–1.5 octaves (Field, 1987). Moreover, roughly 200 Gabor-like filters with different orientations, sizes, and offsets can represent the small patch of natural scene coded by a 12 × 12 array of cone photoreceptors (Olshausen, 2004). This seems like an excessive number of filters, but there are roughly 40 cortical neurons at the first stage for every cone photoreceptor, and this explains why: overrepresentation, as will be further explained, constitutes an efficient code. Thus, theory predicts: (1) an efficient type of receptive field for the first integration in V1; (2) its optimal dimensions; (3) the optimal number and proportion of filter types. These predictions match astonishingly well the actual properties of neurons that integrate the relay cell input (Jones & Palmer, 1987). So how are they wired?

Integrating via *insecure* synapses

The cortical neuron that integrates inputs from a line of thalamic relay cells is the famous "simple cell" (Hubel & Wiesel, 1962). Its dendrites branch within the stratum of brisk-sustained thalamic input and collect from about 30 relay cells with overlapping, linearly aligned receptive fields (Kara & Reid, 2003). This degree of convergence produces the optimal length-to-width ratio (2:1). And, as theory requires, pooling occurs only along one dimension. Orthogonal to that dimension spatial grain is preserved (figure 12.7).

Crucially, the simple cell must not respond to isolated firing of a single relay cell. Rather, it should fire only when the line of relay cells that form its input fire together, and thereby report a linear feature in the scene (Wang et al., 2010). This requires the connection from single relay cells to be weak. Consequently, each relay spike causes only about 3% of the simple cell's spikes. Thus, compared to the thalamus, where the main retinal input causes more than 50% of the output spikes (Weyand, 2007), the relay cell input to a simple cell reduces "synaptic security" by about 20-fold. In short, the synaptic design changes from quasi-secure at the thalamus to insecure at the cortex.

Each relay cell contributes approximately six contacts to a simple cell, so the linear band of 30 relay cells, provides about 180 contacts (Costa & Martin, 2011). To reliably fire one spike the simple cell requires synchronous release, amounting to approximately 100 vesicles within about 10 ms (Frick et al., 2010). Therefore, each synapse needs to release one vesicle with p ~

0.5. Retinal spikes occurring within 5–10 ms of each other are fourfold more likely to excite a simple cell than spikes separated by longer intervals (Kara & Reid, 2003). This design, where 30 quasi-synchronous input spikes trigger one output spike, "sparsifies" the response by 30-fold and signifies an edge.

To encode the predicted range of spatial frequencies, simple cells integrate parallel bands of relay cells that share the same orientation but are wired alternately for bright or dark contrasts. This expands integration in the dimension orthogonal to the orientation bands, while preserving spatial resolution (figure 12.7). Alternating bands are arranged side by side. Simple cells express up to four bright domains arranged variously with similar numbers of dark domains to produce over 30 different patterns—and more, considering a rarer class of "periodic" cells with up to six bright and seven dark domains (Mullikin et al., 1984; Palmer & Davis, 1981). Thus, it is easy to imagine that several hundred types would be needed.

The relay cell axon terminal needs to provide sufficient active zones to allow an insecure connection with all the many simple cells that are needed for optimal tiling of natural scenes. This works out: the terminal expresses about 3,000 active zones (figure 12.5) and contributes about six contacts per simple cell; therefore, it can supply about 500 neurons. Some contacts are reserved for inhibitory neurons, which constitute about 10% of cells in this stratum, but this leaves sufficient contacts for several hundred simple cells of different orientations and bandwidths. This is roughly the number of neurons predicted to be optimal; thus the synaptic numbers and cell numbers are consistent with each other and also with predictions from image-processing theory.

Circuit for the Gabor filter

Recall that the two-dimensional Gabor function optimally encodes space and spatial frequency, extracting the maximum mutual information given the statistical properties of natural images (Field, 1987; Okajima, 1998a, 1998b).

The function, defined as a plane wave localized by a Gaussian envelope, is represented by a formidable equation. However, like the difference-of-Gaussians function of retinal ganglion cells, it is computed as a *primitive*, that is, directly, via clever balancing of excitatory and inhibitory weights. A key difference between the relay cell's difference-of-Gaussians profile and the simple cell's Gabor profile is the depth of the surround. Whereas the difference-of-Gaussians is "sombrero-like," the Gabor has a far deeper

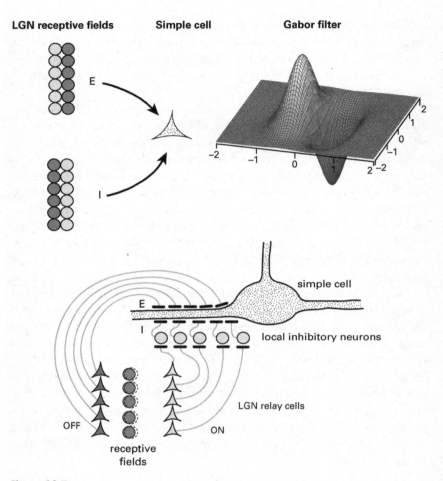

LGN receptive fields   **Simple cell**   **Gabor filter**

Figure 12.7
Relay cell inputs to cortical simple cell directly establish its Gabor weighting function. Upper: This particular Gabor filter has two spatial bands. One is excited (E) by a bright bar and inhibited (I) by a dark bar. The adjacent band is excited by a dark bar and inhibited by a bright bar. This produces a receptive field (weighting function) that closely resembles a Gabor filter that efficiently encodes inputs from 30 thalamic neurons, further reducing the mean spike rate and increasing bits per spike (Kumbhani et al, 2009). **Lower:** Circuit for one band of the Gabor. Dark bar initiates spikes in a row of OFF thalamic relay cells that project together, minutely preserving their nearest neighbor relationships, to deep layer 4. These OFF neurons with circular, difference-of-Gaussian receptive fields converge with excitatory contacts onto the cortical neuron. Each relay cell contributes only a few contacts to the cortical neuron, so its functional connection is quite insecure. One spike from the relay cell has

very low probability of firing the simple cell. However, a bar that synchronously fires the *row* of OFF relay cells reliably fires the simple cell. A cospatial row of ON relay neurons excites cortical interneurons that inhibit the simple cell. Therefore, as the dark bar excites the simple cell ("push"), it also disinhibits ("pull"). This push–pull circuit recombines information that was divided by rectification at the bipolar cell output (figure 11.13). At this stage both components of the signal can be efficiently encoded by the same cell because the coding is now sufficiently sparse. LGN, lateral geniculate nucleus. Upper image, reprinted with permission from Palmer et al (1991).

"brim" that could not be achieved simply by summing difference-of-Gaussians relay cell receptive fields.

For a simple cell to exhibit a Gabor function it recombines the complementary halves of the original cone response (ON and OFF) that were segregated for efficient transmission. Now a bright contrast excites the simple cell via a band of ON relay cells that release glutamate; simultaneously, it *disinhibits* the simple cell by suppressing a cospatial band of OFF relay cells that (via interneurons) release GABA onto the simple cell (figure12.7; Ferster, 1988).[9] This "push–pull" design, recombining both halves of the cone signal, restores the full dynamic range of responses to contrast (Hirsch, 2003). Moreover, by exploiting both excitatory and inhibitory reversal potentials, the circuit extends the linear response range for contrast and rejects noise. These advantages are well-known and thoroughly exploited by designers of electronic circuits.

Why postpone this recombining of information that was divided by rectifying in the retina (figure 11.13)? Because here at last the coding is sparse enough for both halves of the original signal to be processed efficiently by the same cell.

In emphasizing how V1's main computation utilizes Gabor filters, we have slighted numerous complexities—including nonlinear mechanisms at the cellular level and various aspects of local cortical circuitry (Douglas & Martin, 2009). In fact, half a century after Hubel and Wiesel proposed the model for orientation specificity, the questions of *exactly* how it occurs, and *exactly* what makes it so robust to variations in luminance and contrast, are still studied (Priebe & Ferster, 2012). Final answers await a detailed understanding of cortical connectivity (Bock et al., 2011). Meanwhile, we consider some implications of this design.

Advantages of sparse coding
The linear Gabor filter, by optimally representing space and spatial frequency, substantially sparsifies the input. Moreover, as the eye flicks about,

regions beyond the immediate simple cell receptive field are stimulated by complex image structures that activate various nonlinear circuits within V1 to further sparsify the simple cell's response (Vinje & Gallant, 2002; Gallant, 2004; Herikstad et al., 2011). Thus, a simple cell is largely silent, and the response, being usually zero, is highly redundant. Furthermore, since many simple cells are required for each patch of visual space, and most are silent—the degree of redundancy begins to seem possibly worrisome—because, after all, how much idle capacity should be tolerated?

It might seem puzzling that, following the considerable investment at early stages to reduce redundancy, the fifth-stage neuron expands it so tremendously. Yet this is unavoidable with sparse coding: it conserves expensive action potentials but expands unused capacity. However, under the law of diminishing returns (figure 3.6), it is more energy efficient to represent information in many neurons that fire at very low mean rates than in a few neurons firing at high mean rates. Indeed, there is an optimum sparseness that maximizes energy efficiency, and this depends on the ratio between the fixed cost of installing and maintaining a neuron and the additional signaling cost of firing a spike (Levy & Baxter, 1996).

The rule goes: when neurons are expensive and spikes are cheap, spike frequently to increase information per neuron. However, when neurons are cheap and spikes are expensive, spike *in*frequently to increase information per spike. For the average cortical neuron, spikes are relatively expensive—two spikes per second equals the resting cost (Howarth et al., 2012). This ratio favors low rates, and indeed cortical neurons operate close to the optimum efficiency.

Thus, on the path toward perception, the expansion of redundancy represents progress. Indeed, sparsifying serves several purposes besides increasing energy efficiency (Olshausen & Field, 2004). A sparse representation facilitates feature extraction because against lower background activity, the response to a feature stands out. It is also easier to establish and learn correlations via coincidence detection because fewer spikes produce fewer spurious coincidences.

To summarize, the second-order statistics of natural scenes can be represented optimally by a single family of Gabor filters with specific dimensions. They are instantiated in layer 4 by simple cells, which integrate thalamic input with an "insecure" design. The deep lobes of the Gabor filter's sensitivity profile are achieved by a push–pull design that integrates the cospatial ON and OFF components of the original cone signal. The simple cell response is further sparsified by nonlinear mechanisms. Sparsifying reduces the metabolic cost of signaling—but at the expense of space

(100-fold more cortical neurons than retinal ganglion cells) and cell main-tenance costs. This is why V1 is so large: because of the redundancy caused by sparse coding. Given the large number of simple cells, small gains in efficiency could produce big savings.

Simple cells adapt to improve efficiency

V1 simple cells are not static filters. Rather, they adapt continuously to match their coding to context and this improves their efficiency. A fast-acting control of contrast gain matches sensitivity to the prevailing range of contrast without changing the receptive field's Gabor shape (figure 3.4). Gain is controlled by complex mechanisms in the retina and LGN (Atallah et al., 2012), but more simply in cortex by inhibitory interneurons (Carandini et al., 1997).

The interneurons encode a weighted mean of the activity of the local population of simple cells and make inhibitory synapses on the simple cell's soma, sufficiently distant from the excitatory synaptic inputs on den-drites so as not to disturb the receptive field. Driven by a signal representing the local mean, the inhibitory synapses reduce the simple cell's input resis-tance and shunt its response by a constant proportion. Thus, the gain con-trol performs a *divisive normalization* that scales the simple cell's output signal with respect to the local mean. This adaptive mechanism combines with the adjustments of the receptive field mentioned earlier, whereby details beyond the Gabor receptive field sharpen simple cell responses and suppress background activity.

These forms of adaptation increase efficiency by familiar means. By matching coding to the input distribution (figure 9.10), the contrast gain control ensures that a simple cell's signaling range is used efficiently; by throttling high rates, it maintains sparsity, thus increasing the information per spike. By tuning a cell to the wider scene beyond the receptive field, it further reduces correlations between spikes, thus enhancing spike efficiency and increasing sparsity. But adaptation goes beyond simply concentrating information in spikes; it processes information.

By implementing divisive normalization, the contrast gain control advances pattern recognition (Carandini & Heeger, 2012). During pattern recognition, measures of the stimulus that coded information at lower lev-els (e.g., photon rates in photoreceptors) are discarded at higher levels, as circuits now establish the relationships that define objects (e.g., the edge segment extracted by a simple cell). Divisive normalization discards abso-lute values and establishes relative values. In retina, divisive normalization discarded light level in favor of encoding contrast—an invariant property of reflecting surfaces (chapter 8). In simple cells, divisive normalization replaces contrast with *relative* contrast, and this helps the visual system to generalize. A low-contrast shark in turbid water is just as dangerous as a high-contrast shark in clear water. Indeed, divisive normalization is so advantageous that it is applied in peripheral and central sensory systems across all phyla.

There is another gain control to consider. The higher order mechanisms that direct visual attention to one particular region of the visual field increase the gain of V1 neurons in that region. This executive control ensures that the resources used to enhance signals are only employed when and where they are needed.

Binocular convergence

Just as pathways for negative and positive contrasts rejoin at the simple cell, so do pathways from the two eyes. Each point on an object is captured by two sets of ganglion cells, one in each eye, but from slightly different angles. Because the angle depends on viewing distance, their responses can recombine at a simple cell so that it responds selectively to an object a particular depth in the visual field. This serves stereoscopic depth perception, which helps to rapidly segment figure from ground.

Processing for *stereopsis* begins in layer 4 of V1 where inputs from each eye segregate into distinct, alternating patches. These *ocular dominance patches* exchange wires so that their neurons respond to either eye. The circuit further arranges for neurons to respond especially well to both eyes when an object is at just the right depth. V1 combines inputs from the two eyes efficiently, that is, with least wire, because the corresponding relay neurons were aligned at the thalamic level and projected together (figures 12.1 and 12.2). On the other hand, this strategy for using single neurons to sensitively encode depth requires many neurons, another reason why V1 is large.

Motion

Just as stereopsis rapidly segments a scene into objects and surfaces, so does local motion. Local motion, that is, a scene's second-order spatiotemporal statistics, is encoded at the same stage as stereopsis (simple cell) by using orientation-selective Gabor filters that encode the second-order spatial statistics. Moreover, the mechanisms are similar. To understand the spatiotemporal Gabor filter, note that an image component moving at constant velocity describes a trajectory in space–time. If the spatial axis is defined as perpendicular to the orientation axis of the receptive field (in space–space), and if the individual bright and dark bands are oriented in space–time, a cell can report the speed and direction of motion of an image component. A good match between that motion and the Gabor filter's orientation in space–time gives a good response. Simple cell receptive fields in cat V1 and V2 are oriented like this, and the slope of their bands in space–time predicts their tuning to direction and speed of image components (McLean et al., 1994).

Cortical receptive fields become oriented in space–time apparently via the convergence of lagged and nonlagged relay cells. This could be done efficiently with pairs of relay cells whose receptive fields are displaced by 1/2 center diameter (spatial quadrature) if one is lagged and the other non-lagged (temporal quadrature). Thus, a receptive field oriented in space–time can be assembled from receptive fields that are not oriented in space–time—just as a receptive field oriented in space–space is assembled from nonoriented receptive fields. A simple cell can actually possess both properties (orientation in space–space and in space–time.[10]

How V1 generalizes the Gabor representation

After initially segmenting a scene at the input layer with simple cells, V1 then generalizes the computations across broader regions. It must also generalize across dark and bright contrasts in order to represent disparity and direction of motion independently of whether the image component is bright or dark. To accomplish this, simple cells with receptive fields selective for the same orientation, spatial frequency, and motion—but offset spatially—converge onto another class of cortical neuron—now six neuronal stages beyond the photoreceptors. Hubel and Wiesel recognized this class and called it a "complex" cell (Hubel & Wiesel, 1962). The complex cells, as "generalizers," should provide V1's outputs to other areas—and along with simple cells they do (figure 12.8; Karklin & Lewicki, 2009).[11] Complex cells locate in layers above and below the main layer for wiring up simple cells, thus keeping the simple → complex wires short while arranging the circuitry so that the simple and complex wirings do not mutually interfere (figure 12.9). This reason to stratify circuits is the same as in retina (figure 11.12).

Distributing V1 outputs

How many types of output should there be, and how should they be organized? The governing principle again seems to be *send only what is needed.*

Figure 12.8
Cortical neurons use diverse dendritic patterns to collect particular inputs. Left:
Neuron cell bodies distribute densely in distinct layers. **Right**: Thick relay axons (a,
b) branch densely in layer 4; thin relay axons (c, d) branch sparsely in layers 1–3.
Middle: Neuron types are distinguished by their dendritic patterns. For example, all
cells labeled *1* restrict their dendrites to layers 1 and 2; all neurons labeled *4* send apical dendrites up through layers 1–3; cells labeled *7, 8*, and *9* send an apical dendrite
through all the overlying layers to branch profusely in layer 1. Thus, each type can
select just what it needs from specific layers. All 17 neuron types direct their axons
to the underlying white matter. The axons are omitted here for clarity but are shown
in figure 12.9. Cells were drawn from consecutive slices of mouse somatosensory
cortex. Primary visual cortex (V1) shows a similar pattern. Drawing by R. Lorente de
Nó (1938) and reprinted with permission.

Figure 12.9
Cortical neurons use diverse axonal patterns to distribute particular outputs. Same
neurons as shown in figure 12.8. Layer 4 neurons (4, 5) send axon branches to the
upper layers (1–3) and to deeper layers (5–6). In V1 this separates the circuits for Gabor filters (simple cells) from the circuits that generalize them (complex cells). Cell
8 sends its descending axon into the deeper white matter heading for subcortical
structures. From V1, these would be superior colliculus and pontine clusters (Wang
and McCormick, 1993). Cells 4, 7, and 10 send their descending axons into the upper stratum of white matter, heading for other cortical areas, such as V2. Even this
rich pattern is oversimplified. For example, the projection from V1 to V2 comes from
layer 2, 3, 4A, 4B, 5, and 6. Only layers 1 and layer 4C seem not to provide output
to V2. Further explanation of wiring efficiency is given in chapter 13. Drawing by
R. Lorente de Nó (1938); reprinted with permission.

To do this, V1 employs a superficial output stratum (layers 2–3), and a deep output stratum (layers 5–6). The deep neurons collect exactly what they need by projecting dendrites toward the cortical surface to select inputs from any/all layers (figure 12.8).[12] Diverse neuron types can be efficiently wired to summarize whatever it is that V1 needs to tell a particular lower brain structure. The targets include (1) superior colliculus, (2) LGN, (3) *pontine* cluster that informs the cerebellum via high-rate mossy fibers, and (4) *inferior olive* cluster that informs the cerebellum via low-rate climbing fibers.

These downstream targets are functionally diverse, so each needs somewhat different information. For example, the superior colliculus, tasked with selecting targets for "capture" by a rapid eye movement (chapter 4), needs to know the direction of target motion. Correspondingly, layer 5 output neurons to superior colliculus are directionally selective, as are those to the pontine cluster serving cerebellar areas for visually guided movement (Palmer & Rosenquist, 1974; Wang & McCormick, 1993). The dendritic arbors of these layer 5 corticocollicular and corticopontine neurons are essentially indistinguishable in their sampling of other layers. On the other hand, neuron clusters within the LGN are concerned with an entirely different set of issues, and correspondingly they collect from different neuron types in a different output layer (layer 6). These neurons are not directionally selective; they are diverse, collecting from various layers in different combinations and transmitting over axons of different caliber, each probably targeting a different cluster within the LGN (Katz, 1987; figure 12.10).

Thus, the principle *send only what is needed* is served here via customized cell types that constitute V1's deep-layer outputs. By selecting a subset of the data available in V1, each type of deep-layer output can transmit at a modest spike rate. Lower rates allow thinner axons, so these output messages can reach their targets economically.[13] The selection process recalls that used by diverse ganglion cells to reduce their spikes (figure 11.12). The distribution of fiber diameters for V1 deep-layer outputs is unknown. However, the distribution for sensorimotor cortex closely resembles that of the optic nerve (figure 4.2). Thus, as for retina, the diversity of deep-layer output neurons is efficient in space and energy.

The upper-layer output neurons collect most of their input via ascending axons from the middle layers—the simple cells—plus direct contacts from low-rate relay cells and "comments" from ascending (recurrent) axonal branches of the deep-layer output neurons. The upper-layer outputs project *forward* to other cortical regions and across to the opposite hemisphere. Upper-layer types are also diverse and, like the deep-layer types, they send

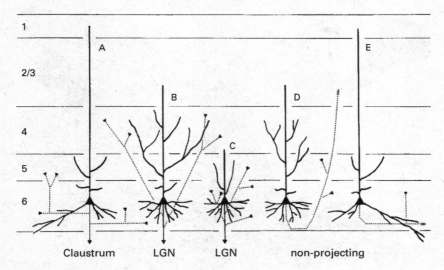

Claustrum   LGN   LGN   non-projecting

Figure 12.10
V1's deep-layer outputs follow the principle *send only what is needed*. Three types of neuron in layer 6 project axons to different subcortical clusters. Their apical dendrites sample different subsets of layers, and within a given layer they sample different volumes with different densities of dendritic membrane. Each type sends recurrent axon branches (dotted lines) to inform different subsets of layers. Only two types of LGN-projecting (LGN, lateral geniculate nucleus) neuron are shown, but there are many more, probably one type for each LGN cluster. Reprinted with modifications and permission from Katz (1987).

only what is needed by the next areas (Sato & Svoboda, 2010). These outputs should be considered as unfinished computations—a continuing conversation with the rest of the cortex—and these are the pathways that we now follow (Felleman & Van Essen, 1991).[14]

Beyond V1

The previous sections concerning the LGN and V1 are based strongly on data from cat. Many points are similar for monkey, though there are certain differences in terminology of V1 layers and possibly in the organization of the LGN, and certainly with respect to color since cat is dichromatic; whereas monkey is trichromatic. Now moving forward to higher levels of processing, most of the work has focused on primate (monkey and human), especially for correlations with behavior.

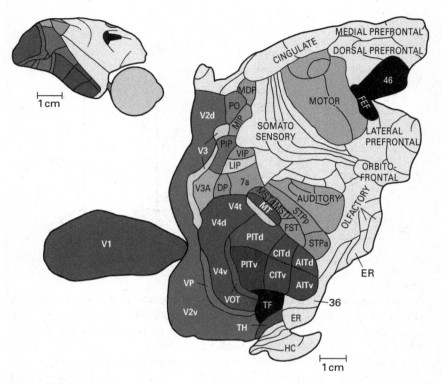

Figure 12.11
Layout of higher visual areas. Upper left: Monkey brain with eye to scale. Visual areas occupy all of the occipital and much of the temporal lobes. **Right**: Cortex flattened to show relationships of visual and other areas. Note: (1) early stage areas (V1, V2) are larger than later specialized stages; (2) areas that interconnect strongly (e.g., V1/V2) are near each other; and (3) just as related neuron clusters group together (figure 12.1), so do related cortical areas. Reprinted with modifications and permission from Van Essen (2004).

Functional architecture of area V2

V1 directs most of its upper-layer output to adjacent cortical areas, mainly V2 (figure 12.11).[15] Adjacency saves wire, as does the layout. Thus, V2's map of the visual field is arranged to be mirror-symmetrical with V1. This allows V1 to deliver its dense representation of the fovea over many short wires and its sparser representation of the periphery by fewer long wires.[16] The reverse layout would be unimaginably wasteful.

V2 is nearly as large as V1 (77%). Its shrinkage compared to V1 seems mostly due to V2's pruning of the far periphery—regions whose essential information has been tapped off at earlier stages and requires no further analysis. For example, information from the full field has been transmitted from V1 to superior colliculus, LGN, pontine nuclei, and so on. So, per square degree of visual space, V2 invests similarly to V1. The reason seems clear: V2 remains a "full service" area, performing the next stages of pattern analysis for all the core pathways, including spatial pattern, color, motion, and stereopsis.

V2 apparently uses all these properties to further segment the image. Moreover, it seems to do so by integrating them and thus needs to reorganize data from V1 into a layout efficient for that purpose. V1 segregated various properties: layers for color, space, and motion; patches for color and luminance; and larger patches for the two eyes. V2 then regroups them into bands (*stripes*) tangential to the cortical surface. For example, maps of motion direction converge as *thick stripes* (Lu et al., 2010). Patches of color and bands of luminance converge as *thin stripes*. This spatially intimate processing of hue and luminance in thin stripes suggests a role for thin stripes in segmenting objects by their surface properties (Wang et al., 2007).

V2 stripes were originally recognized by their denser staining for cytochrome oxidase, and again this reflects higher mean firing rates. V2 thick stripes (motion) collect from high-rate strata in V1 served by the brisk-transient types for motion. V2 thin stripes collect from high-rate patches in V1 served by cells that carry both chromatic and achromatic signals, thus doing "double duty" (Nassi & Callaway, 2009; Lim et al., 2009; Federer et al., 2009).

In short, these patterns of connectivity maintain parallelism established at the cone output: slow to slow, fast to fast, and color to color. Moreover, this pattern holds across various sensory modalities—because it is efficient.

How V2 relates to theories of image segmentation

The hallmark of V1 neurons is their sensitivity to a local edge, especially when it moves. Their push–pull wiring and Gabor weighting makes them efficient at this core task. But to interpret an image, local-edge elements need to be linked as global contours. Humans do this well; indeed when two observers independently assign all the edges in a rich image to particular contours, their agreement is 98% (Geisler et al., 2001).[17] The neural mechanisms for such grouping should match the statistical properties of contours in natural images. Correspondingly, human detection of contours in a complex background is predicted quantitatively by a local grouping rule derived directly from the statistics of "edge co-occurrence." This grouping rule, combined with a simple integration rule (if A goes with B and B goes with C, then A goes with C), links locally grouped contour elements into longer contours.

The rule binds two edge elements if they are more likely to arise from the same physical contour than from different ones. Edge elements are most likely to belong to the same physical contour if they are cocircular, that is, if they are consistent with a smooth, continuous contour. This is the Gestalt principle of *good continuation*. Contour grouping should also recognize the high degree of parallel structure of natural images.

V2 neurons respond as though they are optimized to match these higher order grouping statistics. These neurons are driven by small sets of excitatory neurons with similar Gabor receptive fields but are suppressed by larger sets of neurons with a wide range of different Gabor receptive fields (Willmore et al., 2010). Consequently, a V2 neuron only responds when some of the features that drive the excitatory set of Gabors are present AND most of the features that drive the suppressive set are absent. In this respect V2 provides a key advance toward image segmentation—identifying sets of edges. V2 receptive fields are several times larger in diameter than those in V1, yielding roughly 10-fold larger areas, and like V1 complex cells that provide their input, the V2 responses generalize across the field.

V2 neurons also detect contours that separate a "figure" from the background. This requires a contour enclosing the figure to "capture" the area within. Perceptually, this can be accomplished by integrating contours stereoscopically. It can also be accomplished by hue and surface luminance since both tend to be more similar within a figure than across the boundary. This suggests a role for thin stripes.

Perceptual capture of an area bounded by a contour also occurs rapidly and powerfully with motion. A static object patterned to match the background—leopard in sun-dappled foliage—is damnably difficult to segment. But when the cat moves, the coherent motion of its markings on a static background are rapidly extracted. This suggests a role for thick stripes.

In short, V2 seems to use various streams from V1 to segregate figures from ground. The segmentations are based on spatial, chromatic, stereoscopic, and motion cues. V2 needs all these streams, and they apparently interact. For example, both blob and interblob neurons contribute to V2 thin stripes (Xiao et al., 2003). So this seems like a good reason to keep all the pathways close together in this one large area.

Is V2's coding optimal? One cannot say for sure. It has been proven that the weighting functions of retinal and thalamic neurons are optimal for representing first-order image statistics. And it has been proven that the weighting functions in V1 are optimal for second-order statistics. But higher order image statistics that might allow prediction of optimal encoding have riot been measured. One indicator of third-order statistics—textures— shows that those images with higher information content are more visually salient than those with less information (Tkacik et al., 2010). This hints that cortical mechanisms try to find the richest patterns.

We have reached the edge of the known world for image processing at the millimeter scale. We will continue tracking the visual pathways forward on a coarser scale toward final stages of perception and action because there are surprises. However, before leaving early cortical processes, we consider their generality.

How general are the circuits for early cortical processing?

The structural differences between cortical areas are sufficiently subtle that to recognize the boundaries requires an expert. Moreover, when viewing a cortical slice with individual neurons rendered to stand out from the rest (figures 12.8 and 12.9), even an expert is hard put to identify the area. Thus, the impression is strong that across the expanse of cerebral cortex the same basic circuits repeat and . . . repeat. This certainly suggests similar processing.

Indeed the primary auditory cortex responds sparsely across a variety of stimulus ensembles, including natural sounds. The distribution of firing rates is log normal—resembling the optic nerve distribution—but the mean rates are lower (Hromádka et al., 2008). The auditory system also uses spatially compact, oriented Gabor-like filters. These filters are constructed at a lower level, the central cluster of the inferior colliculus which relays via a thalamic cluster (*medial geniculate nucleus*) to primary auditory cortex (*A1*; figure 12.11). Temporal tuning in the inferior colliculus is about 10-fold faster than V1, but the spectrotemporal Gabor waveforms are nearly identical (Rodríguez et al., 2010; Qiu et al., 2003). Millisecond timing precision established at lower levels is preserved by cortical neurons, allowing them on average 4.3 bits per spike (Kayser et al., 2010). In short, sparse coding— with log-normal distribution of spike rates and mean rates of a few spikes per second seems to be a general property of cerebral cortex (Margrie et al., 2002; Brecht et al., 2004). These low firing rates allow the wires to be fine caliber—which contributes to the structural similarities.

We suggest that sensory cortex is structured similarly across modalities because it performs similar computations: (1) finds correlations in the natural environment, (2) segments the bandwidth to send at lowest acceptable information rates, (3) encodes with optimal weighting functions (e.g., Gabor filters in V1), and (4) regroups for segmentation according to Gestalt rules.

Beyond V2

A new theme for neural investment

The dominant theme for investment has been: match neural resources to the physical distribution of information. For example, because nature contains more negative contrasts than positive, the photoreceptor synapse and all later stages invest more in OFF responses than ON. Also, because nature is structured such that second-order correlations generate $1/f^2$ power spectra, V1 invests in a family of Gabor filters that optimally encode images with that statistical property. And following the statistical distributions of edges, color, and motion in natural scenes, V2 invests in efficient mechanisms to segment figure from ground. The V2 operations are not rigorously proven to be optimal, but at least they follow Gestalt principles that may eventually be proven to be optimal. Thus, V2 appears to complete the process of coding a whole scene efficiently in one area. Indeed, it is the last visual area before the various streams contained within it segregate into separate areas. It is the last area where a lesion causes blindness (Horton & Hoyt, 1991).

But now the theme for neural investment shifts. The new slogan might be "Ask not what the brain can do for natural scenes—ask what the brain can do for the animal." Beyond V2 the brain's task is to *find what matters most*—and *do it quickly*. Accordingly, cortical areas beyond V2 should invest in circuits that rapidly identify what matters for survival and reproduction.

Many things matter. We need to understand a scene's structure, both for changes in viewpoint, novelty, and navigability. We need to identify objects that are useful or dangerous. We need to locate them spatially and perceive their orientation so that the hand in grasping can match its configuration to the object. We need to gauge the speed of moving objects. We need to recognize faces and their emotional expression. As cortical design addresses these needs, what are the governing principles?

Again the engineer's rule applies: specialize. Each need requires a particular computation. Each can be done most cheaply and rapidly by a dedicated circuit—which saves wire by not mingling with the others. Moreover, certain needs require higher temporal frequencies, which being disproportionately expensive, should be served by separate circuits. Some needs may be satisfied cheaply by approximation. For example, in visually guided grasping, the hand can be roughly matched to an object, and then adjusted precisely by feedback from touch receptors (Santello et al., 2013). But other needs demand exquisite precision. For example, particular faces must be met with particular behaviors.

We match a given face, viewed from any angle and lighting, to one of many stored images. This retrieval includes the associated biosketch. Face history must be integrated with the current facial expression to predict an appropriate response. To mistake one face for another could be fatal, as could misreading an expression—the "smooth forehead betokens a hard heart."[18] So we notice every wrinkle, shade, color, and motion—down to the slightest flush and twitch. To read a face at the minimum interpersonal distance requires full acuity—as any myope knows. So, it seems that the idea of a single neuron that only responds to your grandmother was stimulating, but reading her face requires a much greater investment.

How higher cortical areas invest

Visual areas beyond V2 occupy more than half of the territory allotted to vision—more than one quarter of all neocortex. This is an awesome investment; thus to learn how it is structured is key to understanding cortical design. Since V1 and V2 fill most of the occipital lobe, higher visual areas shift forward into parietal and temporal lobes (figure 12.11). Each higher visual area interacts with a distinct set of other cortical areas and, to save wire, tries to locate near them. This determines which lobes contain which areas.

Each area might conceivably reproduce the whole scene using all the components encoded in V2. Yet if nothing were discarded, each area would need the same amount of space as V2, and this would limit the number of stages for processing. But actually, V2 is the last area that contains all information from V1, and thereafter much *is* discarded. This works out because each individual area concerned with a particular high-level task retains all that it needs for that task and rigorously discards the rest. Consequently, behavior can retain the degree of sensitivity set in retina.

The scheme can be imagined as a large-scale version of what occurs in retina. Each of roughly 20 ganglion cell types discards all but a subset of data from the full scene and relays its subset to specific central clusters that need it (figure 12.1). The cortical visual system accomplishes the analogous task: each of roughly 30 areas discards all but a subset of what is contained in V2 and relays its subset to higher areas that need it. This occurs efficiently because V2 has reorganized the full representation especially for this purpose—so particular subsets can be selected to serve particular behaviors.

Three small areas[19] adjacent to V2 select elements of the full *scene* (ignoring particular objects, faces, and scrambled scenes). Each "scene area" has a different purpose; for example, one registers changes in viewpoint and scene novelty, and another serves navigation (Nasr et al., 2011). These scene-selective areas, being small, cannot represent the scene at full resolution, but for many purposes this is unneeded.

Certain areas select from V2 thin stripes and interconnect within the inferior temporal lobe to form the *ventral stream*. Neurons in these areas respond selectively to features such as color, shape, and texture, which are good for *identifying* objects. And as the analysis proceeds anteriorly along the ventral stream, neurons respond to particular objects independently of their orientation and location in the visual field. This step of shedding the object's spatial context and concentrating *what* the object is achieves a tremendous economy. Other areas select from V2 thick stripes and interconnect within the superior temporal lobe to form the *dorsal stream*. Neurons in these areas respond selectively to features such as depth, direction, and speed of motion that are good for localizing *where* objects are (Ungerleider & Pasternak, 2004). Shedding the static aspects of "what" in this stream offers further economy.[20]

The discarding/selecting process continues along these streams for "what" and "where" until various small areas contain highly specific information that is both compact in space and energy and also behaviorally useful without further processing. Each area can then be placed near the ones that need it. For example, grasping requires knowing an object's location and its orientation. This subset of visual information is extracted by a particular *grasp area* (one of several) that locates right where it is needed—in the posterior parietal lobe—to connect via short wires with somatosensory areas that guide grasping (Fattori et al., 2012). The organization of the "what" stream for vision is exemplified by the cortical representation of faces.

Faces

We need to identify a face from any viewpoint in any light. Moreover, we need its expression—the different meanings conveyed by a wrinkled brow, curled lip, or averted eyes. Acute tilt carries meaning, but an upside-down face is hard to recognize, and the cues to its emotional expression are more difficult still (Thompson, 1980). These tasks we manage at a glance—less than 200 ms. So it should be no surprise that the ventral stream would contain an area that responds preferentially to faces.

But it certainly was a surprise to learn that the ventral stream contains not one, but six *face areas* (Freiwald & Tsao, 2010; Tsao & Livingstone, 2008). Within these areas *face cells* vastly predominate (>90%), with a few neurons responding to other round objects such as a clock face and fruit. Early along the ventral stream, the face areas respond to faces viewed from particular angles, but they converge on a far anterior area (*AM*), the last and largest, where neurons respond to a face in any view. AM contains a small subset of cells that respond to only a few particular faces invariant to view, and a region in medial temporal lobe, one stage higher than AM, contains cells that are still more specific. So possibly, the end point of this processing system is indeed a small number of grandmother-like cells (Quiroga et al., 2005; Viskontas et al., 2009).

Thus, to recognize a face requires multiple areas, interconnecting millions of neurons that rapidly refine the representation and carve away all else. As at earlier stages along the visual pathway, sparseness can be used to concentrate information. By summing across many, sparse-coding cortical neurons, behavioral sensitivity can be enhanced by up to threefold compared to single neuron sensitivity (Cohen & Newsome, 2009). The summation probably enhances speed as well (Price & Born, 2010).

The largest face area in the macaque monkey occupies 16 mm^2. This amounts to one tenth the area of V1, and the six face areas together probably amount to one fifth, which is 2.5% of the monkey cortex. Humans use facial expression to express mood and intention, but also to dissemble and mislead, which monkeys cannot manage, so it seems a safe bet that the human cortex would invest even more in face analysis. The design is affordable because all else is discarded, but still the investment is huge. This emphasizes the change in outlook: early stages, gathering what will be needed by all central processors, must operate economically by matching lower order natural statistics. But later stages, finding high-order correlations that represent the nuanced expression of a particular face at speeds that promote survival (Cheney & Seyfarth, 2007), require investment in a dedicated system.

The most anterior face area is purely visual. The information it contains can be used to assess attitude and mood. However, this requires a connection to a neural complex, the *amygdala*, which compares the visual image to a template—like a mug shot—that identifies its probable significance. In effect, the amygdala imbues the image with emotional content.[21] The most anterior cortical face area and amygdala colocate near the hippocampus, which calls forth the stored template image along with its biographical sketch. Each area is critical, and together they must work fast. Indeed the amygdala responds to high-level facial information even before that information is consciously perceived (Freeman et al., 2014). Thus, the proximity of these areas certainly reflects efficient design. Moreover, each reactivation of the stored "file" provides an opportunity to update the face and the bio-sketch (chapter 14).

The dorsal and ventral streams that arise in the posterior temporal lobe eventually reach prefrontal cortex, where they terminate in adjacent areas. These areas serve *working memory*—holding separately in mind both a face *and* its location after it disappears from view. The dorsal and ventral streams finally converge on individual prefrontal neurons that encode both identity and location. The temporal sequence has been established: by 170 ms, emotional content has been processed in the amygdala; by 190 ms, gaze direction with pointing has been processed in parietal and supplementary motor areas. By 200 ms, the final integration of emotion, gaze, and gesture has occurred in premotor cortex, and you are prepared to respond adaptively to another person's intention (Conty et al., 2012). Any design that colocalizes mechanisms for emotional identification, working memory, and long-term storage must save wire.

Disconnection syndromes

The design that generates highly specialized processing leads, when an area is damaged or disconnected, to families of bizarre syndromes. For the ventral stream these include specific impairments of knowledge (*agnosias*). One can lose the ability to recognize faces (*prosopagnosia*) while still recognizing objects—and vice versa. One can lose the ability to recognize color (*achromatopsia*) so that scenes are perceived but only in black and white. One may retain the ability to recognize faces but lose the ability to read their emotional expression or their emotional history—leading to the unshakable conviction that a long-term, intimate companion is an impostor (*Capgras syndrome*).

One can also lose the ability to recognize words (*alexia*), leading to the peculiar condition of being able to write a sentence but then being unable to read it. This suggests a visual area specialized to recognize words (Wandell et al., 2012; Wandell, 2011). But does that imply a cortical area that remained blank until the invention of written language (Carreiras et al., 2009)? This *visual word form area* will be discussed further in chapter 14.

For the dorsal stream there are impairments of action (*apraxias*), such as misreaching or the inability to draw a picture. One can lose the ability to recognize motion (*akinetopsia*), making it dangerous to step down from a curb. If an object area (ventral stream) becomes disconnected from language areas, one loses the ability to name the object while retaining knowledge of its utility (food, tool). Moreover, dorsal stream areas can still direct its proper use. Because high-level cortical areas connect strongly to brain regions for error correction, such as striatum and cerebellum (chapter 4), damage to the corresponding parts of those regions can cause their own peculiar disconnection syndromes (Schmahmann & Pandya, 2008).

These syndromes seem bizarre because they contradict our quotidian experience of perceptual unity. Yet, they are the inevitable consequence of the design principle *complicate*. Our coherent experience of a scene resembles a silk-screen print or a color photograph assembled for the final image from successive layers. Its unity remains mysterious. We can only say that all the necessary information reaches the frontal lobe rapidly and as cheaply in space and energy as is consistent with that speed. Although we have focused on primate cortex for its extreme specialization, mouse cortex already shows connectional nodes corresponding to dorsal and ventral streams (Wang et al., 2012).

Parallels in auditory design

The auditory system also invests heavily in extracting higher order features. Indeed, beyond early auditory areas that establish sparse coding, there are ventral and dorsal streams for "what" and "where," including multiple areas for comprehending language and music (Rauschecker, 2012). These two forms of communication share similar distributions of low-order correlations but differ so greatly at higher orders that computations on the original vibratory patterns are accomplished most efficiently in different hemispheres, left for language and right for music (Purves et al., 2012).

Early auditory areas, like those for vision, supply compactly coded sound patterns to various smaller areas downstream that specialize to recognize words and streams of words as language. Eventually, they register both surface meaning and the emotional content. As for vision, sounds critical for survival, such as animal alarm calls, voices, and rushing water, are identified rapidly.

Given the importance of voices to primates, one might expect dedicated cortical areas for ... grandmother's voice. Indeed, two areas have been identified in humans by fMRI (Belin et al., 2000), and recordings in monkey confirm clusters of single neurons that respond selectively to conspecific voices (Perrodin et al., 2011). These areas are located anteriorly in the temporal lobe—fairly near to the highest level faces areas. The face and voice areas, if not directly interconnected, probably converge on another area, so their proximity saves wire.

Language areas, like the face areas, converge toward the amygdala (LeDoux, 2000), frontal lobe, and hippocampus—for the same reasons. The language areas generally locate in one hemisphere—which reduces wire and conduction time. The corresponding cortical areas in the opposite hemisphere are then free to use the same low-level structures but can then build different high-level structures to identify notes and patterned streams of notes—as music.

The brain in evolving systems for language and music confronts a critical design question. Given that much of what humans need to hear is either language or music, and that both are generated by the brain itself, what sound frequencies should be used? Why, for example, do we not converse in shrill tones and emit the rarer and briefer shrieks of joy or alarm at lower frequencies? One factor may be that high-frequency shrieks deliver information at a higher rate and stand out against a background noise. Another is that the chosen distribution of frequencies, mostly low plus infrequent bursts of high, is many-fold cheaper to generate and process (figure 4.8).

Many of the high-level areas mentioned here, such as the visual word form area, have been identified only recently. The reason is that, being highly specialized, they are small, so their discovery awaited critical improvements to the spatial resolution of *functional magnetic resonance imaging* (fMRI). It now appears that the lower size limit of specific cortical areas may be about 1 cm^2 (Wandell, 2011).

Disconnections of everyday life: Naming

Sigmund Freud, pondering more than a century ago on the "psychopathology of everyday life," suggested that a failure to recall the name of a familiar face may reflect a particular unconscious motivation. Certainly. But independently of unconscious motivations, these failures become more frequent with age. Moreover, it becomes harder to recall the names of common objects—even those we have no reason to resent.

Efficient design requires images of faces and objects to be identified and stored in high-level *visual* areas whereas their names are identified and stored in high-level *auditory* areas. Although a face may be identified visually in 200 ms, to find and attach the name requires querying a high-level auditory area—an additional step and thus additional time. Why this slows with age we cannot say—except that it's just like everything else.

Conclusions

This chapter began by asking where and how the retina's varied representations of a scene finally serve behavior and whether these lines from retina are ultimately reintegrated. It also asked how levels beyond the retina follow principles of neural design. Here, in brief, are some conclusions:

1. Certain stringently reduced versions of the scene, such as slow changes of light intensity or local direction of motion, serve simple behaviors at lower levels (midbrain, hypothalamus, etc.). These behaviors would not be improved by comparing current value to stored ones, nor would storing current values benefit future decisions. Stringently filtered data, essential only for the moment, may not warrant transmission to the highest levels, nor storage for any longer than *E. coli* stores the last encountered glucose concentration (see chapter 2).[22]

2. Richer versions of the scene from nonstringent filters gather in a complex of clusters. Three retinal lines that divide the spatial range from coarse to fine and the temporal range from slow to fast (figure 11.25) sort out in the LGN to prepare for transmission to V1 (figure 12.1). These lines that were subdivided in retina remain so at the LGN level, as does their further subdivision into ON and OFF. The multiple lines from the two eyes also remain separate, but in the LGN they neatly interleave. These representations together contain most of the useful information in a scene that warrants further analysis by cortex.

3. V1's simple cells integrate various lines that were separated for economical transmission to construct Gabor filters that optimally integrate space and spatial frequency. Overlapping ON and OFF receptive fields now combine by cooperative action: simultaneously turning on excitation ("push") and turning off inhibition ("pull"). Neighboring receptive fields sum along one dimension to integrate space, but not along the other—to preserve acuity. However, the lines carrying different spatiotemporal scales and color remain segregated in different layers, whose different energy needs are evident in their different levels of cytochrome oxidase expression. These patterns carry forward to V2, which assembles the next stage of filters that segment the scene in behaviorally relevant ways: distinguishing figure from ground by stereopsis and computing which edges/lines belong together. The middle temporal visual area (*MT*) begins to segment scenes by relative motion. Beyond this basic level of segmentation and grouping elements, the cortex subdivides into more specialized representations of "what" (ventral stream) and "where" (dorsal stream).

4. As these streams flow forward, the pattern of cortical investment is revealed. Areas for recognizing objects and faces become more specific. A face, once perceived, must be compared to stored data to determine if it is familiar; if unfamiliar and useful, it should be stored for future reference. Some inkling of what matters most for human survival is given by the patterns of cortical investment. What is stored long term are images, or edited versions of images, that carry lessons, and conclusions that can be used to predict what might happen and how to respond.

This chapter led deeply into the brain's wiring and noted some features that promote economy of space and energy. But these examples were somewhat anecdotal, whereas this important topic warrants systematic treatment. That is the next chapter.

13 Principles of Efficient Wiring

All of the various conformations of the neuron and its components are simply morphological adaptations governed by laws of conservation for time, space, and material.

—Santiago Ramón y Cajal (1909) (edited for brevity)

Earlier chapters noted that signal processing begins on the nanometer scale with the diffusion of transmitter in the synaptic cleft. Diffusion through the extracellular space allows the contents of one synaptic vesicle (~4,000 molecules) to reach multiple postsynaptic sites. Thus, the chemical signal is broadcast wirelessly, requiring no extra space or energy. Moreover, on this spatial scale (≤ 1 µm) it travels fast. But signals transmitted rapidly beyond about 1 µm must go by wire which, as Cajal asserted, is governed by certain "laws of conservation."

Cajal inferred these laws by noticing various cases where designs save wire. As shown in figure 13.1, for example, axons branch at an acute angle rather than a right angle (**Y** vs. **T**). Also they tend to leave the parent neuron at the point nearest to their destination. Such savings may seem inconsequential, but they add up: the 5 µm saved at 10 axonal branchings by each of 10^{10} cerebrocortical neurons reduces wire by 500 km.[1] Similarly, the cerebellar neurons in figure 13.1 reduce their irreducibly fine wire by 250 km.

Cajal asserted, as though his theory were insufficiently bold, that these conservation laws "should be immediately obvious to anyone thinking about them" (Ramón y Cajal, 1909; edited for brevity). Yet, such observations fall far short of a theory to explain the architecture of neurons and neuronal aggregates at all scales. The conservation laws should explain why neurons branch, what limits their individual volumes, their aggregate volumes in local circuits, and their fractional volumes (wire vs. synapses). The same laws should explain specialized substructures of local circuits, such as cortical layers, columns, stripes, pinwheels, and barrels—plus the structure

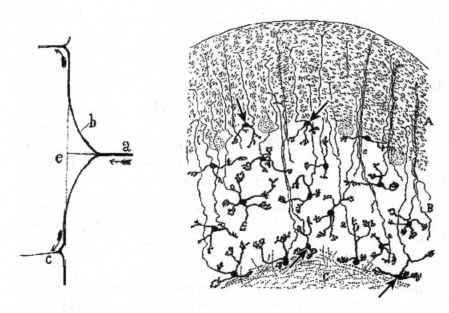

Figure 13.1
Axons exploit many small opportunities to save wire. Left: Axons branch as a Y.
Thus, each branch takes the hypotenuse of a right triangle rather than the sum of
the other two sides. For a 3, 4, 5 right triangle, the saving would be 1.4-fold. **Right**:
When a cell body is near its target layer (*A*), the axon exits from the cell body, but
when the cell body is located far from its target layer, the axon exits from a dendrite
nearest to the target layer. This example shows cerebellar granule cells, the brain's
most abundant cell type. Reprinted (with arrows added to right figure) from Ramón
y Cajal (1909).

of tracts. At a still larger scale, the laws should explain distinctive patterns
of cortical folding, the relationships between cortical areas, and hemispheric specialization.

Cajal missed one key law, conserve metabolic energy. This is easily for-
given because his major work preceded studies of brain metabolism, and
probably he would have accepted the addition as a friendly amendment.
Space and energy are strongly correlated, as we have noted, so any general
theory of efficient wiring must consider energy as well.

This chapter compares three structures: retina, cerebellar cortex, and
cerebral cortex (figure 13.2). These structures share certain features, for
example, all are layered and laid out as maps. Yet, they differ so greatly in
size, form, and function that one wonders how they could possibly be

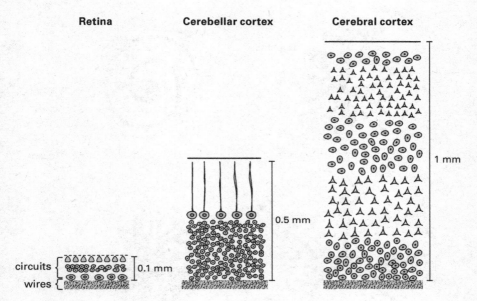

Retina   **Cerebellar cortex**   **Cerebral cortex**

Figure 13.2
Retina, cerebellum, and cerebral cortex differ greatly in thickness, shape, volume, and many other features, but their wiring follows the same principles. Although cerebral cortex folds in larger mammals, especially primates and cetaceans (dolphins and whales) to gain more area, folding is not intrinsic to the microscopic layout as it is in cerebellum.

governed by the same rules? The key is to realize that they share a profound biophysical constraint. Thus, we begin with that constraint and the rules that it engenders. If the same rules can illuminate all three designs, then the generality of Cajal's conservation laws for time, space, material, and energy will indeed become "immediately obvious."

Biophysical constraint on efficient wiring

The core constraint on all circuit design is the substantial and irreducible electrical resistance of neuronal cytoplasm. Resistance attenuates the spread of signals along a neural cable. It also limits the rate at which the membrane capacitance charges, reducing the speed with which signals travel and spreading them out in time (figure 7.7). Consequently, cytoplasmic resistance sets an upper bound to the length of dendrites. Resistance can be reduced by increasing cable diameter (d), but spread of voltage in space and time increases only as $\sqrt{d}$. This causes a law of diminishing returns: a

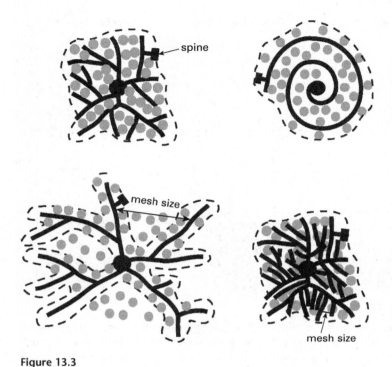

Figure 13.3
Dendritic arbors minimize conduction distance and conduction delay. Upper left:
The optimal arbor is compact, symmetrical, and of optimal mesh. All synapses are
near enough for potential contact on dendrite or spine. Conduction distances and
delays are minimized. **Upper right:** Dendritic arbor has same total length and mesh
but longer conduction distances and times. **Lower left:** Mesh is coarser than "spine-
reach" zone (dashed lines), so dendrites must lengthen to obtain the same number
of potential contacts. **Lower right:** Mesh size is too fine for all potential contacts, so
dendrites occupy excessive space. Modified and reprinted with permission from Wen
& Chklovskii (2008); see also Panico & Sterling (1995).

dendrite, to double its conduction distance and/or halve its conduction
delay, must quadruple its diameter, thus increasing its volume 16-fold.

To conserve volume, dendrites should stay thin. Being thin, they must
stay short, which they typically do by branching symmetrically and compactly about the cell body (figure 13.3; Wen & Chklovskii, 2008). When a
neuron's function requires more synapses, for example, to increase S/N or
sample more inputs, dendrites lengthen (chapter 11). But then to avoid
attenuating and delaying the signal, the dendrites must also thicken,
increasing volume as d^2. But increasing neuronal volume reduces neuronal

packing density. The greater distance between neurons requires longer, and therefore thicker, wires. Thus, the strategy of compensating for attenuation and delay by increasing dendritic diameter becomes self-defeating when the volume fraction of wiring (dendrites + axons) exceeds 3/5 (Chklovskii et al., 2002). This wiring fraction limits the extent to which a single neuron can integrate synaptic inputs; therefore, dendrites are thin and seldom more than 1 mm long.

The same constraint, low conductivity of cytoplasm, limits the length of unmyelinated axons in a local circuit. Their finest diameter (~0.15 μm) conducts at roughly 0.5 mm per millisecond, velocity increasing as $\sqrt{d}$. To double the conduction distance without increasing delay, the axon must double its conduction velocity. As for dendrites, this requires quadrupling diameter, thus increasing volume by 16-fold. Volume rising as d^4 soon outstrips the axon's ability to compensate for time delays; therefore, axons, like dendrites, are constrained to a wire fraction (dendrites + axons) of roughly 3/5 of the total volume. This fraction is optimal—it minimizes delays—and it roughly matches the fraction measured in circuits (Chklovskii et al., 2002).[2]

The "three fifths rule" also sets the optimal fraction for synapses. Synapses are irreducibly small, containing just sufficient vesicles for their job with little safety factor (figure 13.10; Sterling & Freed, 2007). More synaptic space would force the wire fraction away from its optimum. So for every synapse enlarged or added in a mature brain, another is shrunk or deleted (chapter 14). The three-fifths rule determines the branching of dendrites and axons and forces a custom layout for each circuit function (figure 13.4). Retina, cerebellar cortex, and cerebral cortex are all subject to the same constraints; therefore, their local circuits are similar in scale (~1 mm or less) and obey the three fifths rule.[3] However, their tasks are different—which requires different connectivity and thus different optimal layouts.

Here we define their three connectional tasks and, for each, an optimal layout. Then we compare the actual neural structures.

Task 1: Connect a dense array to a sparse array with modest divergence and convergence

Consider 10 densely packed neurons connecting all-to-all to five more sparsely distributed neurons (figure 13.4). The most efficient layout (least wire) is for each presynaptic neuron to send one axon to the sparse array and then diverge to all five neurons via multiple branches. Similarly, it is most efficient for each postsynaptic neuron to send a single dendrite toward the dense array and collect all 10 inputs on branches. Branched axons and

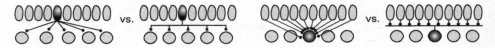

Figure 13.4
Layout for connecting dense array to sparse array with low divergence and convergence: branched axons and dendrites minimize wire. **Left**: One presynaptic neuron diverges to contact five postsynaptic neurons. **Right**: Ten presynaptic neurons converge to contact one postsynaptic neuron. Modest divergence and convergence between arrays conserve wire by interfacing planar meshworks in a single stratum. In this layout the number of potential contacts equals actual contacts.

dendrites serve the same layout goal: they establish a given degree of connectivity with the least wire and thus the least delay.

In two dimensions the optimal layout is for both arrays to tile the same space with matching, cospatial meshworks (figure 13.4). Such interlacing establishes contiguity between the two arrays with no waste by either one. Elements that are not required for this circuit (cell bodies, extraneous wires, and tracts) are excluded because they are obstacles that force a longer path, like a tree across a hiking trail. This type of layout is exemplified by the retina.

Retina
The retinal circuit, about 100 μm thick, sharply segregates cell bodies from synaptic circuits (figure 13.5). This prevents cell bodies from interfering with the layout, and because the circuit connects in a simple sequence, the cell bodies locate near their sites of connection. Thus, photoreceptor and horizontal cell bodies are nearest to the outer synaptic layer; amacrine and ganglion cell bodies are nearest to the inner synaptic layer, and bipolar neurons that connect to both layers are in the middle. The retina's output axons (from ganglion cells) need to travel in the plane of the retina to reach the optic nerve. They require additional space but avoid disrupting the circuit layout by using a separate layer beyond the circuitry.

The inner synaptic layer is thick enough to accommodate processes from approximately 70 neuron types. Each type connects only with a modest subset of other types to form a distinctive circuit that repeats across the retina. Consequently, each circuit is a monolayer of intertwining processes, ≤2 μm thick. The inner synaptic layer comprises a stack of about a dozen of these irreducibly thin circuits that avoid mutual interference. The dendrites of the output cell type with the broadest field span about 0.5 mm and thus cover about 0.2 mm^2,[4] similar to the dendrites of the output neurons of cerebellar and cerebral cortex (figure 13.5).

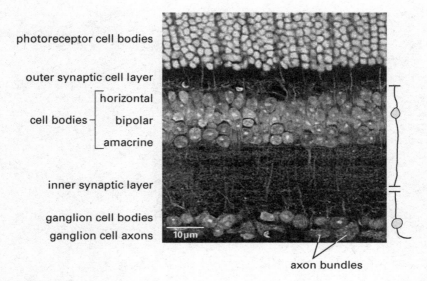

photoreceptor cell bodies

outer synaptic cell layer

cell bodies ⎯ ⎡ horizontal
             ⎢ bipolar
             ⎣ amacrine

inner synaptic layer

ganglion cell bodies
ganglion cell axons

10 µm

axon bundles

Figure 13.5
Retinal layers minimize wire by segregating cell bodies and axon bundles from synaptic circuits. The outer synaptic layer connects to the inner synaptic layer via processes that descend vertically and then branch horizontally at a specified depth to contact particular types of process that ascend vertically and branch horizontally at the same level. Output axons gather in bundles beneath the circuits to avoid interfering. This vertical section through mouse retina omits the photoreceptor outer regions; full thickness is about 200 µm. Photo is modified and reprinted with permission from Masland (2012).

Input axons form a loose mesh laid out as a thin (1–2 µm) stratum, and ganglion cell dendrites do the same (Panico & Sterling, 1995; Chklovskii, 2000). Their meshes match, maximizing points of contiguity, all potential sites for synapses (figure 13.6). For bipolar-to-ganglion cell circuits, the ratio of actual synapses to potential synapses approaches unity (Sterling et al., 1988). In this layout of matched meshes there is no need for dendritic spines, either as "reachers" or "spacers" (see below, figures 13.7 and 13.11); therefore, with some exceptions, they are not used.[5]

Task 2: Connect a dense array to a sparse array with extreme divergence and convergence

A circuit needing extreme divergence and convergence cannot use the retinal layout. The dendritic and axonal arbors would overlap extensively, causing them to interfere with each other and increase path length. Instead,

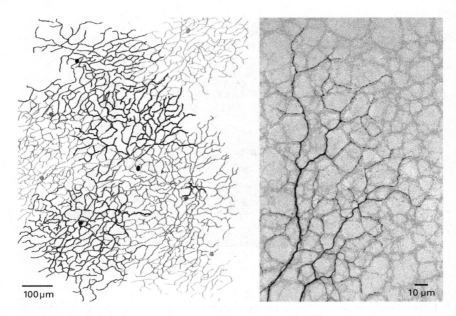

Figure 13.6
Coarse ganglion cell dendritic mesh evenly tiles and efficiently matches a fine axonal meshwork. Left: Eight directionally selective ganglion cells of the ON-OFF type and same preferred direction tile the retina with an even, planar meshwork. Only the OFF meshwork is shown here; the ON meshwork is similar but coarser (see chapter 11). **Right:** Dendrites of one directionally selective ganglion cell (dark branches) neatly cofasciculate with the network of starburst processes (pale), thus establishing many contacts with the least possible wire. All directions are represented within the starburst network, but this ganglion cell selects inhibitory contacts only from processes that respond to motion in the null direction (see figure 11.24). Reprinted with modification and permission from Vaney (1994) and Vaney & Pow (2000).

the optimal layout is for the dense array to use straight, unbranched axons and run them orthogonally through finely branched, planar dendritic meshworks of the sparse array (figure 13.7). This type of layout is used by the cerebellum and by other central structures as well (Oertel & Young, 2004; Bell et al., 1997).

Cerebellar cortex
The cerebellar cortex is roughly 0.5 mm thick, and the unit circuit includes about 10^5 neurons in a volume of about 0.6 mm^3.[6] The scale resembles, in any given dimension, the retina's two-dimensional circuit, reflecting their

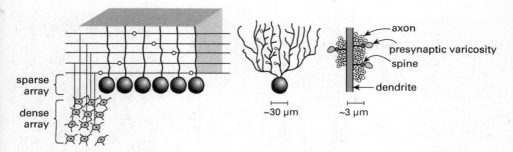

Figure 13.7
Layout for connecting dense array to sparse array with extreme convergence and divergence and many potential synapses. Rectilinear axons run orthogonally to a fine two-dimensional dendritic meshwork with contacts to spines. **Left:** Dense array of tiny neurons sends fine axons vertically to course orthogonally to sparse array of large neurons. Each axon traverses many dendritic arbors, allowing economical divergence via single contacts. **Middle:** Large neuron shown perpendicular to the fine parallel axons. The extensive dendritic arbor allows economical convergence of single contacts from many axons. **Right:** Dendrite projects spines to reach synaptic varicosities from passing axons. This creates space for many potential synaptic contacts. Potential synapses in this layout are fivefold greater than actual synapses.

shared size constraint. The layout resembles what is optimal for a dense-to-sparse array with extreme divergence and convergence. In cerebellum, the dense elements making straight axons are granule cells; the sparse elements intersecting these with coplanar dendrites are Purkinje neurons (compare figure 13.7 vs. figure 13.8). Beyond this obvious correspondence of actual to ideal layout, there are a host of strange and subtle features to be explained.

Unlike retina, the cerebellar cortex mingles neurons with synaptic circuits (figure 13.9). Moreover, it uses a different layering sequence: nearest to the white matter is the input layer; then comes the output layer; and finally, on top, the intermediate layer. Unlike retina, the smallest cerebellar neurons are at the bottom and the largest are in the middle. The output tract from cerebellar cortex is short, just a few millimeters (compared to tens and hundreds of millimeters for retina and cerebral cortex). Finally, unlike retina, which is flat as a pancake, and unlike the cerebral cortex, which is raised like a layer cake and smooth in most species (see below), the cerebellar cortex of all species is corrugated like a tin roof. Moreover, the corrugations are oriented—perpendicular to the long axis of body and brain. All these oddities save wire.

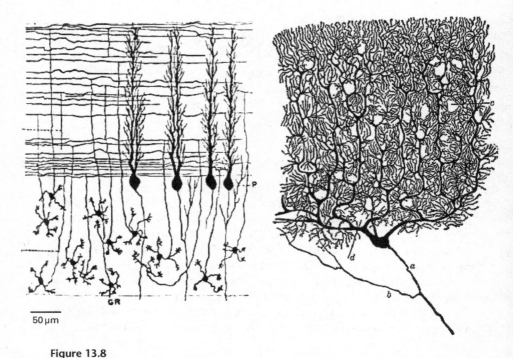

50 µm

Figure 13.8
A layout for extreme connection. Left: Cerebellar wiring viewed parallel to a fold (*folium*). Granule cell axons course parallel to the folium and orthogonal to the Purkinje arbor. **Right:** Purkinje cell's planar dendritic arbor viewed perpendicular to the folium (human). The dendritic mesh fills the plane evenly except for holes occupied by various neuron cell bodies. Left reprinted from Cunningham's *Textbook of anatomy*, by Daniel John Cunningham, published in 1913 by William Wood. Right reprinted from Ramón y Cajal (1909).

Mixing cell bodies with local circuits at the input layer minimizes tracking thick wires through the circuits. The most numerous input axon (mossy fiber) fires at high rates and so is thick; therefore, it should deliver its message immediately upon emerging from the white matter. Its target is the granule cell, which to pack densely, is tiny with thin, short dendrites. A layout like retina that separates cells from synapses would require longer, thicker dendrites. Granule cells reduce the mean spike rate by more than 10-fold, thereby allowing their axons to be thin—essential to their extreme divergence and convergence onto the Purkinje neurons.

Neurons in the intermediate layer are relatively sparse and so do not greatly interfere with the wiring (figure 13.8). Therefore, it is economical to

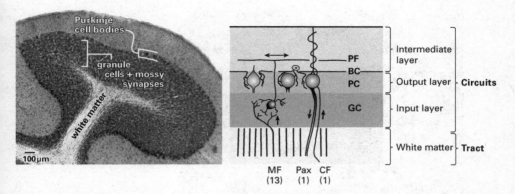

Figure 13.9
Cerebellar layers minimize wire by mixing neuron cell bodies with synaptic circuits.
Left: Large output neurons locate between input and intermediate layers. **Right:** One
type of input axon (mossy fiber) is thick and numerous, so upon emerging from
white matter, it ends deep. The other type of input axon (climbing fiber) is thin
and sparse, so it traverses the deep layer to reach the intermediate synaptic layer.
The single type of output axon (Purkinje cell) is thicker than the mossy fiber but
far sparser (ratio of mossy fibers to Purkinje cells exceeds 13; Llinás et.al., 2004),
so spaced is saved by locating the output layer above the input layer and tracking the output axons through it to the white matter. PF, parallel fiber; BC, basket
cell; PC, Purkinje cell; GC, granule cell; CF, climbing fiber, Pax, Purkinje axon; MF,
mossy fiber. Left, mouse reprinted with permission from http://brainmaps.org/ajax-
viewer.php?datid=62&sname=086&vX=-47.5&vY=-22.0545&vT=1. © The Regents of
the University of California, Davis campus, 2014.

place them where needed. Similarly, a key input, the climbing fiber, is
sparse, one for each Purkinje cell (figure 13.9). It is also thin, so it is efficient
to track the climbing fiber from white matter through the input and output
layers. Purkinje cell axons are thickest of all (figure 13.10), but they are
sparser by more than 10-fold than mossy fibers, so to track them down
through the input layer to the white matter is least bad.

In short, the layout of cerebellar cortex mixes cell bodies with circuits to
minimize total wire. It also arranges the sequence of layers so that the thinnest traversing axons are the longest and the thickest traversing axons are
the sparsest. This allows a granule-to-Purkinje convergence of nearly
200,000, roughly 20-fold greater than for any other circuit,[7] an upper bound
to neural connectivity. To achieve this, both granule and Purkinje neurons
are forced toward various biophysical limits, as follows.

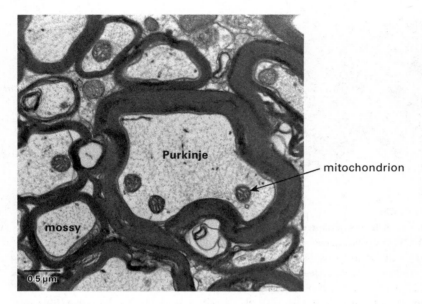

Figure 13.10
Purkinje cell axon is thick, heavily myelinated, and rich in mitochondria. Reprinted
from Perge et al. (2012) with permission.

Granule cell

The granule cell is about as small as a mammalian neuron can be.[8] The cell
body needs just enough synthetic capacity to supply four short dendrites
plus the brain's finest axon (figure 13.11). This axon (parallel fiber), slightly
exceeding 0.2 μm in diameter,[9] is just thick enough to accommodate a few
microtubules for transport and to prevent excessive channel noise due to
high internal resistance (chapter 7). Channel noise causes fluctuations in
conduction velocity and thus in spike arrival time. Such temporal jitter
accumulates with distance, which may limit the axon's useful length. However, the axon branches as a T, thus halving the jitter at each end and possibly doubling its useful length (~5 mm).

The synapse from a parallel fiber is as small as it can be. The axon, passing near a Purkinje cell spine, dilates by fivefold to form a varicosity just
large enough to house a pool of about 480 synaptic vesicles (Xu-Friedman
et al., 2001). These suffice to replenish presynaptic docking sites for a mean
release rate of several vesicles per second, a recycling time of about 1 minute, plus a modest safety factor (Sterling & Freed, 2007). Space is allotted to
mitochondria in only half of the synaptic dilations, as is also true for fine

axons in cerebral cortex (Shepherd & Harris, 1998). The dilation's volume is proportional to its radius cubed whereas the axon's volume is proportional to its smaller radius squared, so the varicosity occupies 33-fold greater volume than an equivalent length of axon.[10] In short, the synapse of a parallel fiber, though irreducibly small, occupies a considerable volume.

A parallel fiber produces one varicosity per 5 μm (Napper & Harvey, 1988; Xu-Friedman et al., 2001). This increases the fiber's total volume (axon + varicosity) over that distance by fivefold.[11] Thus, even the irreducibly small volume of a low-rate synapse (<3 Hz) is so space-greedy that it constrains the number of synapses that can be accommodated under the three-fifths rule. One is also reminded that a higher mean firing rate would require a thicker parallel fiber and a still larger varicosity. Axons of similar caliber elsewhere in the brain should produce similar dilations with similar spacing, and indeed, fine axons in cerebral cortex also produce one varicosity per roughly 5 μm (Braitenberg & Schüz, 1998).

The parallel fiber economizes to some extent on membrane area. To produce about 1,000 active zones, it forms about 800 dilations, that is, 1.25 active zones per dilation. The larger varicosities with 2 or more active zones have a smaller ratio of surface area to volume, thereby reducing the surface area per active zone. Reducing membrane reduces the cost of resting potentials, which dominates the cerebellar energy budget (Howarth et al., 2012) and, by reducing capacitive load, also reduces the cost of action potentials.

Purkinje cell

The Purkinje cell is among the brain's largest neurons (figure 13.8). Its cell body, more than 30 μm in diameter, supports a planar dendritic arbor that extends more than 250 μm across a folium and 250 μm up to the cortical surface. Dendritic branches are just thick enough to support the required length and time constants. The dendritic arbor is also as compact as it can be and still accommodate the passage of parallel fibers. Dendritic spines reaching out for their contacts allow parallel fibers to maintain their straight trajectories (figure 13.11). This saves about 1 μm per synapse, which amounts to 0.2 mm of wire for each Purkinje neuron and thus 3 km of wire for the whole cerebellum (Nairn et al., 1989).

The spines serve another key function: they serve as "spacers" for the tightly compacted Purkinje arbor to allow through-passage for many parallel fibers. Only some of the parallel fibers make contact as they intersect the arbor, but all course within the "spine-reach zone." Thus, every passing parallel fiber is a potential contact—a fiber that could under the right

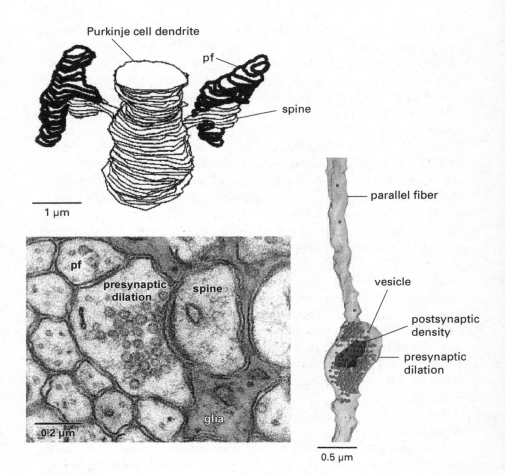

Figure 13.11
Purkinje cell spines minimize wire. Upper left: Purkinje cell dendrite projects two spines that are contacted in passing by parallel fibers. Reconstructed from serial electron micrographs (Harris & Stevens, 1988). **Lower left**: Cross section through irreducibly fine parallel fibers (pf). One fiber is sliced through its presynaptic dilation that contacts a Purkinje cell spine en passage. **Right**: Parallel fiber with presynaptic dilation. Reconstruction. Reproduced with permission from Xu-Friedman, Harris, & Regehr (2001).

circumstances form a synapse. The abundance of potential contacts endows the cerebellar cortex with the opportunity to efficiently alter its connections (chapter 14). This contrasts with retinal ganglion cells where, as noted, every potential contact is actualized.

Efficiency at the cerebellar output

The Purkinje cell axons form what may be the brain's most expensive output tract. The cell's high mean rate (~40 Hz) requires a thick axon with a high mitochondrial volume fraction. For example, a Purkinje axon costs 25-fold more energy than a ganglion cell axon,[12] and Purkinje cells outnumber ganglion cells by about 10-fold, so their total axonal cost is nearly 250-fold greater. A Purkinje cell axon may be cheaper than a vestibular axon because its firing rate is lower by half (figure 4.6), but Purkinje axons are 100-fold more numerous. Given the expense, efficient design would keep Purkinje axons short and would reduce both firing rates and axon numbers before broadcasting their message far beyond the cerebellum.

Indeed, most Purkinje axons project only a few millimeters—down to neuron clusters deeper in the cerebellum (figure 13.12). These deep neurons fire at roughly half the rate of Purkinje cells (Thach, 1968), which reduces their axon diameter and energy use by at least fourfold.[13] Moreover, there is a net convergence that reduces their numbers by 25-fold (Person & Raman, 2012)[14]—for an overall savings in space and energy of at least 100-fold. The synaptic design employed in this integrative step resembles that of other step-down sites, such as thalamus and mossy to granule cell: multiple release sites and enhancement of spillover by glial encapsulation (chapters 4, 7, and 12; Telgkamp et al., 2004). The rate steps down further by about fivefold in the thalamic nuclei that integrate cerebellar signals before relay to motor cortex (chapter 4). Consequently spikes arrive at motor cortex at similar rates to sensory cortex, a feature essential to the shared layout across the whole cortical expanse.

Similarities in striatum

The striatum resembles the cerebellum in using large numbers of neurons that fire at high mean rates over very thick wires. This is a costly design; therefore, as in cerebellum, the striatum optimizes component placement and reduces neuron numbers and rates.

The striatum collects from large expanses of cerebral cortex, especially from anterior regions (figure 4.10). The striatum locates just beneath the cortex and expands anteriorly. The striatum narrows down, wedge-like, to converge on a smaller component (*globus pallidus*) that also fires at high rates (~40 Hz) but reduces the number of output axons by 50- to 100-fold. The next stage of this loop is a neuron cluster in the *substantia nigra* that maintains high firing rates (~70 Hz in primate). But this region projects to the same thalamic clusters as the cerebellar output where the spike rates are stepped down before the final return relay to motor cortex.

In short, both the cerebellum and the striatum are systems for error correction that compute with high-rate neurons. Just why they use high rates remains to be clarified, but what *is* clear is that both systems concentrate their messages by reducing axon numbers and firing rates, thus conserving space and energy over long-distance relays.

Cerebellar macrostructure serves efficient wiring

Efficient layout for the granule-Purkinje cell circuit restricts Purkinje cells to a monolayer and requires their planar arbors to stack in rows (figure 13.8). This design, to accommodate sufficient Purkinje neurons, leads to extreme cortical folding. The *folia* align perpendicular to the brain's long axis and are so finely elaborated as to present in cross section a fern-like pattern (figure 13.12). This so expands cerebellar surface area that in human it exceeds that of the cerebral cortex by 20-fold. Cerebellar surface area per tissue volume exceeds that of cerebral cortex by more than 140-fold. This striking macroscopic feature, as we now explain, is required to accomplish the optimal layout at the microscopic scale (figures 13.7 and 13.8).

Purkinje cells are 6- to 15-fold more numerous than retinal ganglion cells.[15] Moreover, they are on average more than threefold greater in diameter and so occupy 10-fold greater cross-sectional area. Thus, overall, the Purkinje cell monolayer requires 150-fold more area than a ganglion cell monolayer.

Retinal ganglion cells over small regions of high density can stack vertically to avoid folding, but this would be inefficient for Purkinje cells. Stacking would require longer dendrites to reach the synaptic layer and would increase the numbers of thick Purkinje cell axons that track through the underlying granule cell layer and disturb its local wiring. Because stacking the cell bodies would increase wire length, a monolayer is more efficient; this needs a large surface area, and therefore dense folding is inevitable (figure 13.12).

The folia align perpendicular to the body's long axis, allowing the trunk to be represented along the midline with limbs extending bilaterally (figure 13.13). This preserves the body map with least wire, and the map itself saves wire (chapter 3). Cerebellar regions that serve the vestibular system locate near the brain-stem vestibular clusters, thus shortening the direct

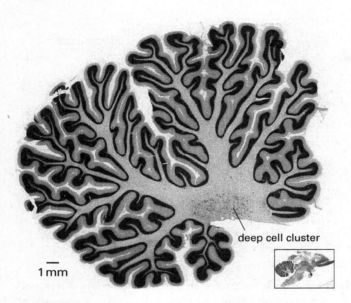

Figure 13.12
Cerebellar folia are essential for optimal layout. Slice shown here, parallel to long axis of body and brain (inset), shows fern-like macroscopic structure. Given a design that restricts Purkinje cell bodies to a monolayer, this fine folding expands the surface area to accommodate more Purkinje cells. Purkinje cell axons converge upon a modest cluster of neurons deep in the white matter. This greatly reduces the number of spikes leaving cerebellum. This section from macaque is reprinted with permission from http://brainmaps.org/ajax-viewer.php?datid=3&sname=0548. © The Regents of the University of California, Davis campus, 2014.

Purkinje cell outputs to that system (figure 13.13). In short, all these macroscopic features—dense folding, orientation of the folds, and topographic layout—serve efficient wiring.

Task 3: Connect many dense arrays to many other dense arrays with moderate divergence and convergence, preserving a high ratio of potential to actual synapses

This task is characteristic for cerebral cortex. To encode and learn complex patterns (chapters 4, 12, and 14), each cortical neuron must access many possible combinations of axonal inputs. The *connectivity repertoire* depends on how many different axons can enter the dendritic spine-reach zone and thus potentially make synaptic contact (figure 13.7; Wen et al., 2009).

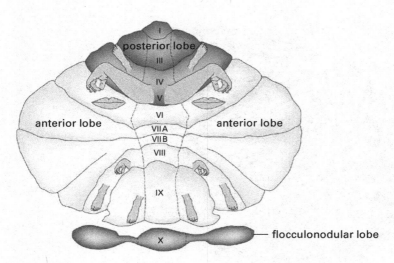

Figure 13.13
Bilateral symmetry of topographic body maps on cerebellum save wire. *Flocculonodular lobe* serves brainstem vestibular circuits that lie just beneath it. Reprinted with permission from Manni & Petrosini (2004).

Because multiple contacts from one axon to one neuron shrink the repertoire, contacts should be sparse. An optimal layout will produce the largest repertoire for a given dendritic cost, and because dendritic arbors differ in size, the optimal layout should work across scales (Wen et al., 2009).

The dendritic arbor that satisfies these specifications is three-dimensional with straight dendrites arranged symmetrically about the cell body (figure 13.14). The arbor is sparse; therefore, straight axons traversing it from all angles will rarely make multiple contacts—except for close neighbors whose axons branch densely near the origin of the dendritic arbor, which is also dense (Markram et al., 1997; Silver et al., 2003). In this layout the connectivity repertoire is set by the dendritic arbor's three-dimensional shape. And by tailoring its arbor for each cortical layer, a neuron can select different repertoires (figure 13.14).

This simple scheme allows the many types of neuron present in each cortical layer to establish their unique connectivity repertoires at least dendritic cost. Consequently, each type delivers a customized signal to a particular target. Targets include specific neuron types in other layers, in other cortical areas, in the opposite hemisphere, and in various subcortical areas, such as striatum, thalamus, midbrain, and clusters that supply the

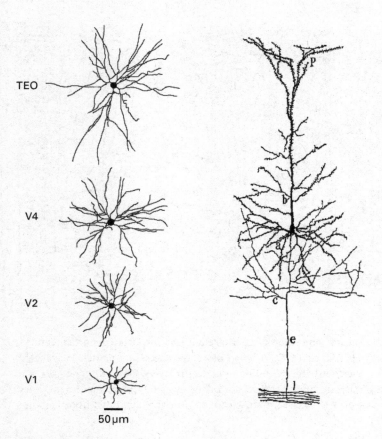

Figure 13.14
Dendritic arbors that maximize connectivity repertoire are sparse in three dimensions and similar across scales. Left: Basal dendritic arbors of pyramidal cells from four visual areas (monkey). Sparsity in three dimensions is far greater than apparent in this two-dimensional projection parallel to the cortical surface. The connectivity repertoire per dendritic length is the same across scales. **Right**: Pyramidal neuron viewed perpendicular to cortical surface. Dendritic arbor assumes a different shape in each layer (a, b, P) and therefore selects a different repertoire from the axons in that layer. The pyramid cell's axon (e) runs straight to the white matter, minimizing wire and local interference; its collateral branches (C) run straight, maximizing the number of dendritic arbors they can encounter with least wire. Figure is modified and reprinted with permission from Wen et al. (2009); pyramidal neuron reprinted from Ramón y Cajal (1909).

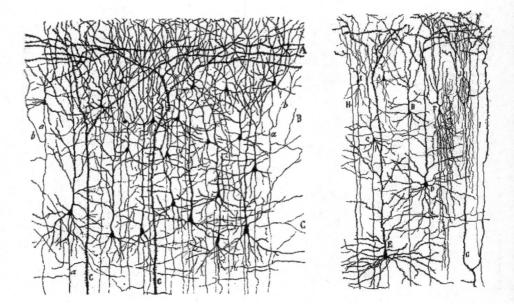

Figure 13.15
Layout to maximize connectional repertoire appears similar across areas. Left: Visual cortex, upper layers (A–C). a, axon descending to white matter; b, collateral branches; C, thick dendrites ascending from deeper layers. **Right:** Motor cortex, also upper layers, showing similar layout. Both drawings are from human infant, a few days postnatal on left, and a month postnatal on right. Reprinted from Ramón y Cajal (1909).

cerebellum (chapters 4 and 12). Overall, the number of types with different connectional repertoires and different targets can be on the order of 100.

This scheme is general in that it works wherever a circuit needs to provide many types with a rich connectivity repertoire.[16] This could help explain why the cortex looks so similar across different areas (figure 13.15).

Cerebral Cortex
The cerebral cortex is roughly 1 mm thick, twice the thickness of cerebellar cortex, and instead of three layers, there are at least six (figure 13.16). The neuron types are more diverse by more than 100-fold. Whereas the cerebellar cortex gathers two excitatory inputs into one synaptic layer and integrates them with one stereotyped output neuron (Purkinje cell), the cerebral cortex sends diverse excitatory inputs to all layers, integrates them with diverse cell types, and broadcasts a multiplicity of outputs to a multiplicity

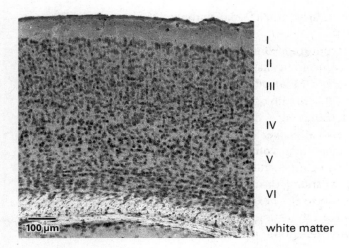

I

II

III

IV

V

VI

100 µm   white matter

Figure 13.16
Cerebral cortex mixes cell bodies and synaptic circuits. Dark cell bodies are surrounded by pale regions of circuitry (dendrites, axons, synapses). Largest cell bodies with thicker axons are sparse and located deep, near white matter, whereas the smaller cell bodies with thinner axons are dense and located high, far from white matter. See also figure 13.17. Mouse, same section as figure 13.9.

of targets (figures 12.8 and 12.9). Whereas cerebellar axonal and dendritic arbors are linear or flat (one- or two-dimensional), the cerebral axonal and dendritic arbors tend to be dense near their origin and spread out with distance (three-dimensional). Thus, the circuit layout matches in many respects what is optimal for high connectivity repertoires and a high ratio of potential to actual contacts.

Because cerebral circuits are built from three-dimensional components, the basic design does not require folding. Indeed the cerebral cortex in most species is smooth.[17] The columnar unit structure described in chapter 12 extends about 0.5–1 mm in all dimensions, encompassing on the order of 10^5 neurons (Braitenberg & Schüz, 1998). Thus, the unit circuits of cerebellar and cerebral cortex, despite glaring differences, use similar numbers of neurons and occupy similar volumes.

The unit circuit is designed to maximize the number of potential connections that can be made within its volume. For densely connecting the cerebral three-dimensional circuit, branching axon arbors with en passant contacts reduce wire volume by 300-fold, and dendritic arbors reduce volume by another 50-fold. By increasing a neuron's reach, dendritic spines further reduce wire volume by approximately threefold.

Thus, volume is reduced 45,000-fold without losing potential synapses (Chklovskii, 2004).

Clearly, this layout to maximize connectivity repertoire is efficient, but can it serve all cortical circuits? Cortical neurons need to detect specific sorts of correlation: in V1, an edge; in V2, a corner; in ventral stream areas, a face (chapter 12). This involves integrating across progressively wider expanses but still using single, insecure synapses so that each type can preserve a sparse signal. Moreover, cortical neurons in all these areas need the capacity to change their connections—to resculpt their circuitry—in response to shifts in correlated input (chapter 14). These needs are served by progressively expanding the characteristic reach of a neuron's basal dendrites from one area to the next going forward (figure 13.14). This allows progressively broader integration by individual neurons and maintains their connectivity repertoire per dendritic length across scales (Wen et al. 2009).

Cortical interfaces (input/output tracts)

The cerebral cortex takes its main input from thalamic clusters that have substantially sparsified the signal (chapter 12). For example, input rates to visual cortex are scaled-down from the optic nerve rates (figure 11.26), so the fiber diameters are scaled down correspondingly.[18] By continuing the relay with a log-normal distribution of axon diameters (figure 4.2), the thalamocortical input tract conserves space and energy. Additional space is saved by efficient interfacing at the cortex.

Thalamocortical axons fire at far lower rates than the cerebellum's primary inputs, so they are thinner. Consequently, unlike the cerebellum's thick mossy fibers, which must terminate immediately upon exiting the white matter (figure 13.9), thalamocortical axons can track through the deep cortical layers to arborize in the middle and upper layers (figures 12.6 and 13.17). The thinnest thalamic axons reach the uppermost cortical layer, taking the longest course through gray matter, whereas the thicker ones branch in the middle, taking the shortest course (figure 12.8).

The smallest pyramidal neurons locate in the upper layers so their thin axons can track downward through the circuitry with least interference (figure 13.17). Then comes a stratum of larger pyramidal neurons with thicker axons, but the cells are sparser, also reducing interference. Consistent with fine axon caliber, the synaptic dilations suggest low release rates, being smaller and with sixfold fewer vesicles than for the cerebellum's parallel fibers. Pyramidal axons, upon entering the white matter, turn to course tangentially toward other cortical areas. Many of the fine axons from the

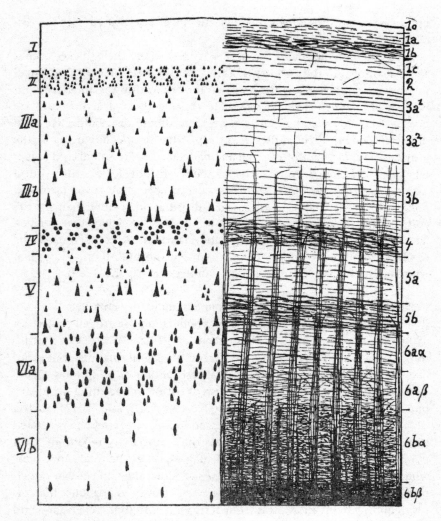

Figure 13.17
Cerebral cortex reduces wire by placing smaller neurons with finer axons near the surface and larger neurons with thicker axons near the white matter. Left: Distribution of neuron cell bodies. **Right:** Distribution of myelinated axons. There are exceptions: for example, larger neurons in layer IIIb lie above smaller neurons in layer IV. However, the smaller neurons distribute densely (like cerebellar granule cells) and receive many thick axons (like cerebellar mossy fibers); whereas the larger neurons distribute sparsely (like cerebellar Purkinje neurons). So this exception also minimizes wire. This feature of cerebellar design repeated in cerebral design suggests that it is a motif. Reprinted from Braitenberg & Schüz (1998), who reprinted it from Bailey & von Bonin (1951).

upper layers form the *corpus callosum*, the huge tract that couples the two hemispheres.

Larger neurons reside in the deeper layers—where their thicker output axons least disturb the circuitry. This reverses the cerebellar layering (small neurons deep), but it satisfies the rule for optimizing the interface with white matter. Deep neurons distribute to subcortical structures and are diverse (chapters 4 and 12). Therefore, they should distribute their axon diameters log-normally, a prediction supported by measurements of the long descending tract from frontal and parietal cortex to spinal cord (chapters 3 and 4). Cerebral white matter is itself layered, the finer corticocortical axons coursing superficial to thicker corticosubcortical axons (Schmahmann & Pandya, 2008), so the overall distribution is finer than for the corticospinal tract. Such a distribution would draw less energy than cerebellar white matter, which contains predominantly thick axons. Indeed, the capillary supply (strongly correlated with glucose utilization) to cerebral white matter is sparser by nearly half (Borowsky & Collins, 1989).

Cortical circuit: diverse types with minimal redundancy and wire

Roughly 80% of cortical neurons are pyramidal and nearly all project an axon into the white matter (Braitenberg & Schüz, 1998; Kevan Martin, unpublished, 2013). Since, as noted, there is a log-normal distribution of diameters, we also expect a log-normal distribution of firing rates (figure 11.26; Perge et al., 2012). This is consistent with reported firing rate distributions in cortex (Koulakov et al., 2009; Rust & DiCarlo, 2012). Such a highly skewed distribution of diameters and firing rates could be generated efficiently with a layout that optimizes the connectivity repertoire. Each pyramidal cell type, integrating a particular aspect of the total input—just what is needed by its downstream targets—follows the same design principles as retina (send only what is needed; send slowly as possible). But cortex, requiring many more arrays, much richer connectivity repertoires, and needing to add and delete connections, requires a different design.

Whereas a retinal bipolar neuron diverges to only about 3 ganglion cells, mostly of one type, a cortical pyramidal neuron diverges locally to more than 5,000 other neurons, mostly pyramidal. This design expands the diversity of pyramidal cell types (i.e., neurons with different connections) by at least an order of magnitude. And, whereas a retinal bipolar axon provides multiple contacts to a given ganglion cell (for high S/N and high redundancy), a pyramidal axon provides few contacts to a given pyramidal neuron (Braitenberg & Schüz, 1998; for low S/N and low redundancy). To efficiently execute this circuit for high diversity/low redundancy requires three dimensions, plus specific features for the axonal and dendritic arbors.

Pyramidal axons, besides descending to white matter, branch locally but loosely and travel straight for relatively long distances (Stepanyants et al., 2004). The local branches form synaptic dilations at roughly 5-μm intervals, so these intracortical axons over their total length of about 20–40 mm produce 4,000–8,000 synapses (Braitenberg & Schüz, 1998). These axons run without specific orientation (unlike cerebellar parallel fibers). Pyramidal cell apical dendrites tend to run vertically, each type selecting one or more particular layers in which to expand its arbor (figures 12.8, 12.9, and 13.14). The apical arbor remains sparse and open (compare to the compact and dense Purkinje arbor). Moreover, pyramidal dendrites are spiny, so besides their roughly 10^4 actual inputs, their extensive spine-reach zone provides space for many potential inputs (Chklovskii, 2004).

Thus, the design of the pyramidal cell circuit: straightish axons that course independently of the trajectories of sparse dendrites (Braitenberg & Schüz, 1998). This ensures that a given axon will rarely contribute more than one synapse to a given neuron. Thus, all the excitatory synapses are insecure (as required for extensive integration—see chapter 12). Synaptic redundancy (even for potential contacts) approaches zero, which is efficient because redundancy would reduce a pyramidal neuron's potential to accept other inputs. In short, the cortical circuit maximizes a pyramidal neuron's actual and potential connectional repertoire per unit of cost of its dendrites, axons, and synapses, which seems fitting for neurons tasked with learning to recognize "suspicious coincidences" (Barlow, 1989). This design holds for pyramidal neurons in different cortical areas across a substantial range of scales (Wen et al., 2009).

Thus, the cortical circuit approaches more closely than any other brain circuit the "all to all" limit of connectivity. In doing so it thoroughly exploits every available strategy of efficient wiring within a circuit: multiterminal axons running straight through sparse, open dendritic trees that branch in three dimensions and are studded with spines—all placed in strata that maximize their opportunities to donate and collect what is needed. Yet, there remains to discuss one more key feature.

Inhibitory contributions to the cortical circuit

Each cortical layer contains 10%–30% inhibitory neurons. These connect in specific patterns to pyramidal neurons, further enhancing diversity of pyramidal cells by sculpting their output messages. Inhibitory neurons comprise at least 14 distinct neurochemical categories: different expression patterns of neuropeptides, potassium channels, calcium-binding proteins, and so on (Gonchar et al., 2007). They also comprise more than half a dozen physiological categories: fast-firing, irregular firing, fast adapting, and so on (Burkhalter, 2008). Also, they express distinctive axon arbors that are generally confined to the layer of origin. A multiparameter space that displayed all known differences, such as used in retina (Sterling, 2004a), would probably reveal more than 100 types (Petilla Interneuron Nomenclature Group, 2008; Gonchar et al., 2007).

Although the connections are incompletely known, the inhibitory circuits clearly depart from the pyramidal cell design. In particular, their design limits their connectional repertoires: (1) many types have denser, more compact dendritic and axonal arbors; (2) the axons are shorter than pyramidal axons and provide only 10% of cortical synapses; (3) arbors tend not to orient vertically or cross many layers; (4) axons, instead of being straight, are tortuous, with short branches that follow the course of their dendritic targets (Stepanyants et al., 2004); (5) axons do not contact spines but rather dendrites and cell bodies; and (6) certain GABA types, the *chandelier cells*, specifically target the initial segment of particular pyramidal axons. Whether this synapse enhances or suppresses spiking is uncertain[19], but either way, it contacts a known choke point.

Moreover, these circuits place many synapses where they will be most potent (proximal dendrite, cell body, axon initial segment). So although 90% of synapses to a pyramidal neuron are excitatory, the less numerous inhibitory synapses being targeted exert substantial influence. The inhibitory neurons serve diverse functions: constrict time windows for coincidence detection; sharpen response tuning; and maintain circuit stability against the avalanche of mutually excitatory connections (Burkhalter, 2008). These design features recall the inner retinal circuit and suggest that these neurons might be usefully viewed as the amacrine cells of the cortical circuit, that is, three-dimensional neurons tasked to carve away information unneeded by a particular pyramidal output in order to maintain economical spike rates.

Certain inhibitory types produce high-rate spike bursts (>50 Hz), and, since spikes are expensive, one wonders whether this violates the energy constraint? In fact, the fast-spiking neuron probably costs less than a pyramidal cell. It is more compact and receives fewer excitatory inputs, which are costly. Also, the axon is roughly 10-fold shorter than a pyramidal axon, so each spike costs less.[20] Only 20% of all cortical neurons are inhibitory, and fast-spiking neurons are a subset, so fast-spiking inhibitory neurons are a small item in the overall budget. Finally, a fast-spiking inhibitory neuron reduces energy consumed by the far larger and more numerous pyramidal neurons and thus pays for itself.

Certain GABA-ergic interneurons in hippocampus project for longer distances and fire tonically at high rates. Correspondingly, these neurons use thick axons with thick myelin, resembling that of the Purkinje cell (Jinno et al., 2007). The point is that circuit design may require departures from the standard design, but the bill for space, materials and energy must inescapably be paid.

Columnar organization

Cortical neurons, besides segregating into layers, also segregate into columns, neurons stacked perpendicular to the cortical surface. Neurons within a column share the same topographic location on the map of whatever sense is being served, plus they also share particular receptive field properties. For example, all neurons in a V1 column respond to the same stimulus orientation or direction of motion, and all neurons in an S1 column respond to the same type of somatic receptor, such as skin pressure or joint angle. Columnar segregation is achieved by orienting the apical dendrites of pyramidal neurons vertically to traverse the layers and by also giving their axon collaterals (branches off the main axon) a net vertical orientation (figures 12.8 and 12.9).

The circuit's function is to advance the assembly of small patterns delivered to the middle layers into larger patterns in the layers above and below and further sparsify the coding (chapter 12). These functions are served efficiently by columns in three respects: (1) topographic maps save wire; (2) placing components (cell bodies) near each other saves wire; and (3) giving axons and dendrites similar orientation allows them to connect with less wire. Thus, columnar organization seems designed to execute the coding requirements discussed in chapter 12 with the least time, space, and energy—by now completely obvious.

Efficient wiring at larger scales

The upper limit of scale for dendritic computing in local circuits is about 1 mm. But brain organization extends to larger scales by nearly two orders of magnitude, and one expects these scales to follow the same conservation laws. Moreover, one expects the execution of these laws to follow the same design principles: maintain the map; place components optimally; reduce wire (Knudsen et al., 1987; Chklovskii & Koulakov, 2004).

It follows that whenever two neurons share significant input, they should stay close together. If they are separated, the shared input will require more wire to reach them. This explains the omnipresence of topographic maps, which extend smoothly for millimeters and centimeters (chapters 4 and 12; Allman, 1999; Allman & Kaas, 1974; Mitchison, 1991; Swindale, 1996). It also explains why, when several "higher" cortical areas collect from a particular "lower" cortical area, they tend to surround its region of highest spatial resolution (fovea, fingertip) because this allows the most numerous connections to be short.

Why subdivide cerebral cortex into many distinct areas, rather than just go with one "superarea"? To prevent the wires for different circuits from sterically interfering with each other—for this would increase wire. Most generally, if two sets of neurons connect primarily *within* their set, it saves wire to separate the sets (Mitchison, 1991; Chklovskii & Koulakov, 2004). Where multiple areas at a given hierarchical level need to exchange connections, the areas with the densest connectivity should locate close together; areas with sparser connectivity should locate farther apart, next to the areas with which they connect most strongly. This explains the arrangement of 11 cortical areas in frontal lobe (Klyachko & Stevens, 2003).

To summarize, wire minimization so far explains: (1) why cortical areas exist; (2) why they contain smooth maps; and (3) why related areas arrange themselves in particular patterns. These features in a large brain occupy a scale of centimeters. But what about features *within* an area on the scale of millimeters? We have explained why V1 contains orientation columns; now, why stripes and patches? The answers are the same.

Wherever one set of neurons needs to integrate its signals with those from a different set, the two sets should be close together. For example, simple cells in V1 integrate inputs from a patch of lateral geniculate neurons that serve one eye. This creates a set of neurons at the cortical input layers which are dominated by that eye, plus a complementary set of neurons dominated by the other eye (chapter 12). The two sets should integrate because corresponding points in each eye "see" the same point in the visual field. Their integration halves the wire needed to relay the integrated signal to higher cortical areas.

The optimal arrangement is to place ocular dominance regions for the two eyes next to each other so that they can exchange connections using the shortest wires. The dominance regions might form either stripes or patches, and which scheme is better depends on the degree of dominance. Where most neurons in a region favor one eye, parallel stripes would use the least wire. But where fewer than 40% of neurons are dominated by one eye, a patchy arrangement would be favored. The actual arrangements follow these predictions (figure 13.18). Therefore, the principle "minimize

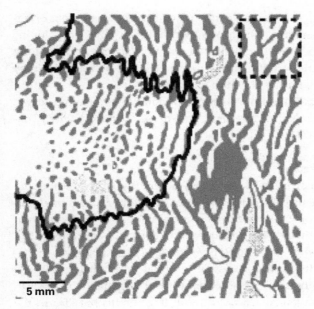

Figure 13.18
Ocular dominance distribution in macaque V1 matches theory of optimal wiring.
Gray and white regions contain neurons dominated by, respectively, the left and
right eye. The large gray spot is the representation of the optic disc. Black line indicates predicted transition between *patchy* and *stripy* patterns, based on the fraction of
left-eye neurons being about 40% (as averaged over an area equivalent to the boxed
region shown in the upper right). Original data were taken from Horton & Hocking
(1996); this image is reprinted from Chklovskii and Koulakov (2004) with permission.

wire" explains why ocular dominance columns exist and why they are
either patchy or striped.

But why do stripes representing central vision run parallel to the horizontal meridian of the visual field whereas stripes representing peripheral
vision run concentrically? This too saves wire. Because the eyes are separated horizontally, they have disparate points of view. Their cortical representations of a given object are also separated horizontally on the cortical
map. Combining these disparate representations allows stereoscopic vision
(chapter 12). This can be done with least wire for central vision when the
stripes run horizontally across V1 and for peripheral vision when they run
concentrically—the pattern that is observed.

V1's orientation columns are grouped to minimize wire. Their layout
problem resembles that for ocular dominance, but instead of two categories

(right eye, left eye), there are many, defined by the preferred stimulus orientations. Preferred orientation changes smoothly across adjacent columns except at a few points where it changes abruptly, a fracture. Also, for long stretches the orientation columns run in parallel, but occasionally they converge to a point to form a pinwheel (Ohki et al., 2006).

All of these features—smoothly changing orientation preferences, parallel stripes, plus occasional fractures and pinwheels—minimize wire (Chklovskii & Koulakov, 2004).

Hemispheric specialization

Information in human brain is processed differently by corresponding regions in the two hemispheres. For example, left auditory areas predominate in processing spoken language whereas right auditory areas predominate for music even though the sound patterns for voice and music span the same spatiotemporal spectrum (Purves et al., 2012). Similarly, orthographic symbols (writing) are processed by visual areas on the left side whereas pictorial symbols are processed on the right (chapter 14). Specialized computations are lateralized not only in large brains but also in the small brains of songbirds (Moorman et al., 2012). Clustering areas for specialized processing minimizes wire and reduces computing delays; the results can then be communicated between hemispheres by extremely fine axons.

The great tract interconnecting the cerebral hemispheres uses extremely fine axons. About 30% are unmyelinated with the most frequent diameter less than 0.2 μm, resembling olfactory axons and cerebellar parallel fibers (figures 4.6 and 13.11). The distribution of diameters, both unmyelinated and myelinated, is log normal—like many central tracts—and constant across a 1,000-fold difference in brain size (rat to horse; Caminiti et al., 2013). Larger animals use some thicker axons—too few to be evident in the distribution—but occupying such disproportionate space as to constrain their numbers (Wang et al., 2008). Thus, the results of local cortical computing are relayed between hemispheres mostly via axons that are irreducibly fine and slow (Ringo et al., 1994). Consistent with this, deficits from severing the corpus callosum are hardly noticeable without special testing (Sperry, 1981).

Pressure to compute locally seems to have driven hemispheric specialization.[21] Auditory patterns for spoken language and music are processed differently by corresponding auditory areas in the two hemispheres; and so are visual patterns for writing and pictures. Separating the circuits for these distinctly specialized functions allots each sufficient computing space, min-imizes mutual interference, and reduces the need for high-rate interhemi-spheric traffic.

Energy costs

Circuits

Having considered the fundamental basis for Cajal's "laws" for conserving time, space, and materials, we turn now to the newer "law" for conserving metabolic energy. This is relevant because every design feature that reduces wire volume correspondingly reduces metabolic costs and thus saves energy for processing information. For scale, consider that human kidney and heart each consume 1.7-fold more energy per gram per day than human brain. So brain is expensive tissue, but not the most expensive (McClave & Snider, 2001). Across different brain regions energy cost varies by up to 3.5-fold, auditory structures being most costly because their neurons need short time constants to process high temporal frequencies.

Most energy in neural circuits goes for pumps to recharge ionic batteries against currents that run them down. In retina, several neuron types depo-larize tonically and therefore must continually extrude sodium and cal-cium. This includes photoreceptors, horizontal cells, and to some degree, ON bipolar cells. All of these elements strongly express pumps, such as the sodium–potassium pump (figure 13.19). Amacrine and OFF bipolar neurons depolarize phasically, allowing them to express the pumps at lower levels. Ganglion cell bodies, and particularly the axons, admit sodium as a large, phasic current during the action potential and therefore strongly express the sodium–potassium pump (figure 13.19).

Recall that when a ganglion cell fires a spike, 90% of the visual informa-tion that entered the neural circuit at the cone terminals has been discarded (figure 11.2). It is sobering to recall that for a ganglion cell merely to double its information rate (bits per second) would require it to increase its spike rate 10-fold, increasing space and energy by 100-fold (figure 11.25). This illustrates vividly why the retina must process images: so that ganglion cells can send only what is needed.

The ATP consumed by ion pumps is produced mostly via respiratory processes in mitochondria that use the enzyme cytochrome oxidase. There-fore, relative energy costs within a neural tissue can be visualized from the distribution of that enzyme, which also correlates with glucose uptake and capillary density (Borowsky and Collins, 1989; Weber et al., 2008). As

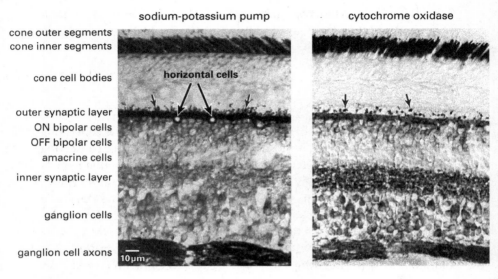

Figure 13.19

**Distribution of the energy demanding sodium-potassium pump is matched distri-
bution of the energy producing enzyme, cytochrome oxidase.** **Left:** Distribution of
sodium-potassium pump follows known requirements for ion pumping: cone inner
segments, rod terminals (short arrows), cone synaptic terminals, horizontal cell bodies, and ON bipolar cells. The latter, being less rectified than OFF bipolar cells, are
more tonically depolarized and so require more pumps. OFF bipolar and amacrine
cell bodies are not tonically active and therefore need fewer pumps. **Right:** Distribution of cytochrome oxidase matches the distribution of pumps. Arrows point to
mitochondria expressing the enzyme in cone terminals of the outer synaptic layer.
Inner synaptic layer expresses cytochrome oxidase to support intense synaptic activity; ganglion cell bodies and axons express the enzyme to support action potentials
Primate retina just beyond the fovea. Reprinted with permission from Wong-Riley
et al. (2010).

expected, cytochrome oxidase matches the pumps (figure 13.19). Its
greater expression in ON bipolar cells (compared to OFF cells) reflects
the ON cells' incomplete rectification, illustrating that rectification conserves energy.

Relative energy costs in cerebellum are evident in the expression of cytochrome oxidase, as shown in figure 13.20. Cytochrome oxidase is strong for
high-rate inputs, the large mossy fiber terminals; and it is strong for high-
rate outputs, the large Purkinje cells. Cytochrome oxidase is weak for low-
rate inputs, the climbing fibers; and it is weak for low-rate neurons, the
granule cells.

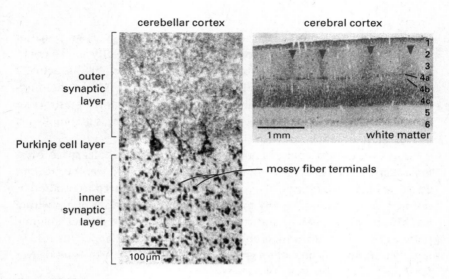

Figure 13.20
**Distribution of cytochrome oxidase identifies energetically expensive compo-
nents. Left:** Mossy fiber terminals fire at high mean rates and form numerous contacts with granule cell dendritic claws (see figure 13.1). Purkinje cells also fire at
high mean rates. Granule cells (pale areas surrounding mossy terminals) fire high-
frequency bursts, but are mostly silent, as are their pale axons in the outer synaptic
layer. **Right:** Striate cortex. Input layers (1–4) express more cytochrome oxidase than
the sparser coding neurons in deep output layers (5, 6). See figures 12.2, 12.6, and
12.8. Both images are from macaque and reprinted with permission from Hevner &
Wong-Riley (1993).

Relative energy costs in striate cortex are also evident in the expression
of cytochrome oxidase (figure 13.20). Expression is strong for high spike
rates at the input layers; and it is weak for low spike rates in the output layers, both superficial and deep (chapter 12).

Cerebellar and cerebral cortex differ remarkably in how they spend their
allowance. The cerebellar circuit accepts a high-rate input with large, energetically expensive synapses and delivers a single type of high-rate output.
In between, it uses tiny low-rate neurons. In contrast, the cerebral circuit
accepts low-rate inputs (compared to cerebellum); then it reduces the rates
still further. Finally the cerebral circuit uses multiple types of output, each
of which retains only what is needed for its particular purpose (figure
12.10). Costs of spiking are similar for cerebellar and cerebral cortex, both
about 20% of their budgets; but overall the cerebral cortex costs 1.7-fold
more (Howarth et al., 2012).

Tracts

Central tracts use only about one third as much energy as gray matter (Sokoloff et al., 1977). Although one might expect that nearly all of a tract's space and energy would go to axons, considerable space (30%) is actually used by glial astrocytes, which also contain more than 70% of the mitochondria (figure 13.21). The volume fraction of mitochondria in astrocytes is nearly twice that in axoplasm (Perge et al., 2009). What is accomplished by this substantial investment?

Each spike dumps a pulse of potassium into the extracellular space, causing a sharp, local rise in potassium concentration. This reduces the gradient across axonal membranes. The extracellular potassium concentration is restored in two phases: rapidly by a glial sodium–potassium pump with a low affinity for potassium and then more slowly by the axonal sodium–potassium pump with a high affinity (Ransom et al., 2000). This role in rapidly restoring the potassium gradient may be what the astrocytes deliver for their considerable share of scarce resources.

End of Story?

Our Introduction asked, how can a human brain with no more volume or power than a laptop be so much smarter than a supercomputer? The long answer, as promised, has been this book. We have deliberately tried not to explain exactly how it all works. Far too much is unknown, and we did not want to present speculative models. Our intent was to organize what is known of the brain's functional architecture in terms of principles that underlie computational efficiency.

The retina's broad computational tasks seem fairly clear, but those of the cerebellum remain to be determined (Alvarez-Icaza & Boahen, 2011). The cerebral cortex might be better understood computationally than cerebellum, but its local circuits are far more obscure. Still, as this chapter has explained, all three structures closely follow the design principle *minimize wire*.

All three structures share a deep biophysical constraint: the substantial and irreducible electrical resistance of neuronal cytoplasm. Cytoplasmic resistance prevents neural wires from being any finer than they are. It also prevents local circuits from being any more voluminous than they are because that would cause excessive conduction delays. This constraint on local circuit volume drives efficient layout—which in turn specifies equal lengths of dendrite and axon, an optimum proportion of wire, and an optimal upper bound on synapse volume.

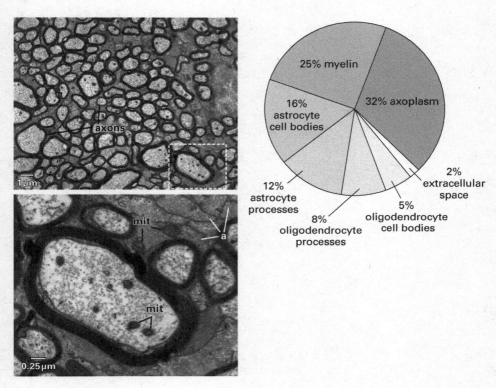

Figure 13.21
How the optic nerve allocates space and energy capacity. Upper left: Astrocytes occupy the space between axons, nearly 30% of the total. **Lower left**: Astrocyte processes (a) contain approximately 70% of the mitochondria (mit). The volume fraction of mitochondria in astrocyte cytoplasm (3.2%) is nearly twice that in myelinated axons. **Right**: Space budget. Reprinted with permission from Perge et al. (2009).

The optimizations governing design of microcircuits have been seen to also govern design of their macro-organization and their interfacing with global connections. Thus, from the subnanometer scale (protein folding) to the meter scale (long tracts)—a 10^{10} range—design follows the same principles. The territory staked out in the Introduction has now been largely covered, sketchily perhaps, but to a degree consistent with another introductory promise, brevity. Yet, one key principle remains: brain efficiency requires the capacity to learn and store new information. This is the topic of the next and final chapter.

That man should be "happy" is not included in the plan of "Creation." What we call happiness comes from the sudden satisfaction of needs, and it is from nature only possible as an episodic phenomenon.
—Sigmund Freud (1930) (edited for brevity)

Freud, writing this comment two decades after publication of Ramón y Cajal's *Histology of the Nervous System*, might have wondered: must not man's steady-state unhappiness and episodic happiness be generated by the brain? Might it not therefore belong to neural design?

This chapter will explain that for a brain to be efficient it must learn. Learning is a principle of brain design. Moreover, for learning itself to be efficient, it must follow an optimal schedule of effort versus reward. The schedule, as it turns out, requires that rewards be episodic and unpredictable (Glimcher, 2011). Thus, our average unhappiness interrupted randomly by moments of happiness is no mere affliction; rather it is part of the design. Of course, what we self-referentially conceive as *humans'* unhappy state is equally the unhappy state of the mollusk, the fly, and the honeybee (Menzel, 2012; Mayford et al., 2012; Caroni et al., 2012). All animals suffer unhappiness because learning is design, and this particular design for learning is optimal.

"Learn" as a principle of design

"Learn" belongs to a broad principle that is actually a continuum: *adapt, match, learn, and forget* (chapter 3). This principle is what distinguishes biology from traditional engineering. An automobile is at its best when it rolls off the assembly line. Use simply wears it down. Moreover, a car built to meet certain environmental conditions is hard to modify when conditions change. But suppose a city car could respond to a rough road by thickening its tires, stiffening its springs, and respecifying its gear ratios—that would be a hot item.

Yet for biological design, adaptive responses are the essence. Each fresh experience helps to prepare for a future need. Where skin sustains rough wear, it thickens to callus; where muscle and bone sustain mechanical tension, they strengthen. Likewise in brain, adjustments begin the instant that sufficient evidence accumulates at the environmental interface, as demonstrated by the rapid adaptation of sensory receptors and circuits to changes in stimulus statistics (chapters 8, 9, and 11). Indeed it seems that wherever one looks, current experience is continually being used to update circuits in order to improve future performance.

These adjustments occur in all brains, from small (*C. elegans*, chapter 2) to large (this chapter). They occur in all systems, from sensory to motor, and at all levels, from spinal cord and cerebellum (figures 7.16 and 7.17) to cerebral cortex (this chapter). Moreover, adjustments are made at all levels of circuit organization, from adding, removing, and modifying the structure of a protein molecule at a specific site in a neuron to adding and removing synapses and even entire neurons.

In the adult human brain—whose circuits until recently were considered as physically immutable as a Toyota—neurons are both recruited from existing circuits and manufactured de novo to meet new demands.

Consider, for example, the hippocampus, which is tasked with storing maps to guide navigation. London taxi drivers, who visually navigate a complex cityscape, expand their hippocampal gray matter (Maguire et al., 2000). This occurs gradually with practice over several years. Piano tuners, who acoustically navigate a complex soundscape, also gradually expand their hippocampal gray matter (Teki et al., 2012), along with auditory areas in the temporal and frontal lobes. White matter also expands, suggesting that the new circuits formed to process and store information send it out over new or updated wires.

To adapt its capacity to changing demands, the hippocampus in adults continually generates new neurons. Many die back, but some newborn neurons are recruited during sustained practice to build new circuits (Chancey et al., 2013)—adding new synapses and new wire. These circuits apparently serve newly formed memories since deleting the new neurons shortly after learning also deletes the new memories (Arruda-Carvalho et al., 2011). The hippocampus also expands because memories are stored by strengthening synapses, which requires them to enlarge. Nor is the hippocampus unique: cortical motor areas expand with motor practice (Xu et al., 2009; Yang et al., 2009; Landi et al., 2011); a visual area (*visual word form area*) expands with literacy; and areas critical to social functioning expand with size of social network (Sallet et al., 2011; Bickart et al., 2012).

Thus, the brain follows a core principle that serves all biological systems: use current conditions to predict future needs and revise circuits accordingly. During every sort of "neural exercise," the brain readjusts physically across timescales from milliseconds to decades, and across spatial scales from nanometers to centimeters. Just as physical exercise sculpts particular muscles and bones, so neural exercise sculpts particular brain circuits.

This biological perspective defines learning as a structured channeling of information from present to future (Varshney et al., 2006). Current motor behavior can improve future motor skills; current perception and cognition can improve future perceptual and cognitive skills; and current behavioral choices can improve future choices. But learning, besides being a design principle, is also a neural function. Therefore, it must be constrained (like all neural functions) by space, time, and energy. Learning must obey all the principles that have been proposed for other aspects of neural design.

Principles for the design of learning

An important constraint on learning is space. The adult brain is jammed with circuits and tracts, and it cannot expand. So learning must conserve space with a design that: (1) is spatially specific; (2) stores only what is needed; (3) stores only for as long as needed (i.e., selectively forgets); (4) stores and retrieves information at the site where it is processed; (5) optimizes the units of storage (size and number); and (6) optimizes a "teaching signal" for the real world, which is an environment rich in small surprises.

Specificity

In humans spatial relationships are processed in the right hemisphere. Correspondingly, in taxi drivers it is the right hippocampus that expands—and not the whole structure—only the posterior region. Tonal relationships are also processed by the right hemisphere. Correspondingly in piano tuners, it is also the right hippocampus that enlarges—but only the anterior region. Thus, the principle *complicate* extends to learning: each circuit reconfigures for its specific task.

A design that expands a neural circuit following practice ought also to retract the circuit when it is persistently unused. Indeed, taxi drivers studied from the start of training across years of practice initially expanded the hippocampus and, upon retirement, restored it to normal (Woollett & Maguire, 2011; Woollett et al., 2009; Woollett & Maguire, 2012).

In practicing one skill over years, we neglect innumerable others. Thus, working taxi drivers perform below average on tests of short-term visual recall. But again, retirement restores performance to normal. Skills decay rapidly when supplanted by other activities, which is why professional athletes, musicians, and surgeons need to constantly practice.

Store only what is needed
An animal whose body grows continually throughout life also continues to grow its brain. For example, as a fish enlarges throughout adulthood, so does its eye. The retina adds new neurons and new circuits in concentric rings; it adds new ganglion cells that send axons to connect to new circuits in the midbrain. Retinal neurons that at first serve the periphery gradually serve more centrally, so as the map expands, it must continually revise itself by breaking and reforming all the connections (Easter & Stuermer, 1984; Reh & Constantine-Paton, 1984). This degree of plasticity allows the animal to regenerate a complete retina from stem cells (pigment epithelium) and grow new optic axons that find their proper central targets (Sperry, 1963; Stone, 1960).

We fantasize that humans might gain this capacity to regenerate an eye and all its circuits and tracts. However, a rigid skull is not the place to unleash brain growth. Therefore mammalian design for learning must restrain growth and refrain from adding new circuits unless it is also prepared to make space by pruning others.[1]

Store information only for as long as needed
The adult brain stores countless items and procedures on innumerable timescales. A glance at the printed number of our hotel room suffices to store it for the period of our stay. But by the time we reach the airport, it is gone. On the other hand, we store the physiognomies of friends, family, children, grandchildren—and scores of casual acquaintances—along with their physical, intellectual, and emotional status. This prodigious number of rich patterns we retain for months and years without practice. Otherwise, we could not exclaim upon reencounter, "How tall you've grown! How deep your voice has become . . ." nor suppress exclamations such as "Yikes! How bald and gray . . ."

At each reencounter the brain promptly revises the stored pattern. Thus, after a year's absence, we recognize a toddler as our grandchild but thereafter cannot recall the infant. We may recognize a friend from our youth but recall his original appearance only from a photo. Like Orpheus, we may store the image of a lover in full bloom, but the instant we gaze upon her once more, that image is irretrievably lost. This "auto-delete" matches the probability that the updated image will best serve future interactions. Thus, early drafts are well discarded.

Why then do we retain apparently trivial facts from our youth, such as the lineup of our favorite sports team, a rock lyric, and so on? Would not space for new information be cleared by discarding great chunks of now useless data? Perhaps, if they were separable, like cream from milk, but such facts that now seem trivial were not learned in an instant, but rather through repeated exposure and practice. They were woven into a rich context and stamped with an emotion. You might have heard the baseball lineup repeated 100 times across a season on the radio as you worked at a project with your father. The broadcasts may be coupled with the memory of his aroma, and a memory of his pleasure at the task and at your participation. Such memories, once contextualized, emotionally stamped, and repeated, are nearly ineradicable.

Why? Adult character emerges through integrating diverse experience, which is needed for prompt and constructive responses to diverse and subtle situations. For example, it tells us in each new context when to speak and when to hold our tongue. It informs our posture, facial expression, gestures, choice of words, and tone of voice—all to better serve the task at hand. Without such integration, it is hard to behave effectively. We are too soon to anger or too late; we are too generous or too stingy, and so on.

Over decades we continually resift our web of experience seeking to recover new insights and deeper understandings. To facilitate this process, we invest diversely: meditation, prayer, music, literature, theater, drugs, dreams, and psychotherapy. The issues to be integrated shift across the life cycle, and the process continues into old age.[2] In short, the brain devotes huge resources to long-term storage of facts and experience because they are raw materials for the continual resculpting of neural circuits that improves our behavior.

The brain *does* delete genuine trivia—facts and experience lacking context and stamped weakly by emotion—like the number of your hotel room. But the team lineup learned with your father and the rock lyric learned with your first kiss are not trivia; they are essential threads, the warp and weft of your neural web. Deprived of this fabric, we become profoundly disoriented—as reported, for example, by individuals with severe retrograde memory loss following treatment with electroconvulsive shock (Donohue, 2000).[3]

Retaining this fabric, we gradually deepen our understanding of people, events, and situations. We gradually improve our ability to perceive, judge, and act in keeping with deepened understanding. We gradually improve control of emotions to better direct each action engendered by reason and knowledge. And we gradually improve intellectual control over actions engendered by emotion. In mature form, these capacities, which define "wisdom," occupy considerable space in the human brain—presumably because they increase biological fitness. Simpler versions are observed in baboon societies—where they have been documented to increase fitness (Cheney & Seyfarth, 2007; Seyfarth et al., 2012). This suggests a rule in design of learning across phyla, including mollusk and honeybee: delete true trivia, but retain contextualized material for long-term integration. In short, *store only what is needed.*

Store and retrieve information without adding wire

The obvious site to compactly store information is at the synapse.[4] Storage occurs by changing its transfer "weight," that is, its ability to excite or inhibit a postsynaptic neuron. Since the synapse is the key site for processing information, storing it there avoids additional wire for relay. Moreover, information stored directly at a synapse can be retrieved directly—also avoiding additional wire. In short, as we peruse a blueprint of brain design, we should not seek a special organ for "information storage"—it is stored, as it should be, in every circuit.

Synaptic sensitivity needs to shift across all timescales. Rapid, brief shifts can improve the efficiency with which information is coded and processed. When the average signal weakens, vesicle release sensitivity tunes up; when the average signal strengthens, release sensitivity turns down (Abbott & Regehr, 2004; Goldman et al., 2002; Yang & Xu-Friedman, 2012; Silver, 2010). These short-term adjustments control gain to maintain coding in a sensitive region of a neuron's input/output curve; moreover, they increase efficiency by reducing redundancy and by avoiding high mean rates (chapters 3, 9, 11, and 12).

This short-term synaptic plasticity fits the broad definition of learning: use present information to adjust a circuit, in order to improve future performance. On the relatively fast timescales of environmental fluctuation, these changes need to be rapid, brief, and cheap. Consequently, they are accomplished by chemistry—by changes in presynaptic calcium concentration that alter release probability and the size of the releasable vesicle pool, and by modifying the sensitivities of signaling proteins by, for example, phosphorylating postsynaptic receptors. On the other hand, being short-term, they require no restructuring and thus no additional space for storage. However, when signal statistics change for longer times, circuits begin to reconfigure.

Sculpt circuits to match stable changes in bottom-up statistics

When input statistics to a particular circuit change stably for a few days in an adult animal, longer term changes are initiated. Synapses that were dormant are strengthened, and newly dormant synapses are depressed. For example, restrict visual input to one eye, and within days the geniculocortical synapses for that eye are expanding in V1's ocular dominance stripes. Simultaneously synapses for the deprived eye are retracting.[5] Or, shift the wavelength of chromatic input to one eye with a tinted contact lens for a few days, and color perception shifts for several weeks (Neitz et al., 2002). Sew two fingers together so that their surfaces function as one, and the primary somatosensory cortex (S1) soon resculpts the sensory map of the hand from five fingers to four (Clark et al., 1988).

Such sculpting of local circuits by a stable change of input statistics occurs most rapidly and completely during a *critical period* of juvenile development, corresponding to the time of the circuit's initial assembly. A kitten, during the first months after eye opening, develops ocular dominance stripes in V1 where neurons in a given stripe respond mainly to one eye (chapters 12 and 13). During this critical period, masking one eye for a few hours reduces responses of cortical neurons to that eye. Masking for a few weeks causes the LGN axon arbors for that eye to shrivel, thereby freeing synaptic territory for expansion of axon arbors from the normal (Hubel et al., 1977; Tomita et al., 2012). This capacity to resculpt is retained by the adult brain. For example, three days of masking in an adult mouse suffices to induce substantial resculpting of its thalamocortical synapses and a shift in ocular dominance (Coleman et al., 2010).

The economics seem clear. Short-term changes of input statistics (a passing cloud, a shift from field to forest) cause circuits in retina, LGN, and V1 to *adapt* via short-term synaptic facilitation and depression—which is cheap. Long-term changes of input (masking an eye) cause circuits in V1 to *resculpt*. This requires dismantling and disposal of used materials and synthesis of fresh ones—which is more costly and harder to reverse. Consequently, adaptation spans milliseconds to minutes while resculpting begins within days and continues for months to years. But what about intermediate-term changes that span hours to weeks? Should a five-day hike in a forest rewire our V1 orientation columns to favor vertical? Apparently so, since five days of intermittent exposure to an oriented grating induces orientation-specific, *long-term potentiation* (LTP).[6]

Resculpting a particular circuit must balance probable gains against costs. Decisions on when to initiate, how far to go, and how long are probably settled case by case for each circuit via natural selection. Some areas are readily sculpted during a critical period. But other areas maintain perma-nently high levels of plasticity and the capacity to grow out new axons. Such permanent plasticity may be expensive, but the payoff can be huge. This suggests another principle in the design of learning: certain brain areas should specialize in maintaining high levels of plasticity. This allows them to learn subsets of arbitrary patterns, as exemplified by a cortical area in humans that serves literacy.

Special circuits learn certain sets of images

The ventral occipitotemporal cortex of the left hemisphere contains an area recently dubbed the *visual word form area* (VWFA).[7] During early childhood, this area serves to recognize objects and symbols—a house and a drawing of a house. In preliterate cultures that is all it does. However, around age 7 in cultures that teach reading, this area starts responding more strongly to written words than to drawings of objects. Across cultures, it is always *this* area and no other that learns writing, whether it is alphabetic or logo-graphic.[8] So something about the circuitry of this cortical area suits it above all others to this task of storing arbitrary squiggles.

Initially, a written word is "sounded out"—that is, the visual symbol links temporally to the voiced sound. That sound, of course, already has linguistic meaning. But this slow process would never get us through *War and Peace*. Facile reading requires storing thousands of orthographic sym-bols in some format that allows rapid access.[9] Over months and years of practice with thousands of words, this area prunes away its responses to checkerboards and faces, and it gradually captures new territory at its boundary with the fusiform face area. Thus, there are territorial trade-offs, but the benefit is huge: the ability to read vastly extends an individual's information storage capacity by moving it beyond the skull.

The growth in selectivity for written words versus drawn objects does not occur spontaneously but requires tutelage. Therefore, it must be initi-ated top-down from higher cortical areas. Moreover, responses to horizon-tal strings of writing and even horizontal checkerboards are strengthened in early visual areas, such as V1 (figure 14.1). Responses to horizontal check-erboards are stronger for the left V1, which is closest to the VWFA. This suggests a function for the reciprocal pathways from higher visual areas down to lower ones: they tune up the lower areas for line segments that have attained symbolic importance.

Chapter 12 noted that cortex invests heavily in specialized areas in order to rapidly identify something critical—a face, a voice, an object. Now we see that the same rule applies to storage of myriad arbitrary symbols and

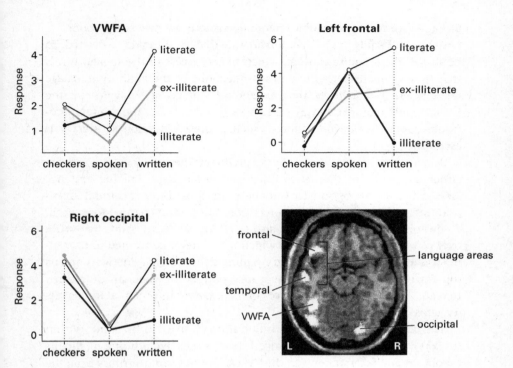

Figure 14.1
VWFA responds to orthographic squiggles when we learn to read. Plots indicate where activation was modulated by literacy during exposure to written sentences relative to rest. Plots report activation to visual checkers, spoken and written language relative to rest. Horizontal slice shows activation in literates of VWFA and left temporofrontal language areas. Right occipital areas are also modulated by literacy—a top-down effect. "ex-illiterate" = illiterate adult who learned to read. Modified and reprinted with permission from Dehaene et al. (2010).

their rapid recall. In reading, the fixations between saccades last about 200 ms. This time interval suffices to transfer the low-level visual image of a word string up to the VWFA and evoke recognition. Also, within this interval, the recalled message relays forward to language areas to recall the meaning of this string and to integrate it with preceding word strings to comprehend the larger message. In short, the text fragment captured during each fixation is processed sufficiently to drive the next saccade: reading does not allow unprocessed fragments to stack up in a buffer (Rayner, 1998). With all this to be accomplished, dwell time in the VWFA can use only a fraction of the 200-ms interval—so visual recall of words must be fast.

Storage area for orthographic images needs to grow new connections
Orthographic squiggles in isolation are true trivia and should be deleted. To be stored, they require context, which means access to the spoken words that they symbolize. This requires connections to the auditory language network, so one expects that expanding storage of squiggles by the VWFA would require two-way traffic between the VWFA and the temporo-frontal language network. Indeed, such pathways develop as literacy is established.

A written sentence viewed by a nonliterate person does not activate either the VWFA or the network for spoken language. But the sentence viewed by a literate person activates both. In short, a recent cultural invention (literacy) captures a long evolved communication channel for speech. To do so, an established cortical area (VWFA) apparently sends new wires over centimeters to an area with which it had never connected until reading was invented. VWFA's forward coupling to the language network and its top-down resculpting of early visual areas both occur when an adult learns to read. Thus, this modest patch of cortex preserves its potential for plasticity across the whole life cycle.

The only downside to growth is that it needs space in the skull. Within the VWFA, as word storage advances, there is selective pruning of circuits for objects and faces; moreover, the VWFA captures some territory from the fusiform face area at their shared boundary, and as that occurs, there are decrements in memory for faces (Dehaene et al., 2010). So, again, plasticity that improves one function causes another to decay. Again, memory capacity is constrained by space. Thus, the design of learning must include rules for deciding what to store.

Empty the trash
In learning words, it would be inefficient to store every voiced sound, for that would include throat clearings, stutters, and so on. The memory bank would soon fill with trash; therefore, to conserve space, true words need sorting from noise. In this regard, the brain should accumulate evidence that a particular utterance belongs in the lexicon, and this implies repetition. But how many repetitions should be required? Over what time span? And where should they be stored? The cortex has clear rules, as we now recount. The rules certainly seem prudent, but whether they are optimal is unknown.

For example, when a word already stored in the lexicon, such as "pipe," is voiced, it evokes a specific voltage signature about 100 ms after the last phoneme ("p"). This voltage pattern is stable as "pipe" is repeated amid

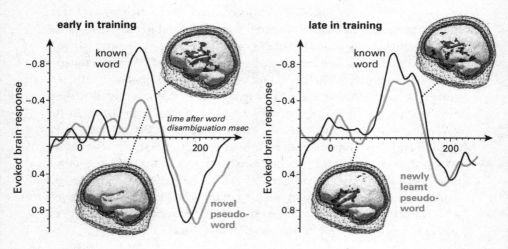

Figure 14.2

Adding a spoken word to the lexicon. Black trace shows that the evoked response to a known word ("pipe") is stable across repetitions and stably activates the perisylvian language areas. Gray trace shows that the evoked response to a new pseudo-word ("pite") initially differs from the response to the known word and evokes no response in the language areas. However, following repetition over minutes, both its evoked response and its localization resemble those of the known word. Electrical responses and maps are modified and reprinted with permission from Shtyrov et al. (2010).

other sounds 160 times over 14 minutes, and it localizes over a large part of the perisylvian language area (figure 14.2; Shtyrov et al., 2010).

When a word not stored in the lexicon, such as "pite," is voiced, its evoked voltage initially differs from that evoked by the known word; moreover it does not localize to the language areas (figure 14.2). However, as "pite" is repeated, its signature eventually resembles that for the known word, and so does its localization. After 160 repetitions over 14 minutes, a pseudo-word has been admitted to the lexicon and stored within the language complex, especially in the ventral "what" region of the auditory stream. Pite is a pseudo-word because it lacks semantic meaning, yet its voltage looks like a word and it is stored along with true words. Apparently, words are initially stored without their meanings—probably because meaning is not what first distinguishes them from noise; rather it is repetition.

These language areas couple to the VWFA and apparently activate it by top-down signals. That is, as a voiced word activates the language areas, it instructs the VWFA to store the corresponding orthographic squiggle. And with top-down instruction VWFA grows connections to the language areas that are needed to attach meaning to the squiggles and to read them aloud.

In summary, designs for learning store information at the sites where it is learned and from whence it can be directly recalled. We have noted a range of examples: arbitrary sound images, such as pseudo-words, are learned over minutes at auditory language sites; arbitrary visual images, such as orthographic squiggles, are learned on a similar timescale in the VWFA. Arbitrary motor tasks during long practice (months) reorganize cerebellum and also primary motor cortex (Karni et al., 1995; Matsuzaka et al., 2007; Kleim et al., 2004). To master two languages instead of one requires more space in the auditory cortex (Ressel et al., 2012). What these designs share is repetition. Repetition is one important way that the brain discovers what is worth storing. But why does storing information require space, and how is this need reflected in the neural designs for learning?

Cellular design for efficient information storage

As noted, information is stored and retrieved at synapses. Storage occurs by increasing synaptic *weight*, that is, its contribution to firing the postsynaptic neuron. Space for synapses is already constrained by competing needs for local wires and long tracts, and cortical synapses are already as small as they can be (chapter 13), so memory capacity cannot increase by shrinking them further. In fact, to increase its synaptic weight, a synapse must *enlarge*. The presynaptic terminal enlarges to accommodate more vesicles and more active zones for release. The postsynaptic structure, typically a dendritic spine, also enlarges to accommodate more transmitter receptors, more synaptic scaffold proteins, and more regulatory proteins. When new active zones are added presynaptically, they are matched postsynaptically by new spines.

When more information is stored at a postsynaptic site, the neuron has more to convey to other sites, and this requires expanding its synaptic output—first by enlarging the active zones, and then by adding more of them. This requires more resources to be ferried from the cell body down the axon, which requires a thicker axon (figures 4.6, 11.9, 11.12, and 11.25). Thus, the costs of learning are not confined to the initial synapse but propagate through the system.[10]

One might wonder if memory storage could be improved by using larger synapses. A larger synapse can store and recall more information because it has a higher S/N, but this requires more space and materials. In fact, the storage capacity per unit volume is greatest when information is stored by many small, noisy synapses—as used in cerebral and cerebellar cortex. The reasons are familiar: information capacity increases in proportion to the number synapses and as the log of synaptic S/N (Varshney et al., 2006). Energy costs rise linearly with the number of synapses but increase quadratically with S/N. Thus it is more efficient to store information at many small synapses of low S/N. Just as information is transmitted more efficiently at low rates, it is stored more efficiently in synapses of low capacity.

Several design features increase the storage efficiency of a low-capacity synapse (Varshney et al., 2006; Sterratt et al., 2011). First, in a population of synapses, positive and negative weights are balanced, thus eliminating redundancy and allowing storage of only what is needed. Second, storage proceeds with sparse firing, so a given synapse less frequently changes weight in response to two or more different input patterns. Thus sparsity reduces interference between stored patterns. Third, synapses are prevented from continually increasing their weights—for that would require continually increasing S/N and thus synapse size. So setting an upper bound to synaptic weight holds storage capacity to the efficient regime of small, low-S/N synapses.

Synaptic weight is limited by three mechanisms: (1) synapses are physically constrained from greatly expanding—the space is already occupied; (2) *Hebbian synapses* (see below) increase weight when activated right before an action potential in their postsynaptic neuron, but they also decrease weight when activated right after an action potential; consequently, when synaptic potentials and action potentials occur at random; that is, when there are no correlations to learn, there is no net change in weight; and (3) synaptic plasticity is regulated homeostatically (Turrigiano, 2011).

Homeostatic regulation occurs by mechanisms that sense a neuron's average firing rate, for example, by sensing calcium accumulation from voltage-gated channels at synapses and elsewhere. This signal couples to protein circuits that control the numbers of ion channels in the membrane. When mean spike rate increases, channels are removed to reduce synaptic currents, and when spike rate falls, channels are added. The mechanism is homeostatic because it stabilizes the sums of synaptic weights (positive and negative) to maintain an efficient firing rate.

Homeostatic mechanisms adjust the rates of synaptic receptor turnover such that the weight of each synapse is reduced/increased by a constant proportion. This normalization preserves the relative differences in weight established by stored patterns so that memories are preserved, costs are bounded, and efficiency is maintained. Thus, local homeostatic synaptic plasticity allows learning's broader circuits to keep operating in an efficient regime. Unsurprisingly, synaptic homeostasis is slow; otherwise it would erase memories as they form. Having indicated that memory systems at the synaptic level follow familiar design principles, we now consider the design of learning at the molecular level.

Molecular design of short-term memory

Cortical mechanisms, like memory mechanisms in general (see below), adhere to a familiar principle: start with chemistry and protein folding because they are cheap, compact, and fast. So a memory begins with a quantum of glutamate binding to its receptors on the spine of a cortical dendrite. The receptors change conformation to admit an inward flood of depolarizing cations.

The key trigger for memory formation is a transient rise in calcium concentration, and this occurs through various means. If the spine expresses the GluR2 subunit, calcium will enter directly through the channel; or if the spine expresses voltage-gated calcium channels, calcium will enter during depolarization; or if the spine expresses the NMDA-type glutamate receptor (NMDAR), calcium will enter through that channel if glutamate binding coincides with a strong depolarization, such as occurs when the neuron fires (chapter 7).

The spine head expresses about 10 glutamate receptors and about 10 voltage-gated calcium channels (Nimchinsky et al., 2004; Sabatini & Svoboda, 2000). These set the magnitude of the calcium surge and its variability. For a calcium surge to initiate synaptic strengthening, it should be brief and local: a concentration sufficient to bind a protein at the synapse. This occurs with a brief channel opening because the cytoplasmic calcium concentration rises steeply at the channel and decays steeply over nanometers (chapters 7, 10, and 11). The calcium flux should be small enough for a pump or exchanger to promptly restore the initial concentration. A single channel would be too noisy and inject too little calcium; 10 channels would improve S/N sufficiently to match S/N of presynaptic release, but more might flood the synapse. So the numbers of channels could well be optimized to store the present signal for the future (Varshney et al., 2006).

Each mechanism for elevating calcium offers something different, as will be explained. But simply to initiate information storage, the means doesn't matter, only the result: when calcium ions bind to a specific kinase (CaM kinase II) attached to the inner face of the postsynaptic membrane, the enzyme changes conformation and begins to phosphorylate the glutamate receptors. Their conformation is thereby altered (chapter 5) so that their next binding of glutamate will cause a larger inward current, and this potentiation increases the synaptic weight. If glutamate quanta recur in the same pattern, all the synapses that were potentiated together will respond strongly together and thus recall the original pattern.[11]

This design, simply tweaking the conformation of about 10 receptor proteins to increase the synaptic weight, is irreducibly fast and cheap (chapter 5). But although the change in weight far outlasts the initiating calcium transient, it nonetheless decays. Calcium is pumped out over milliseconds, the kinase inactivates over seconds, and the phosphorylated glutamate receptors are retrieved from the membrane over and hour or so by endocytosis (chapter 7). As the phosphorylated receptors are replaced with non-phosphorylated receptors from the spine's cytoplasmic pool, the potentiated synaptic current returns to normal. This decay of initial storage is a critical design feature because it prevents long-term storage of events that turn out to be fleetingly correlated and/or of little consequence. This storage mode, *early long-term potentiation*, resembles a "speed date"—little invested, soon forgotten—unless something special warrants a longer look.

A longer look is provided when the behavior that caused the initial potentiation is repeated. One important signal that encourages a behavior to repeat is dopamine—which is released within specific brain regions. When dopamine is released near a potentiating spine, it binds to D1 receptors that trigger a G-protein cascade that activates multiple kinases. Each kinase plays a specific role in extending LTP and converting it to a stable store. Where the dopamine originates and what structures its release will be discussed later.

Molecular design of long-term memory

In one well-studied example, the next step recruits the kinase PKMζ by initiating its translation from messenger RNA parked nearby in the dendrite (Sacktor, 2011; Lisman et al., 2011). PKMζ lacks a regulatory subunit, so it is constitutively active and begins to phosphorylate various proteins that regulate endocytosis. This shifts the balance between receptor insertion and retrieval, stably increasing the number of receptors in the postsynaptic membrane. Signals also travel backward across the synapse, announcing a stable postsynaptic expansion and triggering a presynaptic expansion to match (Tokuoka & Goda, 2008; Vitureira et al., 2012). In these ways, pre- and postsynaptic mechanisms extend synaptic potentiation for several hours while additional proteins are being synthesized. This is LTP.[12]

Meanwhile, dopamine has also recruited another kinase within the spine, PKA, which triggers signaling pathways to the dendrite and cell nucleus that now produce a host of fresh plasticity-related proteins. These are recruited by the spines that were "tagged" during the initial potentiation (Kandel, 2012; Mayford et al., 2012). The new proteins stabilize the potentiated state well beyond 4 hours, doubling the spine's inward current to a single transmitter quantum and also doubling its volume (Govindarajan et al., 2011). Plasticity-related proteins then arrive from the cell body to stabilize the spine's potentiation indefinitely and also induce new spines at the tagged site. Nearby *non*potentiated spines are induced to regress. Thus, spines clustered along a stretch of dendrite come to share a stable record of the past (figure 14.3).

For dopamine to convert early LTP to late LTP, its timing at the synapse need not be critical. (But see Arbuthnott & Wickens, 2007.) Dopamine can stabilize events that occurred many minutes before or minutes after its release. So to initiate long-term storage, dopamine need merely create a broad "penumbra" that can capture and prolong any early LTP before it finally decays[13] (Lisman et al., 2011). This raises an important design issue: how many synapses are needed to stabilize a given memory and where should they be located?

The number of synapses involved in laying down a specific memory (i.e., a particular association of events) has not been determined—we know too little about how patterns that are memorized are read into specific sets of synapses and then read out.[14] However, there is the intriguing possibility that synapses clustering on a single dendrite cooperate to lay down memories. Where late LTP is mild, single spines may not reach threshold for stabilization, but over a roughly 70-μm stretch of dendrite, they share plasticity-related proteins, and so all stabilize together (figure 14.4). Thus, a dendrite could be a compact memory module.

This design allows for efficient recall. Whereas EPSCs in spines located on different dendritic branches sum linearly, EPSCs in multiple spines along the same branch sum *supra*linearly to cause a sharply timed, local dendritic spike (Gasparini & Magee, 2006; Makaral et al., 2009). This design, which uses just the right set of potassium channels, allows recall of 10–100 independent patterns with minimal mutual interference. Thus, the principles *complicate circuits* and *minimize wire* could extend down to the integrative level of a dendritic segment.

In short, memory could be stored when about 100 irreducibly small synapses on spines clustered on a dendritic segment stably double their excitatory currents. This stably doubles their use of constrained resources—energy and space—so the brain invests cautiously, waiting for evidence that a particular memory will be important. The evidence accumulates from a cycle:

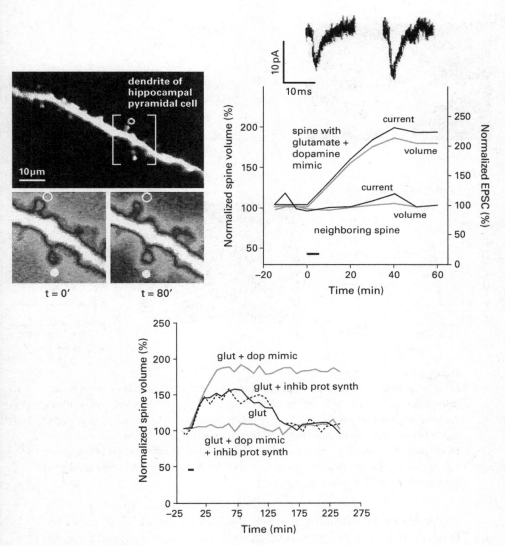

Figure 14.3
Early long-term potentiation (LTP) and late LTP both occur at a single spine. Left:
Closely spaced glutamate pulses are delivered to one spine on the dendrite of a hippocampal pyramidal cell (•) but not its neighbor (o). After 80 minutes the stimulated
spine has enlarged. **Right:** Brief glutamate stimulus (bar) plus a dopamine mimic
to activate a spine's PKA doubles spine volume and postsynaptic current over the
course of an hour. **Lower:** Glutamate alone potentiates the spine (early LTP). Early
LTP does not require protein synthesis, but it decays over several hours. Glutamate
+ dopamine mimic prolongs potentiation beyond four hours (late LTP). Late LTP
requires synthesis of new proteins. Modified and reprinted with permission from
Govindarajan et al. (2011).

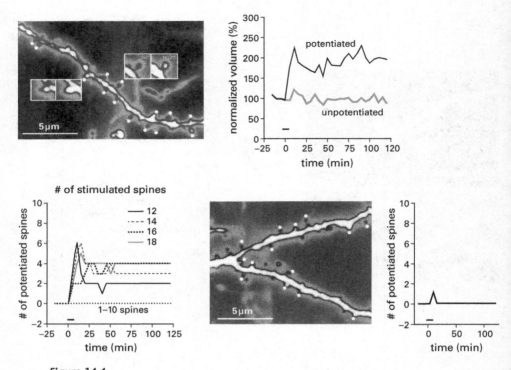

Figure 14.4

Long-term potentiation requires stimulation of more than 10 spines on one dendritic branch. **Upper left:** Fourteen spines stimulated (bright dots) on one dendritic branch. Insets are examples: right spine was potentiated; left spine was not. **Upper right:** Pooled data: spine volumes double. **Lower left:** Potentiation increases when more than 10 spines on same branch are stimulated. **Lower right:** When 14–16 spines are stimulated across branches, they fail to potentiate. Modified and reprinted with permission from Govindarajan et al. (2011).

phasic release of dopamine activates the "reward circuit" that causes the behavior to repeat. Repetition strengthens early LTP at spines and eventually converts them to late LTP. Spine clusters that stabilize simultaneously as separate memory quanta can be widely distributed across brain areas; yet when the stimulus pattern that initially caused storage recurs, they are recruited together. Consequently, what we experience as an integrated memory may be assembled like a jigsaw puzzle.

This account has focused on the roles of dopamine in cortical synaptic plasticity. Other molecules can also initiate long-term storage (Mayford et al., 2012; Gao et al., 2012), but as we now explain, the role of dopamine can

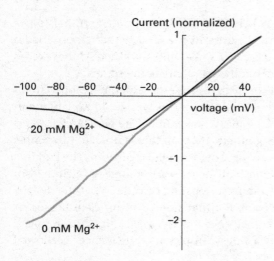

Figure 14.5
NMDA receptor provides a mechanism for Hebbian learning. Current elicited by glutamate binding to the NMDA receptor channel is blocked in the presence of magnesium ion. However, if a glutamate quantum arrives while the neuron is depolarized below about −50 mV by an action potential, the channel is unblocked and a strong current flows, delivering a pulse of calcium. Consequently, the NMDA receptor channel detects the coincident arrival of a glutamate quantum from one source and an action potential from another source. Reprinted with permission from Miyashita et al. (2012).

be followed from cellular mechanisms of plasticity to the broadest principles of learning.

How the design of memory couples to the design of learning

So far we have considered the design of storage. But the point of learning is to store information linked to a particular signal so that it can be recalled. For example, the sound of a word from the stored lexicon recalls its written form. How is such a linkage established at the synapse?

One key is a glutamate receptor of astonishingly clever design. When an NMDA receptor binds a quantum of glutamate, its cation channel admits only a modest current because the channel's mouth is partially blocked by a magnesium ion. However, if a different input excites the neuron to fire a spike, the strong depolarization pops the magnesium from the channel, allowing an inward surge of current carrying calcium (figure 14.5). This pulse of calcium initiates early LTP. Thus, when the printed symbol "pipe" excites spines on a pyramidal dendrite in the VWFA, the response fades quickly. But if the voiced word causes a spike while the visual symbol excites the NMDA receptors, they open fully to potentiate the spines.

This mechanism is termed *Hebbian* because the experimental psychologist D. O. Hebb realized that a mechanism of this type could explain memory storage, and 40 years later, voilá, it was found. The NMDA channel is one important example among others. Here, all that is needed is to couple two inputs temporally to drive a surge of intracellular calcium. The Purkinje cell does this with a different mechanism, releasing a surge of calcium from intracellular stores via a kinase-driven signaling cascade. What the designs all share is a coincidence detector. But that alone is insufficient because it does not indicate whether a coincidence is accidental or significant. This is why synaptic potentiation by a single "suspicious coincidence" is allowed to dissipate.

However, if a coincidence repeats, statistical evidence accumulates that it is no accident, and potentiation grows. What causes repetition? A behavior repeats if it is followed by some rewarding event. If a thirsty lab animal presses a lever by accident and receives a drop of juice, it will likely press again. If a beginning reader views the printed symbol "pipe" and hears it voiced, the coincidence of symbol and sound is suspicious, but repetition is needed to confirm the association. What drives the student to repeat the sound while viewing the symbol? The reward is a teacher's praise—for we thirst for praise as a lab rat thirsts for juice. Another reward is the sense of mastery—which sustains learning when the teacher is long gone.

The brain represents all of these rewards—juice for the rat, praise and mastery for the student—with one internal signal. It is a pulse of dopamine released when neurons clustered deep in the midbrain fire a burst of spikes (figures 4.1 and 4.4). The dopaminergic axons distribute to various sites critical for learning. Some regions, such as prefrontal cortex, select a behavior and evaluate whether it should be repeated. Other regions, such as ventral striatum and hypothalamus, drive the repetition. From these regions an electrode can drive repetition even more potently than common external rewards such as food or sex. Naturally, sites that choose and evaluate connect (via long loops) with the sites that drive repetition (Bromberg-Martin et al., 2010).

To summarize: synapses are potentiated by repetition, and when they are also bathed in dopamine (or a different molecular signal), the combination initiates long-term storage (figure 14.4). At certain sites the same signal drives all the essential components: behavioral repetition, synaptic potentiation, and long-term storage. What makes this "teaching signal" so efficient will be explained later.

Storage in many areas is driven directly by dopamine input. This includes hippocampal maps of spatial and musical landscapes and other areas such as prefrontal cortex, striatum, and amygdala—where decisions to store or not to store are worked out. However, areas closer to the sensory side, such as the VWFA and language areas, do not receive dopamine input, so their synapses require different mechanisms to initiate long-term storage. These posterior areas receive rich feedback connections from forward cortical areas (see later, figure 14.7). This might be another case where complex decisions are computed at higher levels and then communicated back to earlier areas as executive orders ("store"/"don't store"). Corticocortical axons use the same log-normal distribution of diameters noted to be efficient for various other tracts (figure 4.2; Pajevic & Basser, 2013). Which particular chemical signal initiates long-term storage by dendritic spines in VWFA remains to be determined (Caroni et al., 2012; Holtmaat & Svoboda, 2009).

What rules govern choice?

Information is stored and retrieved locally, as we have seen, because that is efficient. Storage can be triggered by repetition and solidified by further repetition. But how does the brain *choose* a particular behavior to repeat and thus to store? Moreover, given that a behavior integrates many elements through a final common motor pathway, how could it possibly be "stored locally"—what would that even mean? Consider the following quotidian example—fictitious in its details but firmly based on current understanding from electrophysiology, neuroanatomy, and fMRI.

Choosing an Exit

London's Paddington Station is a hub with many exits. Its geometry somewhat resembles the Morris maze that tests spatial navigation in rodents. And like rodents, commuters map this space in their hippocampus. Our particular commuter, arriving on the 8:09, steps down, turns right and strides directly to Exit 2. This route he repeats nine days out of ten. He nearly always chooses this exit because he always needs a taxi. Initially he wandered the station in mild bewilderment until he happened upon Exit 2 and was rewarded by a queue of taxis. Taxi drivers gather at Exit 2 because there *they* are rewarded by a queue of commuters. This exemplifies an iron law of learning: what will most likely be rewarded, we repeat (see later, figure 14.8).

But today our commuter pauses at Exit 1, looks to his watch, hesitates again, and finally selects it. What explains this break from routine? Rising late, our commuter skipped breakfast to catch the train, so he's hungry. He knows that Exit 1 leads to a food court and that choosing it will reward him with a snack. Why then does he check his watch and hesitate? Because he was reared to be punctual. In choosing Exit 1, our commuter values the prospect of satisfying his hunger more highly than the prospect of satisfying a family rule.

This leads him to another choice: plump croissant or plain scone at half the price. To choose, he must again compare their subjective values. If our commuter values novelty and savor, one croissant is worth at least two scones, so he will chose the croissant. But now, from some deep memory store, his mother speaks—astonished and disappointed that her son might choose pleasure over price. He selects the scone (Levy & Glimcher, 2012; Smith et al., 2010; Levy & Glimcher, 2012).

Neoclassical economic theory proved mathematically that whenever choices are logically consistent, the chooser behaves as if the alternatives are valued on a common scale. The chooser selects the alternative with the highest value (Levy & Glimcher, 2011). Accordingly, there must be neurons to mediate this behavior—cells whose firing patterns reflect subjective choice value. The subjective scale must integrate myriad factors: both our humble need to eat and our higher need to satisfy the family rules—in this case, punctuality and thrift. Such a scale requires information from all sources to converge for final weighing.

These design specifications are instantiated in a defined region of orbitofrontal cortex (figure 14.6). This region (*area 13* and vicinity) receives convergent input from essentially all sensory modalities and from limbic structures such as the amygdala and accumbens (figure 14.7; Ongur & Price, 2000). The region becomes active during choices; moreover, it contains neurons that code subjective choice value—neurons that can fire more to a croissant than a scone (Levy & Glimcher, 2011; Padoa-Schioppa & Assad, 2008; Kable & Glimcher, 2009). Experiments have yet to include a mother's voice—but one can well imagine that it weighs in regularly from wherever it is stored. If all values are weighed on a single scale, rewards must be distributed to their original contributors to keep them in the game. That is, the family value of thrift must be rewarded somewhere so that it can continue to compete with the taste of a croissant.

In short, even apparently simple choices—to eat or to arrive on time, to select a croissant or a scone—cut deep. That is because some part of the choice often reflects a strongly held value, either instinctual (nature), developmental (nurture), or both. When the commuter chooses a scone, he

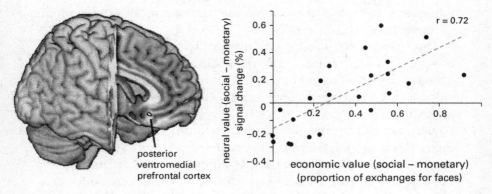

Figure 14.6
Posterior ventromedial prefrontal cortex encodes social value and monetary value on the same scale. Left: Locus in prefrontal cortex where activity correlates with likelihood of economic exchange. **Right:** Within this region, a stronger neurometric signal predicts likelihood of economic exchange: money versus opportunity to view a pretty face. Individuals showing greatest neurometric value for pretty faces readily exchanged money to view more of them whereas individuals showing least neurometric value for pretty faces versus money were less likely to pay for that opportunity. Line represents the least-squares fit to the data points; each point reflects the average response for a participant. Reprinted with permission from Smith et al. (2010).

values his own preference lower on the scale than he values the preference of his mother.

Designs to optimize choice and information storage

The brain addresses the problem of choice with neurons that encode value. A value-coding neuron should optimize its ability to encode different values. This requires a steep curve for response versus reward—just as for neurons at all lower levels (chapters 3, 6, 9, and 11). The value-coding neuron needs to continually reposition this curve along the reward axis in order to match any change in the mean reward distribution (figure 3.4). Moreover, the neuron should flatten the curve if the reward distribution broadens to better encode the wider distribution. About a quarter of the value-coding neurons recorded in orbitofrontal cortex behave this way (Kobayashi et al., 2010; Louie et al., 2011).

Notice that neurons at the highest cortical level implement fundamental economic behavior using the same adaptive design (shifting response curves) as the humblest sensory neurons. This is why we list *adapt* as part of the principle *learn* from which it is distinguished mainly by timescale. As we have noted, any mechanism that uses current inputs to shape future responses is projecting information from the present into the future to change behavior. This is learning.[15]

This chapter asked initially, what information should be stored? By now, one answer seems obvious: store those scenarios that yield better than expected rewards—because those are what bear repeating. Moreover, it is efficient to store the diverse components of a given scenario in the cortical areas where they originated. This is one likely use for the connections that higher areas, such as orbitofrontal cortex, return to areas that provide their inputs (figure 14.7). Top-down orders can be simple: *save* (*facilitate* the synapses) or *delete* (depress the synapses; Ayzenshtat et al., 2012). Simple orders can be conveyed with minimal wire (chapter 4). But to learn which decisions bear repeating, the orbitofrontal areas require a *teaching signal*.

How predictors of value learn the value of their predictions

The most efficient teaching signal for any learner—machine or animal—is computed by the *temporal difference model* (Glimcher, 2011; Schultz, 2011; Montague et al., 1995). The behavioral choice is evaluated by an internal signal that compares the currently experienced reward value to the expected reward value. This signal, reporting the difference between actual and predicted reward, feeds back to the choice mechanism to update the prediction. It says, if better than predicted, "Repeat!" or, if worse than predicted, "Forget it!"

Quantitatively, this *reward prediction error* is the magnitude of the reward signal divided by the standard deviation of its probability distribution—something like a contrast signal divisively normalized (chapter 12; Louie et al., 2011). As learning proceeds, the error diminishes, so the rate of learning declines. When the reward and the prediction are identical, learning does not occur. One hallmark of this optimal teaching signal is its timing shifts. At first, it closely follows the objective reward—delivery of food, sex, money, and so forth. But as suspicion grows that a stimulus predicts an objective reward, the teaching signal diminishes for the actual reward and grows for the predictive signal (figure 14.8).

A signal with all the foregoing properties, shown to be both necessary and sufficient for an optimal teaching signal (Glimcher, 2011), occurs in a defined region of the ventral midbrain (Schultz, 2010). Neurons clustered there show exactly these properties, firing tonically at low mean rates and then phasically when a behavior is unexpectedly well rewarded. When a behavior is rewarded, but no better than expected, the spike burst does not occur. The dynamic range for this critical teaching signal seems

Cortical cascades to prefrontal
cortex

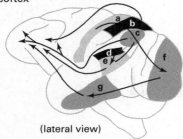

(lateral view)

"Limbic" cascades to prefrontal
cortex

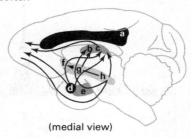

(medial view)

Prefrontal back to cortex

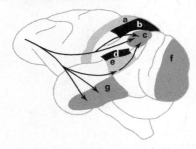

Prefrontal back to "limbic" structures

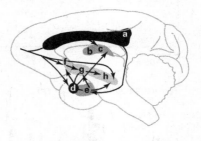

a. primary somatosensory
b. secondary somatosensory
c. inferior parietal lobule (multimodal)
d. primary auditory
e. secondary auditory
f. primary visual
g. secondary visual

a. cingulate gyrus
b. anterior thalamic nucleus
c. dorsomedial thalamic nucleus
d. amygdala
e. hippocampus
f. septum/accumbens
g. midbrain "limbic" area

Figure 14.7
**Prefrontal cortex integrates inputs from neocortical and "limbic" structures—and
feeds back to both.** The forward pathways allow prefrontal cortex to compare various possible choices on the same scale (see figure 14.6). The feedback pathways allow
the integrated decisions to tell the lower areas what information to store. Diagram
shows brain of macaque monkey. Modified from Sterling (2004b) and reprinted with
permission.

dopamine neuron responds to "primary" reward
(liquid and food)

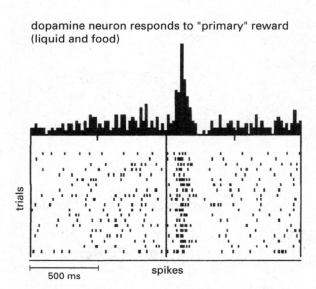

response to light or sound that predicts primary reward

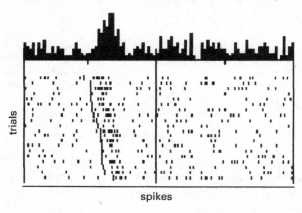

Figure 14.8
Properties of the temporal-difference teaching signal. Upper: Dopamine neuron
fires phasically following primary rewards. **Lower:** As the animal learns that a signal
(light or sound) predicts a primary reward, phasic activation declines for the primary reward and grows for the predictive signal. Black histograms show accumulated
spikes across trials. Reprinted with permission from Schultz (2010).

astonishingly small: a tonic rate of about 4 Hz rises phasically to about 40 Hz, but only for approximately 100 ms, thus four spikes. It may seem surprising that joy could depend on such a slender signal, but this dynamic range is similar to retinal ganglion cells (chapter 11).

Teaching neurons need to distribute their signal widely: to all the areas where value is represented, where choices are made, and where actions are selected. This includes orbitofrontal cortex, dorsal striatum (*caudate*) for the motor components of choice, and ventral striatum (*nucleus accumbens*) for the emotional components (Asaad & Eskandar, 2011; Hare et al., 2008). As noted, these areas can then relay executive orders cheaply by fine axons.

To be clear, many sorts of surprising external and internal rewards—much of what the heart might desire but does not expect—are represented in the brain by the magnitude of this signal and its timing. It is a final common pathway for learning, just as motor neurons are the final common pathway for motor behavior. This design is essential to valuing all choices on a single scale: were we to reward taste, thirst, and sex via separate systems, choosing would be even more of a problem than it already is. Such considerations suggest an economical neural design that we term "quasi-wireless" for a reason now to be explained.

How the teaching signal reaches its forward targets

The cheapest route to inform large volumes of neural tissue would be to broadcast a neuroendocrine signal via the circulation, that is, wireless signaling (chapter 4). However, that would be too slow; moreover, it would smear the critical temporal differences. On the other hand, to implement extremely sharp timing would require that a "teaching synapse" contact every learning synapse—at a huge cost in space.

But the teaching signal does not require extremely sharp timing. It can be roughly 100-fold broader than, for example, an auditory signal. Consequently, the teaching signal can be sent by wire to within a few micrometers of some learning synapses and there release a well-timed pulse of chemical transmitter into the extracellular space. Transmitter is allowed to diffuse over micrometers to reach multiple synapses. Diffusion over a few micrometers allows sharp enough timing, and the spatial blurring from diffusion doesn't matter because only synapses that are already potentiated will respond (figure 14.3).

A single axon delivers this teaching signal far and wide—to prefrontal cortex, hippocampus, dorsal and ventral striatum, amygdala, and hypothalamus (figure 14.9). This requires a huge terminal arbor with up to 10^6 dilations containing vesicles filled with transmitter (Glimcher, 2011;

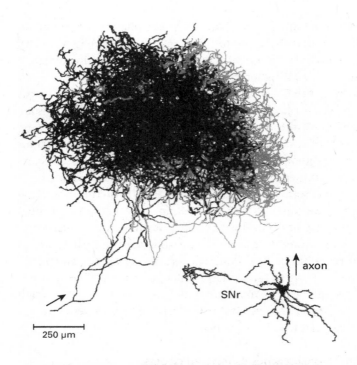

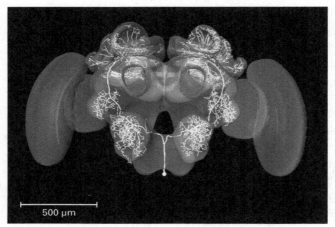

Figure 14.9
Teacher axons employ dense arbors to broadcast signal to large volumes of tissue.
Upper: Axon arbor of a dopamine neuron projecting from substantia nigra (SNr) to
dorsal striatum (Matsuda et al., 2009; rat). The arbor supplies dopamine to a volume
that includes 75,000 neurons. Each point in the target is "covered" by about 150
overlapping input arbors, and thus all dendritic spines lie within 3 μm of a dopami-

nergic synaptic dilation. Total wire length of arbor is about 0.5 m, and human arbor is nearly 10-fold greater. Dopamine arbors in frontal cortex are less dense (Matsuda et al., 2012). **Lower:** Axon arbor of octopamine neuron (honeybee) projecting bilaterally to multiple regions that encode reward for olfactory learning. Like dopamine neurons in the ventral tegmentum of the mammalian brain, this neuron codes for a prediction error: it decreases its response to an expected reward but increases its response to an unexpected reward. Reprinted with permission from Menzel (2012).

Schultz, 2007). This is 100 to 1,000-fold more active zones than are expressed by standard neuron types which require a large cell body and a thick axon to sustain them. This neuron uses a fairly slow action potential— just fast enough to support the degree of spike-timing precision required for efficient teaching. The slow action potentials also match the final stage of wireless transmission, that is, diffusion in the extracellular space across 3–10 μm over tens of milliseconds.

To deliver this teaching signal to large volumes of neural tissue requires in human on the order of 200,000 neurons per hemisphere[16] (Damier et al., 1999). This is about one fifth as many neurons as contribute to the optic nerve. The tasks are essentially the opposite. Whereas ganglion cells need to signal independently and preserve a fine spatial map, the "teacher neurons" need to send one message synchronously throughout the hemisphere. When the same teaching signal is broadcast to all synapses, those that have been potentiated during a repeated behavior will all strengthen together. To promote synchrony, the teacher neurons couple strongly with gap junctions. That allows synchronous release of a transmitter from about 10^{11} synaptic dilations[17] distributed densely and uniformly, an efficient way to synchronously bathe large volumes.

The mammalian neurotransmitter released to signal reward-prediction error is dopamine.[18] So the teaching signal required for efficient learning is a brief pulse of dopamine delivered by a burst of spikes at roughly 40 Hz over 100–500 ms at key sites for subjective value and choice. The teaching signal encourages a behavior to repeat if the behavior delivers a positive reward-prediction error. This design couples the dopamine pulse to a mechanism that causes a subjective pulse of "well-being."[19] Whether, besides humans, other animals also feel this pulse, we cannot say, but their reward-seeking behaviors certainly suggest it.

Neuroscientists were formerly reluctant to think about subjective feelings. However, if we accept that firing by certain orbitofrontal neurons encodes flavor, punctuality, and thrift according to their relative subjective values, we might as well accept that dopamine's sculpting of their synapses is associated with a moment of subjective "happiness." This is what we often seek during choice and learning: a pulse of subjective well-being. The visual image of a croissant is of no particular interest, nor is the raw gustatory information from a bite. It is what they elicit: a signal that delivers a pulse of dopamine to striatal and orbitofrontal neurons.

Mysteries still abound. For example, a thirsty animal enjoys its earned juice reward even though it was anticipated, and the commuter enjoys his scone even though it is what he expected. In these cases the dopamine neurons do not fire because the reward-prediction error = 0. These subjective pleasures must depend on other intrinsic signals, obvious candidates being endogenous opioids and cannabinoids.

Downsides to the "temporal difference" design for learning

Can't get no satisfaction
Temporal-difference learning is, from a theoretical perspective, a highly efficient design (Glimcher, 2011). Moreover, the brain executes it efficiently via dopamine neurons that fire brief bursts of slow action potentials to deliver, quasi-wirelessly, a pulse of subjective well-being that encodes a positive surprise as the teaching signal. This works brilliantly in natural environments because they are unpredictable and therefore provide diverse small surprises and frequent pulses of subjective well-being. But modern environments are crafted to be highly predictable. For the temporal-difference design, this creates two huge problems.

First, because the teaching signal needs to be brief, the sense of well-being—what Freud termed "happiness"—*must* be episodic. This is bearable when the environment delivers small reward-prediction error signals with some frequency. However, as the environment grows more predictable, sources of these signals shrink, and the intervals between episodes of satisfaction lengthen. The efficient design for learning in an unpredictable environment becomes in a predictable environment—zoo or modern city—an efficient design for existential angst.

Live with artificial light in a temperature-controlled building, and there will be few surprises from weather or climate. Shop in a supermarket where the goods are always the same; eat prepared foods designed to always taste the same. Instead of playing a sport, or an instrument, or having sex—play a DVD of someone else doing it. Be surprised by an "action film," but soon it is routine. Repetition shrinks surprise and thus the pulse of well-being.

The second huge problem is that the teaching signal makes us so clever that we find ways to "fool" it. We identify natural products, such as alcohol, nicotine, opiates, and cannabinoids that release dopamine in greater than quotidian amounts (Cheer et al., 2004). We find and distill other products, such as cocaine, that inhibit the dopamine transporter, thereby increasing and prolonging dopamine's elevated concentration in the extracellular space. Both methods intensify the sense of well-being and hold it at a plateau—for a while. But the reward-prediction error signal is all about surprise—so the system includes numerous mechanisms that adapt. This requires progressively stronger stimuli to achieve the same sense of well-being. So we initially fool the system but ultimately fool ourselves—as exemplified by Freud, who was doubly addicted—to nicotine and cocaine.

Given this bit of neural design, the best we can do for comfort is to find sources of small surprise that can deliver brief pulses. On vacation we reencounter surprises from Nature: sunrise, sunset, night sky; warm breeze when we are cool and cool breeze when we are warm; sports, music, crafts, dance, sex. The learning system built upon small, episodic surprises drives us to keep them coming whereas urban life "protects" us from them, leaving a few vicarious sources that prove to be poor substitutes.

Nearly any behavior can act like a drug—eating, running, and gambling can all drive the reward-prediction error signal—some to more constructive effect than others. However, reliance on a single behavior to drive the reward-prediction error signal, like drugs, gradually requires increasing doses to achieve the same effect.

In short, the downside to this brilliantly engineered system is that it evolved to work in environments rich in surprise (figure 14.10). Narrow the range of surprises, and the teaching system locks onto whatever source of reward is readily available. This defines *addiction*—a dysfunctional behavior that develops inexorably when a system designed for surprise is deprived of it. Nothing wrong with the system—just that it was engineered for a different set of conditions. Modern societies address this so poorly. Instead of fighting narco-wars, we should be developing more surprising environments and richer ensembles of activities.

This account is far from complete. For example, what encodes disappointment? Since a positive reward signal delivers a pulse of well-being, does a negative signal that accompanies failure deliver a pulse of unwell-being? What causes single-trial learning, avoidance learning, and motor learning? What strengthens synapses in brain areas devoid of dopamine axons? This chapter, to reduce space and energy, has been circumscribed. It has gathered and integrated what amount to interim reports from highly focused investigators, and the elephant they are exploring is large. Nevertheless, a few conclusions seem warranted.

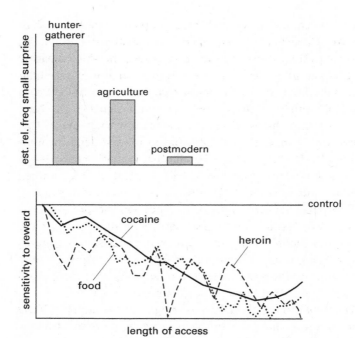

Figure 14.10
Discontents of civilization. Upper: Premodern life provides diverse sources of small surprise that deliver positive reward-prediction error signals. Postmodern life shrinks the possibilities for surprise, making it a challenge to find them. **Lower:** Prolonged access to a source of well-being reduces sensitivity to its rewarding effect. Note that the curve for rich food is identical to those for heroin and cocaine (rat). Modified from Kenny (2011).

Some conclusions

Recycle space

To store information requires space, so ultimately an adult brain can hold only a limited number of memories. The brain solves this problem by recycling the space. In an adult brain there can be no learning without equal forgetting, and this is why *forget* belongs to the design principle that includes *learn*.

The brain winnows all memories. Fresh ones with little promise of future reward, it deletes wholesale. Those with greater promise it saves, but gradually discards inessential details. So the brain continually prunes and resculpts its stories—by continually pruning and resculpting its synaptic connections.

This design—recycling of brain space—involves matching synaptic plasticity, that is, the capacity to resculpt, to local need. For example, synaptic plasticity is high in the VWFA during childhood, which is critical for rapidly learning to read, and it is high in motor areas when it becomes critical to learn fine motor skills. Plasticity eventually tapers off but persists to some degree, which is why a pensioner can learn a new musical instrument or a new language. But as resculpting slows with age, early memories and habits tend to stabilize, and new ones are established more slowly. Consequently, we grow "set in our ways," and old dogs grow harder to teach.

Specializing

Learning is designed to process, store, retrieve, and resculpt information at the same site: the synapse. Therefore, learning follows the basic patterns of cortical specialization. Moreover, specialization extends to the individual hemispheres: languages are learned primarily on the left and music primarily on the right, motor skills (handedness) are learned commonly on the right, and so on. Motor asymmetry is marked in humans—where it can exploit direct connections from motor cortex to individual motor neurons. This pathway is essential for fine control of individual fingers and effective use of our uniquely opposable thumb. The direct corticomotorneuronal connections are most highly developed in humans (Kuypers, 1981; Rathelot & Strick, 2009). Thus, even if other apes had the patience to practice the violin, they would lack the essential fine motor control.[20]

The brain's tendency toward specialized areas for different skills extends to the pool of brains within a community. Because one brain has limited capacity and requires constant practice to maintain its particular skills, an efficient community will support specialization across brains: butcher, baker, doctor, spiritual leader. This can only work, of course, if people manage to cooperate. That requires kindnesses that encourage trust and suspicions that thwart trickery.[21] That simple rules of fairness and reciprocity are shared across cultures suggests that they belong to a neural design that supports specialization of brains within a community (e.g., Gintis & Fehr, 2012; Baumgartner et al., 2009; Fehr & Camerer, 2007; Camerer & Fehr, 2006).

Education

Each brain expresses different splice variants of myriad synaptic proteins that serve learning and plasticity. Moreover, each person experiences different education, different life events. Also, each person behaves differently and so experiences different feedback. Every interaction between two people, even a "simple" handshake, changes both brains in varied and unfathomable ways. The rich interactions between varied synaptic proteins and varied experiences give each brain a unique functional architecture, and this causes each person to learn differently.

Given that brains differ at birth and differ more with every new experience, uniform education cannot be optimal. Yet, we typically confine 30 children to desks in a small room for many hours per day, insisting that they "pay attention" and not fidget. About 5% of these children, whose brains by dint of genes and experience respond poorly to such confinement, we diagnose with a mental health disorder (*attention deficit hyperactivity disorder*) whose peak onset coincides with matriculation. More than half of the children exhibiting this "disorder" are treated with drugs with the same properties as methamphetamine and cocaine (Low, 2012). This is a Procrustean solution: prescribing drugs of abuse, instead of discovering what activities calm and engage the child.

Given the brain's limited storage capacity, it matters how we fill it. So what should we teach and how? For students to grow facile at accessing and evaluating information, they must do it for themselves—discover what they do best, what encourages them to repeat—to practice. Probably the best we can do is to provide diverse opportunities: a walk in the woods, a playground, a garden, a chisel, a cello. Instruction should include basic rules for safety and some pointers on observing, recording, evaluating, and integrating. None of these except for safety are truly teachable, because there is magic to it—like the conductor drawing out a symphony with a baton.

The best a teacher can do beyond providing opportunities is to give feedback that encourages continued, mindful practice. That is why a few principles can be as helpful as a compendium, and that is what this book has attempted.

15 Summary and Conclusions

Summary

This book opened by asking, how can the brain be far smarter than a super-computer yet consume 100,000-fold less space and energy? Brain and computer are governed by mathematical and physical laws specifying that the costs to capture, process, send, store, and retrieve information rise disproportionately for higher information rates. One reason for the brain's efficiency is that it has more ways to stay on the linear region of this curve. Those ways constitute principles of neural design.

To restrain its overall signaling rate, a communication channel should try to fill its capacity with more information and less noise. All neural processes have a noisy component: transmitter molecules arrive stochastically; ligands bind, channels open, vesicles fuse, and spikes fire—all stochastically. To distill information from stochastic processes, all levels use the same strategy: sum n correlated events and thereby improve S/N as $\sqrt{n}$.

This core strategy, which operates at all levels from molecules to behavior, has a drawback: cost rises faster than the benefit. Although S/N can improve 10-fold with only 100 events, the next factor of 10 requires 10,000 events. Because costs rise linearly, whereas improvements rise as the square root, there soon arises a point of diminishing returns where improving S/N by spatial and temporal summation becomes uneconomical. Greater efficiencies for a communication channel require additional design strategies.

One strategy is to clear room for more information on a channel's limited capacity by reducing redundancy. Events arriving closely in space or time carry similar information, so integrating more ion channels, synapses and spikes tends to fill the capacity without adding information. Neural designs at all levels combat this tendency by using some form of inhibition to subtract the mean and transmit only the difference signal. Where individual elements in an array sum their inputs broadly to improve S/N their overlap increases redundancy within the array. In that case, elements restrain their individual S/N to maximize total information sent by the array, as exemplified by an array of retinal ganglion cells coding the retinal image. This illustrates that a design goal at the single neuron level (improve S/N) is integrated with another goal (reduce redundancy) so that an array can maximize its total information.

The theoretical optima for spatial and temporal summation to improve S/N and the theoretical optima for lateral inhibition to reduce redundancy change according to input signal and noise. Consequently, efficient designs should adapt both processes to the statistics of signal and noise. Moreover, adaptation should start as soon as there is sufficient statistical evidence that the mean has indeed changed. Adaptation to enhance efficient use of channel capacity is widely employed by the brain, and many instances are known to approach optimality. They all serve the principle *send only what is needed*.

Another route to restraining a neuron's information rate is to sculpt the message in order to send its downstream user only what it needs. Since there are various downstream users, this strategy requires multiple, parallel pathways. Each communication channel, by sending less information, restrains its own rate, thus economizing on space and energy.

Multiple parallel channels also allow better matching of capacity to the natural distribution of information, and in some cases the match is shown to be optimal. For example, a retina sending bright and dark visual contrasts separately over ON and OFF channels optimally matches the distribution of channels to the distribution of bright and dark in natural scenes. It also halves the information per channel, reducing the cost of transmission by fourfold. Moreover, the benefits carry across many subsequent stages. Thus, routing information via parallel communication channels serves the principle *send at the lowest possible rate*.

Compute at the cheapest level

The mathematical rules mentioned specify the general costs of transmitting and processing information, so they apply to all levels, from molecules to tracts and areas. These rules do not indicate the relative costs of different levels, yet those differences are substantial. Physics sets a lower bound to the energy needed to code 1 bit: $\sim 1\ k_B T$ joules. This lower limit is approached by the coding of 1 bit as a change in protein conformation.

It follows that when neurons capture information efficiently via changes in protein conformation, they should use proteins to process the information efficiently. For this, protein molecules are linked chemically, by binding and catalysis, to form a circuit. With each protein molecule changing the signal in a specific way, the circuit computes.

Computing with such a chemical circuit is efficient. A protein molecule is irreducibly small and operates close to thermodynamic limits and, by changing conformation allosterically, it computes as a finite-state machine. Efficiency is maximized by containing circuits in complexes and small compartments. Delays and noise from diffusion are reduced, and small compartments achieve high concentrations with relatively few molecules. High concentrations speed processing by accelerating reactions and by allowing low binding affinities.

Chemical computing with small numbers of molecules increases noise—from diffusion, stochastic binding, and so on. Compartmentalization and complexing help, but ultimately S/N can only be improved by using more molecules to carry the same signal. This is done most efficiently by *matching* the number of molecules to the signal that they carry. A reliable signal requires many molecules to conserve S/N, but when S/N is low, fewer molecules can be used, thereby avoiding wasteful excess capacity.

Chemical computation has a serious disadvantage. Although chemistry over short distances can be fast—chemical signals diffuse 1 μm in a millisecond—it cannot operate rapidly over longer distances. That requires electrical signals, which transmit passively 1,000-fold farther, 1 mm in a millisecond. To travel still farther, signals amplified as spikes course throughout the brain and body up to 100 mm in a millisecond. But these speeds over these distances cost more energy and space. Thus, wherever possible, neurons follow the principle *compute with chemistry*.

Costs of speed over distance

Electrical signaling for speed over distance is limited by the resistance of a single ion channel, membrane capacitance, the resistance of a neuron's internal cytoplasm and ion channel noise. All four limitations can be reduced by opening more channels. But in contrast to a protein signaling chemically, an ion channel's electrical signal is expensive. For example, the cost of opening a sodium channel by changing its conformation is only about 25 $k_B T$ joules. But one millisecond's worth of current flowing through that channel, for a power gain of $\times 1,000$, admits 6,000 sodium ions. These cost 2,000 ATP (equivalent to 50,000 $k_B T$) to pump out, which is 2,000 times the cost of a G protein signaling chemically.

Electrical signaling also costs substantially more space. Chemical signaling is wireless, but electrical signals travel along wires, such as the dendrites and axons that occupy more than 50% of a brain's volume. Supplying energy also takes space. The sodium–potassium pump works relatively slowly: a pump protein molecule extrudes 3 sodium ions every 5 ms. Consequently, keeping pace with one sodium channel requires 10,000 pump molecules. The pumps are powered with ATP, and although the mitochondria that produce ATP pack their oxidative proteins densely, mitochondria cannot be packed too tightly inside a neuron. Consequently it takes 5 μm^3 of a neuron's internal volume to power one open sodium channel. Thus, although a neuron could easily pack its membrane with more than 10^3 channels μm^{-2}, the additional membrane area for pumps and cytoplasmic volume for mitochondria set stringent limits. Consequently, opening more channels requires more space, so neurons are larger and wires longer.

Responding to these constraints on energy and space, designs follow the principle *make neural components irreducibly small*. A smaller neuron needs fewer ion channels to produce a signal of given amplitude because it has less membrane to charge and polarize. Thus, with space and energy costs rising as (diameter)^2, axons are encouraged to shrink, but eventually they hit a lower bound set by channel noise. As axon diameter falls below 0.2 μm, a single sodium channel depolarizes the membrane to threshold, so channel noise generates spontaneous action potentials that would overwhelm communication. Other key components, such as synapses, are also small, and they exploit chemistry and mechanics at the nanoscale to compute with least energy and space. They too approach limits set by noise.

Irreducibly thin axons conduct at about 1 mm per millisecond, matching the timescale at which local circuits compute. Therefore, circuits that rely on such fine axons to conserve resources must limit their spatial extent. Indeed, local circuits in cerebral cortex and cerebellar cortex, which together account for most of the brain's gray matter, rely on the finest axons. Consequently, the upper bound to the extent of their local circuits is on the order of a millimeter.

In short, neural structures on each scale are irreducibly small. This starts on the nanoscale with single protein molecules which can shrink no further without loss of functionality. Then protein molecules assemble into circuits inside synapses, dendrites and axons. These irreducibly fine neural structures set the possibilities for neural wiring at the millimeter scale—six orders of magnitude above the proteins. The irreducibly small neural components set an upper bound to spike rates. The thinnest axons support the lowest mean firing rates which convey the lowest information rates. Thus, these circuits stay low on the information-versus-cost curve.

Note that neural designs often satisfy several principles at once. Proteins are irreducibly small, compute by chemistry, reduce wire by sending chemical signals, and can transmit efficiently by sending at very low rates. Axons can be irreducibly thin, transmit with the fewest possible channels, and send slowly as possible to stay low on the rate-versus-cost curve. But what additional principles govern how a neuron and a neuronal array should spend the available resources?

Analogue is cheapest

Analogue signals are cheapest for several reasons: (1) Output can continuously match input, reducing waste. (2) Analogue is well suited for chemistry, and chemistry is cheapest. (3) Analogue computing is direct and therefore completes a basic operation in the fewest steps. (4) Analogue processing avoids the physical constraint of dividing time into discrete intervals—so it can operate at high information rates.

The transition from cheap chemistry to costly electrical processing preserves efficiency by continuing in analogue mode. Thus, a graded EPSP of 1.5 mV costs 100-fold less per wire length than an all-or-none spike of 150 mV. Therefore, it pays to integrate many graded electrical events before sending a spike. Spikes are unavoidable because they alone provide speed over longer distances, and they become efficient when noise accumulates in analogue circuits. Taking a threshold by triggering an action potential reestablishes sufficiently high S/N that the brain can track a single action potential across many levels and rely on it for behavior. So, neural designs follow the principles *compute directly with analogue primitives* and *combine analogue and pulsatile processing.*

Concentrate resources

A key design issue at all levels is how to distribute computation. Physics dictates that to maximize efficiency, signals should be concentrated in space and time. Thus, it pays to spatially concentrate chemical transmitter molecules in a small packet and to temporally concentrate their delivery to a spatially concentrated cluster of postsynaptic receptor proteins. The resulting current, when integrated in space and time with other currents at the axon initial segment, causes a few temporally concentrated spikes. Thus, neurons are typically silent for substantial intervals between concerted spike bursts. At the next level, the neuronal array, activity is also most efficiently encoded when it is concentrated in a small proportion of the neurons.

The consequence of these "designs for concentration" is that fast electrical pulses, costing 1,000-fold more energy than the chemical signals that caused them, occur relatively rarely and in relatively few neurons. This restrains the mean spike rate in an array. Thus, information is concentrated at every scale: a few transmitter puffs open a few ligand-gated ion channels that allow a few analogue currents that sum to cause a few spikes in a few active neurons; and with this design information is sent and received efficiently. Certain cases are even shown to be optimal, but all follow the principle *sparsify*.

Specialize

To be efficient, neural designs at all scales specialize. Single protein molecules specialize to match speed and sensitivity to the properties of their inputs. For example, the sodium channel serving a thin, low-rate axon differs from the sodium channel serving thick, high-rate axons. Protein circuits specialize to match their properties and ratios to each other for appropriate speed and S/N.

Synapses specialize presynaptically, for example, by using different voltage-gated calcium channels and different calcium sensors to achieve particular sensitivity and gain. Synapses specialize postsynaptically, for example, by using transmitter receptor proteins of different binding affinity, different cooperativity, different selectivity for ions, and different abilities to activate G proteins. By varying these parameters, a synapse can match its transfer function to provide optimal sensitivity and S/N, plus an I/O function that computes directly across the necessary timescale. Neural circuits specialize their connections for the same reason, to use resources more efficiently. In short, a key contribution to the brain's ability to stay low on the rate-versus-cost curve is its immense capacity to specialize—its obeisance to the principle *complicate*.

Save wire

An action potential, to travel much faster than the graded signal that caused it, relies on two large, countervailing currents. And despite their optimal matching, a spike costs about 100-fold more than the graded signal. Moreover, this initial high cost repeats along the length of an axon. This leads naturally to the principle *minimize wire*.

To minimize wire volume, it is efficient to reduce diameter because volume falls as d^2. But small diameters are limited to low firing rates whereas certain axons require high firing rates to serve their particular function. Where thick axons are needed, they often project for relatively short distances and then converge on relay neurons that step down their rates and numbers. However, even irreducibly thin axons are costly, so there is pressure to economize on length; and there are many ways to do so.

On the finer scales, neurons form spines to maximize connectivity and minimize volume, and both dendrites and axons branch to save wire. Ascending in scale, neurons with the strongest interconnections locate closest to each other, and in some cases, such as the fly lamina cartridge of 16 neurons and the *C. elegans* brain of 302 neurons, neurons locate optimally to save wire. Neurons form orderly maps to save wire by reducing mutual interference that would increase path length. Neuronal circuits segregate into specific clusters for the same reason; they form layers to minimize interference from input and output wires. Cortical areas with strongest interconnections locate closest together to minimize wire length, and they signal to distant circuits with reduced instructions to minimize the number and caliber of wires.

Efficiency from prediction

Many of the efficiencies mentioned so far apply to time intervals in which the statistics of signals and wiring are essentially fixed. However, as time advances, probabilities change, leading to mismatches and inefficiencies. This occurs at all levels from proteins to behavior. The best way to deal with changing probabilities is to predict what they are most likely to be in the next instant and the next . . . and then use those predictions to readjust all the matches in timely and cost-effective fashion.

Consequently, circuits at all levels—from molecules to large-scale neuron arrays—adapt their input/output curves to maintain efficiency across huge changes in dynamic range. Each level efficiently matches the next. Such adaptive adjustments span a broad range of timescales. Some are fast and cheap: a synapse can adapt in within milliseconds by altering the number of vesicles at docking sites and changing the conformation of receptor proteins; and it can adapt over minutes by altering the number of receptor proteins. Other adjustments are slower and more costly: a neuron can adapt over tens of minutes by changing the concentration of its protein circuits and over hours by synthesizing new postsynaptic receptors and inserting them stably into postsynaptic membrane. A neuron can adapt over days to years by adding new synapses and pruning old ones. Finally, a brain can adapt by adding and removing neurons.

Of all these adjustments, the more stable and costly are classed as "learning." Neuroscience is still learning about learning; yet it is already evident that learning—stable changes in functional architecture of synapses in response to their signaling history—is ubiquitous in the brain, from early sensory stages to cortex and final motor outputs. Moreover, learning follows mathematical rules that deliver the greatest reward for a given investment. Thus, the process of learning that improves brain efficiency is itself efficient. These processes constitute the final principle of neural design: *adapt, match, learn, and forget.*

Conclusions

If the principles suggested for neural design are correct and not mere slogans, they should apply across a range of spatial and temporal scales. Indeed, we have noted their traces across a 10^9 range of spatial scales—from nanometers (protein folding, structure of ion channels) to micrometers (structure of synapses and local circuits) to a meter (structure of long tracts). Ditto for a 10^{13} range of temporal scales—from microseconds (transmitter diffusion) to decades (human memory). If the principles are correct, they should apply across brain regions and across brains of diverse species—and they do.

Principles for neural design should apply to related devices. And they do, especially regarding energy-efficient electronics (Sarpeshkar, 2010, 2012). Two neural principles, *compute directly with analogue primitives* and *mix analogue and pulsatile* (digital), were adopted from electronics (Sarpeshkar, 1998). Four more principles, *send at the lowest rate, send only what is needed, sparsify,* and *match and adapt,* have been derived independently. Two more principles, *minimize wire* and *make components irreducibly small,* are standard practice in chip design. Of the last two, *complicate* comes from mechanical engineering (chapter 1); *compute with chemistry* derives from biology and is just now entering electronics (Akyildiz et al., 2008; Sarpeshkar, 2010, 2014). In short, the 10 principles of neural design seem well grounded in the physics, chemistry, and mathematics of information.

Our coverage in applying the 10 design principles has been selective— some might say "spotty"—for we chose not to write a textbook. Instead, we constructed a close "interpretive reading" of neural design: a brain's functional architecture viewed from the perspective of information theory and biophysics. With some exceptions these perspectives do not reveal what any bit of brain architecture "does."

But these perspectives do explain many "whys" of the design and provide tools to interrogate any system. Are the axons thick or thin? Are they uniform or mixed? How are the layers arranged? Do neurons use lateral inhibition to eliminate redundancy? What is optimized when neurons adapt and match? So viewed, each structure and mechanism suggests how much information might be sent and how fast. So these principles can serve as a guide to the perplexed.

Could our brain do better?

Neuroscientists are frequently asked at parties, "Is it true that we use only 10% of our brain?" People wonder if they might make more efficient use of the brain that natural selection has provided them. The design principles suggest not: we seem to fully use all of our brain. For example, memories are stored at synapses, which always fill their allotted space. Even if some fraction of synapses were held in reserve, the reserve pool belongs to the overall design. So our memory banks are effectively full, and new memories can be stored only by pruning old ones.

It is true that most areas are relatively silent until called. This is evident in the shifting patterns of cerebral blood flow visualized by fMRI. Moreover, when an area *is* called, only a fraction of its neurons respond; therefore, activity increases by just a few percent. This may strike the questioner as tragically wasteful: if we used more of our neurons more of the time, would not the brain compute more effectively?

But, of course, using neurons sparsely is efficient design. Neurons may not be used to their full effect, but space, materials, and energy are. Indeed, sparse usage is so important that the brain devotes special resources to regulate which neurons should turn up, and which should turn down. Such neural mechanisms for *attention* and *mind wandering* are essential for the brain to follow the core principles of efficient design. Although we have omitted to describe these mechanisms in any detail, they certainly wield the baton (Cohen & Maunsell, 2009, 2011; Ni et al., 2012; Purcell et al., 2012; Gruberger et al., 2011).

If the neuroscientist concedes that the brain does not work for very long at peak capacity, the questioner visibly brightens: yes, too bad that the mind tires with intense mental effort; too bad that we need to sleep—such a waste! But it resembles the too bad that we cannot run at top speed for very long or run a marathon every day. The brain, like the body, must repair, revise, and replenish. So it is a fantasy that we could be much better than we are. Of course, brain, like body, can improve with exercise—but up to definite limits. Beyond these, exercise in one mental domain necessarily comes at the expense of others.

Could natural selection do better?

Another party question concerns brain efficiency: "Is it true we could evolve a more efficient brain if natural selection were allowed more time?" Again, information theory and physics suggest not. Many neural computations already operate near the lower bounds on space and energy costs defined by thermodynamics. Recall that the lower bound on space per bit of information is approached by single protein molecules that could be no smaller and still stably change conformation. Recall that the theoretical lower bound on energy per bit of information is approached when a protein molecule changes conformation.

Recall also that proteins coupled in circuits are often matched for best possible trade-offs of speed, S/N, and sensitivity. And when parallel protein circuits specialize to improve performance under different conditions, the same protein types are often retained but modified slightly to tweak performance appropriate to the condition. Thus, the rod specialized for processing starlight, and the cone specialized for daylight, both use opsin coupled to a cascade of other proteins with feedbacks. Rod and cone opsins differ slightly, as do their G proteins, phosphodiesterases, kinases, cyclases, and so on. Thus, information processing by proteins and circuits of proteins seems already to have approached maximum efficiency, and one cannot expect much improvement at that level.

Cellular structures also operate near their lower physical bounds for space and energy efficiency. Recall, for example, that axons approach the lower limit on diameter set by channel noise and microtubule size. Thus, the thinnest axons are as economical as can be. Dendrites also approach a lower limit on caliber. However, the finest axons and dendrites support only low information rates, so functions that require higher information rates need thicker axons with more ion channels and microtubules, plus thicker dendrites with more synapses.

But the returns on these investments are limited by mathematical and physical laws that all produce diminishing returns. Natural selection cannot skirt the statistics of channel opening and vesicle release, nor can it tunnel under the cable equations that describe membrane space and time constants, nor has it evolved metallic conductors to increase length constants, nor can it outrun Shannon's equations that explain why higher information rates cost disproportionately more. In short, natural selection can certainly improve a neuron's performance by investing more, but for a given cost, it cannot do a whole lot better.

At the next level of structure, neural circuits try to minimize wire length. We have seen that the retinal circuit succeeds in laying out 20 parallel circuits of modest convergence and divergence with the least possible wire. The cerebellar circuit also succeeds in laying out a quite different circuit of tremendous convergence and divergence with the least possible wire. Spinal motor circuits do the same, and if one accepts the idea that cerebral cortex circuits try to maximize potential connections, its local circuits also minimize wire. Thus, circuits that together constitute a substantial proportion of total gray matter are wired about as efficiently as possible.

Myelinated tracts occupy up to about 40% of brain volume, and that space is also used efficiently. Axon diameters in white matter distribute lognormally, with most axons near the lower bound on diameter and a few with larger diameters. This requires local circuits to accomplish intense computations so that long tracts, such as fornix, corticospinal tract, and corpus callosum, can guide them with a "reduced instruction set." Space could be saved, of course, by shifting a long tract's distribution toward finer fibers or by amputating its tail of thicker ones, but this would impair performance, not improve it.

Circuits at all levels from molecules to large-scale neuron arrays adapt their input/output curves to optimize gain across huge changes in dynamic range; moreover, the different levels efficiently match each other. Given that these adaptive adjustments are initiated the instant that there is sufficient statistical evidence to direct a change, this cannot improve. Adaptations on longer timescales use mechanisms that are classed as "learning"— stable changes in synaptic structure and function. These changes are ubiquitous, so learning appears to serve every function. Moreover, since it follows mathematical rules that deliver the greatest reward given investment, the algorithm cannot be improved.

In short, the processing, transmitting, and storing of information is efficient on all levels: from proteins and protein circuits to neurons and neural circuits, and thence to neural systems and behavior. In certain cases, optimal designs can be followed stepwise across spatiotemporal scales and across hierarchical levels, for example, from photoreceptors through retina and V1 to the level of visual behavior. Indeed, when discriminating a small brief spot, sensitivity is set by neural noise measured at the ganglion cell output; noise associated with processing across the many subsequent stages leading to a perceptual decision do not reduce performance. Optimality is proven only for a limited number of examples, but nothing has been encountered so far to suggest that brains built of protein circuits could become a great deal more efficient.

Natural selection has driven fly and mammal in strikingly different directions for 450 Myr, allowing the fly roughly 100-fold more generations per year than human. Yet, on all comparable measures the neural components, processing, and layouts of fly and mammal are equally efficient. So it seems that with regard to brain design, natural selection has followed different routes but reached the same destination. This suggests that the core principles of neural design were established already in their shared ancestors. If we evolve for another 450 Myr, we will certainly look different; we will think and behave differently because we will have continued to be shaped by natural selection. But our brains will probably be no more efficient.

Last question

The brain is effective in allowing its owner to economically explore a wider world that contains more resources. The brain has taken animals from soup to soil, thence to land, sea, and air—and now to outer space. The brain allows a human to repair a heart and even to do it robotically from half a world away—or, alternatively, to identify and assassinate an enemy. This raises what is perhaps the biggest question. Will our efficient brain—our 12-watt bundle of cleverness—irreversibly overexploit all of nature and extinguish our species along with many others? This question neuroscience cannot answer.

Principles of Neural Design

Compute with chemistry
- Bits per joule approaches lower bound set by thermodynamic limit.
- Bits per liter reaches lower bound set by protein structure.
- Signals are fast at short distances.
- Computation is direct.

Compute directly with analogue primitives
- Analogue completes a basic operation in fewer steps than digital.
- Analogue is well suited to chemical and electrical computing.

Combine analogue and pulsatile processing
- Analogue processes information at high rates.
- Analogue electrical signals are cheaper than pulses.
- But stochasticity at all stages (vesicle release, ligand binding, channel opening) accumulates noise.
- Therefore, compute locally in analogue; threshold to restore S/N, and send noise-resistant pulses.

Sparsify
- Signal with proteins in small clusters.
- Release vesicles in brief bursts.
- Fire spikes in brief bursts.
- Maximize information per array for least space and energy: optimize fraction of active neurons; optimize S/N vs redundancy.

Send only what is needed
- Reduce noise and redundancy.
- Sculpt message for downstream users.
- Reduce number of signals to save energy and space.

Send at the lowest acceptable rate
- Higher rates cost disproportionately more.

Minimize wire

- Space and energy decrease as length and $(\text{diameter})^2$.
- Small diameter allows few bits per second.
- Slowest signals can use zero wire (neuromodulators, hormones).
- Shorter wires reduce processing time.

To shorten wire:

- Organize neurons in maps;
- Within a map segregate computations in parallel circuits.
- Separate circuits in layers, columns, stripes, barrels.
- Arrange maps to interconnect with least wire
- Connect neurons by matching their axonal and dendritic meshworks.
- Reduce instruction set to send long distance.

Make neural components irreducibly small

- Smaller reaction vessel allows faster chemistry with fewer molecules.
- Lower membrane capacitance charges with smaller current.
- Nanoscale molecular components allow smaller axons and synapses.

Complicate

- Specialize molecules to match signal properties (match molecular binding affinity to temporal bandwidth, protein stability to photon energy).
- Specialize neural circuits to match task (rod circuit for starlight, cone circuit for daylight).
- Optimize across levels, from molecules to neural circuits.

Adapt, match, learn, and forget

- Adapt output capacity to predicted range of inputs.
- Match capacity across levels (symmorphosis).
- Learn in order to improve future predictions.
- Forget in order to preserve storage capacity.

Notes

Introduction

1. According to Darwin, "the best definition of a high standard of organization is the degree to which the parts have been specialized, and natural selection tends toward this end, inasmuch as the parts are thus enabled to perform their functions more efficiently" (Darwin, 1859, slightly edited for brevity).

2. Anticipatory regulation has been termed allostasis (stability through change) to distinguish it from homeostasis (stability through constancy) (Sterling, 2012).

3. Here we feel in good company. Santiago Ramón y Cajal stated a century ago:

. . . we must state that such terms as goals, designs, and refinements, do not indicate an intentional or preconceived plan in the evolution of nature; they are only variations and adaptations that have prevailed because of their utility in the struggle for survival. (Ramon y Cajal, 1909) (edited for brevity)

And more than 150 years ago, Darwin wrote:

It has been said that I speak of natural selection as an active power or Deity; but who objects to an author speaking of the attraction of gravity as ruling the movements of the planets? Everyone knows what is meant and is implied by such metaphorical expressions; and they are almost necessary for brevity. So again it is difficult to avoid personifying the word Nature; but I mean by Nature, only the aggregate action and product of many natural laws, and by laws the sequence of events as ascertained by us. With a little familiarity such superficial objections will be forgotten. (Darwin, 1859)

Chapter 1

1. The following sections draw heavily on two wonderful books by G. L. Glegg, *The Design of Design* (Glegg, 1969/2009a) and *The Selection of Design* (Glegg, 1972/2009b).

2. A comprehensive account of design is given by Pahl, Beitz, Feldhusen, and Grote (2007, chapters 1, 2, 3, and 7).

3. This history of the Model T design draws on Casey (2009).

Chapter 2

1. There is overlap in that some small multicellular organisms can be smaller than a large unicellular organism and live a shorter time. For example, rotifers are among the smallest multicellular animals and are about the same size as *Paramecium*. The smallest insects, parasitic wasps, are only slightly larger.

Chapter 3

1. This occurs via hormones detected in the brain by the *subfornical organ*—a circumventricular organ (Fry et al., 2007). The gut produces additional satiety hormones, such as cholecystokinin (CCK) and neuropeptide Y (NPY) that are also detected in brain.

2. Antidiuretic hormone (ADH), also called vasopressin.

3. Later chapters explain more. For now consider some numbers: macaque brain, which commonly is treated as a surrogate for human, is smaller by about 16-fold in weight and neuron number. The number of cortical areas expands from small mammals to human by about 10-fold: 20 in mouse and 200 in human. The breakdown in macaque is 30–40 areas for vision, 15–20 areas for hearing, 15–20 areas for tactile sensing, and more than 10 areas for motor control (Kaas, 2008).

4. Certain small neurons, such as spinal Renshaw cells, cerebellar granule cells (chapter 7), and cortical inhibitory neurons, fire at high rates in brief bursts. However, their mean rates over seconds and minutes are low.

Chapter 4

1. There are exceptions, for example, the tiny vole whose high metabolic rate requires frequent foraging and whose life therefore departs from the standard circadian rhythm (van der Veen et al., 2006).

2. In vertebrate neuroanatomy a neuron cluster is called "nucleus"—but since that can be confused with a cell nucleus, and the term doesn't apply to insect neuroanatomy, we use "cluster."

3. "Sham rage" is easily elicited when the cortex is removed—so the complex pattern does not require the cortex, but without the cortex, the pattern is poorly directed since it has no access to the critical sensory processors.

4. Certain neuron clusters in this network have intrinsic circadian oscillators that require entrainment by SCN but not continuous signaling.

5. Another example is the *area postrema* (figure 4.4) which locates near the brainstem generator for respiratory patterns and senses blood pH. Collecting this data wirelessly, it communicates via short wires to the respiratory center, providing key information for use in regulating blood levels of CO_2. The area postrema has various additional functions.

6. See Chung and Coggeshall (1983). This work shows that the ventral funiculus (serving local motor circuits) has sevenfold more myelinated than unmyelinated fibers whereas the lateral funiculus (serving local sensory circuits) has more unmyelinated than myelinated fibers.

7. Of course, larger mammals use mostly fine fibers but also need a few thick ones (3–22 µm in humans) for transiently high information rates and high conduction velocity. Nevertheless, the diameter distributions of the human and rodent pyramidal tract are nearly identical but for a long tail toward larger diameters.

8. The rule of thumb for calculating diameter of myelinated axons from conduction velocity and vice versa is 1 µm ~ 6 m/s.

9. This assumes that cross sections of ventral horn and hypothalamic pattern generators are equal, that hypothalamic pattern generators extend for 10 mm, and that the spinal cord extends for 1 m.

10. This differs from the standard slogan: "Neurons that fire together wire together." That slogan refers to how the connections are established during development and learning, whereas our point refers to the principle *minimize wire*.

11. These are temperature receptors gathered to form an imaging organ of low spatial resolution. These receptors are served by the trigeminal nerve that innervates the skin on our face, cornea, teeth, and so on (Newman & Hartline, 1981)

12. This is for human, but the range in octaves is similar for guinea pig (54 Hz–50 kHz), whose auditory axons are shown in figures 4.6 and 4.8.

13. This occurs also for tactile exploration with finger tips and tongue.

14. A mechanism that integrates across sensors should weight most strongly the sense with the most reliable information. In daylight this would be vision, but in twilight, as vision becomes noisier, the mechanism relies more on sound and touch (Burge et al., 2010).

15. From retina, inferior colliculus, spinal, and brainstem somatosensory clusters. Thus, they have undergone some processing, but nothing like what has occurred between V1 and frontal/parietal cortex.

16. Projection is from deep superior colliculus to mediodorsal nucleus of thalamus to frontal eye field (Sommer & Wurtz, 2006).

17. The *dorsal column nuclei*.

18. The thalamic pulvinar and mediodorsal nucleus reduce rates from superior colliculus by 2–3 times (Berman & Wurtz, 2010). The ventral anterior and ventrolateral thalamic nuclei reduce rates from the deep cerebellar nuclei and striatum by as much as fivefold (D. Contreras and M. Farries, private communications).

19. For example, they implement similar spatial, chromatic, and temporal sampling and express similar numbers of ganglion cells projecting to the brain (Sterling, 2004a).

20. Mammal and insect occupy the same taxonomic levels: class Mammalia belongs to the phylum Vertebrata, and class Insecta belongs to the phylum Arthropoda. Mouse, monkey, and human belong to different orders: Rodentia and Primate; fly, bee, and locust also belong to different orders: Diptera, Hymenoptera, and Orthoptera.

21. In a delayed match-to-sample task, a bee flies into a y-maze and has to choose between two colors, one of which is rewarding. The bee learns to associate a color displayed at the maze entrance with the color of the reward inside the maze. In the symbolic delayed match-to-sample task, the bee shows that she can generalize. Whatever pattern is displayed at the maze entrance (e.g., horizontal stripes), once inside she heads for it to collect her reward.

Chapter 5

1. Some wireless messengers, such as steroid hormones, are nonpolar and therefore penetrate the cell membrane and the nuclear membrane to bind a protein within the cell nucleus, changing its conformation and so delivering its information.

2. This pleased Shannon's employer, Bell Telephone. Bell prospered immensely, and Shannon had the satisfaction of seeing his mathematical theory revolutionize information technology.

3. *Power spectrum* plots the power of each frequency against frequency.

4. Allostery serves many functions besides information processing and in fact mediates most of a cell's vital processes. For example, it moves molecular motors, pumps ions across membranes, reads DNA, converts energy from sugar and oxygen into ATP. Allostery is, therefore, a cornerstone of the brain's winning technology, cell and molecular biology. See Monod (1971) and Alberts et al. (2008).

5. Wikipedia gives the following definition. "A finite-state machine (FSM) or finite-state automaton (plural: automata), or simply a state machine, is a mathematical model of computation used to design both computer programs and sequential logic circuits. It is conceived as an abstract machine that can be in one of a finite number of states. The machine is in only one state at a time; the state it is in at any given

time is called the current state. It can change from one state to another when initiated by a triggering event or condition; this is called a transition. A particular FSM is defined by a list of its states, and the triggering condition for each transition."

6. When a bead of sweat drops to the floor from an author's brow its kinetic energy is equivalent to more than 10^{12} bits.

Chapter 6

1. We routinely practice arithmetic and algebra, but how our brains use circuits of identified neurons to generate this behavior remains a mystery.

2. There are cases where noise is helpful, but, more often than not, noise is deleterious (McDonnell & Ward, 2011).

3. The hydrolysis of 2,000 ATP molecules to ADP releases $2,000 \times 25\ k_B T$ joules $= 5 \times 10^4\ k_B T$ joules. The current carried by the 6,000 sodium ions passing through the channel delivers $2.4 \times 10^4\ k_B T$ joules of energy.

4. The durations of sodium current, potassium current, sodium inactivation, refractory period, and of the action potential vary among neuron types by several milliseconds to satisfy particular functional requirements (e.g., Carter & Bean, 2009). The values given here are typical.

5. The human genome contains at least 232 genes for the subunits that form ion channels (Nayak et al., 2009).

6. One alternative design, using pumps located in other cells to create an electrical driving force, is used by invertebrate mechanosensors (from the simple *Hydra* to more complicated forms), and by the hair cells of the inner ear. Here pumps in the stria vascularis create and maintain the electrical potential in the scala media that drives transduction current into hair cells.

7. Not all brains adopt this complete package. In chapter 2 we saw that *C. elegans*, which is relatively small and slow, has synapses but no sodium action potentials.

Chapter 7

1. James Eberwine estimates that that there are approximately five- to eightfold more protein species than mRNA species (due to posttranslational modifications). That would be 50,000 to 80,000 different types of proteins, each at a differing abundance.

2. Order of magnitude estimate based on 3.3×10^9 ATP per second for a cortical pyramidal cell firing at 4 Hz (Attwell and Laughlin 2001) and transmitting 8 bits per spike.

3. Certain electrical synapses can also be more complicated: some rectify to serve as a diode, respond to neuromodulators, undergo long-term potentiation, and so on (Pereda, 2014).

4. We have added an estimate for the cost of SNAREs. Assuming that each of the four SNARE proteins is phosphorylated × 5, and that three SNAREs release their energy to discharge a vesicle, the total is 60 ATP. This small number is realistic. Attwell and Laughlin (2001) suggested a minimum cost of 10 ATP based on Siegel (1993).

5. This basic picture applies to a variety of transmitters: glutamate, GABA, acetylcholine, epinephrine, dopamine, and serotonin. All use vesicles in this size range; all use specialized vesicular transporters for loading, and all use low affinity receptors at conventional synapses. All transmitters also use high affinity receptors for greater distances and greater times, as will be explained.

6. Calcium enters only if the channel includes a particular subunit (*GluR2*), as will be further explained, serving as a chemical messenger.

7. Vesicular glutamate concentration does vary across the vesicles at a synapse; this causes variations in cleft concentration which are a significant source of noise. See Wu et al. (2007).

8. When receptors are removed from or added to a cluster by endocytosis or exocytosis, the unit change is about 20 receptors.

9. These abbreviations stand for long chemical names of the molecules whose binding competition with glutamate led to the discovery of the two receptor types.

10. Distance for signal amplitude to reach $1/e = 37\%$.

11. Because spines are so small, it is difficult to estimate neck resistance, and published values range from 1 MΩ to 500 MΩ. This wide range of resistances in one type of neuron (CA1 pyramidal cells in mouse hippocampus) is associated with large proportional changes in neck length and diameter. These variations suggest that neck resistance is actively adjusted to change the amplitude of the EPSP in the head and the current delivered via the neck by 20-fold or more (Tønnesen et al., 2014).

12. In insects the cell body is separated from synapses by a long, naked neurite. The spike initiation site is on the primary neurite, near the point at which it becomes the axon (Burrows, 1996). Certain invertebrate neurons have more than one spike initiation site (Zecevic, 1996). In vertebrates spikes are initiated at the axon's initial segment, roughly 10–100 μm from the point at which the axon leaves the cell body (Debanne et al., 2011; Kole and Stuart, 2012).

13. This is something like *E. coli*'s permease informing its DNA that lactose has arrived and new proteins are needed to exploit it (figure 2.2).

Chapter 8

1. Finer spatial detail could be gained by longer exposure, but since photons follow Poisson statistics, doubling the image quality (S/N) would require quadrupling the exposure—to nearly a second. This strategy works for a slow-moving amphibian, but not for a fast-moving mammal.

2. Convergence varies with species (reviewed in Sterling, 2004a).

3. This paper reported a low value for the mouse rod S/N and thus a high value for false negatives. Subsequent studies with improved recordings find a higher S/N that would relax the threshold and give fewer false negatives (Cangiano et al., 2012).

4. The proportion declines at highest intensities as rod photopigment bleaches and recovers slowly whereas cone photopigment regenerates rapidly (Yin et al., 2006; Borghuis et al., 2009).

5. Following exposure to bright light, it takes more than 20 minutes to replenish all rod opsins with cis chromophore.

6. Although cones pack densely in the primate fovea, the fovea occupies such a small area that the overall ratio is as stated.

7. R* has sites for phosphorylation, but under physiological conditions it is inactivated by arrestin within 20 ms, without phosphorylation (Yau & Hardie, 2009).

8. Many nocturnal insect photoreceptors make longer microvilli to catch more photons, but all microvilli have approximately the same diameter. To guide photons efficiently, the waveguide has a rounded cross section. Consequently, a few microvilli are less than 1 µm long (figure 8.9).

9. Waveguiding sets the same limit to the diameter of a foveal cone inner segment (reviewed in Sterling, 2004a).

10. A fly photoreceptor also uses a light-adaptation mechanism that acts within seconds to reduce the number of photons reaching its microvilli. Pigment granules are moved up against the photosensitive waveguide. Here they reduce the photon flux in the waveguide 100-fold by absorbing light as it is internally reflected at the waveguide's boundary (strictly speaking, they attenuate the evanescent wave). Because the pigment granules are moved mechanically to attenuate light, then by analogy with the human pupil, the fly's mechanism is referred to as a longitudinal pupil. Reviewed by Hardie (1984).

11. Calcium shifts the blowfly curve by 2 log units, mainly at lower light levels. At higher light levels the longitudinal pupil shifts the curve 2 more log units.

12. Neither a cone's bit rate nor its energy consumption has been measured. We assume that a cone uses similar energy as a rod midway through its dynamic range.

13. Efficiency also increases with light level because the input S/N improves, but this has less effect on efficiency than gain control.

14. Bandwidth is defined conventionally as the frequency at which the fraction of transmitted power is $1/\sqrt{2}$. In other words, signal amplitude is halved. This frequency is well below the true cutoff frequency, above which no detectable signal is transmitted. As a rule of thumb, the cutoff occurs at the frequency at which transmission reduces signal amplitude by 99%. When an input signal at this frequency has its maximum amplitude, a contrast of 1 is reduced to 1%, which is close to the threshold for detection by the brain.

15. The contrast of a dark target is increased by making the background brighter and this is achieved by increasing sensitivity to the background.

Chapter 9

1. These percentages are for the housefly *Musca* (Strausfeld, 1976, tables 5.1–5.6). They will be similar in the slightly larger blowfly, *Calliphora*.

2. R7 and R8 form a narrower waveguide, so they receive slightly fewer photons.

3. See also chapter 3, "Ramón y Cajal's Hunting Case Watch" in Strausfeld (2012), for a beautifully illustrated account of these exquisite neural circuits.

4. To illustrate this point, Laughlin et al. (1998) observe that this could be done by operating the system linearly and tuning each synapse to deliver its own band of temporal frequencies.

5. Examples include the bipolar cells and ganglion cells of vertebrate retina, neurons in the mammalian lateral geniculate nucleus (chapters 11 and 12). Neurons in the olfactory glomeruli of vertebrates and insects also remove redundancy with lateral inhibition.

6. H. Keffer Hartline (1903–1983) was awarded the Nobel Prize in 1967 for discovering how sensory neurons implement lateral inhibition. He made these discoveries in an advantageous model system that evolved 400,000,000 years ago, the compound eye of the horseshoe crab, *Limulus*.

7. Predictive coding removes the background signal which in daylight is 3 times the mean amplitude of the response to contrast.

8. Effects that can be attributed to voltage-gated channels, fast voltage transients and high-frequency noise, have been observed in photoreceptor terminals (Weckström et al., 1992).

9. An efficient molecular mechanism, desensitization of the histamine-gated chloride channels, acts postsynaptically to reduce signal amplitude without employing

extra synapses (Skingsley, Laughlin, & Hardie, 1995), but its contributions to predictive coding and circuit efficiency have not been determined.

10. If another potential mechanism for generating lateral inhibition presynaptically exists—namely, lateral-inhibitory interneurons that synapse onto photoreceptors— then the mechanism must be weak and slow. The connectome demonstrates that a photoreceptor terminal receives fewer than five synapses, at most one from L2 and the rest from neurons that spread laterally, the amacrines and the processes of L4. These two neurons are well placed to mediate lateral interactions between cartridges, but their roles are unknown.

11. Reducing low frequencies is a winning strategy because information increases as $\log_2(S/N)$. Consequently, when a high frequency signal and a low frequency signal are reduced by the same proportion, each loses the same number of bits. However, because the low frequency signal's amplitude is larger, its reduction frees up more dynamic range.

Chapter 10

1. By comparison, a rod photosensor expresses 4×10^7 receptor molecules in order to capture a substantial fraction of the arriving photons.

2. *Merkel's disk*, of different morphology and located more superficially, transmits lower frequencies (5–40 Hz) that we sense as light touch. Recent studies suggest that the epithelial Merkel's cells themselves can release a transmitter to evoke spikes in the sensory neuron. So the story for skin sensors continues. See Maksimovic et al. (2013).

3. The skin's mechanical arrangements are relatively few—mostly named for their discoverers, Italian and German histologists of the late nineteenth century— including Ruffini, Pacini, Meissner, and Merkel. However, the possibilities for mechanical amplification and filtering, rectification, and so on are limitless and are exploited by every organism: wind receptors, hair bending, and so forth.

4. Processing also occurs prior to a synapse, for example, by voltage-sensitive and calcium-sensitive potassium channels in the hair cell membrane. These contribute to fast and strong adaptive mechanisms that remove redundancy from the voltage that drives synaptic calcium channels. Thus, the synapse is part of a system that uses several prominent mechanisms to ensure that the right bits are encoded as spikes.

5. Although the cone synapse uses similar numbers of synaptic ribbons, they are uniform. The cone synapse, like the hair cell, performs custom filtering before digitizing, but because it can tolerate temporal delays, its spikes can be postponed to a second processing stage, greatly reducing the spike rates.

6. This account omits numerous complexities of the vestibular system. There are different organs with different mechanical couplings and varied species of ion channel—for example, multiple isoforms of voltage-gated calcium and sodium channels. Correspondingly there are different synaptic arrangements—but they all show the basic pattern described here. See Eatock et al. (2008) and Eatock and Songer (2011). For a start at sorting out these complexities using information theory, see Sadeghi et al. (2007).

Chapter 11

1. The human eye uses 4.5 megapixels (cones) concentrated in a central fovea. Mars rover uses 96 megapixels (charge-coupled device sensors) elongated in a "streak" (4,000 pixels high × 24,000 pixels wide) arranged to pan horizontally.

2. It is not images that are stored, but their lessons. One hundred milliseconds of raw photoreceptor data would be 10 Mbits; in 1 minute, 60 Mbits; in an hour, 3,600 Mbits = 3.6 Gbits. The brain could not store this, so it stores the analysis—the conclusions and the lessons after processing by temporal and frontal areas. What we recall from a visual image, no matter how vivid and raw it may seem, has been shaped and reshaped profoundly. That is why an "eyewitness" is so often wrong (Neisser, 1997).

3. In dim light images are impoverished—starved for photons—so the retina uses various mechanisms to extend summation in both space and time. The design principles are the same (Sterling, 2004a).

4. This includes cat, monkey, and human versus about 20 μm for "avascular" retinas, such as rabbit and guinea pig. Birds have finer spatial vision and thus finer optics and a denser cone array; this requires more circuitry and thus a thicker retina—therefore, a special apparatus (*pecten*) to supply energy and remove waste.

5. This chapter answers these questions for mammals. It includes data from human, monkey, cat, rabbit, guinea pig, and mouse. Because the retina's fundamental plan is identical across these species, we will not burden the reader by noting species in every case. A determined reader can track them down via the citations.

6. Approximately 100 glutamate pulses per 100 ms from about 20 active zones per cone and about 1,000 cone terminals. $\sqrt{10^5} > 300$.

7. It is sobering to realize that Cajal, who brilliantly launched modern studies of the retina, totally rejected the hypothesis of intercellular coupling; moreover, he repeatedly ridiculed its proponents. Yet the retina's first three neural integrative mechanisms use intercellular coupling (gap junctions): rod–cone, cone–cone, and horizontal cell–horizontal cell. There is a lesson: in denying the existence of something smaller than you can actually resolve, at least be polite because you could well be wrong.

8. The calculation: about 300 docking sites per terminal releasing about 600 vesicles s^{-1} gives about 2 vesicles/site s^{-1}. That amounts to 0.2 synaptic vesicles released per site for a 100-ms interval (Sterling 2004a; Sterling & Freed, 2007; DeVries et al., 2006).

9. Haverkamp et al. (2000) give the ribbon numbers. The calculation: S/N ~ 2.2 × √20 ~ 10. Strong stimuli can cause steep voltage changes that can synchronize release and thereby transmit a higher information rate than by simply modulating a tonic rate (DeVries, 2000; Rabl et al., 2005).

10. Note that the inner hair cell's postsynaptic knob is near to voltage-gated sodium channels, but the cone bipolar postsynaptic dendrites are farther away, about 100 μm. This allows the cone bipolar to integrate via multiple mechanisms before the next round of analogue-to-discrete recoding.

11. High rate: about 100 receptor clusters for a midget bipolar cell; lower rate: 60, 35, and 25 receptor clusters for several nonmidget types (Haverkamp et al., 2001b).

12. For primate, see Dacey (2004); for mouse, see Kim et al. (2010). Smallest mouse ganglion cell dendritic field is about 55 μm and largest is about 350 μm

13. The discovery of ganglion cell structural and functional diversity has extended over half a century. The following references are offered as highlights: Lettvin et al. (1959); Barlow and Levick (1965); Enroth-Cugell and Robson (1966); Cleland and Levick (1974); Boycott and Wassle (1974); Stone and Fukuda (1974); Kolb et al. (1981); Rodieck and Brening (1984); Amthor et al. (1989); Rockhill et al. (2002); van Wyk et al. (2006); Zhang et al., (2012).

14. Every mammalian retina exhibits this basic pattern. Some show subtle variants. For example, the wide-field, brisk-transient type in cat (alpha/Y) collects only from the fast-transient bipolar type (Freed & Sterling, 1988), but the narrow-field, brisk-sustained type (beta/X) collects from three bipolar types in a specific ratio: 4:2:1 (Cohen & Sterling, 1992). The brisk-sustained type thus assembles its bandwidth by combining inputs from different bipolar types in specific proportions. Brisk-sustained types have not been identified in mouse (though they are present in rabbit, guinea pig, and rat), but mouse bipolar cells certainly release glutamate in sustained manner, and brisk-transient ganglion cells exhibit sustained responses (Borghuis et al., 2014).

15. In fact, the smallest, slowest type (local-edge) branches at the junction of OFF and ON strata, collecting most contacts from the slowest OFF bipolars but some also from the slowest ON bipolars. This uses the bright contrasts, which are sparse on this spatial scale, to fill out the information capacity of the lowest rate ganglion cell. Thus, the local-edge type matches the proportion of dark and bright contrasts available on that fine scale, which contains insufficient slow ON information to justify a separate array.

16. Mouse, about 3 mm; cow, about 30 mm eye diameter.

17. For the AII's role in spatially summing rod bipolar inputs, a much broader field would better, and the AII cells achieve this through intense coupling via many large gap junctions (Sterling et al., 1988).

18. Primate retina uses several more with larger fields to capture higher temporal frequencies. Thus, a brisk-transient type with twofold wider dendritic field (Crook et al., 2011) and a second type of blue–yellow cell, also tuned to lower spatial frequencies (Schein et al., 2004).

19. Cone photovoltage (analogue) → voltage-gated calcium channels and cone synaptic quanta (discrete) → bipolar voltage (analogue) → voltage-gated Ca channels and bipolar quanta (discrete) → ganglion cell membrane voltage (analogue) → spikes to brain (discrete, i.e., pulsatile).

Chapter 12

1. Experts on the LGN will notice that this treatment draws on cat, rodent, and primate but omits various details to allow a readable narrative on our theme: efficiency of design.

2. The calculation: brisk-sustained ganglion cell axon delivers about 1,600 active zones to LGN; 25% go to inhibitory interneurons, leaving 600 for standard and 600 for lagged relay cells. Of the 600 contacts to a brisk-sustained relay cell, half (300) are from the private line ganglion cell and the rest from about five overlapping ganglion cells, each contributing 60 contacts. The private line ganglion cell that delivers only 300 contacts to its main target diverges the rest to relay cells that it overlaps.

3. This paper shows how this works for a second type of Y cell in lamina C with slightly different properties than the main Y cell in the A laminae.

4. There is a history of research on "blindsight." This is demonstrable visual behavior that remains in the absence of V1. Subjects may deny seeing anything but show behaviorally that they do see. Given the nearly two dozen pathways to lower brain regions, and the rule to carry out the behavior at lowest possible level, this should be no surprise. But one should keep in mind that "blindsight" is too rudimentary to support independent navigation.

5. Experts will recognize that our treatment simplifies. It ignores differences between species and avoids discussing points that are not yet fully resolved. Therefore, we consider this interpretation of V1 design as provisional.

6. Primate has its own terminology: the high-rate neurons (b-t) are termed M and supply upper layer 4 (primate 4Cα) whereas medium-rate neurons (b-s) are termed P and supply lower layer 4 (primate 4A, 4Cβ). The lowest-rate relay neurons are termed K and supply upper layers (1–3). Cat has a different terminology, but the same pat-

terns hold: high-rate (Y) terminates in upper layer 4; medium-rate (X) deeper in layer 4; lowest rate (W) in layers 2–3.

7. Valverde Salzmann et al. (2012). This paper shows strong responses to red–green isoluminant stimuli, but not to blue–yellow. However, studies of other primate species report blue–yellow signals as well.

8. Economides et al. (2011). This paper shows that patch neurons respond to achromatic as well as chromatic stimuli at higher than average rates.

9. A dark contrast that excites an adjacent line of OFF relay cells to "push," simultaneously disinhibits by suppressing a cospatial line of ON relay cells that drive a different inhibitory input to "pull."

10. Essentially identical mechanisms occur in V1 and V2 (cat). V2's larger spatial receptive fields allows tuning for threefold higher velocities. This explanation of early motion coding was provided by L. A. Palmer.

11. Learning has been suggested as a mechanism for complex cell generalization (Karklin & Lewicki, 2009).

12. This account is simplified. For example, the full V1 → V2 projection comes from layers 2–3, 4A, 4B, 5, and 6. Only layers 1 and 4C ignore V2 (Sincich & Horton, 2005).

13. This contrasts with high-rate outputs from cerebellar cortex that require a stage of convergence to relay neurons in order to reduce numbers and rates (chapter 4).

14. Just as the LGN relay neurons are gated by feedback from V1, each cortical area also receives reciprocal projections from the higher order cortical area(s) to which it projects (Felleman & Van Essen, 1991).

15. The upper layers also project to the opposite hemisphere via the giant commissure, the *corpus callosum* (chapter 13).

16. All connections between cortical areas are reciprocal. Upper layers forward (V1 → V2) and lower layers the reverse.

17. Our account here of Gestalt grouping draws on this paper.

18. Bertold Brecht, "To Posterity #1."

19. These are the *parahippocampal place area, transverse occipital sulcus*, and *retrosplenial cortex* (Nasr et al., 2011).

20. This oversimplifies V2 and its connections—for example, we have omitted "pale stripes" and their connections. But the basic theme is captured here. We too are shedding detail!

21. Early exposure to faces of one race, excluding all others, stably enhances sensitivity to recognizing emotional expression of that race and enhances amygdala

activity to out-group faces. Thus, an "early warning system" to rapidly recognize and trigger alarm to out-group faces appears to originate at least partly in neural development (Telzer et al., 2013).

22. We recognize that unsuspected connections continue to be discovered as methods for tracing grow more sensitive. A new connection would not lead us to reject the principle *send only what is needed* but rather to investigate what value accrues.

Chapter 13

1. For perspective, adult human nervous system has of the order of 1.5×10^6 km of axon, so the granule cell saving is 0.03% of total wire.

2. Braitenberg and Schüz (1998) report 35% for each but may have used somewhat different criteria. This fraction is intended as a rule of thumb—to emphasize that a constraint exists.

3. In figure 13.5, the thickness of the retinal synaptic layers (inner + outer) ÷ the sum of those layers + the cell body layers (inner nuclear + ganglion cell layer) = 3/5; and in figure 13.9, the thickness of outer synaptic layer + 30% of granule cell layer occupied by synapses ÷ total thickness is approximately 3/5.

4. Certain specialized amacrine types contribute long axons (>1 mm). The cells and axons are sparse; nevertheless, because the axons are so long, they contribute a locally dense meshwork (Dacey, 1989).

5. Spines are used by a few specialized types, such as the dopamine amacrine cell—a hint that, unlike most retinal synapses, it might participate in stable plastic changes (see chapter 14).

6. Rat: 0.3 mm flanking Purkinje cell bodies × 5-mm parallel fiber length = 0.3 mm^2 × 0.37 mm thick = 0.55 mm^3

7. Convergence/divergence = 200 for cerebellar circuit (200,000/1,000), but convergence/divergence is about 10 for cerebral pyramidal neuron (10,000/1,000). A few giant pyramids may converge 60,000 synapses according to Braitenberg and Schüz.

8. Granule cell is about 7 μm in diameter, the same as a red blood cell, but the latter reduces its volume by ejecting the nucleus. This limits the lifetime of the red blood cell to about 45 days whereas the granule cell lasts for the life of the animal.

9. This irreducibly thin axon is confined to a low mean spike rate, and a key task of the mossy-fiber-to-granule-cell connection to reduce spiking to a rate that the parallel fiber can support.

10. This is 4/3 times the ratio between radii squared (axon radius = 0.15 μm, varicosity radius = 0.75 μm.

11. Ratio of parallel fiber (pf) volume to varicosity volume: pf radius squared $(0.15 \ \mu m)^2 \times$ length $(6.25 \ \mu m) = 0.141 \ mm^3 \div 4/3 \times$ varicosity radius cubed $(0.75)^3 = 0.563 \ \mu m^3$; pf length excluded from varicosity $= 6.25 \ \mu m - 0.3 \ \mu m$ (diameter of varicosity) $= 5.95 \ \mu m$; volume of that part of pf $= 0.225 \times 5.95 = 0.134 \ \mu m^3$. So, pf $=$ varicosity $= 0.134 + 0.563 = 0.696$; ratio $= 4.95$.

12. Sevenfold greater mitochondrial volume fraction $\times$ 3.5-fold greater cytoplasmic volume (Perge et al., 2012).

13. If the mitochondrial volume fraction is lower in the deep neurons, this would reduce energy costs by an additional factor.

14. Some Purkinje axons do leave the cerebellum for the vestibular neuron cluster, which, being nearby, also keeps the axons short.

15. Cat: $\sim 1.2 \times 10^6$ Purkinje cells versus $\sim 0.2 \times 10^6$ ganglion cells. Human: $\sim 15 \times 10^6$ Purkinje cells versus 1×10^6 ganglion cells.

16. The number of synapses converging on one pyramidal neuron is about 10^3 to 10^4. Thus, the divergence and convergence of each type of pyramidal cell exceed that of a retinal ganglion cell but are far less than for a Purkinje cell.

17. Folding only becomes prominent in higher primates and dolphins (Cetacea).

18. Mean diameter of retino–LGN axons: brisk-sustained (X) $= 2.0 \ \mu m$; brisk transient (Y) $= 3.0 \ \mu m$ (Sur et al., 1987). Mean diameter of LGN–area 17 axons: brisk-sustained (x) $= 1.3 \ \mu m$; brisk transient (Y) $= 1.8 \ \mu m$ (Humphrey et al., 1985).

19. Whether GABA enhances or suppresses spiking depends on whether the particular chloride transporter that is expressed holds local E_{Cl} positive or negative to resting potential (figure 7.4).

20. Assuming similar frequency of synaptic outputs for pyramidal and nonpyramidal neurons, given that inhibitory synapses are only 10% of total.

21. Recognition that that the right hemisphere in humans serves unique functions and is not simply a "spare part" is relatively recent. For example, the 1943 edition of *Human Neuroanatomy*, the text by Strong and Elwyn, stated, "in man the higher cortical functions are vested principally in one cerebral hemisphere, the left one in right-handed individuals—lesions of the other hemisphere producing as a rule no recognizable disturbances" (see Bogen, 1969).

Chapter 14

1. A professor of ichthyology famously justified his reluctance to learn students' names, saying, "Whenever I learn a name, I forget a fish." This has been considered merely an excuse for laziness, but perhaps it is justified by neural design. The professor was David Starr Jordan, first president of Stanford University.

2. For a sense of the different stages, consult the works of Erik Erikson (e.g., Erikson & Erikson, 1997). For an integration of the neuroscience of memory and personal history, see Kandel (2006).

3. Conversely, when large quantities of facts and experience are obliterated, as in the retrograde memory loss of Korsakoff's syndrome, the context offered by religious practice, music, or gardening may remain to knit the raveled web. See Oliver Sacks's "The Lost Mariner" (in Sacks, 1985).

4. The idea that learning occurs by changing connections and adding new ones dates back more than a century. Cajal was not the first to suggest this, but he made many relevant observations and many comments as well. See DeFelipe (2006).

5. This shaping of neural circuits to match their anticipated long-term input statistics (*plasticity*) serves the same design goals as familiar types of learning. Moreover, it uses the same mechanisms, as will later be explained.

6. Visual stimulation activates postsynaptic NMDA receptors that, under control of the enzyme PKMζ, cause insertion of the GluR1 subunit into the postsynaptic GluR receptors. This maintains potentiation, which improves contrast sensitivity and spatial acuity of the visual evoked potential (Cooke & Bear, 2010).

7. This section draws on studies over the last decade using fMRI. Summarized critically by Wandell et al. (2012). Also Dehaene et al. (2010).

8. The smallest unit of meaning (morpheme) is the smallest unit of writing—as in Chinese characters.

9. This is true as well for face areas, where face recognition is 80 ms faster for a familiar versus an unfamiliar face (380 ms vs 460 ms) (Ramon et al., 2011).

10. This is shown for motor cortex learning as early changes in the descending tract.

11. This account necessarily simplifies and compresses what is actually a complex and diverse set of mechanisms. To get some idea, consider that CaM kinase has 28 isoforms and constitutes 1%–2% of brain protein.

12. PKMζ also provides a simple mechanism for long-term depression, that is, returning a potentiated synapse to its basal condition. Again it is chemistry: an appropriate stimulus reduces the calcium influx, thereby activating a phosphatase that dephosphorylates PKMζ, restoring the initial balance of receptor cycling and the basal number of receptors.

13. However, in dopamine's role as a "teaching signal" its timing is critical because, as will be discussed, dopamine can be important in encouraging an animal to repeat the behavior.

14. For the better characterized cerebellar circuit, estimates of the numbers of associations made by a Purkinje cell's input synapses vary from 40,000 to 100, depend-

ing on the model used for storage and recall. This is reviewed by Howarth et al., (2010). Promising new techniques for resolving the engram, including those presented here, are critically reviewed by Sakaguchi and Hayashi (2012).

15. This account is certainly oversimplified, for it is based mainly on studies centered on one particular area of medial and orbital frontal cortex whereas the whole region contains 22 areas. Moreover, our commuter has dealt so far only with an inanimate bit of pastry and has not yet even gotten to work.

16. These neurons in the right hemisphere teach music, spatial mapping, and art; in the left hemisphere they teach language, history, and zoology.

17. 10^6 synapses per neuron $\times$ 10^5 neurons. Although a dopamine neuron fires at similar mean rates to a retinal ganglion cell, its cell body is about 10-fold greater in volume and its axon is also thicker. This seems attributable to its need to supply a much larger axon arbor with about 1,000-fold more release sites.

18. Arthropods use the same signal but a different transmitter, octopamine.

19. This captures the experience that follows brief block of reuptake by a snort of cocaine.

20. However, monkeys and chimpanzees do show evidence of hemispheric specialization for face recognition (Dahl et al., 2013).

21. There must be differences in brain structure between a leader and a confidence man, but they are probably subtle.

References

Abbott, L. F., & Regehr, W. G. (2004). Synaptic computation. *Nature, 431,* 796–803.

Aguera y Arcas, B., Fairhall, A. L., & Bialek, W. (2003). Computation in a single neuron: Hodgkin and Huxley revisited. *Neural Computation, 15,* 1715–1749.

Akyildiz, I. F., Brunetti, F., & Blazquez, C. (2008). Nanonetworks: A new communication paradigm. *Computer Networks, 52,* 2260–2279.

Ala-Laurila, P., Greschner, M., Chichilnisky, E. J., & Rieke, F. (2011). Cone photoreceptor contributions to noise and correlations in the retinal output. *Nature Neuroscience, 14,* 1309–1316.

Alberts, B., Johnson, A., Lewis, J., Raff, M., Roberts, K., & Walter, P. (2008). *Molecular biology of the cell.* New York: Garland Science.

Alexander, R. M. (1996). *Optima for animals.* Princeton, NJ: Princeton University Press.

Allada, R., & Chung, B. Y. (2010). Circadian organization of behavior and physiology in *Drosophila. Annual Review of Physiology, 72,* 605–624.

Allman, J. M. (1999). *Evolving brains.* New York: Freeman.

Allman, J. M., & Kaas, J. H. (1974). The organization of the second visual area (V II) in the owl monkey: A second order transformation of the visual hemifield. *Brain Research, 76,* 247–265.

Alvarez-Icaza, R., & Boahen, K. (2011). Deep cerebellar neurons mirror the spinal cord's gain to implement an inverse controller. *Biological Cybernetics, 105,* 29–40.

Amthor, F. R., Takahashi, E. S., & Oyster, C. W. (1989). Morphologies of rabbit retinal ganglion cells with complex receptive fields. *Journal of Comparative Neurology, 280,* 97–121.

Anderson, J. C., & Laughlin, S. B. (2000). Photoreceptor performance and the co-ordination of achromatic and chromatic inputs in the fly visual system. *Vision Research, 40,* 13–31.

Angueyra, J. M., & Rieke, F. (2013). Origin and effect of phototransduction noise in primate cone photoreceptors. *Nature Neuroscience, 16*(11), 1692–1700. Epub 2013 Oct 6.

Arbuthnott, G. W., & Wickens, J. (2007). Space, time and dopamine. *Trends in Neurosciences, 30*(2), 62–69.

Ardiel, E. L., & Rankin, C. H. (2010). An elegant mind: Learning and memory in *Caenorhabditis elegans*. *Learning & Memory, 17*, 191–201.

Arendt, D. (2008). The evolution of cell types in animals: Emerging principles from molecular studies. *Nature Reviews. Genetics, 9*, 868–882.

Arenz, A., Silver, R. A., Schaefer, A. T., & Margrie, T. W. (2008). The contribution of single synapses to sensory representation in vivo. *Science, 321*(5891), 977–980.

Arruda-Carvalho, M., Sakaguchi, M., Akers, K. G., Josselyn, S. A., & Frankland, P. W. (2011). Posttraining ablation of adult-generated neurons degrades previously acquired memories. *Journal of Neuroscience, 31*(42), 15113–15127.

Arthur, W. B. (2009). *The nature of technology: What it is and how it evolved*. London: Allen Lane.

Asaad, W. F., & Eskandar, E. N. (2011). Encoding of both positive and negative reward prediction errors by neurons of the primate lateral prefrontal cortex and caudate nucleus. *Journal of Neuroscience, 31*(49), 17772–17787.

Ashmore, J. (2008). Cochlear outer hair cell motility. *Physiological Reviews, 88*, 173–210.

Atallah, B. V., Bruns, W., Carandini, M., & Scanziani, M. (2012). Parvalbumin-expressing interneurons linearly transform cortical responses to visual stimuli. *Neuron, 73*, 159–170.

Attwell, D., & Gibb, A. (2005). Neuroenergetics and the kinetic design of excitatory synapses. *Nature Reviews Neuroscience, 6*, 841–849.

Attwell, D., & Laughlin, S. B. (2001). An energy budget for signaling in the grey matter of the brain. *Journal of Cerebral Blood Flow and Metabolism, 21*, 1133–1145.

Ayzenshtat, I., Gilad, A., Zurawel, G., & Slovin, H. (2012). Population response to natural images in the primary visual cortex encodes local stimulus attributes and perceptual processing. *Journal of Neuroscience, 32*(40), 13971–13986.

Azevedo, F. A. C., Carvalho, L. R. B., Grinberg, L. T., Farfel, J. M., Ferretti, R. E. L., Leite, R. E. P., et al. (2009). Equal numbers of neuronal and non-neuronal cells make the human brain an isometrically scaled-up primate brain. *Journal of Comparative Neurology, 513*, 532–541.

Baden, T., Euler, T., Weckström, M., & Lagnado, L. (2013). Spikes and ribbon synapses in early vision. *Trends in Neurosciences, 36*(8), 480–488.

Bailey, P., & von Bonin, G. (1951). *The isocortex of man*. Urbana: University of Illinois Press.

Balasubramanian, V., & Sterling, P. (2009). Receptive fields and functional architecture in the retina. *Journal of Physiology, 587*, 2753–2767.

Bard, P., & Mountcastle, V. B. (1947). Some forebrain mechanisms involved in expression of rage with special reference to suppression of angry behavior. *Association for Research in Nervous and Mental Diseases, 27*, 362–404.

Bargmann, C. I. (2012). Beyond the connectome: How neuromodulators shape neural circuits. *BioEssays, 34*, 458–465.

Barlow, H. (1989). Unsupervised learning. *Neural Computation, 1*, 295–311.

Barlow, H. B. (1972). Single units and sensation: A neuron doctrine for perception? *Perception, 1*, 371–394.

Barlow, H. B., & Levick, W. R. (1965). The mechanism of directionally selective units in rabbit's retina. *Journal of Physiology, 178*(3), 477–504.

Bartlett, E. L., & Wang, X. (2011). Correlation of neural response properties with auditory thalamus subdivisions in the awake marmoset. *Journal of Neurophysiology, 105*, 2647–2667.

Baumgartner, T., Fischbacher, U., Feierabend, A., Lutz, K., & Fehr, E. (2009). The neural circuitry of a broken promise. *Neuron, 64*(5), 756–770.

Bean, B. P. (2007). The action potential in mammalian central neurons. *Nature Reviews Neuroscience, 8*, 451–465.

Belin, P., Zatorre, R. J., Lafaille, P., Ahad, P., & Pike, B. (2000). Voice selective areas in human auditory cortex. *Nature, 403*, 309–312.

Bell, C., Bodznick, D., Montgomery, J., & Batian, J. (1997). The generation and subtraction of sensory expectations within cerebellum-like structures. *Brain, Behavior and Evolution, 50*(Suppl. 1), 17–31.

Bell, C. C., Han, V., & Sawtell, N. B. (2008). Cerebellum-like structures and their implications for cerebellar function. *Annual Review of Neuroscience, 31*, 1–24.

Bennett, C. H. (1982). The thermodynamics of computation—A review. *International Journal of Theoretical Physics, 21*, 905–940.

Bennett, C. H. (2000). Notes on the history of reversible computation. *IBM Journal of Research and Development, 44*, 270–277.

Berg, H. (1993). *Random walks in biology*. Princeton, NJ: Princeton University Press.

Berg, H. C., & Brown, D. A. (1972). Chemotaxis in *Escherichia coli* analysed by three-dimensional tracking. *Nature, 239*(5374), 500–504.

References

Berman, R. A., & Wurtz, R. H. (2010). Functional identification of a pulvinar path from superior colliculus to cortical area MT. *Journal of Neuroscience, 30,* 6342–6354.

Berntson, A., Smith, R. G., & Taylor, W. R. (2004). Transmission of single photon single photon signals through a binary synapse in mammalian retina. *Visual Neuroscience, 21*(5), 693–702.

Berry, M. J., Warland, D. K., & Meister, M. (1997). The structure and precision of retinal spike trains. *Proceedings of the National Academy of Sciences of the United States of America, 94*(10), 5411–5416.

Bevan, A. T., Honour, A. J., & Stott, F. H. (1969). Direct arterial pressure recording in unrestricted man. *Clinical Science, 36,* 329–344.

Bialek, W. (2012). *Biophysics. Searching for principles.* Princeton: Princeton University Press.

Bickart, K. C., Hollenbeck, M. C., Barrett, L. F., & Dickerson, B. C. (2012). Intrinsic amygdala–cortical functional connectivity predicts social network size in humans. *Journal of Neuroscience, 32*(42), 14729–14741.

Bizzi, E., & Cheung, V. C. K. (2013). The neural origin of muscle synergies. *Frontiers in Computational Neuroscience, 7,* 51.

Bock, D. D., Lee, W.-C. A., Kerlin, A. M., Andermann, M. L., Hood, G., Wetzel, A. W., et al. (2011). Network anatomy and in vivo physiology of visual cortical neurons. *Nature, 471,* 177–183.

Bogen, J. E. (1969). The other side of the brain II: An appositional mind. *Bulletin of the Los Angeles Neurological Society, 34,* 135–162.

Boiko, T. B., Van Wart, A., Caldwell, J. H., Levinson, S. R., Trimmer, J. S., & Matthews, G. (2003). Functional specialization of the axon initial segment by isoform-specific sodium channel targeting. *Journal of Neuroscience, 23*(6), 2306–2313.

Borghuis, B. G., Ratliff, C. P., Smith, R. G., Sterling, P., & Balasubramanian, V. (2008). Design of a neuronal array. *Journal of Neuroscience, 28*(12), 3178–3189.

Borghuis, B. G., Sterling, P., & Smith, R. G. (2009). Loss of sensitivity in an analog neural circuit. *Journal of Neuroscience, 29*(10), 3045–3058.

Borghuis, B. G., Looger, L. L., Tomita, S., & Demb, J. B. (2014). Kainate receptors mediate signaling in both transient and sustained OFF bipolar cell pathways in mouse retina. *Journal of Neuroscience 34*(18), 6128–6139.

Borowsky, I. W., & Collins, R. C. (1989). Metabolic anatomy of brain: A comparison of regional capillary density, glucose metabolism, and enzyme activities. *Journal of Comparative Neurology, 288,* 401–413.

Borycz, J. A., Borycz, J., Kubów, A., Kostyleva, R., & Meinertzhagen, I. A. (2005). Histamine compartments of the *Drosophila* brain with an estimate of the quantum content at the photoreceptor synapse. *Journal of Neurophysiology, 93,* 1611–1619.

Bowling, D. L., Gill, K., Choi, J., Prinz, J., & Purves, D. (2010). Major and minor music compared to excited and subdued speech. *Journal of the Acoustical Society of America, 127*(1), 491–503.

Bowling, D. L., Sundararajan, J., Han, S., & Purves, D. (2012). Expression of emotion in Eastern and Western music mirrors vocalization. *PLoS ONE, 7*(3).

Boycott, B. B., & Wässle, H. (1974). The morphological types of ganglion cells of the domestic cat's retina. *Journal of Physiology, 240*(2), 397–419.

Boyd, I. A., & Davey, M. R. (1968). *Composition of peripheral nerves.* New York: Harcourt Brace/Churchill Livingstone.

Braitenberg, V., & Schüz, A. (1998). *Cortex: Statistics and geometry of neuronal connectivity.* Berlin: Springer-Verlag.

Branco, T., & Häusser, M. (2010). The single dendritic branch as a fundamental functional unit in the nervous system. *Current Opinion in Neurobiology, 20,* 494–502.

Bray, D. (1995). Protein molecules as computational elements in living cells. *Nature, 376,* 307–312.

Bray, D. (2009). *Wetware: A computer in every living cell.* New Haven: Yale University Press.

Brecht, M., Schneider, M., Sakmann, B., & Margrie, T. W. (2004). Whisker movements evoked by stimulation of single pyramidal cells in rat motor cortex. *Nature, 427,* 704–710.

Bromberg-Martin, E. S., Matsumoto, M., & Hikosaka, O. (2010). Dopamine in motivational control: Rewarding, aversive, and alerting. *Neuron, 68,* 815–834.

Brooks, R. A. (1990). Elephants don't play chess. *Robotics and Autonomous Systems, 6,* 3–15.

Buisman, H. J., ten Eikelder, H. M. M., Hilbers, P. A. J., & Liekens, A. M. L. (2008). Computing algebraic functions with biochemical reaction networks. *Artificial Life, 15,* 5–19.

Bullock, T. H., Orkand, R., & Grinnell, A. D. (1977). *Introduction to nervous systems.* San Francisco: W.H. Freeman.

Burge, J., Ernst, M. O., & Banks, M. S. (2008). The statistical determinants of adaptation rate in human reaching. *Journal of Vision, 8*(4), 20.1–19.

Burge, J., Girshick, A. R., & Banks, M. S. (2010). Visual–haptic adaptation is determined by relative reliability. *Journal of Neuroscience, 30*, 7714–7721.

Burkhalter, A. (2008). Many specialists for suppressing cortical excitation. *Frontiers in Neuroscience, 2*, 155.

Burkhardt, D. A., Fahey, P. K., & Sikora, M. A. (2006). Natural images and contrast encoding in bipolar cells in the retina of the land- and aquatic-phase tiger salamander. *Visual Neuroscience, 23*, 35–47.

Burns, M. E., & Pugh, E. N., Jr. (2013). Visual transduction by rod and cone photoreceptors. In J. S. Werner & L. M. Chalupa (Eds.), *The new visual neurosciences* (pp. 7–18). Cambridge, MA: MIT Press.

Burr, D. (2004). Eye movements: Keeping vision stable. *Current Biology, 14*(5), R195–R197.

Burris, C., Klug, K., Ngo, I. T., Sterling, P., & Schein, S. (2002). How Muller glial cells in macaque fovea coat and isolate the synaptic terminals of cone photoreceptors. *Journal of Comparative Neurology, 453*, 100–111.

Burrows, M. (1996). *The neurobiology of an insect brain*. Oxford: Oxford University Press.

Burton, B. G. (2002). Long-term light adaptation in photoreceptors of the housefly, *Musca domestica. Journal of Comparative Physiology, A, 188*, 527–538.

Burton, B. G., & Laughlin, S. B. (2003). Neural images of pursuit targets in the photoreceptor arrays of male and female houseflies *Musca domestica. Journal of Experimental Biology, 206*, 3963–3977.

Burton, B. G., Tatler, B. W., & Laughlin, S. B. (2001). Variations in photoreceptor response dynamics across the fly retina. *Journal of Neurophysiology, 86*, 950–960.

Büschges, A., Scholz, H., & El Manira, A. (2011). New moves in motor control. *Current Biology, 21*, R513–R524.

Calkins, D. J. & Sterling, P. (1999). Evidence that circuits for spatial and color vision segregate at the first retinal synapse. *Neuron, 24*(2), 313–321.

Calkins, D. J., Tsukamoto, Y., & Sterling, P. (1996). Foveal cones form basal as well as invaginating contacts with diffuse ON bipolar cells. *Vision Research, 36*(21), 3373–3381.

Callaway, E. M. (2005). Structure and function of parallel pathways in the primate early visual system. *Journal of Physiology, 566*(Pt 1), 13–19.

Calvert, P. D., Strissel, K. J., Schiesser, W. E., Pugh, E. N., Jr., & Arshavsky, V. Y. (2006). Light-driven translocation of signaling proteins in vertebrate photoreceptors. *Trends in Cell Biology, 16*(11), 560–568.

Camerer, C. F., & Fehr, E. (2006). When does "economic man" dominate social behavior? *Science, 311*(5757), 47–52.

Caminiti, R., Carducci, F., Piervincenzi, C., Battaglia-Mayer, A., Confalone, G., Visco-Comandini, F., et al. (2013). Diameter, length, speed, and conduction delay of callosal axons in macaque monkeys and humans: Comparing data from histology and magnetic resonance imaging diffusion tractography. *Journal of Neuroscience, 33*(36), 14501–14511.

Cangiano, L., Asteriti, S., Cervetto, L., & Gargini, C. (2012). The photovoltage of rods and cones in the dark-adapted mouse retina. *Journal of Physiology, 590*(16), 3841–3855.

Carandini, M., & Heeger, D. J. (2012). Normalization as a canonical neural computation. *Nature Reviews Neuroscience, 13*, 51–62.

Carandini, M., Heeger, D. J., & Movshon, J. A. (1997). Linearity and normalization in simple cells of the macaque primary visual cortex. *Journal of Neuroscience, 17*, 8621–8644.

Carandini, M., Horton, J. C., & Sincich, L. C. (2007). Thalamic filtering of retinal spike trains by postsynaptic summation. *Journal of Vision, 7*(14), 1–11.

Caroni, P., Donato, F., & Muller, D. (2012). Structural plasticity upon learning: Regulation and functions. *Nature Reviews Neuroscience, 13*, 478–490.

Carr, C. E., & Boudreau, R. E. (1993). An axon with a myelinated initial segment in the bird auditory system. *Brain Research, 628*, 330–334.

Carr, C. E., & Soares, D. (2002). Evolutionary convergence and shared computational principles in the auditory system. *Brain, Behavior and Evolution, 59*, 294–311.

Carreiras, M., Seghier, M. L., Baquero, S., Estévez, A., Lozano, A., Devlin, J. T., & Price, C. J. (2009). An anatomical signature for literacy. *Nature, 461*, 983–986.

Carter, B. C., & Bean, B. P. (2009). Sodium entry during action potentials of mammalian neurons: Incomplete inactivation and reduced metabolic efficiency in fast-spiking neurons. *Neuron, 64*, 898–909.

Casey, R. H. (2009). The Model T turns 100! American Heritage.com, reprinted from *Invention & Technology, 23*.

Chakrapani, S., & Auerbach, A. (2005). A speed limit for conformational change of an allosteric membrane protein. *Proceedings of the National Academy of Sciences of the United States of America, 102*, 87–92.

Chancey, J. H., Adlaf, E. W., Sapp, M. C., Pugh, P. C., Wadiche, J. I., & Overstreet-Wadiche, L. S. (2013). GABA depolarization is required for experience-dependent synapse unsilencing in adult-born neurons. *Journal of Neuroscience, 33*(15), 6614–6622.

Chanda, M. L., & Levitin, D. J. (2013). The neurochemistry of music. *Trends in Cognitive Sciences, 17*, 179–193.

Cheer, J. F., Wassum, K. M., Heien, M. L. A. V., Phillips, P. E. M., & Wightman, R. M. (2004). Cannabinoids enhance subsecond dopamine release in the nucleus accumbens of awake rats. *Journal of Neuroscience, 24*(18), 4393–4400.

Chen, B. L., Hall, D. H., & Chklovskii, D. B. (2006). Wiring optimization can relate neuronal structure and function. *Proceedings of the National Academy of Sciences of the United States of America, 103*, 4723–4728.

Cheney, D. L., & Seyfarth, R. M. (2007). *Baboon metaphysics: The evolution of a social mind*. Chicago: University of Chicago Press.

Cheng, S. M., & Carr, C. E. (2007). Functional delay of myelination of auditory delay lines in the nucleus laminaris of the barn owl. *Developmental Neurobiology, 67*(14), 1957–1974.

Cherniak, C. (1995). Neural component placement. *Trends in Neurosciences, 18*, 522–527.

Chiang, A.-S., Lin, C.-Y., Chuang, C.-C., Chang, H.-M., Hsieh, C.-H., Yeh, C.-W., et al. (2011). Three-dimensional reconstruction of brain-wide wiring networks in *Drosophila* at single-cell resolution. *Current Biology, 21*, 1–11.

Chichilnisky, E. J., & Kalmar, R. S. (2002). Functional asymmetries in ON and OFF ganglion cells of primate retina. *Journal of Neuroscience, 22*, 2737–2747.

Chittka, L., & Niven, J. (2009). Are bigger brains better? *Current Biology, 19*, R995–R1008.

Chklovskii, D. B. (2000). Optimal sizes of dendritic and axonal arbors in a topographic projection. *Journal of Neurophysiology, 83*(4), 2113–2119.

Chklovskii, D. B. (2004). Synaptic connectivity and neuronal morphology: Two sides of the same coin. *Neuron, 43*, 609–617.

Chklovskii, D. B., & Koulakov, A. A. (2004). Maps in the brain: What can we learn from them? *Annual Review of Neuroscience, 27*, 369–392.

Chklovskii, D. B., Schikorski, T., & Stevens, C. F. (2002). Wiring optimization in cortical circuits. *Neuron, 34*, 341–347.

Chow, D. M., Theobald, J. C., & Frye, M. A. (2011). An olfactory circuit increases the fidelity of visual behavior. *Journal of Neuroscience, 31*, 15035–15047.

Chowdhury, S., & Chanda, B. (2012). Estimating the voltage-dependent free energy change of ion channels using the median voltage for activation. *Journal of General Physiology, 139*, 3–17.

Christian, C. A., Herbert, A. G., Holt, R. L., Peng, K., Sherwood, K. D., Pangratz-Fuehrer, S., et al. (2013). Endogenous positive allosteric modulation of GABA(A) receptors by diazepam binding inhibitor. *Neuron, 78*(6), 1063–1074.

Chua, L. O. (1971). Memristor—The missing circuit element. *IEEE Transactions on Circuit Theory, CT-18*(5), 507–519.

Chung, K., & Coggesshall, R. E. (1983). Propriospinal fibers in the rat. *Journal of Comparative Neurology, 217*, 47–53.

Chung, K. Y., Rasmussen, S. G. F., Liu, T., Li, S., DeVree, B. T., Chae, P. S., et al. (2011). Conformational changes in the G protein Gs induced by the β2 adrenergic receptor. *Nature, 477*, 611–615.

Clark, S. A., Allard, T., Jenkins, W. M., & Merzenich, M. M. (1988). Receptive fields in the body-surface map in adult cortex defined by temporally correlated inputs. *Nature, 332*(6163), 444–445.

Clark, B. A., & Cull-Candy, S. G. (2002). Activity-dependent recruitment of extrasynaptic NMDA receptor activation at an AMPA receptor-only synapse. *Journal of Neuroscience, 22*(11), 4428–4436.

Cleland, B. G., & Levick, W. R. (1974). Brisk and sluggish concentrically organized ganglion cells in the cat's retina. *Journal of Physiology, 240*(2), 421–456.

Cognigni, P., Bailey, A. P., & Miguel-Aliaga, I. (2011). Enteric neurons and systemic signals couple nutritional and reproductive status with intestinal homeostasis. *Cell Metabolism, 13*, 92–104.

Cohen, E., & Sterling, P. (1990). Demonstration of cell types among cone bipolar neurons of cat retina. *Philosophical Transactions of the Royal Society of London, Series B, 330*(1258), 305–321.

Cohen, E., & Sterling P. (1992). Parallel circuits from cones to the on-beta ganglion cell. *European Journal of Neuroscience, 4*(6):506–520.

Cohen, M. R., & Maunsell, J. H. (2009). Attention improves performance primarily by reducing interneuronal correlations. *Nature Neuroscience, 2*, 1594–1600.

Cohen, M. R., & Maunsell, J. H. (2011).When attention wanders: How uncontrolled fluctuations in attention affect performance. *Journal of Neuroscience, 31*, 15802–15806.

Cohen, M. R., & Newsome, W. T. (2009). Estimates of the contribution of single neurons to perception depend on timescale and noise correlation. *Journal of Neuroscience, 29*, 6635–6648.

Coleman, J. E., Nahmani, M., Gavornik, J. P., Haslinger, R., Heynen, A. J., Erisir, A., & Bear, M. F. (2010). Rapid structural remodeling of thalamocortical synapses parallels experience-dependent functional plasticity in mouse primary visual cortex. *Journal of Neuroscience, 30*(29), 9670–9682.

Collins, C. E., Lyon, D. C., & Kaas, J. H. (2005). Distribution across cortical areas of neurons projecting to the superior colliculus in New World monkeys. Anat Record Part A 285A:619–627.

Conty, L., Dezecache, G., Hugueville, L., & Grèzes, J. (2012). Early binding of gaze, gesture, and emotion: Neural time course and correlates. *Journal of Neuroscience, 32,* 4531–4539.

Cooke, F., & Bear, M. F. (2010). Visual experience induces long-term potentiation in the primary visual cortex. *Journal of Neuroscience, 30*(48), 16304–16313.

Creed, R. S., & Sherrington, C. S. (1926). Observations on concurrent contraction of flexor muscles in the flexion-reflex. *Proceedings of the Royal Society of London. Series B, 100,* 258–267.

Crocker, A., & Sehgal, A. (2010). Genetic analysis of sleep. *Genes & Development, 24,* 1220–1235.

Crook, J. D., Manookin, M. B., Packer, O. S., & Dacey, D. M. (2011). Horizontal cell feedback without cone type-selective inhibition mediates "red-green" color opponency in midget ganglion cells of the primate retina. *Journal of Neuroscience, 31*(5), 1762–1772.

Crook, J. D., Packer, O. S., Troy, J. B., & Dacey, D. M. (2013). Synaptic mechanisms of color and luminance coding: Rediscovering the X–Y-cell dichotomy in primate retinal ganglion cells. In J. S. Werner & L. M. Chalupa (Eds.), *The new visual neurosciences* (pp. 123–143). Cambridge, MA: MIT Press.

Crowell, J., Banks, M. S., Anderson, S. J., & Geisler, W. S. (1988). Physical limits of grating visibility: Fovea and periphery. *Investigative Ophthalmology & Visual Science* (Suppl. 29), 139.

Cunningham, D. J. (1913). *Cunningham's text book of anatomy.* New York: William Wood.

Cuttle, M. F., Hevers, W., Laughlin, S. B., & Hardie, R. C. (1995). Diurnal modulation of photoreceptor potassium conductance in the locust. *Journal of Comparative Physiology. A, Neuroethology, Sensory, Neural, and Behavioral Physiology, 176,* 307–316.

Dacey, D. (2004). Origins of perception: Retinal ganglion cell diversity and creation of parallel visual pathways. In M. S. Gazzaniga (Ed.), *The cognitive neurosciences III* (pp. 281–301). Cambridge, MA: MIT Press.

Dacey, D. M. (1989). Axon-bearing amacrine cells of the macaque monkey retina. *Journal of Comparative Neurology, 284*(2), 275–293.

Dacey, D. M., & Petersen, M. R. (1992). Dendritic field size and morphology of midget and parasol ganglion cells of the human retina. *Proceedings of the National Academy of Sciences of the United States of America, 89,* 9666–9670.

da Costa, N. M., & Martin, K. A. C. (2011). How thalamus connects to spiny stellate cells in the cat's visual cortex. *Journal of Neuroscience, 31,* 2925–2937.

Dahl, C. D., Rasch, M. J., Tomonaga, M., & Adachi, I. (2013). Laterality effect for faces in chimpanzees *(Pan troglodytes). Journal of Neuroscience, 33*(33), 13344–13349.

Dallos, P. (2008). Cochlear amplification, outer hair cells and prestin. *Current Opinion in Neurobiology, 18*(4), 370–376.

Damier, P., Hirsch, E. C., Agid, Y., & Graybiel, A. M. (1999). The substantia nigra of the human brain. I. Nigrosomes and the nigral matrix, a compartmental organization based on calbindin D_{28k} immunohistochemistry. *Brain, 122,* 1421–1436.

Darwin, C. (1859). *On the origin of species by means of natural selection, or the preservation of favoured races in the struggle for life.* London: John Murray.

Darwin, C. (1881). *The autobiography of Charles Darwin, 1809–1882.* With original omissions restored/edited with appendix and notes by his granddaughter, Nora Barlow. Reprinted in 1958 by Collins (London).

Dastjerdi, M., Weyand, T. G., Dong, D. W. (2003). The spatiotemporal receptive field (STRF) properties of the lateral geniculate geniculate nucleus (LGN) in the awake cat during free-viewing of natural time-varying images. *Society for Neuroscience Abstracts* 229.3.

Dastjerdi, M., Weyand, T. G., & Dong, D. W. (2007). Paired recording reveals temporal decorrelation of retinal inputs in the thalamus. *COSYNE, 7,* 45.

Dastjerdi, M. M., Weyand, T. G., & Dong, D. W. (2011). Temporal differentiation underlying synaptic transmission in the thalamus during natural viewing. (unpublished manuscript).

Daugman, J. G. (1985). Uncertainty relation for resolution in space, spatial frequency, and orientation optimized by two-dimensional visual cortical filters. *Journal of the Optical Society of America. A, Optics and Image Science, 2,* 1160–1169.

Davenport, C. M., Detwiler, P. B., & Dacey, D. M. (2007). Functional polarity of dendrites and axons of primate A1 amacrine cells. *Visual Neuroscience, 24*(4), 449–457.

Davenport, C. M., Detwiler, P. B., & Dacey, D. M. (2008). Effects of pH buffering on horizontal and ganglion cell light responses in primate retina: Evidence for the proton hypothesis of surround formation. *Journal of Neuroscience, 28*(2), 456–464.

Debanne, D., Campanac, E., Bialowas, A., Carlier, E., & Alcaraz, G. (2011). Axon physiology. *Physiological Review, 91,* 555–602.

de Bono, M., & Bargmann, C. I. (1998). Natural variation in a neuropeptide y receptor homolog modifies social behavior and food response in *C. elegans*. *Cell, 94*, 679–689.

de Bono, M., & Maricq, A. V. (2005). Neuronal substrates of complex behaviors in *C. elegans*. *Annual Review of Neuroscience, 28*, 451–501.

DeFelipe, J. (2006). Brain plasticity and mental processes: Cajal again. *Nature Reviews Neuroscience, 7*, 811–817.

Dehaene, S., Pegado, F., Braga, L. W., Ventura, P., Nunes Filho, G., Jobert, A., et al. (2010). How learning to read changes the cortical networks for vision and language. *Science, 330*, 1359–1364.

Deleuze, C., David, F., Béhuret, S., Sadoc, G., Shin, H.-S., Uebele, V. N., et al. (2012). T-type calcium channels consolidate tonic action potential output of thalamic neurons to neocortex. *Journal of Neuroscience, 32*(35), 12228–12236.

Demb, J. B., & Singer, J. H. (2012). Intrinsic properties and functional circuitry of the AII amacrine cell. *Visual Neuroscience, 29*(1), 51–60.

de Ruyter van Steveninck, R. R., & Laughlin, S. B. (1996). The rate of information transfer at graded-potential synapses. *Nature, 379*, 642–645.

DeVries, S. H. (2000). Bipolar cells use kainate and AMPA receptors to filter visual information into separate channels. *Neuron, 28*(3):847–856.

DeVries, S. H. (2014). Cone bipolar cells: ON and OFF pathways in the outer retina. In J. S. Werner & L. M. Chalupa (Eds.), *The new visual neurosciences* (pp. 53–62). Cambridge, MA: MIT Press.

DeVries, S. H., & Baylor, D. A. (1997). Mosaic arrangement of ganglion cell receptive fields in rabbit retina. *Journal of Neurophysiology, 78*(4), 2048–2060.

DeVries, S. H., Li, W., & Saszik, S. (2006). Parallel processing in two transmitter microenvironments at the cone photoreceptor synapse. *Neuron, 50*, 735–748.

DeVries, S. H., Qi, X., Smith, R., Makous, W., & Sterling, P. (2002). Electrical coupling between mammalian cones. *Current Biology, 12*, 1900–1907.

Dhingra, N. K., Freed, M. A., & Smith, R. G. (2005). Voltage-gated sodium channels improve contrast sensitivity of a retinal ganglion cell. *Journal of Neuroscience, 25*, 8097–8103.

Dhingra, N. K., & Smith, R. G. (2004). Spike generator limits efficiency of information transfer in a retinal ganglion cell. *Journal of Neuroscience, 24*, 2914–2922.

Dickinson, M. H., & Tu, M. S. (1997). The function of dipteran flight muscle. *Comparative Biochemistry and Physiology Part A: Physiology, 116*, 223–238.

Dong, D. W., Weyand, T. G., & Usrey, W. M. (2005). The spatiotemporal receptive field of lateral geniculate neurons during free-viewing of natural time-varying images in the rhesus monkey. *Society for Neuroscience Abstracts*, 591.5.

Donohue, A. B. (2000). Electroconvulsive therapy and memory loss: A personal journey. *Journal of ECT*, *16*, 133–143.

Dougherty, R. F., Koch, V. M., Brewer, A. A., Fischer, B., Modersitzki, J., & Wandell, B. A. (2003). Visual field representations and locations of visual area V1/2/3 in human visual cortex. *Journal of Vision*, *3*, 586–598.

Douglas, R. J., & Martin, K. A. C. (2009). Inhibition in cortical circuits. *Current Biology*, *19*, 398–402.

Doya, K. (2000). Complementary roles of basal ganglia and cerebellum in learning and motor control. *Current Opinion in Neurobiology*, *10*(6), 732–739.

Dror, R. O., Dirks, R. M., Grossman, J. P., Xu, H., & Shaw, D. E. (2012). Biomolecular simulation: A computational microscope for molecular biology. *Annual Review of Biophysics*, *41*, 429–452.

Dubs, A., Laughlin, S. B., & Srinivasan, M. V. (1981). Single photon signals in fly photoreceptors and first order interneurones at behavioral threshold. *Journal of Physiology*, *317*, 317–334.

Dunbar, R. I. M., & Shultz, S. (2007). Evolution in the social brain. *Science*, *317*, 1344–1347.

Duncan, G., Rabl, K., Gemp, I., Heidelberger, R., & Thoreson, W. B. (2010). Quantitative analysis of synaptic release at the photoreceptor synapse. *Biophysical Journal*, *98*, 1–9.

Dworak, M., McCarley, R. W., Kim, T., Kalinchuk, A. V., & Basheer, R. (2010). Sleep and brain energy levels: ATP changes during sleep. *Journal of Neuroscience*, *30*, 9007–9016.

Easter, S. S., Jr., & Stuermer, C. A. (1984). An evaluation of the hypothesis of shifting terminals in goldfish optic tectum. *Journal of Neuroscience*, *4*(4), 1052–1063.

Eatock, R. A., & Songer, J. E. (2011). Vestibular hair cells and afferents: Two channels for head motion signals. *Annual Review of Neuroscience*, *34*, 501–534.

Eatock, R. A., Xue, J., & Kalluri, R. (2008). Ion channels in mammalian vestibular afferents may set regularity of firing. *Journal of Experimental Biology*, *211*, 1764–1774.

Eckert, R. (1972). Bioelectric control of ciliary activity. *Science*, *176*, 473–481.

Economides, J. R., Sincich, L. C., Adams, D. L., & Horton, J. C. (2011). Orientation tuning of cytochrome oxidase patches in macaque primary visual cortex. *Nature Neuroscience*, *14*, 1574–1580.

Eggermann, E., Bucurenciu, I., Goswami, S. P., & Jonas, P. (2012). Nanodomain coupling between Ca^{2+} channels and sensors of exocytosis at fast mammalian synapses. *Nature Reviews Neuroscience, 13*, 7–21.

Eisenmann, D. M. Wnt signaling (June 25, 2005), *WormBook*, ed. The *C. elegans* Research Community, WormBook, doi/10.1895/wormbook.1.7.1, http://www.wormbook.org.

Emes, R. D., Pocklington, A. J., Anderson, C. N. G., Bayes, A., Collins, M. O., Vickers, C. A., et al. (2008). Evolutionary expansion and anatomical specialization of synapse proteome complexity. *Nature Neuroscience, 11*, 799–806.

Enroth-Cugell, C., & Robson, J. G. (1966). The contrast sensitivity of retinal ganglion cells of the cat. *Journal of Physiology, 187*(3), 517–552.

Erikson, E. H., & Erikson, J. M. (1997). *The life cycle completed*. New York: W.W. Norton.

Fain, G. L., Hardie, R., & Laughlin, S. B. (2010). Phototransduction and the evolution of photoreceptors. *Current Biology, 20*(3), R114–R124.

Faisal, A. A., & Laughlin, S. B. (2007). Stochastic simulations on the reliability of action potential propagation in thin axons. *PLoS Computational Biology, 3*(5), e79.

Faisal, A. A., White, J. A., & Laughlin, S. B. (2005). Ion-channel noise places limits on the miniaturization of the brain's wiring. *Current Biology, 15*, 1143–1149.

Famiglietti, E. V., Jr. (1970). Dendro-dendritic synapses in the lateral geniculate nucleus of the cat. *Brain Research, 20*, 181–191.

Fang-Yen, C., Wyart, M., Xie, J., Kawai, R., Kodger, T., Chen, S., Wen, Q., & Samuel, A. D. T. (2010). Biomechanical analysis of gait adaptation in the nematode *Caenorhabditis elegans*. *Proceedings of the National Academy of Sciences of the United States of America, 107*, 20323–20328.

Farris, S. M. (2011). Are mushroom bodies cerebellum-like structures? *Arthropod Structure & Development, 40*, 368–379.

Farris, S. M., & Schulmeister, S. (2011). Parasitoidism, not sociality, is associated with the evolution of elaborate mushroom bodies in the brains of hymenopteran insects. *Proceedings. Biological Sciences, 278*, 940–951.

Fattori, P., Breveglieri, R., Raos, V., Bosco, A., & Galletti, C. (2012). Vision for action in the macaque medial posterior parietal cortex. *Journal of Neuroscience, 32*, 3221–3234.

Federer, F., Ichida, J. M., Jeffs, J., Schiessl, I., McLoughlin, N., & Angelucci, A. (2009). Four projection streams from primate V1 to the cytochrome oxidase stripes of V2. *Journal of Neuroscience, 29*, 15455–15471.

Federer, F., Williams, D., Ichida, J. M., Merlin, S., & Angelucci A. (2013). Two projection streams from macaque V1 to the pale cytochrome oxidase stripes of V2. *Journal of Neuroscience, 33*(28), 11530–11539.

Fehr, E., & Camerer, C. F. (2007). Social neuroeconomics: The neural circuitry of social preferences. *Trends in Cognitive Science, 11*(10), 419–427.

Félix, M. A., & Braendle, C. (2010). The natural history of *Caenorhabditis elegans*. *Current Biology, 20*, R965–R969.

Felleman, D. J., & Van Essen, D. C. (1991). Distributed hierarchical processing in the primate cerebral cortex. *Cerebral Cortex, 1*, 1–47.

Fernández-Chacón, R., & Südhof, T. C. (1999). Genetics of synaptic vesicle function: Toward the complete functional anatomy of an organelle. *Annual Review of Physiology, 61*, 753–776.

Ferster, D. (1988). Spatially opponent excitation and inhibition in simple cells of cat visual cortex. *Journal of Neuroscience, 8*, 1172–1180.

Field, D. J. (1987). Relations between the statistics of natural images and the response properties of cortical cells. *Journal of the Optical Society of America. A, Optics and Image Science, 4*, 2379–2394.

Field, G. D., & Rieke, F. (2002). Nonlinear signal transfer from mouse rods to bipolar cells and implications for visual sensitivity. *Neuron, 34*, 773–785.

Fischbach, P. K.-F., & Dittrich, A. P. M. (1989). The optic lobe of *Drosophila melanogaster*. I. A Golgi analysis of wild-type structure. *Cell and Tissue Research, 258*, 441–475.

Franklin, K. B. J., & Paxinos, G. (1996). *The mouse brain in stereotaxic coordinates*. San Diego: Acadamic.

Freed, M. A. (2000). Parallel cone bipolar pathways to a ganglion cell use different rates and amplitudes of quantal excitation. *Journal of Neuroscience, 20*(11), 3956–3963.

Freed, M. A. (2005). Quantal encoding of information in a retinal ganglion cell. *Journal of Neurophysiology, 94*, 1048–1056.

Freed, M. A., & Liang, Z. (2010). Reliability and frequency response of excitatory signals transmitted to different types of retinal ganglion cell. *Journal of Neurophysiology, 103*(3), 1508–1517.

Freed, M. A., Smith, R. G. & Sterling, P. (2003). Timing of quantal release from the retinal bipolar terminal is regulated by a feedback circuit. *Neuron, 38*(1), 89–101.

Freed, M. A., & Sterling, P. (1988). The ON-alpha ganglion cell of the cat retina and its presynaptic cell types. *Journal of Neuroscience, 8*, 2303–2320.

Freeman, R. L. (1999). *Fundamentals of telecommunications*. New York: John Wiley & Sons.

Freeman, J. B., Stolier, R. M., Ingbretsen, Z. A., & Hehman, E. A. (2014). Amygdala responsivity to high-level social information from unseen faces. *Journal of Neuroscience, 34*, 10573–10581.

Freiwald, W. A., & Tsao, D. Y. (2010). Functional compartmentalization and viewpoint generalization within the macaque face-processing system. *Science, 330*, 845–851.

French, M. (1994). *Invention and evolution*. Cambridge: Cambridge University Press.

Freud, S. (1930). *Civilization and its discontents*. Trans. James Strachey. Reprint: New York: W.W. Norton [1962].

Frick, A., Sakmann, B., & Helmstaedter, M. (2010). Cell type-specific thalamic innervation in a column of rat vibrissal cortex. *Cerebral Cortex, 20*, 2287–2303.

Friedrich, R. W., & Laurent, G. (2001). Dynamic optimization of odor representations by slow temporal patterning of mitral cell activity. *Science, 291*, 889–894.

Fry, M., Hoyda, T. D., & Ferguson, A. V. (2007). Making sense of it: Roles of the sensory circumventricular organs in feeding and regulation of energy homeostasis. *Experimental Biology and Medicine, 232*, 14–26.

Fu, J., DiPatrizio, N. V., Guijarro, A., Schwartz, G. J., Li, X., Gaetani, S., et al. (2011). Sympathetic activity controls fat-induced oleoylethanolamide signaling in small intestine. *Journal of Neuroscience, 31*, 5730–5736.

Fu, Y., Kefalov, V., Luo, D. G., Xue, T., & Yau, K. W. (2008). Quantal noise from human red cone pigment. *Nature Neuroscience, 11*, 565–571.

Gallant, J. L. (2004). Neural mechanisms of natural scene perception. In L. Chalupa & J. Werner (Eds.), *The visual neurosciences*. Cambridge, MA: MIT Press

Gao, Z., van Beugen, B. J., & De Zeeuw, C. I. (2012). Distributed synergistic plasticity and cerebellar learning. *Nature Reviews Neuroscience, 13*, 619–635.

Garrigan, P., Ratliff, C. P., Klein, J. M., Sterling, P., Brainard, D. H., & Balasubramanian, V. (2010). Design of a trichromatic cone array. *PLoS Computational Biology, 6*(2), e100067, pp. 1–17.

Gasparini, S., & Magee, J. C. (2006). State-dependent dendritic computation in hippocampal CA1 pyramidal neurons. *Journal of Neuroscience, 26*(7), 2088–2100.

Gauthier, J. L., Field, G. D., Sher, A., Shlens, J., Greschner, M., Litke, A. M., & Chichilnisky, E. J. (2009). Uniform signal redundancy of parasol and midget ganglion cells in primate retina. *Journal of Neuroscience, 8*, 4675–4680.

Geisler, W. S. (1989). Sequential ideal-observer analysis of visual discriminations. *Psychological Review, 96*, 267–314.

Geisler, W. S., Perry, J. S., Super, B. J., & Gallogly, D. P. (2001). Edge co-occurrence in natural images predicts contour grouping performance. *Vision Research, 41*, 711–724.

Gigerenzer, G. (2008). *Gut feelings.* London: Penguin Books

Gintis, H., & Fehr, E. (2012). The social structure of cooperation and punishment. *Behavioral and Brain Sciences, 35*(1), 28–29.

Glegg, G. L. (1969/2009a). *The design of design.* Cambridge: Cambridge University Press.

Glegg, G. L. (1972/2009b). *The selection of design.* Cambridge: Cambridge University Press.

Glickstein, M. (2014). *Neuroscience: A historical introduction.* Cambridge, MA: MIT Press.

Glimcher, P. W. (2011). Quantification of Behavior Sackler Colloquium: Understanding dopamine and reinforcement learning: The dopamine reward prediction error hypothesis. *Proceedings of the National Academy of Sciences of the United States of America, 108*(Suppl 3), 15647–15654.

Goldberg, J. M. (2000). Afferent diversity and the organization of central vestibular pathways. *Experimental Brain Research, 130*, 277–297.

Goldman, M. S., Maldonado, P., & Abbott, L. F. (2002). Redundancy reduction and sustained firing with stochastic depressing synapses. *Journal of Neuroscience, 22*, 584–591.

Gonchar, Y., Wang, Q., & Burkhalter, A. (2007). Multiple distinct subtypes of GABAergic neurons in mouse visual cortex identified by triple immunostaining. *Frontiers in Neuroanatomy, 1*, 3.

Gonzalez-Bellido, P. T., Wardill, T. J., & Juusola, M. (2011). Compound eyes and retinal information processing in miniature dipteran species match their specific ecological demands. *Proceedings of the National Academy of Sciences of the United States of America, 108*, 4224–4229.

Gould, S. J. (1992). *The panda's thumb: More reflections in natural history.* New York: W.W. Norton.

Gould, S. J., & Lewontin, R. C. (1979). The spandrels of San Marco and the Panglossian paradigm: A critique of the adaptationist programme. *Proceedings of the Royal Society of London, Series B, 205*, 581–598.

Goulding, M. (2009). Circuits controlling vertebrate locomotion: Moving in a new direction. *Nature Reviews Neuroscience, 10*, 507–518.

Govindarajan, A., Israely, I., Huang, S.-Y., & Tonegawa, S. (2011). The dendritic branch is the preferred integrative unit for protein synthesis-dependent LTP. *Neuron, 69*(1), 132–146.

Grande, G., & Wang, L.-Y. (2011). Morphological and functional continuum underlying heterogeneity in the spiking fidelity at the calyx of Held synapse in vitro. *Journal of Neuroscience*, *31*, 13386–13399.

Grant, L., Yi, E., & Glowatzki, E. (2010). Two modes of release shape the postsynaptic response at the inner hair cell ribbon synapse. *Journal of Neuroscience*, *30*, 4210–4220.

Graydon, C. W., Cho, S., Diamond, J. S., Kachar, B., vonGersdorff, H., & Grimes, W. N. (2014). Specialized postsynaptic morphology enhances neurotransmitter dilution and high-frequency signaling at an auditory synapse. *Journal of Neuroscience*, *34*, 8358–8372.

Graydon, C. W., Cho, S., Li, G.-L., Kachar, B., & von Gersdorff, H. (2011). Sharp Ca^{2+} nanodomains beneath the ribbon promote highly synchronous multivesicular release at hair cell synapses. *Journal of Neuroscience*, *31*(46), 16637–16650.

Green, D. G. (1970). Regional variations in the visual acuity for interference fringes on the retina. *Journal of Physiology*, *207*, 351–356.

Grice, E. A., Kong, H. H., Conlan, S., Deming, C. B., Davis, J., Young, A. C., NISC Comparative Sequencing Program, Bouffard, G. G., Blakesley, R. W., Murray, P. R., Green, E. D., Turner, M. L., & Segre, J. A. (2009). Topographical and temporal diversity of the human skin microbiome. *Science*, *324*, 1190–1192.

Grimes, W. N., Zhang, J., Graydon, C. W., Kachar, B., & Diamond, J. S. (2010). Retinal parallel processors: More than 100 independent microcircuits operate within a single interneuron. *Neuron*, *65*(6), 873–885.

Grodd, W., Hülsmann, E., Lotze, M., Wildgruber, D., & Erb, M. (2001). Sensorimotor mapping of the human cerebellum: fMRI evidence of somatotopic organization. *Human Brain Mapping*, *13*(2), 55–73.

Grosmaitre, X., Santarelli, L. C., Tan, J., Luo, M., & Ma, M. (2007). Dual functions of mammalian olfactory sensory neurons as odor detectors and mechanical sensors. *Nature Neuroscience M10*(3), 348–354.

Gross, O. P., Pugh, E. N., Jr., & Burns, M. E. (2012). Spatiotemporal cGMP dynamics in living mouse rods. *Biophysical Journal*, *102*, 1775–1784.

Gruberger, M., Ben-Simon, E., Levkovitz, Y., Zangen, A., & Hendler, T. (2011). Towards a neuroscience of mind-wandering. *Frontiers in Human Neuroscience*, *5*, 56.

Guilding, C., Hughes, A. T., Brown, T. M., Namvar, S., & Piggins, H. D. (2009). A riot of rhythms: Neuronal and glial circadian oscillators in the mediobasal hypothalamus. *Molecular Brain*, *27*, 2–28.

Güldner, F.-H. (1983). Numbers of neurons and astroglial cells in the suprachiasmatic nucleus of male and female rats. *Experimental Brain Research*, *50*, 373–376.

Halassa, M. M., & Haydon, P. G. (2010). Integrated brain circuits: Astrocytic networks modulate neuronal activity and behavior. *Annual Review of Physiology, 72*, 335–355.

Hallermann, S., Fejtova, A., Schmidt, H., Weyhersmüller, A., Silver, R. A., Gundelfinger, E. D., & Eilers, J. (2010). Bassoon speeds vesicle reloading at a central excitatory synapse. *Neuron, 68*(4), 710–723.

Hallermann, S., & Silver, R. A. (2013). Sustaining rapid vesicular release at active zones: Potential roles for vesicle tethering. *Trends in Neurosciences, 36*(3), 185–194.

Han, Se., Sundararajan, J., Bowling, D. L., Lake, J., & Purves, D. (2011). Co-variation of tonality in the music and speech of different cultures. *PLoS One 6*(5).

Hardie, R. C. (1984). Functional organization of the fly retina. *Progress in Sensory Physiology, 5*, 1–79.

Hardie, R. C. (1989). A histamine-activated chloride channel involved in neurotransmission at a photoreceptor synapse. *Nature, 339*, 704–706.

Hardie, R. C., & Franze, K. (2012). Photomechanical responses in *Drosophila* photoreceptors. *Science, 338*, 260–263.

Hardie, R. C., & Postma, M. (2008). Phototransduction in microvillar photoreceptors of *Drosophila* and other invertebrates. In R. Masland & T. D. Albright (Eds.), *The senses, a comprehensive reference. Volume 1, Vision I* (pp. 77–130). San Diego: Academic Press.

Hardie, R. C., & Raghu, P. (2001). Visual transduction in *Drosophila*. *Nature, 413*, 186–193.

Hare, T. A., O'Doherty, J., Camerer, C. F., Schultz, W., & Rangel, A. (2008). Dissociating the role of the orbitofrontal cortex and the striatum in the computation of goal values and prediction errors. *Journal of Neuroscience, 28*(22), 5623–5630.

Harnett, M. T., Xu, N.-L., Magee, J. C., & Williams, S. R. (2013). Potassium channels control the interaction between active dendritic integration compartments in layer 5 cortical pyramidal neurons. *Neuron, 79*, 516–529.

Harris, J. J., & Attwell, D. (2012). The energetics of CNS white matter. *Journal of Neuroscience, 32*(1), 356–371.

Harris, J. J., Jolivet, R., & Attwell, D. (2012). Synaptic energy use and supply. *Neuron, 75*, 762–777.

Harris, K. M., & Stevens, J. K. (1988). Dendritic spines of rat cerebellar purkinje cells: Serial electron microscopy with reference to their biophysical characteristics. *Journal of Neuroscience, 8*, 4455–4489.

Harris-Warrick, R. M., & Marder, E. (1991). Modulation of neural networks for behavior. *Annual Review of Neuroscience, 14*, 39–57.

Häusser, M., Raman, I. M., Otis, T., Smith, S. L., Nelson, A., du Lac, S., et al. (2004). The beat goes on: Spontaneous firing in mammalian neuronal microcircuits. *Journal of Neuroscience, 24,* 9215–9219.

Haverkamp S., Grünert U., & Wässle H. (2000). The cone pedicle, a complex synapse in the retina. *Neuron, 27,* 85–95.

Haverkamp, S., Grünert, U., & Wässle, H. (2001a). Localization of kainate receptors at the cone pedicles of the primate retina. *Journal of Comparative Neurology, 436,* 471–486.

Haverkamp, S., Grünert, U., & Wässle, H. (2001b). The synaptic architecture of AMPA receptors at the cone pedicle of the primate retina. *Journal of Neuroscience, 21*(7), 2488–2500.

Heck, D. H., De Zeeuw, C. I., Jaeger, D., Khodakhah, K., & Person, A. L. (2013). The neuronal code(s) of the cerebellum [Review]. *Journal of Neuroscience, 33*(45), 17603–17609.

Hedwig, B. (2000). Control of cricket stridulation by a command neuron: efficacy depends on the behavioral state. *Journal of Neurophysiology, 83,* 712–722.

Heidelberger, R., Thoreson, W. B., & Witkovsky, P. (2005). Synaptic transmission at retinal ribbon synapses [Review]. *Progress in Retinal Eye Research. 24*(6), 682–720.

Heinze, S., & Reppert, S. M. (2011). Sun compass integration of skylight cues in migratory monarch butterflies. *Neuron, 69,* 345–358.

Herikstad, R., Baker, J., Lachaux, J.-P., Gray, C. M., & Yen, S.-C. (2011). Natural movies evoke spike trains with low spike time variability in cat primary visual cortex. *Journal of Neuroscience, 31,* 15844–15860.

Herculano-Houzel, S. (2011). Scaling of brain metabolism with a fixed energy budget per neuron: implications for neuronal activity, plasticity and evolution. *PLoS one 6,* e17514.

Herman, M. A. (2006). Hermaphrodite cell-fate specification. The *C. elegans* Research Community, WormBook, doi/10.1895/wormbook.1.39.1, http://www.wormbook.org/.

Hess, W. (1949). The central control of the activity of internal organs. In *Nobel Lectures, Physiology or Medicine 1942–1962.* Amsterdam: Elsevier Publishing Co. http://www.nobelprize.org/nobel_prizes/medicine/laureates/1949/hess-lecture.html.

Hevner, R. F., & Wong-Riley, M. T. T. (1993). Mitochondrial and nuclear gene expression for cytochrome oxidase subunits are disproportionately regulated by functional activity in neurons. *Journal of Neuroscience, 13*(5), 1805–1819.

Hille, B. (2001). *Ion channels of excitable membranes.* Sunderland, MA: Sinauer.

Hirasawa, H., & Kaneko, A. (2003). pH changes in the invaginating synaptic cleft mediate feedback from horizontal cells to cone photoreceptors by modulating Ca^{2+} channels. *Journal of General Physiology, 122*, 657–671.

Hirsch, J. A. (2003). Synaptic physiology and receptive field structure in the early visual pathway of the cat. *Cerebral Cortex, 13*, 63–69.

Hjelmfelt, A., Weinberger, E. D., & Ross, J. (1991). Chemical implementation of neural networks and Turing machines. *Proceedings of the National Academy of Sciences of the United States of America, 88*, 10983–10987.

Hodgkin, A. L., & Huxley, A. F. (1952). A quantitative description of membrane current and its application to conduction and excitation in nerve. *Journal of Physiology, 117*, 500–544.

Holtmaat, A., & Svoboda, K. (2009). Experience-dependent structural synaptic plasticity in the mammalian brain. *Nature Reviews Neuroscience, 10*(9), 647–658.

Homann, J., & Freed, M. A. (2012). Optimal weighting of inhibition and excitation to an off alpha ganglion cell. *ARVO Meeting Abstracts, 53*, 6917.

Hornstein, E. P., Verweij, J., & Schnapf, J. L. (2004). Electrical coupling between red and green cones in primate retina. *Nature Neuroscience, 7*(7), 745–750.

Horridge, G. A., & Meinertzhagen, I. A. (1970). The accuracy of the patterns of connexions of the first- and second-order neurons of the visual system of *Calliphora. Proceedings of the Royal Society of London, Series B, 175*, 69–82.

Horton, J. C., & Hocking, D. R. (1996). Intrinsic variability of ocular dominance column periodicity in normal macaque monkeys. *Journal of Neuroscience, 16*(22), 7228–7239.

Horton, J. C., & Hoyt, W. F. (1991). Quadrantic visual field defects: A hallmark of lesions in extrastriate (V2/V3) cortex. *Brain, 114*, 1703–1718.

Hoshi, H., Tian, L.-M., Massey, S. C., & Mills, S. L. (2011). Two distinct types of ON directionally selective ganglion cells in the rabbit retina. *Journal of Comparative Neurology, 519*(13), 2509–2521.

Howard, J. (2001). *Mechanics of motor proteins and the cytoskeleton.* Sunderland, MA. Sinauer.

Howard, J., Blakeslee, B., & Laughlin, S. B. (1987). The intracellular pupil mechanism and photoreceptor signal: Noise ratios in the fly *Lucilia cuprina. Proceedings of the Royal Society of London, Series B, 231*, 415–435.

Howarth, C. (2014). The contribution of astrocytes to the regulation of cerebral blood flow. *Frontiers in Neuroscience, 8*(103), 1–9.

Howarth, C., Gleeson, P., & Attwell, D. (2012). Updated energy budgets for neural computation in the neocortex and cerebellum. *Journal of Cerebral Blood Flow and Metabolism, 32*, 1222–1232.

Howarth, C., Peppiatt-Wildman, C. M., & Attwell, D. (2010). The energy use associated with neural computation in the cerebellum. *Journal of Cerebral Blood Flow and Metabolism, 30*, 403–414.

Hromádka, T., Deweese, M. R., & Zador, A. M. (2008). Sparse representation of sounds in the unanesthetized auditory cortex. *PLoS Biology, 6*, e16.

Hsu, A., Smith, R. G., Buchsbaum, G., & Sterling, P. (2000). Cost of cone coupling to trichromacy in primate fovea. *Journal of the Optical Society of America. A, Optics, Image Science, and Vision, 17*, 635–640.

Huang, H. J., Kram, R., & Ahmed, A. A. (2012). Reduction of metabolic cost during motor learning of arm reaching dynamics. *Journal of Neuroscience, 32*, 2182–2190.

Hubel, D. H., & Wiesel, T. N. (1962). Receptive fields, binocular interaction and functional architecture in the cat's visual cortex. *Journal of Physiology, 160*, 106–154.

Hubel, D. H., Wiesel, T. N., & LeVay, S. (1977). Plasticity of ocular dominance columns in monkey striate cortex. *Philosophical Transactions of the Royal Society of London, Series B, 278*(961), 377–409.

Huber, T., & Sakmar, T. P. (2011). Escaping the flatlands: New approaches for studying the dynamic assembly and activation of GPCR signaling complexes. *Trends in Pharmacological Sciences, 32*, 410–419.

Hudspeth, A. J. (2005). How the ear's works work: Mechanoelectrical transduction and amplification by hair cells. *Comptes Rendus Biologies, 328*, 155–162.

Humphrey, A. L., Sur, M., Uhlrich, D. J., & Herman, S. M. (1985). Projection patterns of individual X- and Y cell axons from the lateral geniculate nucleus to cortical area 17 in the cat. *Journal of Comparative Neurology, 233*, 159–189.

Immonen, E., & Ritchie, M. G. (2011). Animal communication: Flies' ears are tuned in. *Current Biology, 21*, R278–R280.

Ivannikov, M. V., Harris, K. M., & Macleod, G. T. (2010). Mitochondria: Enigmatic stewards of the synaptic vesicle reserve pool. *Frontiers in Synaptic Neuroscience, 2*, 145.

Jennings, H. (1904). *Contributions to the study of the behaviour of lower organisms.* Philadelphia: University of Pennsylvania Press.

Jia, S., Dallos, P., & He, D. Z. Z. (2007). Mechanoelectric transduction of adult inner hair cells. *Journal of Neuroscience, 27*, 1006–1014.

Jinno, S., Klausberger, T., Marton, L. F., Dalezios, Y., Roberts, J. David B., Fuentealba, P., Bushong, E. A., et al. (2007). Neuronal diversity in GABAergic long-range projections from the hippocampus. *Journal of Neuroscience, 27*(33), 8790–8804.

Jacobs, G. H. (2009). Evolution of colour vision in mammals [Review]. *Philosophical Transactions of the Royal Society of London, Series B, 364*(1531), 2957–2967.

Jones, E. G. (2001). The thalamic matrix and thalamocortical synchrony. *Trends in Neurosciences, 24*, 595–600.

Jones, J. P., & Palmer, L. A. (1987). An evaluation of the two-dimensional Gabor filter hypothesis of simple receptive fields in cat striate cortex. *Journal of Neurophysiology, 58*, 1233–1258.

Jortner, R. A., Farivar, S. S., & Laurent, G. (2007). A simple connectivity scheme for sparse coding in an olfactory system. *Journal of Neuroscience, 27*, 1659–1669.

Josephson, E. M., & Morest, D. K. (2003). Synaptic nests lack glutamate transporters in the cochlear nucleus of the mouse. *Synapse, 49*, 29–46.

Julliard, A. K., Chaput, M. A., Apelbaum, A., Aimé, P., Mahfouz, M., & Duchamp-Viret, P. (2007). Changes in rat olfactory detection performance induced by orexin and leptin mimicking fasting and satiation. *Behavioural Brain Research, 183*, 123–129.

Kaas, J. H. (2005). From mice to men: The evolution of the large, complex human brain. *Journal of Biosciences, 30*, 155–165.

Kaas, J. H. (2008). The evolution of the complex sensory and motor systems of the human brain. *Brain Research Bulletin, 75*, 384–390.

Kable, J. W., & Glimcher, P. W. (2009). The neurobiology of decision: consensus and controversy. *Neuron, 63*, 733–745.

Kageyama, G. H., & Wong-Riley, M. (1985). An analysis of the cellular localization of cytochrome oxidase in the lateral geniculate nucleus of the adult cat. *Journal of Comparative Neurology, 242*, 338–357.

Kageyama, G. H., & Wong-Riley, M. T. T. (1984). The histochemical localization cytochrome oxidase in the retina and LGN of the ferret, cat and monkey, with particular reference to retinal mosaics and ON:OFF-center visual channels. *Journal of Neuroscience, 4*, 2445–2459.

Kandel, E. R. (2006). *In search of memory: The emergence of a new science of mind.* New York: Norton.

Kandel, E. R. (2012). The molecular biology of memory: cAMP, PKA, CRE, CREB-1, CREB-2, and CPEB. *Molecular Brain, 5*, 14.

Kandel E. R., Schwartz, J. H., Jessell, T. M., Siegelbaum, S. A., & Hudspeth, A. J. (2012). *Principles of Neural Science* (fifth edition). New York: Elsevier.

Kanjhan, R., & Sivyer, B. (2010). Two types of ON direction-selective ganglion cells in rabbit retina. *Neuroscience Letters, 483*, 105–109.

Kara, P., & Reid, R. C. (2003). Efficacy of retinal spikes in driving cortical responses. *Journal of Neuroscience, 23*, 8547–8557.

Karklin, Y., & Lewicki, M. S. (2009). Emergence of complex cell properties by learning to generalize in natural scenes. *Nature, 457*, 83–86.

Karni, A., Meyer, G., Jezzard, P., Adams, M. M., Turner, R., & Ungerleider, L. G. (1995). Functional MRI evidence for adult motor cortex plasticity during motor skill learning. *Nature, 377*(6545), 155–158.

Katz, L. C. (1987). Local circuitry of identified projection neurons in cat visual cortex brain slices. *Journal of Neuroscience, 7*, 1223–1249.

Katz, L. C., Burkhalter, A., & Dreyer, W. J. (1984). Fluorescent latex microspheres as a retrograde neuronal marker for in vivo and in vitro studies of visual cortex. *Nature, 310*, 498–500.

Katz, P. S. (2011). Neural mechanisms underlying the evolvability of behaviour. *Philosophical Transactions of the Royal Society of London, Series B, 366*, 2086–2099.

Kayser, C., Logothetis, N. K., & Panzeri, S. (2010). Millisecond encoding precision of auditory cortex neurons. *Proceedings of the National Academy of Sciences of the United States of America, 107*, 16976–16981.

Kenny, P. J. (2011). Reward mechanisms in obesity: New insights and future directions. *Neuron, 69*, 664–679.

Kier, C. K., Buchsbaum, G., & Sterling, P. (1995). How retinal microcircuits scale for ganglion cells of different size. *Journal of Neuroscience, 15*(11), 7673–7683.

Kim, M. H., Li, G. L., & von Gersdorff, H. (2013). Single Ca^{2+} channels and exocytosis at sensory synapses. *Journal of Physiology, 591*(13), 3167–3178.

Kim, I. J., Zhang, Y., Meister, M., & Sanes, J. R. (2010). Laminar restriction of retinal ganglion cell dendrites and axons: Subtype-specific developmental patterns revealed with transgenic markers. *Journal of Neuroscience, 30*, 1452–1462.

Kirschfeld, K. (1967). Die projektion der optischen umwelt auf das raster der rhabdomere im komplexauge von Musca. *Experimental Brain Research, 3*, 248–270.

Klaassen, L. J., Fahrenfort, I., & Kamermans, M. (2012). Connexin hemichannel mediated ephaptic inhibition in the retina. *Brain Research, 1487*, 25–38.

Kleim, J. A., Hogg, T. M., VandenBerg, P. M., Cooper, N. R., Bruneau, R., & Remple, M. (2004). Cortical synaptogenesis and motor map reorganization occur during late, but not early, phase of motor skill learning. *Journal of Neuroscience, 24*(3), 628–633.

Klyachko, V. A., & Stevens, C. F. (2003). Connectivity optimization and the positioning of cortical areas. *Proceedings of the National Academy of Sciences of the United States of America, 100*, 7937–7941.

Knudsen, E. I., Lac, S., & Esterly, S. D. (1987). Computational maps in the brain. *Annual Review of Neuroscience, 10*, 41–65.

Kobayashi, S., Pinto de Carvalho, O., & Schultz, W. (2010). Adaptation of reward sensitivity in orbitofrontal neurons. *Journal of Neuroscience, 30*(2), 534–544.

Koch, C. (2012). *Consciousness: Confessions of a romantic reductionist.* Cambridge, MA: MIT Press.

Koch, C. (1999). *Biophysics of computation: Information processing in single neurons.* Oxford: Oxford University Press.

Koch, K., McLean, J., Berry, M., Sterling, P., Balasubramanian, V., & Freed, M. A. (2004). Efficiency of information transmission by retinal ganglion cells. *Current Biology, 14*, 1523–1530.

Koch, K., McLean, J., Segev, R., Freed, M. A., Berry, M. J., II, Balasubramanian, V., & Sterling, P. (2006). How much the eye tells the brain. *Current Biology, 16*, 1428–1434.

Kolb, H., Nelson, R., & Mariani, A. (1981). Amacrine cells, bipolar cells and ganglion cells of the cat retina: A Golgi study. *Vision Research, 21*(7), 1081–1114.

Kole, M. H. P., & Stuart, G. J. (2012). Signal processing in the axon initial segment. *Neuron, 73*, 235–247.

Korn, H., & Axelrad, H. (1980). Electrical inhibition of Purkinje cells in the cerebellum of the rat. *Proceedings of the National Academy of Sciences of the United States of America, 77*(10), 6244–6247.

Koshland, D. E., Goldbeter, A., & Stock, J. B. (1982). Amplification and adaptation in regulatory and sensory systems. *Science, 217*, 220–225.

Koulakov, A. A., Hromádka, T., & Zador, A. M. (2009). Correlated connectivity and the distribution of firing rates in the neocortex. *Journal of Neuroscience, 29*(12), 3685–3694.

Kumbhani, R. D., Nolt, M. J., & Palmer, L. A. (2007) Precision, reliability, and information-theoretic analysis of visual thalamocortical neurons. *Journal of Neurophysiology, 98*(5), 2647–2663.

Kuypers, H. G. J. M. (1981). Anatomy of the descending pathways. In J. M. Brookhart & V. B. Mountcastle (Eds.), *The nervous system II, handbook of physiology* (pp. 597–666). Bethesda: American Physiological Society.

Lamb, T. D. (2013). Evolution of phototransduction, vertebrate photoreceptors and retina. *Progress in Retinal and Eye Research, 36*, 52e119.

Lamb, T. D., & Pugh, E. N., Jr. (2006). Phototransduction, dark adaptation, and rhodopsin regeneration. The Proctor Lecture. *Investigative Ophthalmology & Visual Science, 47*, 5138–5152.

Landauer, R. (1996). Minimal energy requirements in communication. *Science, 272*, 1914–1918.

Landi, S. M., Baguear, F., & Della-Maggiore, V. (2011). One week of motor adaptation induces structural changes in primary motor cortex that predict long-term memory one year later. *Journal of Neuroscience, 31*(33), 11808–11813.

Laughlin, S. (1981). A simple coding procedure enhances a neuron's information capacity. *Zeitschrift fur Naturforschung. Section C. Biosciences, 36*, 910–912.

Laughlin, S. B. (1974). Neural integration in the first optic neuropil of dragonflies. III. Transfer of angular information. *Journal of Comparative Physiology, 92*, 377–396.

Laughlin, S. B. (1992). Retinal information capacity and the function of the pupil. *Ophthalmic & Physiological Optics, 12*, 161–164.

Laughlin, S. B. (1994). Matching coding, circuits, cells, and molecules to signals: General principles of retinal design in the fly's eye. *Progress in Retinal and Eye Research, 13*, 165–196.

Laughlin, S. B. (1996). Matched filtering by a photoreceptor membrane. *Vision Research, 36*, 1529–1541.

Laughlin, S. B. (2010). The optic lamina of fast flying insects as a guide to neural circuit design. In G. M. Shepherd, & S. Grillner (Eds.), *Handbook of brain microcircuits* (pp. 404–415). Oxford: Oxford University Press.

Laughlin, S. B. (2011). Energy, information and the work of brain. In R. Levin, S. B. Laughlin, C. De La Rocha, & A. Blackwell (Eds.), *Work meets life* (pp. 39–67). Cambridge, MA: MIT Press.

Laughlin, S. B., & Osorio, D. (1989). Mechanisms for neural signal enhancement in the blowfly compound eye. *Journal of Experimental Biology, 144*, 113–146.

Laughlin, S. B., Howard, J., & Blakeslee, B. (1987). Synaptic limitations to contrast coding in the retina of the blowfly *Calliphora*. *Proceedings of the Royal Society of London, Series B, 231*, 437–467.

Laughlin, S. B., & Weckström, M. (1993). Fast and slow photoreceptors—A comparative study of the functional diversity of coding and conductances in the Diptera. *Journal of Comparative Physiology, A, 172*, 593–609.

Laughlin, S. B., de Ruyter van Steveninck, R. R., & Anderson, J. C. (1998). The metabolic cost of neural information. *Nature Neuroscience, 1*, 36–41.

Laurent, G. (2002). Olfactory network dynamics and the coding of multidimensional signals. *Nature Reviews Neuroscience, 3*, 884–895.

LeDoux, J. E. (2000). Emotion circuits in the brain. *Annual Review of Neuroscience, 23*, 155–184.

Leeper, H. F., & Charlton, J. S. (1985). Response properties of horizontal cells and photoreceptor cells in the retina of the tree squirrel, *Sciurus carolinensis*. *Journal of Neurophysiology, 54*(5), 1157–1166.

Leibovic, K. N., & Moreno-Diaz Jr., R. (1992). Rod outer segments are designed for optimum photon detection. *Biological Cybernetics, 66,* 301–306.

Lennie, P. (2003). The cost of cortical computation. *Current Biology, 13,* 493–497.

Leskov, I. B., Klenchin, V. A., Handy, J. W., Whitlock, G. G., Govardovskii, V. I., Bownds, M. D., Lamb, T. D., Pugh, E. N. Jr. & Arshavsky, V. Y. (2000). The gain of rod phototransduction: Reconciliation of biochemical and electrophysiological measurements *Neuron, 27*(3), 525–537.

Lettvin J. Y., Maturana H. R., McCulloch W. W., & Pitts W. H. (1959). What the frog's eye tells the frog's brain. *Proceedings of the Institute of Radio Engineers, 47,* 1940–1951.

Levinthal, D. J., & Strick, P. L. (2012). The motor cortex communicates with the kidney. *Journal of Neuroscience, 32,* 6726–6731.

Levy, D. J., & Glimcher, P. W. (2011). Comparing apples and oranges: Using reward-specific and reward-general subjective value representation in the brain. *Journal of Neuroscience, 31*(41), 14693–14707.

Levy, D. J., & Glimcher, P. W. (2012). The root of all value: A neural common currency for choice. *Current Opinion in Neurobiology, 22*(6), 1027–1038.

Levy, W. B., & Baxter, R. A. (1996). Energy efficient neural codes. *Neural Computation, 8,* 531–543.

Li, H., Zhang, Z., Blackburn, M. R., Wang, S. W., Ribelayga, C. P., & O'Brien, J. (2013). Adenosine and dopamine receptors coregulate photoreceptor coupling via gap junction phosphorylation in mouse retina. *Journal of Neuroscience, 33,* 3135–3150.

Li, W., & DeVries, S. H. (2004). Separate blue and green cone networks in the mammalian retina. *Nature Neuroscience, 7,* 751–756.

Liang, J., & Freed, M. A. (2010). The On pathway rectifies the Off pathway of the mammalian retina. *Journal of Neuroscience, 30,* 5533–5543.

Liang, Y., Fotiadis, D., Filipek, S., Saperstein, D. A., Palczewski, K., & Engel, A. (2003). Organization of the G protein coupled receptors rhodopsin and opsin in native membranes. *Journal of Biological Chemistry, 278,* 21655–21662.

Liebman, P. A., Parker, K. R. & Dratz, E. A. (1987). The molecular mechanism of visual excitation and its relation to the structure and composition of the rod outer segment. *Annual Review of Physiology, 49,* 765–791.

Light, A. C., Zhu, Y., Shi, J., Saszik, S., Lindstrom, S., Davidson, L., et al. (2012). Organizational motifs for ground squirrel cone bipolar cells. *Journal of Comparative Neurology, 520,* 2864–2887.

Lightner, M. (2011). Performance-yield trade-offs in work in manufactured and living system: Design centering look at the world of work. In R. Levin, S. B.

Laughlin, C. De La Rocha, & A. Blackwell (Eds.), *Work meets life* (pp. 69–95). Cambridge, MA: MIT Press.

Lim, H., Wang, Y., Xiao, Y., Hu, M., & Felleman, D. J. (2009). Organization of hue selectivity in macaque V2 thin stripes. *Journal of Neurophysiology, 102,* 2603–2615.

Lisman, J., Grace, A. A., & Duzel, E. (2011). A neoHebbian framework for episodic memory; role of dopamine-dependent late LTP. *Trends in Neurosciences, 34,* 536–547.

Llinás, R. R., Walton, K. D., & Lang, E. J. (2004). Cerebellum. In G. M. Shepherd (Ed.), *The synaptic organization of the brain* (pp. 271–310). New York: Oxford University Press.

Lockery, S. R. (2011). The computational worm: Spatial orientation and its neuronal basis in *C. elegans. Current Opinion in Neurobiology, 21,* 782–790.

Loewenstein, W. R., & Mendelson, M. (1965). Components of receptor adaptation in a Pacinian corpuscle. *Journal of Physiology, 177*(3), 377–397.

Lorente de Nó, R. (1938). Architectonics and structure of the cerebral cortex. In J. F. Fulton (Ed.), *Physiology of the nervous system* (pp. 291–330). New York: Oxford University Press.

Louie, K., Grattan, L. E., & Glimcher, P. W. (2011). Reward value-based gain control: Divisive normalization in parietal cortex. *Journal of Neuroscience, 31*(29), 10627–10639.

Low, K. (2012). How common is ADHD? About.com Guide. Updated September 1.

Lu, H. D., Chen, G., Tanigawa, H., & Roe, A. W. (2010). A motion direction map in macaque V2. *Neuron, 68,* 1002–1013.

Lukasiewicz, P. D., & Shields, C. R. (1998). Different combinations of GABAA and GABAC receptors confer distinct temporal properties to retinal synaptic responses. *Journal of Neurophysiology, 79*(6), 3157–3167.

Mackenzie, D. (2005). Take it to the limit. *New Scientist, 187,* 38–41.

MacNeil, M. A., Heussy, J. K., Dacheux, R. F., Raviola, E., & Masland, R. H. (1999). The shapes and numbers of amacrine cells; matching of photofilled with golgi-stained cells in the rabbit retina and comparison with other mammalian species. *Journal of Comparative Neurology, 413,* 305–326.

Macosko, E. Z., Pokala, N., Feinberg, E. H., Chalasani, S. H., Butcher, R. A., Clardy, J., & Bargmann, C. I. (2009). A hub-and-spoke circuit drives pheromone attraction and social behaviour in *C. elegans. Nature, 458,* 1171–1175.

Maguire, E. A., Gadian, D. G., Johnsrude, I. S., Good, C. D., Ashburner, J., Frackowiak, R. S., & Frith, C. D. (2000). Navigation-related structural change in the hippocampi of taxi drivers. *Proceedings of the National Academy of Sciences of the United States of America, 97*, 4398–4403.

Makaral, J. K., Losonczy, A., Wen, Q., & Magee, J. C. (2009). Experience-dependent compartmentalized dendritic plasticity in rat hippocampal CA1 pyramidal neurons. *Nature Neuroscience, 12*(12), 1485–1487. Epub 2009 Nov 8.

Maksimovic, S., Baba, Y., & Lumpkin, E. A. (2013). Neurotransmitters and synaptic components in the Merkel cell-neurite complex, a gentle-touch receptor. *Annals of the New York Academy of Sciences, 1279*, 13–21.

Manni, E., & Petrosini, L. (2004). A century of cerebellar somatotopy: A debated representation. *Nature Reviews Neuroscience, 5*, 241–249.

Manookin, M. B., Beaudoin, D. L., Ernst, Z. R., Flagel, L. J., & Demb, J. B. (2008). Disinhibition combines with excitation to extend the operating range of the OFF visual pathway in daylight. *Journal of Neuroscience, 28*, 4136–4150.

Marcelja, S. (1980). Mathematical description of the responses of simple cortical cells. *Journal of the Optical Society of America, 70*, 1297–1300.

Marder, E., & Bucher, D. (2001). Central pattern generators and the control of rhythmic movements. *Current Biology, 11*, R986–R996.

Margolis, D. J., Gartland, A. J., Euler, T., & Detwiler, P. B. (2010). Dendritic calcium signaling in ON and OFF mouse retinal ganglion cells. *Journal of Neuroscience, 30*(21), 7127–7138.

Margrie, T. W., Brecht, M., & Sakmann, B. (2002). In vivo, low-resistance, whole-cell recordings from neurons in the anaesthetized and awake mammalian brain. *Pflugers Archiv, 444*, 491–498.

Markram, H., Lubke, J., Frotscher, M., Roth, A., & Sakmann, B. (1997). Physiology and anatomy of synaptic connections between thick tufted pyramidal neurones in the developing rat neocortex. *Journal of Physiology, 500*, 409–440.

Masland, R. H. (2012). The neuronal organization of the retina. *Neuron, 76*, 266–280.

Masse, N. Y., Turner, G. C., & Jefferis, G. S. X. E. (2009). Olfactory information processing in *Drosophila*. *Current Biology, 19*, R700–R713.

Matić, T., & Laughlin, S. B. (1981). Changes in the intensity-response function of an insect's photoreceptors due to light adaptation. *Journal of Comparative Physiology, A, 145*, 169–177.

Matsuda,W. (2012). Imaging of dopaminergic neurons and the implications for Parkinson's disease. In P. Wellstead & M. Cloutier (Eds.), *Systems biology of Parkinson's disease* (pp. 1–17). New York: Springer.

Matsuda, W., Furuta, T., Nakamura, K. C., Hioki, H., Fujiyama, F., Arai, R., & Kaneko, T. (2009). Single nigrostriatal dopaminergic neurons form widely spread highly dense axonal arborizations in the neostriatum. *Journal of Neuroscience, 29*(2), 444–453.

Matsuzaka, Y., Picard, N., & Strick, P. L. (2007). Skill representation in the primary motor cortex after long-term practice. *Journal of Neurophysiology, 97*, 1819–1832.

Matthews, G., & Fuchs, P. (2010). The diverse roles of ribbon synapses in sensory neurotransmission. *Nature Reviews Neuroscience, 11*, 812–822.

Matthews, G., & Sterling, P. (2008). Evidence that vesicles undergo compound fusion on the synaptic ribbon. *Journal of Neuroscience, 28*(21), 5403–5411.

Mayford, M., Siegelbaum, S. A., & Kandel, E. R. (2012). Synapses and memory storage. *Cold Spring Harbor Perspectives in Biology, 4*(6), 1–22.

McClave, S. A., & Snider, H. L. (2001). Dissecting the energy needs of the body. *Current Opinion in Clinical Nutrition and Metabolic Care, 4*, 143–147.

McDonnell, M. D., & Ward, L. M. (2011). The benefits of noise in neural systems: Bridging theory and experiment. *Nature Reviews Neuroscience, 12*, 415–426.

McLean, J., Raab, S., & Palmer, L. A. (1994). Contribution of linear mechanisms to the specification of local motion by simple cells in areas 17 and 18 of the cat. *Visual Neuroscience, 11*, 271–294.

Meinertzhagen, I. (1993). The synaptic populations of the flys optic neuropil and their dynamic regulation—Parallels with the vertebrate retina. *Progress in Retinal Research, 12*, 13–39.

Menzel, R. (2012). The honeybee as a model for understanding the basis of cognition. *Nature Reviews Neuroscience, 13*, 758–768.

Mercer, A. J., Rabl, K., Riccardi, G. E., Brecha, N. C., Stella, S. L., Jr., & Thoreson, W. B. (2011). Photoreceptor ribbon synapse channels relative to calcium channels at the location of release sites and calcium-activated chloride. *Journal of Neurophysiology, 105*, 321–335.

Meyer, A. C., Frank, T., Khimich, D., Hoch, G., Riedel, D., Chapochnikov, N. M., Yarin, Y. M., Harke, B., Hell, S. W., Egner, A. & Moser, T. (2009). Tuning of synapse number, structure and function in the cochlea. *Nature Neuroscience, 12*(4), 444–453.

Middleton, F. A., & Strick, P. L. (1996). The temporal lobe is a target of output from the basal ganglia. *Proceedings of the National Academy of Sciences of the United States of America, 93*, 8683–8687.

Miller, P., Zhabotinsky, A. M., Lisman, J. E., & Wang, X.-J. (2005). The stability of a stochastic CaMKII switch: Dependence on the number of enzyme molecules and protein turnover. *PLoS Biol 3*, e107.

Mishchenko, Y., Hu, T., Spacek, J., Mendenhall, J., Harris, K. M., & Chklovskii, D. B. (2010). Ultrastructural analysis of hippocampal neuropil from the connectomics perspective. *Neuron, 67*, 1009–1020.

Mitchison, G. (1991). Neuronal branching patterns and the economy of cortical wiring. *Proceedings of the Royal Society of London, Series B, 245*, 151–158.

Miyashita, T., Oda, Y., Horiuchi, J., Yin, J. C. P., Morimoto, T., & Saitoe, M. (2012). Mg^{2+} block of *Drosophila* NMDA receptors is required for long-term memory formation and CREB-dependent gene expression. *Neuron, 74*(5), 887–898.

Moerel, M., De Martino, F., & Formisano, E. (2012). Processing of natural sounds in human auditory cortex: Tonotopy, spectral tuning, and relation to voice sensitivity. *Journal of Neuroscience, 32*(41), 14205–14216.

Mogensen, J., Williams, G., & Divac, I. (1983). Activity of the auditory system in rats habituated to a test chamber: A 2-deoxyglucose study. *Acta Neurobiologiae Experimentalis, 43*, 283–290.

Molnar, A., Hsueh, H. A., Roska, B., & Werblin, F. S. (2009). Crossover inhibition in the retina: Circuitry that compensates for nonlinear rectifying synaptic transmission. *Journal of Computational Neuroscience, 27*(3), 569–590.

Monod, J. (1971). *Chance and necessity.* New York: Vintage.

Montague, P. R., Dayan, P., Person, C., Sejnowski, T. J. (1995). Bee foraging in uncertain environments using predictive hebbian learning. *Nature, 377*(6551), 725–728.

Moore, C., & Mertens, S. (2011). *The nature of computation.* New York: Oxford University Press.

Moorman, S., Gobes, S. M. H., Kuijpers, M., Kerkhofs, A., Zandbergen, M. A., & Bolhuis, J. J. (2012). Human-like brain hemispheric dominance in birdsong learning. *Proceedings of the National Academy of Sciences of the United States of America, 109*, 12782–12787.

Morgans, C. W., Brown, R. L., & Duvoisin, R. M. (2010). TRPM1: The endpoint of the mGluR6 signal transduction cascade in retinal ON-bipolar cells. *BioEssays, 2*, 609–614.

Moser, T., Neef, A., & Khimich, D. (2006). Mechanisms underlying the temporal precision of sound coding at the inner hair cell ribbon synapse. [Topical Review.] *Journal of Physiology, 576*(1), 55–62.

Mullikin, W. H., Jones, J. P., & Palmer, L. A. (1984). Periodic simple cells in cat area 17. *Journal of Neurophysiology, 52*, 372–387.

Naarendorp, F., Esdaille, T. M., Banden, S. M., Andrews-Labenski, J., Gross, O. P., & Pugh, E. N., Jr. (2010). Dark light, rod saturation, and the absolute and incremental sensitivity of mouse cone vision. *Journal of Neuroscience, 30*(37), 12495–12507.

Nahir, B., & Jahr, C. E. (2013). Activation of extrasynaptic NMDARs at individual parallel fiber–molecular layer interneuron synapses in cerebellum. *Journal of Neuroscience 33*(41), 16323–16333.

Nairn, J. G., Bedi, K. S., Mayhew, T. M., & Campbell, L. F. (1989). On the number of Purkinje cells in the human cerebellum: Unbiased estimates obtained by using the "fractionator." *Journal of Comparative Neurology, 290*(4), 527–532.

Napper, R. M., & Harvey, R. J. (1988). Number of parallel fiber synapses on an individual Purkinje cell in the cerebellum of the rat. *Journal of Comparative Neurology, 274*(2), 168–177.

Nasr, S., Liu, N., Devaney, K. J., Yue, X., Rajimehr, R., Ungerleider, L. G., & Tootell, R. B. H. (2011). Scene-selective cortical regions in human and nonhuman primates. *Journal of Neuroscience, 31*, 13771–13785.

Nassi, J. J., & Callaway, E. M. (2009). Parallel processing strategies of the primate visual system. *Nature Reviews Neuroscience, 10*(5), 360–372.

Nawroth, J. C., Greer, C. A., Chen, W. R., Laughlin, S. B., & Shepherd, G. M. (2007). An energy budget for the olfactory glomerulus. *Journal of Neuroscience, 27*, 9790–9800.

Nayak, S. K., Batalov, S., Jegla, T. J., & Zmasek, C. M. (2009). Evolution of the human ion channel set. *Combinatorial Chemistry & High Throughput Screening, 12*, 2–13.

Neisser, U. (1997). The ecological study of memory. *Philosophical Transactions of the Royal Society of London, Series B, 352*(1362), 1697–1701.

Neitz, J., Carroll, J., Yamauchi, Y., Neitz, M., & Williams, D. R. (2002). Color perception is mediated by a plastic neural mechanism that is adjustable in adults. *Neuron, 35*(4), 783–792.

Nelson, P. (2008). *Biological physics: Energy, information, life.* W. H. Freeman and Co.

Nemenman, I. (2012). Gain control in molecular information processing: Lessons from neuroscience. *Physical Biology, 9*, 026003.

Newman, E. A., & Hartline, P. H. (1981). Integration of visual and infrared information in bimodal neurons in the rattlesnake optic tectum. *Science, 213*(4509), 789–791.

Ni, A. M., Ray, S., & Maunsell, J. H. (2012). Tuned normalization explains the size of attention modulations. *Neuron, 73*, 803–813.

Nicol, D., & Meinertzhagen, I. A. (1982). An analysis of the number and composition of the synaptic populations formed by photoreceptors of the fly. *Journal of Comparative Neurology, 207*, 29–44.

Nieuwenhuys, R., & Nicholson, C. (1969). A survey of the general morphology, the fiber connections, and the possible functional significance of the gigantocerebellum of mormyrid fishes. In R. Llinás (Ed.), *Neurobiology of cerebellar evolution and development* (pp. 107–134). Chicago, IL: American Medical Association.

Nimchinsky, E. A., Yasuda, R., Oertner, T. G., & Svoboda, K. (2004). The number of glutamate receptors opened by synaptic stimulation in single hippocampal spines. *Journal of Neuroscience, 24*(8), 2054–2064.

Niu, J., Ding, L., Li, J. J., Kim, H., Liu, J., Li, H., et al. (2013). Modality-based organization of ascending somatosensory axons in the direct dorsal column pathway. *Journal of Neuroscience, 33*(45), 17691–17709.

Niven, J. E., Anderson, J. C., & Laughlin, S. B. (2007). Fly photoreceptors demonstrate energy-information trade-offs in neural coding. *PLoS Biology, 5*, 828–840.

Niven, J. E., Vähäsöyrinki, M., & Juusola, M. (2003a). Shaker K$^+$-channels are predicted to reduce the metabolic cost of neural information in *Drosophila* photoreceptors. *Proceedings of the Royal Society of London, Series B, 270*, S58–S61.

Niven, J. E., Vähäsöyrinki, M., Kauranen, M., Hardie, R. C., Juusola, M., & Weckström, M. (2003b). The contribution of Shaker K$^+$ channels to the information capacity of *Drosophila* photoreceptors. *Nature, 421*, 630–634.

Normann, R. A., & Perlman, I. (1979). The effects of background illumination on the photoresponses of red and green cones. *Journal of Physiology (Lond.), 286*, 491–507.

Normann, R., & Werblin, F. (1974). Control of retinal sensitivity. I. Light and dark-adaptation of vertebrate rods and cones. *Journal of General Physiology, 63*, 37–61.

Novák, B., & Tyson, J. J. (2008). Design principles of biochemical oscillators. *Nature Reviews. Molecular Cell Biology, 9*, 981–991.

O'Connor, D. H., Huber, D., & Svoboda, K. (2009). Reverse engineering the mouse brain. *Nature, 461*, 923–929.

Oertel, D., & Young, E. D. (2004). What's a cerebellar circuit doing in the auditory system? *Trends in Neurosciences, 27*(2), 104–110.

Oesch, N. W., & Diamond, J. S. (2011). Ribbon synapses compute temporal contrast and encode luminance in retinal rod bipolar cells. *Nature Neuroscience, 14*(12), 1555–1561.

Ohki, K., Chung, S., Karal, P., Hübener, M., Bonhoeffer, T., & Reid, T. C. (2006). Highly ordered arrangement of single neurons in orientation pinwheels. *Nature, 442*, 925–928.

Oishi, K., & Klavins, E. (2011). Biomolecular implementation of linear I/O systems. *IET Systems Biology, 5*, 252–260.

Okajima, K. (1998a). The Gabor function extracts the maximum information from input local signals. *Neural Networks, 11*, 435–439.

Okajima, K. (1998b). Two-dimensional Gabor-type receptive field as derived by mutual information maximization. *Neural Networks, 11*, 441–447.

Okawa, H., Sampath, A. P., Laughlin, S. B., & Fain, G. L. (2008). ATP consumption by mammalian rod photoreceptors in darkness and in light. *Current Biology, 18*, 1917–1921.

Olivares, R., Montiel, J., & Aboitiz, F. (2001). Species differences and similarities in the fine structure of the mammalian corpus callosum. *Brain, Behavior and Evolution, 57*, 98–105.

Olshausen, B., & Field, D. (2004). Sparse coding of sensory inputs. *Current Opinion in Neurobiology, 14*, 481–487.

Olshausen, B. A. (2004). Principles of image representation in visual cortex. In L. Chalupa & J. Werner (Eds.), *The visual neurosciences* (pp. 1603–1615). Cambridge, MA: MIT Press.

Ongur, D., & Price, J. L. (2000). The organization of networks within the orbital and medial prefrontal cortex of rats, monkeys and humans. *Cerebral Cortex, 10*, 206–219.

Packer, O., Hendrickson, A. E., & Curcio, C. A. (1989). Photoreceptor topography of the retina in the adult pigtail macaque (*Macaca nemestrina*). *Journal of Comparative Neurology, 288*(1), 165–183.

Packer, O. S., Verweij, J., Li, P. H., Schnapf, J. L., & Dacey, D. M. (2010). Blue-yellow opponency in primate S cone photoreceptors. *Journal of Neuroscience, 30*(2), 568–572.

Padoa-Schioppa, C., & Assad, J. A. (2008). The representation of economic value in the orbitofrontal cortex is invariant for changes of menu. *Nature Neuroscience, 11*(1), 95–102.

Pahl, G., Beitz, W., Feldhusen, J., & Grote, K. H. (2007). Wallace, K., & Blessing, L. (Trans. Eds.), *Engineering design: A systematic approach* (third edition). Berlin: Springer.

Pajevic, S., & Basser, P. J. (2013). An optimum principle predicts the distribution of axon diameters in normal white matter. *PLoS One, 8*(1), e54095.

Palay, S. L., & Chan-Palay, V. (1974). *Cerebellar cortex*. New York: Springer-Verlag.

Palmer, L. A. & Rosenquist, A. C. (1974). Visual receptive fields of single striate cortical units projecting to the superior colliculus in the cat. *Brain Research, 67*(1):27–42.

Palmer, L. A., & Davis, T. L. (1981). Receptive-field structure in cat striate cortex. *Journal of Neurophysiology, 46*(2), 260–276.

Palmer, L. A., Jones, J. P., & Stepnoski, R. A. (1991). Striate receptive fields as linear filters: Characterization in two dimensions of space. In A. Leventhal (Ed.), *Vision and visual dysfunction, Vol 4: The neural basis of visual function* (pp. 246–265). Boca Raton: CRC Press.

Pammer, L., O'Connor, D. H., Hires, S. A., Clack, N. G., Huber, D., Myers, E. W., & Svoboda, K. (2013). The mechanical variables underlying object localization along the axis of the whisker. *Journal of Neuroscience, 33*, 6726–6741.

Pan, F., & Massey, S. C. (2007). Rod and cone input to horizontal cells in the rabbit retina. *Journal of Comparative Neurology, 500*(5), 815–831.

Panico, J., & Sterling, P. (1995). Retinal neurons and vessels are not fractal but space-filling. *Journal of Comparative Neurology, 361*(3), 479–490.

Park, S., Hwang, H., Nam, S.-W., Martinez, F., Austin, R. H., et al. (2008). Enhanced *Caenorhabditis elegans* locomotion in a structured microfluidic environment. *PLoS ONE, 3*, e2550.

Pavlov, I. (1904). Physiology of digestion. Nobel Lecture. Available at http:// www.nobelprize.org/nobel_prizes/medicine/laureates/1904/pavlov-lecture.html/.

Pereda, A. E. (2014). Electrical synapses and their functional interactions with chemical synapses. *Nature Reviews Neuroscience, 15*, 250–263.

Pereira, S., & van der Kooy, D. (2012). Two forms of learning following training to a single odorant in *Caenorhabditis elegans* AWC neurons. *Journal of Neuroscience, 32*, 9035–9044.

Pérez-Escudero, A., & de Polavieja, G. G. (2007). Optimally wired subnetwork determines neuroanatomy of *Caenorhabditis elegans*. *Proceedings of the National Academy of Sciences of the United States of America, 104*, 17180–17185.

Perge, J. A., Koch, K., Miller, R., Sterling, P., & Balasubramanian, V. (2009). How the optic nerve allocates space, energy capacity, and information. *Journal of Neuroscience, 29*(24), 7917–7928.

Perge, J. A., Niven, J. E., Mugnaini, E., Balasubramanian, V., & Sterling, P. (2012). Why do axons differ in caliber? *Journal of Neuroscience, 32*, 626–638.

Perrodin, C., Kayser, C., Logothetis, N. K., & Petkov, C. I. (2011). Voice cells in the primate temporal lobe. *Current Biology, 21*, 1408–1415.

Person, A. L., & Raman, I. M. (2012). Purkinje neuron synchrony elicits time-locked spiking in the cerebellar nuclei. *Nature, 481*, 502–506.

Petilla Interneuron Nomenclature Group (PING), Ascoli, G. A., Alonso-Nanclares, L., Anderson, S. A., Barrionuevo, G., Benavides-Piccione, R., et al. (2008). Petilla terminology: Nomenclature of features of GABAergic interneurons of the cerebral cortex. *Nature Reviews Neuroscience, 9*(7), 557–568.

Petroski H. (1996). *Invention by design: How engineers get from thought to thing.* Cambridge, MA: Harvard University Press.

Pfeifer, R., & Bongard, J. C. (2006). *How the body shapes the way we think. A new view of intelligence.* Cambridge, MA: MIT Press.

Phillips, R., Kondev, J., & Theriot, J. (2009). *Physical biology of the cell.* New York: Garland Science.

Phillips-Portillo, J., & Strausfeld, N. J. (2012). Representation of the brain's superior protocerebrum of the flesh fly, *Neobellieria bullata*, in the central body. *Journal of Comparative Neurology, 520*, 3070–3087.

Poulet, J. F. A., & Hedwig, B. (2006). The cellular basis of a corollary discharge. *Science, 311*, 518–522.

Poulet, J. F. A., & Hedwig, B. (2007). New insights into corollary discharges mediated by identified neural pathways. *Trends in Neurosciences, 30*, 14–21.

Price, N. S. C., & Born, R. T. (2010). Timescales of sensory- and decision-related activity in the middle temporal and medial superior temporal areas. *Journal of Neuroscience, 30*, 14036–14045.

Priebe, N. J., & Ferster, D. (2012). Mechanisms of neuronal computation in mammalian visual cortex. *Neuron, 75*, 194–208.

Purcell, B. A., Weigand, P. K., & Schall, J. D. (2012). Supplementary eye field during visual search: Salience, cognitive control, and performance monitoring. *Journal of Neuroscience, 32*(30):10273–10285.

Purcell, E. M. (1977). Life at low Reynolds number. *American Journal of Physics, 45*, 3–11.

Purves, D., Augustine, G. J., Fitzpatrick, D., Hall, W. C., LaMantia, A., White, L. E. (2012). *Neuroscience* (fifth edition). Sunderland, MA: Sinauer Associates.

Qiu, A., Schreiner, C. E., & Escabí, M. A. (2003). Gabor analysis of auditory midbrain receptive fields: Spectro-temporal and binaural composition. *Journal of Neurophysiology, 90*, 456–476.

Quallo, M. M., Kraskov, A., & Lemon, R. N. (2012). The activity of primary motor cortex corticospinal neurons during tool use by macaque monkeys. *Journal of Neuroscience, 32*(48), 17351–17364.

Quiroga, R. Q., Reddy, L., Kreiman, G., Koch, C., & Fried, I. (2005). Invariant visual representation by single neurons in the human brain. *Nature, 435*, 1102–1107.

Rabl, K., Cadetti, L., & Thoreson, W. B. (2005). Kinetics of exocytosis is faster in cones than in rods. *Journal of Neuroscience, 25*(18), 4633–4640.

Raczkowski, D., Hamos, J. E., & Sherman, S. M. (1988). Synaptic circuitry of physiologically identified W-cells in the cat's dorsal lateral geniculate nucleus. *Journal of Neuroscience, 8,* 31–48.

Rall, W. (1959). Branching dendritic trees and motoneuron membrane resistivity. *Experimental Neurology, 1,* 491–527.

Raman, I. M., Sprunger, L. K., Meisler, M. H., & Bean, B. P. (1997). Altered subthreshold sodium currents and disrupted firing patterns in Purkinje neurons of *Scn8a* mutant mice. *Neuron, 19,* 881–891.

Ramon, M., Caharel, S., & Rossion, B. (2011). The speed of recognition of personally familiar faces. *Perception, 40,* 437–449.

Ramón y Cajal, S. (1909). *Histologie du système nerveux de l'homme & des vertébrés.* English translation (1995) as *Histology of the nervous system of man and vertebrates* (N. Swanson, & L. Swanson, Trans.). Oxford: Oxford University Press.

Ramón y Cajal, S. (1917). Originally published as *Recuerdos de Mi Vida.* English translation (1937) as *Recollections of my life* (E. H. Craigie & J. Cano, Trans). Cambridge, MA: MIT Press.

Ransom, C. B., Ransom, B. R., & Sontheimer, H. (2000). Activity-dependent extracellular K$^+$accumulation in rat optic nerve: The role of glial and axonal Na$^+$ pumps. *Journal of Physiology, 522,* 427–442.

Rao, R., Buchsbaum, G., & Sterling. P. (1994). Rate of quantal transmitter release at the mammalian rod synapse. *Biophysical Journal, 67*(1), 57–63.

Rao-Mirotznik, R., Buchsbaum, G., & Sterling, P. (1998). Transmitter concentration at a three-dimensional synapse. *Journal of Neurophysiology, 80,* 3163–3172.

Rasmussen, S. G. F., DeVree, B. T., Zou, Y., Kruse, A. C., Chung, K. Y., Kobilka, T. S., et al. (2011). Crystal structure of the β2 adrenergic receptor—Gs protein complex. *Nature, 477,* 549–555.

Rathbun, D. L., Warland, D. K., & Usrey, W. M. (2010). Spike timing and information transmission at retinogeniculate synapses. *Journal of Neuroscience, 30,* 13558–13566.

Rathelot, J.-A., & Strick, P. L. (2009). Subdivisions of primary motor cortex based on cortico-motoneuronal cells. *Proceedings of the National Academy of Sciences of the United States of America, 106,* 918–923.

Ratliff, C. P., Borghuis, B. G., Kao, Y. H., Sterling, P., & Balasubramanian, V. (2010). Retina is structured to process an excess of darkness in natural scenes. *Proceedings of the National Academy of Sciences of the United States of America, 107*(40), 17368–17373.

Ratliff, C. P., & DeVries, S. H. (2011). Different bipolar cell types process information from the cone synapse at different rates [Abstract]. Presented at *Arvo 2011*, Fort Lauderdale, FL, May 1–5.

Rauschecker, J. P. (2012). Ventral and dorsal streams in the evolution of speech and language. *Frontiers in Evolutionary Neuroscience, 4*, Article 7, 1–4.

Rayner, K. (1998). Eye movements in reading and information processing: 20 years of research. *Psychological Bulletin, 124*, 372–422.

Reh, T. A., & Constantine-Paton, M. (1984). Retinal ganglion cell terminals change their projection sites during larval development of *Rana pipiens*. *Journal of Neuroscience, 4*, 442–457.

Reingruber, J., Pahlberg, J., Woodruff, M. L., Sampath, A. P., Fain, G. L., & Holcman, D. (2013). Detection of single photons by toad and mouse rods. *Proceedings of the National Academy of Sciences of the United States of America, 110*(48), 19378–19383.

Renteria, R. C., Tian, N., Cang, J., Nakanishi, S., Stryker, M. P., & Copenhagen, D. R. (2006). Intrinsic ON responses of the retinal OFF pathway are suppressed by the ON pathway. *Journal of Neuroscience, 26*, 11857–11869.

Ressel, V., Pallier, C., Ventura-Campos, N., Díaz, B., Roessler, A., Ávila, C., & Sebastián-Gallés, N. (2012). An effect of bilingualism on the auditory cortex. *Journal of Neuroscience, 32*(47), 16597–16601.

Reynolds, A. M., & Rhodes, C. J. (2009). The Lévy flight paradigm: Random search patterns and mechanisms. *Ecology, 90*, 877–887.

Ribrault, C., Sekimoto, K., & Triller, A. (2011). From the stochasticity of molecular processes to the variability of synaptic transmission. *Nature Reviews Neuroscience, 12*(7), 375–387.

Richards, W. (1982). Lightness scale from image intensity distributions. *Applied Optics, 21*, 2569–2604.

Rieke, F., Bodnar, D. A., & Bialek, W. (1995). Naturalistic stimuli increase the rate and efficiency of information transmission by primary auditory afferents. *Proceedings of the Royal Society of London. Series B, 262*, 259–265.

Rieke, F., Warland, D., de Ruyter van Steveninck, R. R., & Bialek, W. (1997). *Spikes. Exploring the neural code.* Cambridge, MA: MIT Press.

Ringo, J. L., Don, R. W., Demeter, S., & Simard, P. Y. (1994). Time is of the essence: A conjecture that hemispheric specialization arises from interhemispheric conduction delay. *Cerebral Cortex, 4*(4), 331–343.

Rivera-Alba, M., Vitaladevuni, S. N., Mishchenko, Y., Lu, Z., Takemura, S., Scheffer, L., et al. (2011). Wiring economy and volume exclusion determine neuronal placement in the *Drosophila* brain. *Current Biology, 21*, 2000–2005.

Roche, B. (1974). As empty as Eve. *The New Yorker*, Sept 9, pp 84–100.

Rockhill, R. L., Daly, F. J., MacNeil, M. A., Brown, S. P., & Masland R. H. (2002). The diversity of ganglion cells in a mammalian retina. *Journal of Neuroscience, 22*(9), 3831–3843.

Rodieck, R. W., & Brening, R. K. (1983). Retinal ganglion cells: Properties, types, genera, pathways and trans-species comparisons. *Brain, Behavior and Evolution, 23*(3–4), 121–164.

Rodríguez, F. A., Chen, C., Read, H. L., & Escabí, M. A. (2010). Neural modulation tuning characteristics scale to efficiently encode natural sound statistics. *Journal of Neuroscience, 30*, 15969–15980.

Roe, A. W., Garraghty, P. E., & Sur, M. (1989). Terminal arbors of single ON-center and OFF-center X and Y retinal ganglion cell axons within the ferret's lateral geniculate nucleus. *Journal of Comparative Neurology, 288*, 208–242.

Roeder, K. D. (1967). *Nerve cells and insect behavior*. Cambridge, MA: Harvard University Press.

Rose, A. (1974). *Vision: Human and electronic (optical physics and engineering)*. IBM Research Symposia Series.

Rossel, S., & Wehner, R. (1982). The bee's map of the e-vector pattern in the sky. *Proceedings of the National Academy of Sciences of the United States of America, 79*, 4451–4455.

Ruigrok, T. J., Hensbroek, R. A., & Simpson, J. I. (2011). Spontaneous activity signatures of morphologically identified interneurons in the vestibulocerebellum. *Journal of Neuroscience, 31*, 712–724.

Rust, N. C., & DiCarlo, J. J. (2012). Balanced increases in selectivity and tolerance produce constant sparseness along the ventral visual stream. *Journal of Neuroscience, 32*, 10170–10182.

Rutherford, M. A., Chapochnikov, N. M., & Moser, T. (2012). Spike encoding of neurotransmitter release timing by spiral ganglion neurons of the cochlea. *Journal of Neuroscience, 32*(14), 4773–4789.

Sabatini, B. L., & Svoboda, K. (2000). Analysis of calcium channels in single spines using optical fluctuation analysis. *Nature, 408*, 589–593.

Sachdev, R. N., & Catania, K. C. (2002). Receptive fields and response properties of neurons in the star-nosed mole's somatosensory fovea. *Journal of Neurophysiology, 87*(5), 2602–2611.

Sacks, O. (1985). *The man who mistook his wife for a hat*. New York: Summit Books.

Sacktor, T. C. (2011). How does PKMζ maintain long-term memory? *Nature Reviews Neuroscience, 12*, 9–15.

Sadeghi, S. G., Chacron, M. J., Taylor, M. C., & Cullen, K. E. (2007). Neural variability, detection thresholds and information transmission in the vestibular system. *Journal of Neuroscience, 27,* 771–781.

Sakaguchi, M., & Hayashi, Y. (2012). Catching the engram: Strategies to examine the memory trace. *Molecular Brain, 5,* 32.

Sakurai, T. (2007). The neural circuit of orexin (hypocretin): Maintaining sleep and wakefulness. *Nature Reviews, 8,* 171–181.

Sala, C., & Segal, M. (2014). Dendritic spines: The locus of structural and functional plasticity. *Physiological Reviews, 94,* 141–188.

Sallet, J., Mars, R. B., Noonan, M. P., Andersson, J. L., O'Reilly, J. X., Jbabdi, S., et al. (2011). Social network size affects neural circuits in macaques. *Science, 334,* 697–700.

Santello, M., Baud-Bovy, G., & Jörntell, H. (2013). Neural bases of hand synergies. *Frontiers in Computational Neuroscience, 7,* 1–15.

Saper, C. B., Lu, J., Chou, T. C., & Gooley, J. (2005). The hypothalamic integrator for circadian rhythms. *Trends in Neurosciences, 28,* 152–157.

Sarpeshkar, R. (1998). Analog versus digital: extrapolating from electronics to neurobiology. *Neural Computation, 10,* 1601–1638.

Sarpeshkar, R. (2010). *Ultra low power bioelectronics.* Cambridge: Cambridge University Press.

Sarpeshkar, R. (2012). Universal principles for ultra low power and energy efficient design. *IEEE Transactions on Circuits and Systems II: Express Briefs, 59,* 193–198.

Sarpeshkar, R. (2014). Analog synthetic biology. *Philosophical Transactions of the Royal Society, Series A, 372*(2012), 20130110.

Sato, R., & Svoboda, K. (2010). The functional properties of barrel cortex neurons projecting to the primary motor cortex. *Journal of Neuroscience, 30,* 4256–4260.

Saul, A. B. (2008). Lagged cells. *Neuro-Signals, 16,* 209–225.

Savage, G. L., & Banks, M. S. (1992). Scotopic visual efficiency: Constraints by optics, receptor properties, and rod pooling. *Vision Research, 32,* 645–656.

Saviane, C., & Silver, R. A. (2006). Fast vesicle reloading and a large pool sustain high bandwidth transmission at a central synapse. *Nature, 439,* 983–987.

Savtchenko, L. P., & Rusakov, D. A. (2007). The optimal height of the synaptic cleft. *Proceedings of the National Academy of Sciences of the United States of America, 104,* 1823–1828.

Schein, S., Sterling, P., Ngo, I. T., Huang, T. M., & Herr, S. (2004). Evidence that each S cone in macaque fovea drives one narrow-field and several wide-field blue-yellow ganglion cells. *Journal of Neuroscience, 24*(38), 8366–8378.

Schmahmann, J. D., & Pandya, D. N. (2008). Disconnection syndromes of basal ganglia, thalamus, and cerebrocerebellar systems. *Cortex*, *44*, 1037–1066.

Schmitt, L. I., Sims, R. E., Dale, N., & Haydon, P. G. (2012). Wakefulness affects synaptic and network activity by increasing extracellular astrocyte-derived adenosine. *Journal of Neuroscience*, *32*(13), 4417–4425.

Schneider, T. D. (2010). A brief review of molecular information theory. *Nano Communication Networks*, *1*, 173–180.

Schreiber, S., Machens, C. K., Herz, A. V. M., & Laughlin, S. B. (2002). Energy-efficient coding with discrete stochastic events. *Neural Computation*, *14*, 1323–1346.

Schultz, W. (2007). Multiple dopamine functions at different time courses. *Annual Review of Neuroscience*, *30*, 259–288.

Schultz, W. (2010). Dopamine signals for reward value and risk: Basic and recent data. *Behavior and Brain Function*, *6*(24), doi: 10.1186/1744-9081-6-24.

Schultz, W. (2011). Potential vulnerabilities of neuronal reward, risk, and decision mechanisms to addictive drugs. *Neuron*, *69*, 603–617.

Schüz, A., & Palm, G. (1989). Density of neurons and synapses in the cerebral cortex of the mouse. *Journal of Comparative Neurology*, *286*, 442–455.

Schwartz, D. A., Howe, C. Q., & Purves, D. (2003). The statistical structure of human speech sounds predicts musical universals. *Journal of Neuroscience*, *23*(18), 7160–7168.

Schwartz, T. W., & Sakmar, T. P. (2011). Structural biology: Snapshot of a signalling complex. *Nature*, *477*, 540–541.

Sebald, W. G. (1998). *The rings of Saturn* (M. Hulse, Trans.) London: Harvill Press.

Sengupta, B., Faisal, A. A., Laughlin, S. B., & Niven, J. E. (2013). The effect of cell size and channel density on neuronal information encoding and energy efficiency. *Journal of Cerebral Blood Flow and Metabolism*, *33*, 1465–1473.

Sengupta, B., Laughlin, S. B., & Niven, J. E. (2013). Balanced excitatory and inhibitory synaptic currents promote efficient coding and metabolic efficiency. *PLoS Computational Biology*, *9*: e1003263.

Sengupta, B., Laughlin, S. B., & Niven, J. E. (2014). Consequences of converting graded to action potentials upon neural information coding and energy efficiency. *PLoS Computational Biology*, *10*, e1003439.

Sengupta, B., Stemmler, M., Laughlin, S. B., & Niven, J. E. (2010). Action potential energy efficiency varies among neuron types in vertebrates and invertebrates. *PLoS Computational Biology*, *6*, e1000840.

Sengupta, P., & Samuel, A. D. (2009). *Caenorhabditis elegans:* A model system for systems neuroscience. *Current Opinion in Neurobiology*, *19*, 637–643.

Seyfarth, R. M., Silk, J. B., & Cheney, D. L. (2012). Variation in personality and fitness in wild female baboons. *Proceedings of the National Academy of Sciences of the United States of America, 109*(42), 16980–16985.

Shannon, C., & Weaver, W. (1949). *The mathematical theory of communication.* Urbana: University of Illinois Press.

Shaw, S. R. (1975). Retinal resistance barriers and electrical lateral inhibition. *Nature, 255,* 480–483.

Sheehan, M. J., & Tibbetts, E. A. (2011). Specialized face learning is associated with individual recognition in paper wasps. *Science, 334,* 1272–1275.

Shepherd, G. M. (1994). *Neurobiology* (third edition). New York: Oxford University Press.

Shepherd, G. M. (2004). *The synaptic organization of the brain* (fifth edition). New York: Oxford University Press.

Shepherd, G. M. & Harris, K. M., (1998). Three-dimensional structure and composition of CA3→CA1 axons in rat hippocampal slices: Implications for presynaptic connectivity and compartmentalization. *Journal of Neuroscience, 18,* 8300–8310.

Sherman, S. M. (2004). Interneurons and triadic circuitry of the thalamus. *Trends in Neurosciences, 27,* 671–675.

Sherman, S. M., & Guillery, R. W. (2002). The role of the thalamus in the flow of information to the cortex. *Philosophical Transactions of the Royal Society of London, Series B, 357,* 1695–1708.

Sherrington, C. S. (1906). *Integrative action of the nervous system.* New Haven: Yale University Press.

Sherrington, C. S. (1910). Flexion-reflex of the limb, crossed extension-reflex, and reflex stepping and standing. *Journal of Physiology, 40,* 28–121.

Shtyrov, Y., Nikulin, V. V., & Pulvermüller, F. (2010). Rapid cortical plasticity underlying novel word learning. *Journal of Neuroscience, 30*(50), 16864–16867.

Siegel, D. (1993). Energetics of intermediates in membrane fusion: Comparison of stalk and inverted micellar intermediate mechanisms. *Biophysical Journal, 65,* 2124–2140.

Silver, R. A. (2010). Neuronal arithmetic. *Nature Reviews Neuroscience, 11,* 474–489.

Silver, R. A., Lubke, J., Sakmann, B., & Feldmeyer, D. (2003). High-probability uniquantal transmission at excitatory synapses in barrel cortex. *Science, 302,* 1981–1984.

Simpson, J. I. (1984). The accessory optic system. *Annual Review of Neuroscience, 7,* 13–41.

Sincich, L. C., & Horton, J. C. (2005). Input to V2 thin stripes arises from V1 cyto-chrome oxidase patches. *Journal of Neuroscience, 25*(44), 10087–10093.

Sincich, L. C., Horton, J. C., & Sharpee, T. O. (2009). Preserving information in neural transmission. *Journal of Neuroscience, 29*, 6207–6216.

Singer, J. H., Lassová, L., Vardi, N., & Diamond, J. S. (2004). Coordinated multive-sicular release at a mammalian ribbon synapse. *Nat Neuroscience, 7*(8), 826–833.

Skingsley, D. R., Laughlin, S. B., & Hardie, R. C. (1995). Properties of histamine-activated chloride channels in the large monopolar cells of the dipteran compound eye: A comparative study. *Journal of Comparative Physiology. A, Neuroethology, Sensory, Neural, and Behavioral Physiology, 176*, 611–623.

Smith, D. V., Hayden, B. Y., Truong, T. K., Song, A. W., Platt, M. L., & Huettel, S. A. (2010). Distinct value signals in anterior and posterior ventromedial prefrontal cortex. *Journal of Neuroscience, 30*, 2490–2495.

Smith, R. G., Freed, M. A., & Sterling, P. (1986).Microcircuitry of the dark-adapted cat retina: Functional architecture of the rod-cone network. *Journal of Neuroscience, 6*(12), 3505–3517.

Smith, R. G. (1995). Simulation of an anatomically-defined local circuit: The cone-horizontal cell network in cat retina. *Visual Neuroscience, 12*, 545–561.

Smith, R. G., & Sterling, P. (1990). Cone receptive field in cat retina computed from microcircuitry. *Visual Neuroscience, 5*, 453–461.

Sokoloff, L. (1977). Relation between physiological function and energy metabolism in the central nervous system. *Journal of Neurochemistry, 29*, 13–26.

Sokoloff, L., Reivich, M., Kennedy, C., Des Rosiers, M. H., Patlak, C. S., Pettigrew, K. D., et al. (1977). The [^{14}C]deoxyglucose method for the measurement of local cerebral glucose utilization: Theory, procedure, and normal values in the conscious and anesthetized albino rat. *Journal of Neurochemistry, 28*, 897–916.

Sokolowski, M. B. (2010). Social interactions in "simple" model systems. *Neuron, 65*, 780–794.

Sommer, M. A., & Wurtz, R. H. (2004). What the brain stem tells the frontal cortex. I. Oculomotor signals sent from superior colliculus to frontal eye field via mediodor-sal thalamus. *Journal of Neurophysiology, 91*, 1381–1402.

Sommer, M. A., & Wurtz, R. H. (2006). Influence of the thalamus on spatial visual processing in frontal cortex. *Nature, 444*, 374–377.

Sommer, M. A., & Wurtz, R. H. (2008). Brain circuits for the internal monitoring of movements. *Annual Review of Neuroscience, 31*, 317–338.

Song, Z., Postma, M., Billings, S. A., Coca, D., Hardie, R. C., & Juusola, M. (2012). Stochastic, adaptive sampling of information by microvilli in fly photoreceptors. *Current Biology, 22*, 1371–1380.

Sotelo, C. (2003). Viewing the brain through the master hand of Ramon y Cajal. *Nature Reviews Neuroscience, 4*, 71–77.

Sotelo, C. (2008). Viewing the cerebellum through the eyes of Ramón Y Cajal. *Cerebellum, 7*, 517–522.

Sotelo, C. (2010). Camillo Golgi and Santiago Ramon y Cajal: The anatomical organization of the cortex of the cerebellum. Can the neuron doctrine still support our actual knowledge on the cerebellar structural arrangement? *Brain Research Reviews, 66*, 16–34.

Sotelo, C., & Llinás, R. (1972). Specialized membrane junctions between neurons in the vertebrate cerebellar cortex. *Journal of Cell Biology, 53*(2), 271–289.

Sperry, R. W. (1963). Chemoaffinity in the orderly growth of nerve fiber patterns and connections. *Proceedings of the National Academy of Sciences of the United States of America, 50*, 703–710.

Sperry, R. W. (1981). Some effects of disconnecting the cerebral hemispheres. Nobel Lecture.

Spruston, N., Jaffe, D. B., & Johnston, D. (1994). Dendritic attenuation of synaptic potentials and currents: The role of passive membrane properties. *Trends in Neurosciences, 17*, 161–166.

Squire, L. R., Berg, D., Bloom, F., & du Lac, S. (2008). *Fundamental neuroscience*. New York: Academic Press.

Srinivasan, M. V. (2010). Honey bees as a model for vision, perception, and cognition. *Annual Review of Entomology, 55*, 267–284.

Srinivasan, M. V., Laughlin, S. B., & Dubs, A. (1982). Predictive coding—A fresh view of inhibition in the retina. *Proceedings of the Royal Society of London, Series B, 216*, 427–459.

Srivastava, S., Orban, G. A., De Mazière, P. A., & Janssen, P. (2009). A distinct representation of three-dimensional shape in macaque anterior intraparietal area: Fast, metric, and coarse. *Journal of Neuroscience, 29*, 10613–10626.

Stepanyants, A., Tamas, G., & Chklovski, D. B. (2004). Class-specific features of neuronal wiring. *Neuron, 43*, 251–259.

Sterling, P. (2004a). How retinal circuits optimize the transfer of visual information. In J. S. Werner & L. M. Chalupa (Eds.), *The visual neurosciences* (pp. 234–259). Cambridge, MA: MIT Press.

Sterling, P. (2004b). Principles of allostasis: Optimal design, predictive regulation, pathophysiology and rational therapeutics. In J. Schulkin (Ed.), *Allostasis, homeostasis, and the costs of physiological adaptation* (pp. 17–64). Cambridge: Cambridge University Press.

Sterling, P. (2012). Allostasis: A model of predictive regulation. *Physiology & Behavior*, *106*, 5–15.

Sterling, P. (2013). Some principles of retinal design. The Proctor Lecture. *Investigative Ophthalmology and Visual Science*, *54*(3), 2267–2275.

Sterling, P., Cohen, E., Smith, R. G., & Tsukamoto, Y. (1992). Retinal circuits for daylight: Why ballplayers don't wear shades. In F. H. Eeckman (Ed.), *Analysis and modeling of neural systems* (pp. 141–162). New York: Springer.

Sterling, P., & Demb, J. (2004). Retina. In G. M. Shepherd (Ed.), *Synaptic organization of the brain* (pp. 217–270). New York: Oxford University Press.

Sterling, P., & Freed, M. A. (2007). How robust is a neural circuit? *Visual Neuroscience*, *24*, 563–571.

Sterling, P., Freed, M. A., & Smith, R. G. (1988). Architecture of rod and cone circuits to the *On*-beta ganglion cell. *Journal of Neuroscience*, *8*, 623–642.

Sterling, P., & Kuypers, H. G. J. M. (1967). Anatomical organization of the brachial spinal cord of the cat. II. The motoneuron plexus. *Brain Research*, *4*, 16–32.

Sterratt, D., Graham, B., Gillies, A., & Willshaw, D. (2011). *Principles of computational modelling in neuroscience*. Cambridge University Press.

Stockman, A., & Brainard, D. H. (2009). Color vision mechanisms. In M. Bass, C. DeCusatis, J. Enoch, & V. Lakshminarayanan (Eds.), *Handbook of optics, Third edition, Volume iii: Vision and vision optics* (pp. 11.1–11.104). New York: McGraw-Hill.

Stone, L. S. (1960). Regeneration of the lens, iris, and neural retina in a vertebrate eye. *Yale Journal of Biology and Medicine*, *32*, 464–473.

Stone, J., & Fukuda, Y. (1974). Properties of cat retinal ganglion cells: A comparison of W-cells with X- and Y-cells. *Journal of Neurophysiology*, *37*(4), 722–748.

Strausfeld, N. J. (1971). The organization of the insect visual system (light microscopy). I. Projections and arrangements of neurons in the lamina ganglionaris of Diptera. *Zeitschrift Fur Zellforschungs*. *121*, 377–441.

Strausfeld, N. J. (1976). *Atlas of an insect brain*. Berlin: Springer-Verlag.

Strausfeld, N. J. (2012). *Arthropod brains: Evolution, functional elegance and historical significance*. Cambridge, MA: Belknap Press.

Strausfeld, N. J., & Hirth, F. (2013). Deep homology of arthropod central complex and vertebrate basal ganglia. *Science*, *340*, 157–161.

Strauss, R., Krause, T., Berg, C., & Zäpf, B. (2011). Higher brain centers for intelligent motor control in insects. *Intelligent Robotics and Applications, 7102*, 56–64.

Strick, P. L., Dum, R. P., & Fiez, J. A. (2009). Cerebellum and nonmotor function. *Annual Review of Neuroscience, 32*, 413–434.

Strukov, D. B., Snider, G. S., Stewart, D. R., & Williams, R. S. (2008). The missing memristor found. *Nature, 453*, 80–83.

Stuart, G., & Spruston, N. (1998). Determinants of voltage attenuation in neocortical pyramidal neuron dendrites. *Journal of Neuroscience, 18*, 3501–3510.

Südhof, T. C. (2013). Neurotransmitter release: The last millisecond in the life of a synaptic vesicle. *Neuron, 80*, 675–690.

Sun, Y., Olson, R., Horning, M., Armstrong, N., Mayer, M., & Gouaux, E. (2002). Mechanism of glutamate receptor desensitization. *Nature, 417*(6886), 245–253.

Sur, M., Esguerra, M., Garraghty, P. E., Kritzer, M. F., & Sherman, S. M. (1987). Morphology of physiologically identified retinogeniculate X- and Y-axons in the cat. *Journal of Neurophysiology, 58*, 1–32.

Swindale, N. V. (1996). The development of topography in the visual cortex: A review of models. *Network: Computation in Neural Systems, 7*, 161–247.

Szmajda, B. A., & DeVries, S. H. (2011). Glutamate spillover between mammalian cone photoreceptors. *Journal of Neuroscience, 31*(38), 13431–13441.

Taberner, A. M., & Liberman, M. C. (2005). Response properties of single auditory nerve fibers in the mouse. *Journal of Neurophysiology, 93*, 557–569.

Talbot, W. H., Darian-Smith, I., Kornhuber, H. H., & Mountcastle, V. B. (1968). The sense of flutter-vibration: Comparison of the human capacity with response patterns of mechanoreceptive afferents from the monkey hand. *Journal of Neurophysiology, 31*(2), 301–334.

Tatler, B., O'Carroll, D. C., & Laughlin, S. B. (2000). Temperature and the temporal resolving power of fly photoreceptors. *Journal of Comparative Physiology. A, Neuroethology, Sensory, Neural, and Behavioral Physiology, 186*, 399–407.

Taylor, W. R., & Smith, R. G. (2004). Transmission of scotopic signals from the rod to rod-bipolar cell in the mammalian retina. *Vision Research, 44*, 3269–3276.

Teki, S., Kumar, S., von Kriegstein, K., Stewart, L., Lyness, C. R., Moore, B. C. J., et al. (2012). Navigating the auditory scene: An expert role for the hippocampus. *Journal of Neuroscience, 32*, 12251–12257.

Telgkamp, P., Padgett, D. E., Ledoux, V. A., Woolley, C. S., & Raman, I. M. (2004). Maintenance of high-frequency transmission at Purkinje to cerebellar nuclear synapses by spillover from boutons with multiple release sites. *Neuron, 41*, 113–126.

Telzer, E. H., Flannery, J., Shapiro, M., Humphreys, K. L., Goff, B., Gabard-Durman, L., et al. (2013). Early experience shapes amygdala sensitivity to race: An international adoption design. *Journal of Neuroscience, 33*(33), 13484–13488.

Thach, W. T. (1968). Discharge of cerebellar neurons during rapidly alternating arm movements in the monkey. *Journal of Neurophysiology, 31*, 785–797.

Thompson, P. (1980). Margaret Thatcher: A new illusion. *Perception, 9*, 483–484.

Thompson, R. H., & Swanson, L. W. (2003). Structural characterization of a hypothalamic visceromotor pattern generator network. *Brain Research. Brain Research Reviews, 41*, 153–202.

Thoreson, W. B., & Mangel, S. C. (2012). Lateral interactions in the outer retina. *Progress in Retinal and Eye Research, 31*, 407–441.

Tkacik, G., Garrigan, P., Ratliff, C., Milcinski, G., Klein, J. M., Seyfarth, L. H., et al. (2011). Natural images from the birthplace of the human eye. *PLoS ONE, 6*(6), e20409.

Tkacik, G., Prentice, J. S., Victor, J. D., & Balasubramanian, V. (2010). Local statistics in natural scenes predict the saliency of synthetic textures. *Proceedings of the National Academy of Sciences of the United States of America, 107*, 18149–18154.

Tokuoka, H., & Goda, Y. (2008). Activity-dependent coordination of presynaptic release probability and postsynaptic GluR2 abundance at single synapses. *Proceedings of the National Academy of Sciences of the United States of America, 105*, 14656–14661.

Tomita, K., Sperling, M., Cambridge, S. B., Bonhoeffer, T., & Hübener, M. A (2012). Molecular correlate of ocular dominance columns in the developing mammalian visual cortex. *Cerebral Cortex, 23*(11), 2531–2541.

Tong, J., Mannea, E., Aimé, P., Pfluger, P. T., Yi, C.-X., Castaneda, T. R., et al. (2011). Ghrelin enhances olfactory sensitivity and exploratory sniffing in rodents and humans. *Journal of Neuroscience, 31*, 5841–5846.

Tønnesen, J., Katona, G., Rózsa, B., & Nägerl, U. V. (2014). Spine neck plasticity regulates compartmentalization of synapses. *Nature Neuroscience, 17*, 678–685.

Townes-Anderson, E., Dacheux, R. F., & Raviola, E. (1988). Rod photoreceptors dissociated from the adult rabbit retina. *Journal of Neuroscience, 8*(1), 320–331.

Townes-Anderson, E., MacLeish, P. R., & Raviola, E. (1985). Rod cells dissociated from mature salamander retina: ultrastructure and uptake of horseradish peroxidase. *Journal of Cell Biology, 100*(1), 175–188.

Tsai, P. S., Kaufhold, J. P., Blinder, P., Friedman, B., Drew, P. J., Karten, H. J., et al. (2009). Correlations of neuronal and microvascular densities in murine cortex revealed by direct counting and colocalization of nuclei and vessels. *Journal of Neuroscience, 29*, 14553–14570.

Tsao, D. Y., & Livingstone, M. S. (2008). Mechanisms of face perception. *Annual Review of Neuroscience, 31*, 411–437.

Tsukamoto, Y., Smith, R. G., & Sterling, P. (1990). "Collective coding" of correlated cone signals in the retinal ganglion cell. *Proceedings of the National Academy of Sciences of the United States of America, 87*(5):1860–1864.

Turrigiano, G. (2011). Homeostatic synaptic plasticity: Local and global mechanisms for stabilizing neuronal function. *Cold Spring Harbor Perspectives in Biology, 4*, a005736–a005736.

Tyson, J. J., Chen, K. C., & Novak, B. (2003). Sniffers, buzzers, toggles and blinkers: Dynamics of regulatory and signaling pathways in the cell. *Current Opinion in Cell Biology, 15*, 221–231.

Ungerleider, L. G., & Pasternak, T. (2004). Ventral and dorsal cortical processing streams. In L. Chalupa & J. Werner (Eds.), *The visual neurosciences* (pp. 541–562). Cambridge, MA: MIT Press.

Unwin, N. (2013). Nicotinic acetylcholine receptor and the structural basis of neuromuscular transmission: insights from *Torpedo* postsynaptic membranes. *Quarterly Reviews of Biophysics, 46*, 283–322.

Vallet, A. M., Coles, J. A., Eilbeck, J. C., & Scott, A. C. (1992). Membrane conductances involved in amplification of small signals by sodium channels in photoreceptors of drone honey bee. *Journal of Physiology (Lond.), 456*, 303–324.

Vallet, A. M., & Coles, J. A. (1993). Is the membrane voltage amplifier of drone photoreceptors useful at physiological light intensities? *Journal of Comparative Physiology, A, 173*, 163–168.

Valverde Salzmann, M. F., Bartels, A., Logothetis, N. K., & Schüz, A. (2012). Color blobs in cortical areas V1 and V2 of the New World monkey *Callithrix jacchus*, revealed by non-differential optical imaging. *Journal of Neuroscience, 32*(23), 7881–7894.

van der Veen, D. R., Le Minh, N., Gos, P., Arneric, M., Gerkema, M. P., & Schibler, U. (2006). Impact of behavior on central and peripheral circadian clocks in the common vole *Microtus arvalis*, a mammal with ultradian rhythms. *Proceedings of the National Academy of Sciences of the United States of America, 103*(9), 3393–3398.

Van Essen, D. C. (2004). Organization of visual areas in macaque and human cerebral cortex. In L. Chalupa & J. Werner (Eds.), *The visual neurosciences* (pp. 507–521). Cambridge, MA: MIT Press.

van Hateren, J. H. (1992a). A theory of maximizing sensory information. *Biological Cybernetics, 68*, 23–29.

van Hateren, J. H. (1992b). Theoretical predictions of spatiotemporal receptive fields of fly LMCs, and experimental validation. *Journal of Comparative Physiology, 171*, 157–170.

van Hateren, J. H. (1992c). Real and optimal neural images in early vision. *Nature, 360*, 68–70.

van Hateren, J. H., & Laughlin, S. B. (1990). Membrane parameters, signal transmission, and the design of a graded potential neuron. *Journal of Comparative Physiology. A, Neuroethology, Sensory, Neural, and Behavioral Physiology, 166*, 437–448.

van Hateren, J. H., & Schilstra, C. (1999). Blowfly flight and optic flow. II. Head movements during flight. *Journal of Experimental Biology, 202*, 1491–1500.

van Kan, P. L., Gibson, A. R., & Houk, J. C. (1993). Movement-related inputs to intermediate cerebellum of the monkey. *Journal of Neurophysiology, 69*, 74–94.

van Rossum, M. C. W., & Smith, R. G. (1998). Noise removal at the rod synapse of mammalian retina. *Visual Neuroscience, 15*, 809–821.

van Wyk, M., Taylor, W. R., & Vaney, D. I. (2006). Local edge detectors: A substrate for fine spatial vision at low temporal frequencies in rabbit retina. *Journal of Neuroscience, 26*(51), 13250–13263.

Vaney, D. I. (1990). The mosaic of amacrine cells in the mammalian retina. In N. Osborne & G. Chader (Eds.), *Progress in retinal research* (pp. 49–100). New York: Pergamon.

Vaney, D. I. (1994). Territorial organization of direction-selective ganglion cells in rabbit retina. *Journal of Neuroscience, 14*(11), 6301–6316.

Vaney, D. I., & Pow, D. V. (2000). The dendritic architecture of the cholinergic plexus in the rabbit retina: Selective labeling by glycine accumulation in the presence of sarcosine. *Journal of Comparative Neurology, 421*, 1–13.

Vaney, D. I., Sivyer, B., & Taylor, W. R. (2012). Direction selectivity in the retina: Symmetry and asymmetry in structure and function. *Nature Reviews Neuroscience, 13*(3), 194–208.

Varshney, L. R., Chen, B. L., Paniagua, E., Hall, D. H., & Chklovskii, D. B. (2011). Structural properties of the *Caenorhabditis elegans* neuronal network. *PLoS Computational Biology, 7*, e1001066.

Varshney, L. R., Sjöström, P. J., & Chklovskii, D. B. (2006). Optimal information storage in noisy synapses under resource constraints. *Neuron, 52*, 409–423.

Verlinden, H., Vleugels, R., Marchal, E., Badisco, L., Pflüger, H.-J., Blenau, W., et al. (2010). The role of octopamine in locusts and other arthropods. *Journal of Insect Physiology, 56*, 854–867.

Verweij, J., Hornstein, E. P., & Schnapf, J. L. (2003). Surround antagonism in macaque cone photoreceptors. *Journal of Neuroscience, 23*(32), 10249–10257.

Vigeland, L. E., Contreras, D., & Palmer, L. A. (2013). Synaptic mechanisms of temporal diversity in the lateral geniculate nucleus of the thalamus. *Journal of Neuroscience, 33*(5), 1887–1896.

Vinje, W. E., & Gallant, J. L. (2002). Natural stimulation of the nonclassical receptive field increases information transmission efficiency in V1. *Journal of Neuroscience, 22*(7), 2904–2915.

Viskontas, I. V., Quiroga, R. Q., & Fried, I. (2009). Human medial temporal lobe neurons respond preferentially to personally relevant images. *Proceedings of the National Academy of Sciences of the United States of America, 106*, 21329–21334.

Vitureira, N., Letellier, M., & Goda, Y. (2012). Homeostatic synaptic plasticity: From single synapses to neural circuits. *Current Opinion in Neurobiology, 22*, 516–521.

Von der Twer, T., & MacLeod, D. I. A. (2001). Optimal nonlinear codes for the perception of natural colours. *Network: Computation in Neural Systems, 12*, 395–407.

von Greefe, R. (1899). *Die mikroskopische Anatomie des Sehnerven und der Netzhaut.* Liepzig: W. Engelmann.

Wandell, B. A. (2011). The neurobiological basis of seeing words. *Annals of the New York Academy of Sciences, 1224*(1), 63–80.

Wandell, B. A., Rauschecker, A. M., & Yeatman, J. D. (2012). Learning to see words. *Annual Review of Psychology, 63*, 31–53.

Wang, H.-P., Spencer, D., Fellous, J.-M., & Sejnowski, T. J. (2010). Synchrony of thalamocortical inputs maximizes cortical reliability. *Science, 328*, 106–109.

Wang, Q., Sporns, O., & Burkhalter, A. (2012). Network analysis of corticocortical connections reveals ventral and dorsal processing streams in mouse visual cortex. *Journal of Neuroscience, 32*, 4386–4399.

Wang, S. S.-H., Shultz, J. R., Burish, M. J., Harrison, K. H., Hof, P. R., Towns, L. C., et al. (2008). Functional trade-offs in white matter axonal scaling. *Journal of Neuroscience, 28*(15), 4047–4056.

Wang, Y., Xiao, Y., & Felleman, D. J. (2007). V2 thin stripes contain spatially organized representations of achromatic luminance change. *Cerebral Cortex, 17*, 116–129.

Wang, Z., & McCormick, D. A. (1993). Control of firing mode of corticotectal and corticopontine layer V burst-generating neurons by norepinephrine, acetylcholine, and 1 S,3 R-ACPD. *Journal of Neuroscience, 13*(5), 2199–2216.

Wässle, H. (2004). Parallel processing in the mammalian retina. *Nature Reviews Neuroscience, 5*, 747–757.

Wässle, H., Puller, C., Muller, F., & Haverkamp, S. (2009). Cone contacts, mosaics, and territories of bipolar cells in the mouse retina. *Journal of Neuroscience, 29*, 106–117.

Weber, B., Keller, A. L., Reichold, J., & Logothetis, N. K. (2008). The microvascular system of the striate and extrastriate visual cortex of the macaque. *Cerebral Cortex, 18*, 2318–2330.

Weckström, M., Hardie, R. C., & Laughlin, S. B. (1991). Voltage-activated potassium channels in blowfly photoreceptors and their role in light adaptation. *Journal of Physiology, 440*, 635–657.

Weckstrom, M., Juusola, M., & Laughlin, S. B. (1992). Presynaptic enhancement of signal transients in photoreceptor terminals in the compound eye. *Proceedings of the Royal Society of London, Series B, 250*, 83–89.

Weckström, M., & Laughlin, S. (2010). Extracellular potentials modify the transfer of information at photoreceptor output synapses in the blowfly compound eye. *Journal of Neuroscience, 30*, 9557–9566.

Wehner, R. (1987). "Matched filters"—Neural models of the external world. *Journal of Comparative Physiology. A, Neuroethology, Sensory, Neural, and Behavioral Physiology, 161*, 511–531.

Weibel, E. R. (2000). *Symmorphosis: On form and function in shaping life.* Cambridge, MA: Harvard University Press.

Weiner, J. (1994). *The beak of the finch: A story of evolution in our time.* Knopf.

Weiner, J. (1999). *Time, love and memory.* New York: Random House.

Wen, Q., & Chklovskii, D. B. (2005). Segregation of the brain into gray and white matter: A design minimizing conduction delays. *PLoS Computational Biology, 1*, e78.

Wen, Q., & Chklovskii, D. B. (2008). A cost-benefit analysis of neuronal morphology. *Journal of Neurophysiology, 99*(5), 2320–2328.

Wen, Q., Stepanyants, A., Elston, G. N., Grosberg, A. Y., & Chklovskii, D. B. (2009). Maximization of the connectivity repertoire as a statistical principle governing the shapes of dendritic arbors. *Proceedings of the National Academy of Sciences of the United States of America, 106*(30), 12536–12541.

Werner, G., & Mountcastle, V. B. (1965). Neural activity in mechanoreceptive cutaneous afferents: Stimulus-response relations, Weber functions, and information transmission. *Journal of Neurophysiology, 28*(2), 359–397.

Weyand, T. G. (2007). Retinogeniculate transmission in wakefulness. *Journal of Neurophysiology, 98*, 769–785.

White, J. G., Southgate, E., Thomson, J. N., & Brenner, S. (1986). The structure of the nervous system of the nematode *Caenorhabditis elegans. Philosophical Transactions of the Royal Society of London, Series B, 314*(1165), 1–340.

Williamson, M. (2011). *How proteins work.* New York: Garland Science.

Willis, W. D., Jr., & Coggeshall, R. E. (1991). *Sensory mechanisms of the spinal cord.* New York: Plenum Press.

Willmore, B. D. B., Prenger, R. J., & Gallant, J. L. (2010). Neural representation of natural images in visual area V2. *Journal of Neuroscience, 30*, 2102–2114.

Wilson, J. R., Friedlander, M. J., & Sherman, S. M. (1984). Fine structural morphology of identified X- and Y-cells in the cat's lateral geniculate nucleus. *Proceedings of the Royal Society of London, Series B, 221*, 411–436.

Wilson, R. I., & Mainen, Z. F. (2006). Early events in olfactory processing. *Annual Review of Neuroscience, 29*, 163–201.

Wolf, R., Voss, A., Hein, S., Heisenberg, M., & Sullivan, G. D. (1992). Can a fly ride a bicycle? *Philosophical Transactions of the Royal Society of London, Series B, 337*, 261–269.

Wong, K. Y. (2012). A retinal ganglion cell that can signal irradiance continuously for 10 hours. *Journal of Neuroscience, 32*, 11478–11485.

Wong-Riley, M. (2010). Energy metabolism of the visual system. *Eye and Brain, 2*, 99–116.

Woollett, K., & Maguire, E. A. (2011). Acquiring "the knowledge" of London's layout drives structural brain changes. *Current Biology, 21*, 2109–2114.

Woollett, K., & Maguire, E. A. (2012). Exploring anterograde associative memory in London taxi drivers. *Neuroreport, 23*(15), 885–888.

Woollett, K., Spiers, H. J., & Maguire, E. A. (2009). Talent in the taxi: A model system for exploring expertise. *Philosophical Transactions of the Royal Society of London, Series B, 364*(1522), 1407–1416.

Wormbook: The online review of *C. elegans* biology. www.wormbook.org.

Wu, X.-S., Xue, L., Mohan, R., Paradiso, K., Gillis, K. D., & Wu, L.-G. (2007). The origin of quantal size variation: Vesicular glutamate concentration plays a significant role. *Journal of Neuroscience, 27*, 3046–3056.

Wurtz, R. H., McAlonan, K., Cavanaugh, J., & Berman, R. A. (2011). Thalamic pathways for active vision. *Trends in Cognitive Sciences, 15*, 177–184.

Xiao, Y., Wang, Y., & Felleman, D. J. (2003). A spatially organized representation of colour in macaque cortical area V2. *Nature, 421*, 535–539.

Xu, T., Yu, X., Perlik, A. J., Tobin, W. F., Zweig, J. A., Tennant, K., et al. (2009). Rapid formation and selective stabilization of synapses for enduring motor memories. *Nature, 462*, 915–919.

Xu, Y., Dhingra, A., Fina, M. E., Koike, C., Furukawa, T., & Vardi, N. (2012). mGluR6 deletion renders the TRPM1 channel in retina inactive. *Journal of Neurophysiology, 107*(3), 948–957.

Xu, Y., Vasudeva, V., Vardi, N., Sterling, P., & Freed, M. A. (2008). Different types of ganglion cell share a synaptic pattern. *Journal of Comparative Neurology, 507*(6), 1871–1878.

Xue, T., Do, M. T., Riccio, A., Jiang, Z., Hsieh, J., Wang, H. C., et al. (2011). Melanopsin signalling in mammalian iris and retina. *Nature, 479*, 67–73.

Xu-Friedman, M. A., Harris, K. M., & Regehr, W. G. (2001). Three-dimensional comparison of ultrastructural characteristics at depressing and facilitating synapses onto cerebellar purkinje cells. *Journal of Neuroscience, 21*(17), 6666–6672.

Xu-Friedman, M. A., & Regehr, W. G. (2003). Ultrastructural contributions to desensitization at cerebellar mossy fiber to granule cell synapses. *Journal of Neuroscience, 23*(6), 2182–2192.

Xu-Friedman, M. A., & Regehr, W. G. (2004). Structural contributions to short-term synaptic plasticity. *Physiological Reviews, 84*, 69–85.

Yakovenko, S., Krouchev, N., & Drew, T. (2011). Sequential activation of motor cortical neurons contributes to intralimb coordination during reaching in the cat by modulating muscle synergies. *Journal of Neurophysiology, 105*, 388–409.

Yang, G., Pan, F., & Gan, W. B. (2009). Stably maintained dendritic spines are associated with lifelong memories. *Nature, 462*, 920–924.

Yang, H., & Xu-Friedman, M. A. (2012). Emergence of coordinated plasticity in the cochlear nucleus and cerebellum. *Journal of Neuroscience, 32*(23), 7862–7868.

Yau, K.-W., & Hardie, R. C. (2009). Phototransduction motifs and variations. *Cell, 139*, 246–264.

Yeh, C. I., Stoelzel, C. R., & Alonso, J. M. (2003). Two different types of Y cells in the cat lateral geniculate nucleus. *Journal of Neurophysiology, 90*, 1852–1864.

Yin, L., Smith, R. G., Sterling, P., & Brainard, D. H. (2006). Chromatic properties of horizontal and ganglion cell responses follow a dual gradient in cone opsin expression. *Journal of Neuroscience, 26*(47), 12351–12361.

Yu, J., & Ferster, D. (2013). Functional coupling from simple to complex cells in the visually driven cortical circuit. *Journal of Neuroscience, 33*(48), 18855–18866.

Yuste, R. (2013). Electrical compartmentalization in dendritic spines. *Annual Review of Neuroscience, 36,* 429–449.

Zampini, V., Johnson, S. L., Franz, C., Lawrence, N. D., Münkner, S., Engel, J., et al. (2010). Elementary properties of Cav1.3 Ca^{2+} channels expressed in mouse cochlear inner hair cells. *Journal of Physiology, 588*(1), 187–199.

Zecevic, D. (1996). Multiple spike-initiation zones in single neurons revealed by voltage-sensitive dyes. *Nature, 381,* 322–325.

Zhang, Y., Kim, I.-J., Sanes, J. R., & Meister, M. (2012). The most numerous ganglion cell type of the mouse retina is a selective feature detector. *Proceedings of the National Academy of Sciences of the United States of America, 109*(36), E2391–E2398b.

Zheng, L., de Polavieja, G. G., Wolfram, V., Asyali, M. H., Hardie, R. C., & Juusola, M. (2006). Feedback network controls photoreceptor output at the layer of first visual synapses in *Drosophila. Journal of General Physiology, 127,* 495–510.

Ziburkus, J., Bickford, M. E., & Guido, W. (2000). NMDAR-1 staining in the lateral geniculate nucleus of normal and visually deprived cats. *Visual Neuroscience, 17*(2), 187–196.

Index